CHEMISTRY FOR THE PROTECTION OF THE ENVIRONMENT 2

ENVIRONMENTAL SCIENCE RESEARCH

Series Editor:

Herbert S. Rosenkranz
Department of Environmental and Occupational Health
Graduate School of Public Health
University of Pittsburgh
130 DeSoto Street
Pittsburgh, Pennsylvania

Founding Editor:

Alexander Hollaender

Recent Volumes in this Series

A Continuation Order Plan is available for this series. A continuation order will bring delivery of each new volume immediately upon publication. Volumes are billed only upon actual shipment. For further information please contact the publisher.

CHEMISTRY FOR THE PROTECTION OF THE ENVIRONMENT 2

Edited by

Lucjan Pawłowski
Technical University of Lublin
Lublin, Poland

William J. Lacy
Lacy and Associates
Alexandria, Virginia

Christopher G. Uchrin
Rutgers University
New Brunswick, New Jersey

and

Marzenna R. Dudzińska
Technical University of Lublin
Lublin, Poland

Plenum Press • New York and London

Library of Congress Cataloging-in-Publication Data

Chemistry for the protection of the environment 2 / edited by Lucjan Pawłowski ... [et al.].
p. cm. -- (Environmental science research ; v. 51)
"Proceedings of the 10th International Conference for the Protection of the Environment, held September 17-21, 1995, in Lublin, Poland"--T.p. verso.
Includes bibliographical references and index.
ISBN-13:978-1-4612-8037-5 e-ISBN-13:978-1-4613-0405-0
DOI: 10.1007/978-1-4613-0405-0

1. Environmental chemistry--Congresses. 2. Sewage--Purification--Congresses. 3. Waste minimization--Congresses. I. Pawłowski, Łucjan. II. International Conference on Chemistry for the Protection of the Environment (10th : 1995 : Lublin, Poland) III. Series.
TD193.C4845 1996
628--dc20 96-25716
CIP

Proceedings of the 10th International Conference on Chemistry for the Protection of the Environment, held September 17–21, 1995, in Lublin, Poland

ISBN-13:978-1-4612-8037-5

Softcover reprint of the hardcover 1st edition 1996

A Division of Plenum Publishing Corporation
233 Spring Street, New York, N.Y. 10013

10 9 8 7 6 5 4 3 2 1

PREFACE

Over the past two decades, this environmental conference series has emerged to become one of the major international forums on the chemical aspects of environmental protection. The forum is called Chemistry for the Protection of the Environment (CPE). The sponsors of this CPE series have included the Chemical Societies of Poland, France, Belgium, Italy, Egypt, and the U.S.A., the American Institute of Chemical Engineers, the American Society of Testing and Materials, the International Ozone Association, the United Nations Industrial Development Organization, the Ministries of the Environment of Poland, France, Belgium, and Italy, U.S. Agency for International Development, U.S. Environmental Protection Agency, more than twenty universities and institutes of higher learning, and five national academies of sciences.

The first meeting in this series was organized by Prof. Pawłowski and Dr. Lacy in 1976 at the Marie Curie–Sklodowska University in Lublin, Poland. The conference dealt with various physicochemical methodologies for water and wastewater treatment research projects that were jointly sponsored by U.S. EPA and Poland.

The great interest expressed by the participants led the organizers to expand the scope of the second conference, which was also held in Poland in September 1979. The third and enlarged symposium was again successfully held in 1981 in Lublin, Poland. At that time the participating scientists and engineers expressed their desire to broaden the coverage as well as the title of the conference series. The International Committee, approved the title "Chemistry for the Protection of the Environment" and designated that date of the fourth conference, CPE IV, which was convened in September 1983 at the Paul Sabatier University in Toulouse, France, and was hosted and arranged by Prof. A. Verdier. This conference series includes participants from various government agencies, academia, and the private sector, representing industrialized countries as well as emerging nations, both East and West in an independent, nonpolitical forum.

The central goals of CPE are to improve technology transfer and scientific dialogue, thereby leading to a better comprehension of and solution to a broad spectrum of environmentally related problems. The fifth conference was held in September 1985 at the Catholic University in Leuven, Belgium. It was hosted by Profs. A. Van Haute and G. Alaerts. CPE V covered topics dealing with treatment technologies and phenomena related to hazardous waste and the utilization of fossil fuels. It provided an opportunity for interdisciplinary discussions and encouraged the exchange of ideas among international specialists from diverse fields and backgrounds.

Under the leadership of Profs. Mentasti and Sarzanini and with the assistance of Dr. Gennero, CPE VI was held in 1987 at the University of Turin in Italy. Over 150 selected scientific papers and posters were presented to an audience of specialists from 32 nations.

This assemblage comprised in equal measure scientists from Europe, the New World, and developing nations.

CPE VII was convened at the Catholic University in Lublin, Poland in 1989. The exchange of information by approximately 200 scientists and engineers made this a memorable scientific occurrence. The scientific committee selected presenters of high intellectual and technical merit. The distinguished participants of CPE VII included Poland's Minister and Deputy Minister for Environmental Protection, U.S. Scientific Council, Israel's Deputy Minister of the Environment, presidents and vice presidents of five universities, representatives of the academies of sciences for Czechoslovakia, France, Italy, Poland, and the U.S.S.R., as well as many department heads and acclaimed scientists.

CPE VIII was scheduled to convene in Budapest, Hungary in September 1991. But due to international administrative difficulties it was moved to Lublin. Despite this last minute change of venue, the scientific meeting was voted an outstanding success by the participants. The technical presentation were original and informative, the major topics included chemical/physical/biological/treatment technologies, monitoring modeling and risk assessment. There were over 120 attendees and some 90 scientific oral and poster presentations. Through CPE International Committee, Dr. Hartstein of U.S. Dept. of Energy, had the proceedings published in the U.S.A.

CPE IX was held in September 1993 in Alexandria, Cairo/Luxor, Egypt, and included a joint conference with the fourth international symposium on Industry in the Developing World. The issues covered were an extremely successful workshop on industrial pollution prevention and clean technologies plus cooperation and institutional issues. The chairpeople were Drs. Pawłowski, Hamza, Saad, Gallup and Lacy. It was a great success, highly informative. The multi-disciplined technical group from 27 countries (and international organizations affiliated with the global environmental movement) were a captive-like audience aboard a Nile River Cruiser. This atmosphere promoted free, open exchanges and dialogues between all the attendees. Selected papers were published in a volume by the High Institute of Public Health, Alexandria, Egypt.

CPE X, the 20th anniversary meeting, was held in the city of its birth, Lublin, Poland. The spirit of this 20th anniversary not only permeated the program and the international group of experts but was reflected in the various folk music festival, folk dancing and social/cultural programs that were arranged by Dr. Dudzińska. The most musically gifted and talented scientist, chairman of the College of Environmental Engineering, Rutgers University, N.J., Prof. Christopher Uchrin entertained one and all with a Chopin recital plus an operatic rendition.

Interesting, original, and informative papers and posters were presented on the following topics: adsorption, analytical methods, chemical/biological treatment, groundwater studies, ion exchange, modeling, ozonation, photodegradation, risk assessment, sludge treatment, waste minimization, innovative technology, acid rain, and indoor air pollution.

Some of the major benefits for the participants were technology transfer and exchange of innovative and alternative treatment methods and information about activities in other countries related to environmental problems. The meeting was enhanced by the participation of large delegations from both The People's Republic of China and the Taiwan Chinese Republic, with excellent papers.

The venue for the next conference, CPE XI will be selected during the spring 1996 meeting of the International Committee. Dr. Hamza of the University of Alexandria, Egypt expressed an invitation to again serve as host. In this closing remarks at CPE X, Dr. W.J. Lacy stated that during CPE XI to be held in Egypt, all scientists and engineers need to be aware of what is being done in other countries, since even scientists from the smallest developing nation could have the solution to the most vexing environmental problems, and that as long as scientists from around the globe wish to meet in this type of open, technical forum and exchange ideas and information on environmental protection, the Chemistry for Protection of the Environment series will continue.

L. Pawłowski, W.J. Lacy, C.G. Uchrin, and M.R. Dudzińska

CONTENTS

WASTEWATER TREATMENT

WASTEWATER MINIMALIZATION AND UTILIZATION

PATHWAYS OF POLLUTANTS

1

CLEANER PRODUCTION IN THE CHEMICAL INDUSTRY

B. A. Bolto

CSIRO Chemicals and Polymers
Private Bag 10, Rosebank MDC, Clayton, Victoria 3169, Australia

ABSTRACT

Various facets of cleaner production are currently receiving much attention. Initial efforts in better housekeeping and improved end-of-pipe processes have matured. While there is still significant effort being expended on recovering useful materials from wastes, there is now a much sharper focus on modifying manufacturing techniques to avoid or markedly decrease the use of toxic starting materials, by-products and solvents. It is in this latter area that research is encouraged.

1. INTRODUCTION

Cleaner production goes beyond environmentally sustainable activity, as it implies

- a minimisation of waste
- a more efficient use of resources & technologies
- a safer means of achieving an end output [1].

The current trend is to put less effort into pollution control, and a much greater emphasis on pollution prevention by not generating toxic wastes in the first place. The present environmental standards favour existing technologies, since implementing unproven technology risks non-compliance. Hence there is a need to upgrade regulations to promote innovation. We should design for the environment, by devising new technologies which work better and cost less. It is also essential that new accounting tools be developed that incorporate environmental costs and benefits into managerial accounting and capital budgeting practices [2].

End-of-pipe solutions involve large capital expenditure and non-productive plant; the costly problem of disposing of the eventual waste product is not solved. Cleaner production, on the other hand, eliminates waste streams and waste by-products, avoids the use of toxic reagents, intermediates and solvents, and produces cleaner products. It makes use of less energy intensive

Chemistry for the Protection of the Environment 2, edited by Pawłowski et al.
Plenum Press, New York, 1996

processes, achieves lower emissions, and can function on a small scale. Manufacturing in this way is more profitable than dirty or hazardous processes, and there is more effective use of raw materials and capital, along with a safer means of achieving an end output.

By-products formation varies within the industry, as shown below [3]. However, the throughput decreases as the list is descended, so that the actual tonnage of by-product is greatest in the reverse direction by two or three orders of magnitude.

Oil refineries	0.1 kg of by-product per kg of product
Bulk chemicals	<1–5
Fine chemicals	5–50
Pharmaceuticals	25 to >100

Action needs to be prompt, as a more than 50% increase in chemical manufacturing is anticipated in the 1990s, most of it to take place in Asia [3]. To this end, it is noteworthy that the EPA in the US has quadrupled the amount of money available for research on cleaner production. A wide range of issues is involved, from waste minimisation and end-of-pipe treatment to the use of non-toxic solvents and reagents, better catalysts, and more energy efficient processes.

2. WASTE MINIMISATION AND WATER REUSE

Waste minimisation featured strongly in the initial moves by the chemical industry towards cleaner manufacturing. Some modifications carried out by the Australian chemical industry cover better housekeeping and major process improvements. ICI Australia, as Australia's largest chemicals manufacturer, has reduced its wastewater flows and hydrocarbon loads by 88% and 95% respectively at its olefins plant near Sydney; the improvements have led to the disconnection of a number of waste collection drains from the sewerage system. Many other process modifications have been carried out, with a total capital outlay for 22 examples from nine factory sites amounting to A$23.7M, resulting in an overall saving of at least A$5.2M yearly. Some of the changes are concerned with water reuse on site, others with eliminating spillage, minimisation of energy input, new catalysts for cleaner and more efficient reactions, and the recovery of valuable chemicals, such as cyanide, ammonia and ethoxylation heavies [4].

Other firms have been active also. BASF Australia , which manufactures pigments in Melbourne, has installed reversible pumps to return pipeline fills to dispensing tanks, rather than washing out the pipe contents to waste at the end of each shipment [5]. Albright and Wilson at their surfactant manufacturing site in Sydney have substantially reduced pollutant flows to the sewerage system by recycling, reuse and re-blending [6]. Reckitt and Coleman have re-installed bunds at their detergent making location in Sydney, making clean up of spills more difficult than a simple hosing down; this has resulted in the virtual elimination of spills by workers.

Handling procedures have been subjected to general improvements such as:

- bulk transport instead of drums, kegs and barrels, the first step being a multi-use intermediate bulk carrier such as a tanker or shipping container
- the use of dedicated pipelines, as with domestic gas
- magnetic couplings to avoid leaking seals met in centrifugal pumps [1].

Many process improvements have been made. Thus hydrocyclones have been utilised to remove solids more efficiently from an effluent than does filtration; ultrafiltration with membranes rather than evaporation to obtain a concentrated product gives energy savings, as does solar power for evaporation. The introduction of new technology for milling and homogenisation can give similar significant economies[1]. In one example, the original system required 75 kW of power for 2–4 hours to predisperse 1000 litres of a mixture, followed by 75 kW for 2–4 hours for milling. The new arrangement needs only 7.5 kW for 4 hours to predisperse 8000 litres, plus 90 kW to mill.

2.1 Recovery of Useful Materials from Wastes

In the food industry, starch, protein and fats can be separated by flotation provided that palatable, non-toxic additives are used to coagulate and flocculate these useful organics, which can then be sold as animal feed.

Pigment manufacture makes use of titanium dioxide, made by SCM Chemicals Ltd, which operates one of the world's most efficient and clean plants at Kemerton in Western Australia [6]. All wastes from the plant are fully treated and neutralised before disposal as clean salty water to the ocean and as an inert solid waste to landfill. The solid waste has potential for use as road base, paving bricks, soil additive and nutrient absorbent, and research projects on all of these applications are under way.

In hydrogen cyanide manufacture from the catalysed reaction of ammonia and methane, dilute waste solutions containing ammonia can be concentrated by stripping them at high pH and then absorbing the ammonia in ammonium nitrate acidified with nitric acid. The concentrated liquor is then recycled to ammonium nitrate manufacture. Recovery of HCN is similarly done on an acidic solution [4].

In the fertiliser industry solutions are concentrated by evaporation, and the condensed vapours form a wastewater which contains several impurities [7]. Ammonia manufacture results in a condensate which has 100–3000 mg/L of NH_4^+ and a pH of 7–9; carbonate is the main anion. A weak acid ion-exchange resin gives good regeneration efficiency, CO_2 is removed by aeration, and a roughly demineralised water is obtained for reuse. The resin is regenerated with mineral acid, yielding ammonium nitrate.

In the chemical industry ion exchange has been used for the recovery of organic compounds, notably phenol and proteins [7]. Sulfonic acids may be separated from sulfuric acid other than by precipitating the latter as $CaSO_4$ or $BaSO_4$, which creates disposal problems; instead, sulfate may be removed on a strong base anion exchange resin, as has been done for 5-sulfoisophthalic acid.

3. ULTIMATE WASTE DISPOSAL

3.1 Aqueous Wastes

The use of wetlands and the disposal of aqueous wastes to land are options well suited to Australia because of the high evaporation rate. They are much used inland for domestic and agricultural wastes, and have recently been applied to wastes from the chemical industry [6]:

Kemcor Plastics, a manufacturer of polyethylene in Melbourne, has diverted its wastewater from the sewerage system to a lake which has attracted a large number of bird and fish species; they are being monitored and are universally in a very healthy state. Similarly, Hoechst, a Melbourne manufacturer of plastics, pigments and pharmaceuticals has recently developed a wetlands project involving storm water cleaning and retention, which follows wastewater treatment and recycling that saves 60% of their water consumption.

CSBP & Farmers makes superphosphate at Albany on the coast of Western Australia, where in recent years it has modified the production process, improved the site drainage system and installed equipment to treat storm water run off and planted more than 90,000 Australian native trees; these initiatives have reduced phosphorus discharges to the harbour from more than 8,000 kg to less than 50 kg per year. Likewise Dow Chemical in Melbourne has diverted treated wastewater from the sewerage system to a tree plantation of 48,000 native trees which receive water by an advanced drip irrigation system. Rohm and Haas, manufacturing mainly raw materials for paint, is in the process of diverting wastewater at their Geelong site to a plantation of 10,000 native trees via holding ponds developed like natural wetlands.

3.2 Destruction of Non-aqueous Toxic Wastes

There are many technology options for degrading chlorinated organics [8], such as:

- high temperature incineration, an approach not acceptable to the Australian community.
- plasma arc processes, which can be in-line factory units for specific by-products, as will be discussed further below.
- cement kilns utilise the energy value, and can treat large volumes of low level wastes with limestone and shale, or clay and coal/oil at 1450°.
- molten metal systems via oxygen injection, commercialised in the USA and Australia.
- molten slag processes are similar.
- molten salt is used in the same way, but can be operated under reducing or oxidising conditions at 850° and 450° respectively.
- oxidation methods generally cover a number of approaches, such as catalysts for gas phase systems; wet air oxidation; supercritical water oxidation; chemical approaches with UV/ozone or UV/hydrogen peroxide
- hydrogenation produces energy rich materials, and can be catalysed with Pd/C at 55° and 3 atmospheres for 3 hours, or at high temperaure, as in refinery hydrocracking; and from hydrogen donors other than molecular hydrogen
- base catalysed dechlorination has three variations: *K-PEG*, $NaHCO_3$, and catalytic transfer hydrogenation, and will be detailed later
- biological remediation is a benign technology especially suitable for chlorinated organics which are dispersed at low concentrations.

Two of these, the plasma arc and base catalysed dechlorination processes, will be discussed in more detail. A plasma arc technique has been jointly developed by CSIRO and Siddons Ramset and is in operation at Nufarm Limited, a manufacturer of herbicides in Melbourne [9]. It is a waste destruction process which utilises exremely high temperatures (10,000° or more) resulting from the discharge of a large electric current in an inert gas. The superheated cloud of gas or plasma instantaneously converts toxic materials into atomic or ionic forms, and then converts these atoms into simple environmentally benign molecules by subsequent downstream processing. Special features are

- high levels of destruction performance, with performance in excess of 99.9999% being common
- an ability to treat highly concentrated organic liquids and gases
- low capital and operating costs relative to other methods; a 150 kW plant which can treat 1–5 tonnes/day depending on the waste costing A$750,000, with an operating cost generally well below A$3,000 per tonne
- an extremely compact plant, suitable for on-site operation so that the waste does not have to be transported to a centralised facility
- it is capable of being shut down in fractions of a second, thus making the process very safe
- the plant is simple to operate and requires little maintenance.

Base cataysed dechlorination prcesses are available in several forms [8]. The *K-PEG* method employs KOH in poly(ethylene glycol), MW 350, at 250°–340°. The polymeric alkoxide displaces the chlorine by nucleophilic substitution:

$$Ar\text{-}Cl + RO^- \longrightarrow Ar\text{-}OR + Cl^-$$

$$Ar\text{-}OR + H_2O \longrightarrow Ar\text{-}OH + ROH$$

It is effective on highly substituted compounds, but has low reactivity with di- and some trichloro-substituted biphenyls. Sodium and naphthalene have been used in a similar manner in a non-aqueous system. Sodium bicarbonate has been used at 280°–340° for treating soils and sediments; it has been applied to PCBs, lindane, dieldrin and pentachlorophenol, with 2.2 g/l levels lowered to <1mg/l in 0.25 to 1.5 hour.

Ball milling in the presence of quick lime gives a similar effect at room temperature, when a 99.996% destruction of PCBs is claimed [10]; very high local temperatures can be obtained. Thus when 5–10 g of DDT, chlorobenzene or PCBs are milled with reactants such as Mg, Ca or Fe metal or CaO by using 12 mm diameter steel balls in a sealed steel vial placed in a laboratory vibratory mill, there is destruction to the levels shown in Table 1. Quicklime is the most effective reagent for the destruction of PCBs.

Catalytic transfer hydrogenation on the other hand uses a saturated oil as the hydrogen donor [11]. The organochlorine compound is dissolved in the oil to concentrations of up to 30% v/v, in the presence of 1% by volume of a carbon catalyst. Alkali is added in a mole ratio of 3:1 (NaOH:Cl group), and the mixture heated at 300°–340° for 2 hours. The reaction proceeds through hydrogen radicals produced from the donor oil molecules:

$$Ar\text{-}Cl + NaOH + RCH_2\text{-}CH_2R' \longrightarrow Ar\text{-}H + NaCl + RCH{=}CHR' + H_2O$$

Table 1. Destruction of chlorinated organic compounds

	Amount of chlorinated compound remaining by GC-EC, in mg/ L		
Reactant	DDT	Chlorobenzene	PCB
Ca	< 1	50	1,400
Mg	4	< 1	260
Fe	36,000	280	120,000
CaO	4	7	5

Table 2. Costs of Treating PCB-contaminated soil at Guam [8]

Treatment	Cost (1992 US$/ton)
Off-site incineration	2,000-3,320
On-site incineration	2,020
Secure landfill	910
Catalytic hydrogen transfer	245

The advantages of catalytic transfer hydrogenation are

- the organic product is a fuel
- low-cost chemicals are used, eg the H donor
- a cheap, non-toxic catalyst is needed that can be disposed of in treated matrices
- reagent recovery is eliminated
- treatment times are reduced to 1–2 h instead of several hours
- pollutants are destroyed in closed and non-pressurised equipment
- compounds of high and low Cl levels are completely dechlorinated because of the reactivity of the H radicals produced and the absence of any steric effects as in *K-PEG*
- the aqueous residues can be put to sewer
- continuous processing is possible
- it is cost effective and environmentally acceptable, as illustrated in Table 2.

4. SYNTHESES WITHOUT SOLVENTS

The best situation is no solvent at all, if that is at all feasible. It should be asked whether a solvent is really necessary; currently, solvent use is declining at 1.6% per year [12]. There are environmental and cost advantages if a solvent can be avoided. For that to happen

- the starting materials must be pure
- the physical nature of the reactants must permit molecular contact
- any by-products should be volatile.

Ideally, at least one of the reactants should be a liquid at the reaction temperature, and all other reactants dissolve in it [13].

Mechanochemical systems have been devised which ignore the solvent. The use of a ball mill which achieves localised high temperatures has been exploited in a few cases. Sodium α-sodioalkanoates can be prepared in this way [14]. Sodium α-sodioacetate is obtained in better than 90% yield from sodium acetate and sodium metal under nitrogen at 200–230° for 5 hours, thus avoiding the use of sodium amide:

$$2\ CH_3COOH + 2\ Na \longrightarrow 2\ NaCH_2COONa + H_2$$

Chlorinated hydrocarbons can be destroyed by ball mill technology, as described above and detailed in Table 1. There are claims of high impact temperatures.

5. WATER AS SOLVENT FOR ORGANIC REACTIONS

Solvents do have advantages [15], in that they

- slow the rate of fast bimolecular reactions
- prevent the temperature rising above the solution boiling point
- speed up the mixing of reactants
- facilitate the isolation of solids by filtration
- make pyrophoric compounds easier to handle, by means of an inert solvent

Reactions in aqueous media were previously limited to electrochemical processes and aldol condensations. This contrasts with many enzymatic processes, which are largely confined to aqueous media [16]. There are a number of examples now of the use of aqueous solvents for organic syntheses; transition-metal-catalysed reactions in water will be dealt with as a separate topic. Carbon-carbon bond formation is the essence of organic synthesis. There are two classes of organic reactions where water can successfully act as the solvent, the Diels-Alder reaction and the Claisen rearrangement.

5.1 The Diels-Alder Reaction

The reaction involves a thermal cycloaddition between a diene and an activated alkene or alkyne, sometimes catalysed by Lewis acids. A well known example is the reaction of butadiene with maleic anhydride, but this is not carried out in water. This type of reaction can actually be accelerated when water is the solvent, normally at temperatures below 100°, as water brings together the two non-polar substrates via hydrophobic effects. That is, there is a tendency for non-polar species to aggregate in water to decrease the hydrocarbon-water interfacial area [17, 18]. Water also markedly affects the selectivity as well as the rate of reaction. The following reaction of butenone with cyclopentadiene gives the results shown in Table 3 for various solvent conditions:

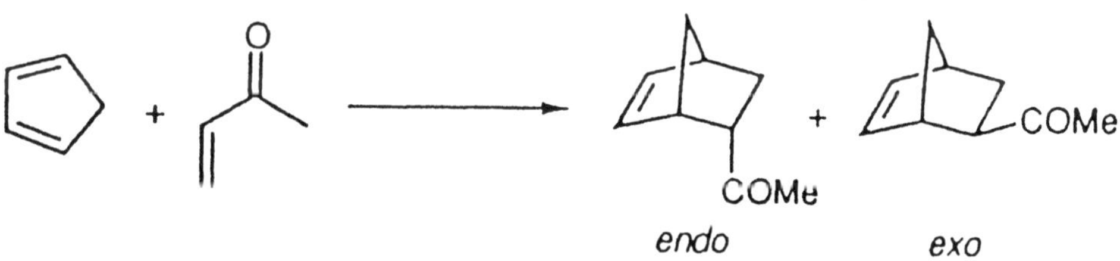

It can be seen that water is preferred over organic solvents as far as the rate of reaction is concerned; LiCl makes the aqueous solution more polar and enhances the hydrophobic effect, giving the fastest reaction rate in the series. The transition-state with the smaller surface

Table 3. Rates and selectivity of a Diels-Alder reaction in different solvents

Solvent	Relative reaction rate	*Endo*/*exo* ratio
iso octane	1	—
Cyclopentadiene	—	3.85
Ethanol	13	8.5
Water	741	21.4
4.9 *M* LiCl	1820	—

area is encouraged when water is the medium; the *endo* form has the more compact stereochemistry and is hence produced before the *exo* form. Aqueous Diels-Alder reactions have been used as key steps in the synthesis of 11-ketotestosterone and gibberellin.

5.2 The Claisen Rearrangement

This is the reaction of phenyl allyl ethers to form allyl substituted phenols, the rearrangement of phenyl allyl ether to 2-allylphenol being the simplest example:

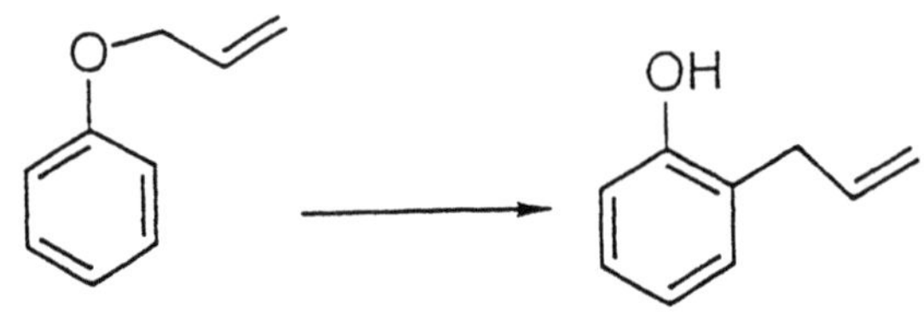

Similar reactions occur with vinyl allyl ethers. Because the volume change of activation has a negative value, water would be expected to have a similar effect to that observed in Diels-Alder reactions [16]. Thus chorismic acid rearrangements are 100 times faster in water than in methanol:

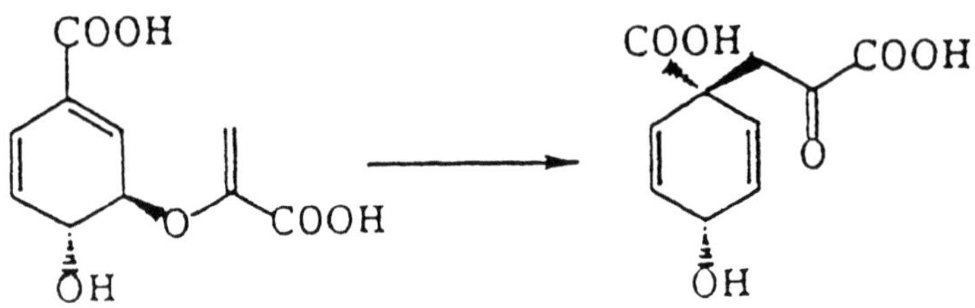

6. ORGANIC REACTIONS IN SUPERHEATED WATER

There are favourable changes in the chemical and physical properties of water at high temperatures and pressures [19, 20]:

- there is a gradual decrease in the dielectric constant (79 at 25°, 33 at 200° and 20 at 300°)
- at 300° the polarity and density are as for acetone at room temperature (density 0.71 g/cm^3)
- organic compounds dissolve more readily
- ionisation is enhanced, with K_w = 11.3 at 200°, suggesting possible promotion of acid- or base-catalysed reactions.

Organic molecules previously considered to be unreactive in water undergo many reactions when superheated. This includes reactions once thought to occur only in the presence of strong acid or base; for example, ethers and esters undergo facile cleavage and hydrolysis. Ionic chemistry predominates, with reactions often facilitated in brine or catalysed by clays, which suggests geothermal analogies. It should be emphasised that the pressures generated in a sealed system are about 4 MPa at 250°; this is well below the supercritical level of water, which is 35 MPa at 385–400°.

6.1 Hydrolysis

The decomposition of condensation polymers can occur in a neutral hydrolytic system in superheated water [21]. Polyethylene terephthalate is converted into its starting materials in less than one hour:

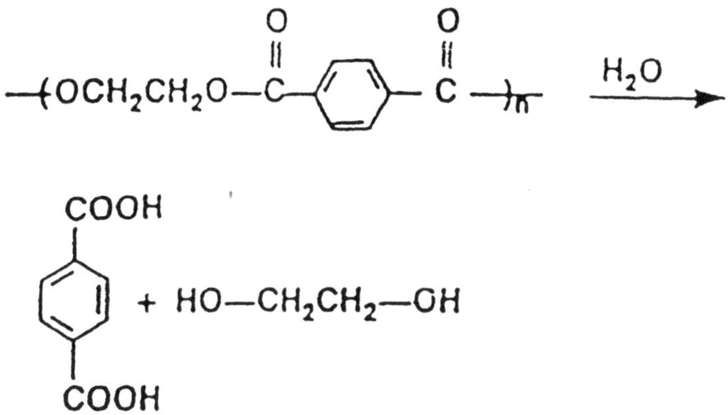

Polycarbonates, nylon 6 and 66 behave similarly, as do polyurethane foams which can be hydrolysed to reusable diamines and glycols, the group R typically being an aliphatic link such as tetramethylene:

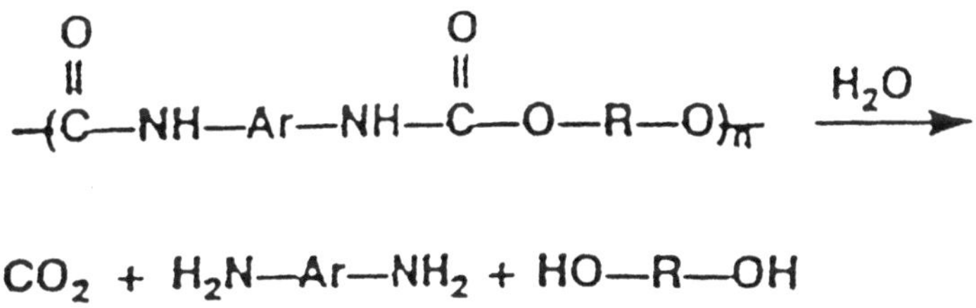

Aromatic acids can be made from nitriles in superheated water; no catalyst is needed but ammonia from the hydrolysis of the intermediate amide assists the hydrolysis. Glycerine is obtained from glycerol triacetate at 180–245° in the absence of a catalyst.

6.2 Bond Cleavage Reactions

Cyclohexyl phenyl compounds split with ring contraction of the cyclohexyl moiety in superheated water[20]:

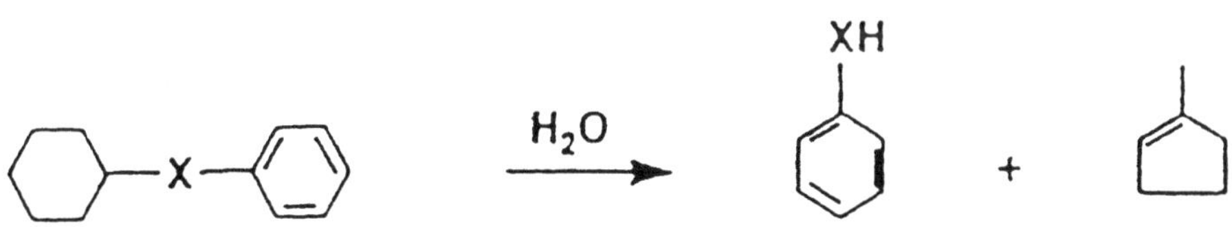

X = O,S,NH

Hence superheated water can sometimes provide a low cost, simpler, cleaner and safer system than that based on organic media. Water is starting to live up to its name as the "universal solvent".

7. CATALYSTS FOR MINIMISING TOXIC REAGENTS

Since the Bhopal disaster there has been a massive reappraisal by chemical manufacturers of the way that they produce, ship and store highly dangerous chemicals [21]. Chemical companies now aim to avoid or minimise the transportation and storage of hazardous and toxic reagents such as phosgene, methylisocyanate, hydrogen cyanide, hydrofluoric acid, hydrochloric acid, chlorine, acrylonitrile, formaldehyde, ethylene oxide and sulfuric acid.

Catalysts are the key to reducing storage and transportation, since they are used in over 90% of the processes carried out in the chemical and petroleum refining industries. Catalysts need to be both highly active and selective for economic viability in making chemicals, polymers and pharmaceuticals.

7.1 Phosgene

Phosgene has been eliminated as a reagent for preparing methylisocyanate, an intermediate in the manufacture of agricultural chemicals, once made as follows:

$$CH_3NH_2 + COCl_2 \longrightarrow CH_3NCO + 2HCl$$

Phosgene can be avoided by a catalytic oxidative-dehydrogenation route [22]:

$$CH_3NH_2 + CO \longrightarrow CH_3NHCHO$$

$$2CH_3NHCHO + O_2 \longrightarrow 2CH_3NCO + H_2O$$

Methylisocyanate can hence be made and converted on site, greatly decreasing the potential for exposure.

7.2 Hydrogen Cyanide

The use of hydrogen cyanide in the manufacture of methyl methacrylate from acetone via the cyanohydrin has been avoided by moving to the catalytic oxidation of isobutene [23]. Nitriles, important as intermediates in the production of amides, carboxylic acids, amines and esters, can now be made without the use of hydrogen cyanide by a new route which involves the catalysed reaction of aldehydes with hydroxylamine [22]:

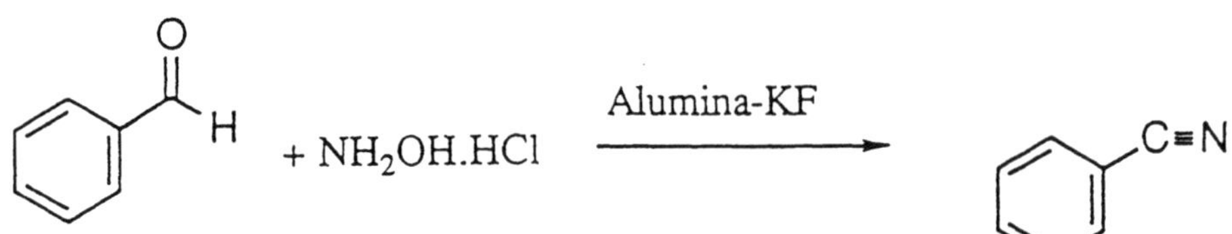

On-site production and consumption of hydrogen cyanide reduces the chance of accidental exposure. This can be achieved by a new process being developed by DuPont [24], which directly converts methane and ammonia to HCN, using a Pt, Rh or Pd catalyst on alumina at 1200°. The yield is high at 95%, there are almost no waste streams, and the process is quick to start up and shut down:

$$CH_4 + NH_3 \longrightarrow HCN + 3H_2$$

The making of nitriles from hydrogen cyanide can be done on-site now by a new nickel phosphite catalysed hydrocyanation of mono-, di-, tri- or polyolefins [22]. Hexamethylenediamine, a nylon intermediate, is an example:

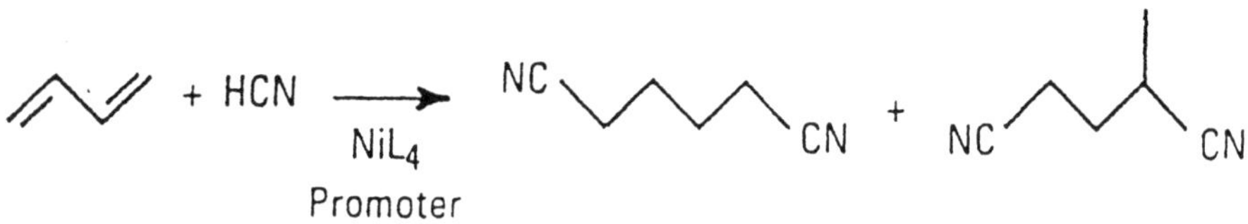

7.3 Tetrahydrofuran

A new DuPont method for the preparation of tetrahydrofuran (THF) from butane has been possible because of three innovations in catalyst technology [25]. A circulating fluidised bed reactor was harnessed for the partial oxidation of *n*-butane to maleic anhydride, from which virtually no by-products are generated. This required the development of attrition-resistant catalyst particles to withstand service in the reactor. Then a highly selective catalyst for the hydrogenation of maleic acid to THF was invented. The method combines significant environmental improvements with greater economy and better product quality. The old method was based on the reaction of acetylene with formaldehyde and proceeded through butene, butadiene and furan.

$$CH_3CH_2CH_2CH_3 \xrightarrow[\text{then hydr. } H_2]{\text{vapor phase oxidn. Pd/Re/C}} \text{HOOC-CH=CH-COOH} \longrightarrow \text{THF}$$

7.4 Styrene

A two-step synthesis of styrene from crude butadiene has been developed by Dow [26]. The first step makes use of a pressurised liquid phase dimerisation of butadiene with a Cu^+ zeolite which gives vinylcyclohexene selectively by what is actually a Diels-Alder reaction. The second converts this to styrene in 92% yield using a catalysed oxidative dehydrogenation:

$$CH_2\text{=CH-CH=}CH_2 \xrightarrow[100°]{Cu^+\text{/zeolite}} \text{vinylcyclohexene} \xrightarrow[400°]{\text{deHn.}} C_6H_5\text{-CH=}CH_2$$

7.5 Liquid Acid Catalysts

Materials such as HF and H_2SO_4 which are hazardous and present corrosion and waste disposal problems have been replaced by new solid acid catalysts in alkylation processes.

7.6 Other New Catalysts

Further extensive work [27] on better catalysts has been aimed at increasing their selectivity, thus decreasing side reactions which produce toxic substances, extending the operating life of catalysts, and developing ecologically safe methods. Thus new fibrous carbon supports for Pd catalysed dechlorination have high surface areas of up to 200 m^2/g and contain Pd hydride complexes; they show activity in the dechlorination of chlorobenzene and other chlorinated aromatics:

$$C_6H_5Cl \longrightarrow C_6H_6 \qquad 95\% \text{ at } 70°$$

Other metal-carbon hydrogenation catalysts give better performance in the hydrogenation of nitro compounds and benzoic acid, and in converting phenol to aniline with ammonia:

$$C_6H_5OH + NH_3 \longrightarrow C_6H_5NH_2 + H_2O$$

A porous silver catalyst for oxidising alcohols has improved metal surface properties and a better resistance to sintering in selective exothermic oxidation of alcohols, including ethylene glycol, butanol, ethanol and methanol:

$$CH_3OH \longrightarrow HCHO \qquad 95\% \text{ at } 500°$$

Finely divided nickel for hydrogenating fat has an active phase of high magnetic susceptibility, allowing the development of a magnetic separation method for the most active component. Cu-V bronzes and dilute hydrogen peroxide can be optimised for the oxidation of anthracene and phenanthrene, a promising variant in reorienting industrial processes because of the harmless reduction products of hydrogen peroxide; this and also nitrous oxide, another ecologically pure oxidising agent in the sense that the reduction product is nitrogen gas, are proposed as ecologically safe methods for the oxidation of organics. Finely divided metal zeolites have been used for the oxidation of benzene to phenol, and the preparation of similar aromatics like naphthols; they can decrease the production time and improve other ecological parameters of industrial processes.

8. CATALYSIS IN AQUEOUS SYSTEMS

When water-soluble catalysts are used in large-scale chemical manufacture product isolation is simpler and the catalyst can be recycled [16]. This is exemplified in transition-metal-catalysed C-C formation reactions. Hydroformylation of propylene is a major industrial process that produces an aldehyde and an alcohol from propylene, CO and H_2 :

$$CH_3CH{=}CH_2 + CO + H_2 \longrightarrow CH_3CH_2CH_2CHO + (CH_3)_2CHCHO$$

It can now be done in aqueous systems with Rh complexes of water-soluble phosphine ligands such as $(C_6H_5)_2P(m\text{-}\ C_6H_4SO_3Na)$, $P(m\text{-}\ C_6H_4SO_3Na)_3$ or $(C_6H_5)_2PCH_2CH_2NMe_3Cl$. The catalyst is practically insoluble in the aldehyde products, so it is easily recovered for reuse.

A catalyst of this type also converts the methanolic group in 5-hydroxymethylfurfural (HMF) to the corresponding acetic acid, leaving the aldehyde group intact as follows, Ø being the furan ring [28]:

$$\text{OHC–Ø–CH}_2\text{OH} \longrightarrow \text{OHC–Ø–CH}_2\text{COOH}$$

This offers the potential of producing fine chemicals from renewable resources in an environmentally friendly way, as HMF is readily available from carbohydrate sources; the acid product is a possible precursor for the corresponding dicarboxylic acid, which may be a promising starting material for condensation polymers of the polyester or polyamide type.

The exact role of water in all of these reactions is not well understood, but the significant advantages are that an aqueous medium is economical to use, and inflammable organic solvents are avoided

9. NON-TOXIC STARTING MATERIALS

One futuristic suggestion is that glucose be used as a starting material instead of benzene. Benzene, a potent carcinogen, comes from petroleum, a non-renewable resource. It is used to make vanillin, nylon 66, L-DOPA, and many phenols. As a first step, it is necessary to introduce oxygen into the molecule. In the production of adipic acid, this gives rise to 10% of the annual world increase in nitrous oxide, an ozone depleting, global warming substance. In catechol and vanillin manufacture, it is necessary to employ concentrated hydrogen peroxide, a highly energetic, corrosive material.

On the other hand, glucose is non-toxic and a renewable resource obtained from plant starch and cellulose. It is already highly oxygenated, so there need be no nitrous oxide formation or peroxide use involved when it is employed as a starting material [29, 30]. Water can be the solvent, at low temperatures, versus organic solvents and elevated temeratures with benzene. Biocatalysis is the key, using intact microbes, genetically engineered by transposing DNA fragments from one microbe into another, giving new biosynthetic pathways. The waste streams would be easily biodegradable, as for domestic sewage.

Manufacture of all of the above substances from glucose is economically competitive except for adipic acid, but the costs of eliminating benzene and nitrous oxide emissions raise the real cost of making this by the old route. However, there are still many scale up and processing problems yet to be solved. A multiple harvested source requiring minimal fertilising and pesticide inputs is preferred as the glucose source, the suggestion being via a type of grass rather than corn [29]. It is very early days, the main problems being that grain processing companies lack experience of chemical markets, and chemical companies lack experience in biocatalysis and renewable resources.

10. ENERGY EFFICIENT PROCESSES

Microwave technology is the main opportunity here. Microwave heating is very efficient relative to conventional methods [31, 32], because energy is absorbed directly by the material being heated, heating is extremely fast, and the temperature of the reaction mixture is more uniform. Microwaves are most commonly used at a frequency of 2450 MHz, when oscillations occur 4.9 x 10^9 times per second. There are two main mechanisms of heat conduction involved:

- ionic conduction, where the conductive agitation and movement of the dissolved ions in the applied electromagnetic field generates heat
- dielectric heating because of the continuing realignment of magnetic dipoles with the electric field component of the radiation.

Compounds with large dipole moments, such as water, ethanol, acetonitrile and dimethylsufoxide heat up very readily. Less polar substances like aromatic and aliphatic hydrocarbons, and highly ordered crystals are poorly absorbing.

A batch reactor based on microwave heating has been developed by CSIRO which holds the chemicals in a pressurised environment, and consists of a small (100 mL) ceramic vessel lined with an inert polymer which is transparent to the microwaves [33]. It can operate at up to 260° and 10 MPa (100 atm). A continuous microwave reactor has also been developed for organic synthesis which has a residence time of 1–2 minutes when the reaction coil is 3 m long and made of Teflon tubing 3 mm internal and 6 mm external diameter, giving a processing rate of 1 L/h [34]; tubing of up to 25 mm internal diameter is feasible, giving up to 100 L/h of reactant flow.

Under the prevailing conditions of high temperature and pressure, water takes on the properties of an organic solvent, behaving at 300° like acetone does at 25°, as has already been mentioned. Many organic compounds which do not normally react in water do so under these conditions, which represent an opportunity for cleaner production. Typical organic syntheses which have been carried out include [32, 35]:

- hydrolysis of tertiary amides in dilute HCl at 200° in a few minutes versus no outcome after several hours by conventional means
- phenylacetamide yield of 72% in 15 minutes at 185° by the reaction of acetophenone with sulfur and aqueous ammonia in isopropanol, the conventional route taking several hours (during which time the amide may be hydrolysed):
- 2-hydroxymethacrylates require heat for their formation from acrylates, but are themselves thermally labile; the conventional route requires several days at ambient temperature, but with microwaves at 160° and rapid quenching a comparable yield is obtained in two minutes:
- isopropylideneglycerol can be prepared cleanly in a microwave reactor by heating glycerol with an excess of acetone and a catalytic quantity of 2-toluenesulfonic acid, giving an 84% yield in 1.2 minutes, whereas the conventional method takes about one day for the same yield:
- the Claisen rearrangement of allyl phenyl ether to 2-allylphenol when carried out in water for 10 minutes at 200° gives 10% conversion, whereas at 240° the conversion is 84%. This is the first successful use of water as solvent for this reaction.

The many other examples studied include oxidation, decarboxylation, elimination, esterification, etherification, amidation, isomerisation and hydrolytic reactions.

11. SMALL-SCALE PRODUCTION

It is desirable to manufacture necessary chemicals on site in an economical way, despite the required quantities being smaller than the huge amounts previously thought necessary to achieve economies of scale. In this way transportation is minimised and accidental exposure risks reduced. There are several areas where advances are being made. Microwave technology along the lines describe above is well suited to the future manufacture of low volume, high value chemicals [31, 32].

Reactive distillation systems have been devised in which the reaction and separation steps are carried out in the same equipment, making for significant capital savings [23]. In the case of equilibrium limited reactions, higher yields are obtained as the reaction is pushed in the desired direction by removal of product from the system.

Membrane reactors are another innovation. During reaction, one of the products is continuously removed through a permeable membrane, so capital savings are again made by combining reaction and separation [23]. Where a normal reactor would require high temperatures to drive the reaction to completion, use of a membrane reactor may allow a lower temperature to be used, with continuous removal of product. By this means energy usage and by-product formation can be markedly reduced. Membranes composed of catalytic materials lead to a combined action which may provide further significant advances in this field [27].

12. CONCLUSIONS

Various facets of cleaner production are currently receiving much attention. It is noteworthy that the initial efforts in better housekeeping and improved end-of-pipe processes have matured. While there is still significant effort being expended on recovering useful materials from wastes, there is now a much better focus on modifying manufacturing techniques to avoid or markedly decrease the use of toxic starting materials, by-products and solvents. It is in this latter area that future research is likely to be concentrated.

13. ACKNOWLEDGMENTS

The input of a number of people has been invaluable; in particular, help and advice from Alan Cope, Russell Mills, Graeme Paul, Chris Strauss and Tsung-Tsan Su are gratefully acknowledged.

14. REFERENCES

1. Paul, G. A. L., Cleaner production in the chemical industry. *Proc. 6th Asian Chemical Congress*, Paper InC-O-4, Manila, 1995.
2. Browner, C. M., EPA's technology innovation strategy. *EPA Journal*, Fall 1994, pp. 9–12.
3. . Su, T-T., Cleaner production: an opportunity for chemical societies. *Proc. 6th Asian Chem. Congress*, Paper InC-K-1, Manila, 1995.
4. Mills, R., ICI Australia, personal communication.
5. Victorian EPA, BASF stretch their dollar further. *Cleaner Production Fact Sheet*, Melbourne, 1994.
6. Michelson, R., Living in the catchment. *Press Release*, Plastics & Chemicals Industry Association, Melbourne, 1994.
7. Bolto, B. A. and Pawlowski, L., *Wastewater Treatment by Ion Exchange*, pp. 53, 103, 114, E & F.N. Spon, London, 1987.
8. Independent Panel on Intractable Waste, Assessment of Management Options. *A Cleaner Australia*, Vol. 2, Canberra 1992.

9. Hawkes, R., The role of Plascon in waste destruction and cleaner production. *Proc. Economic Growth with Clean Production Conf.*, UNIDO / CSIRO, Melbourne, 1994.
10. Rowlands, S. A., Hall, A. K., McCormick, P. G., Street, R., Hart, R. J., Ebekll, G. F. and Donnecker, P., Destruction of toxic materials. *Nature*, **367**, 223, 1994.
11. Rogers, C. J., Kornel, A.and Sparks, H. L., Method for the destruction of halogenated compounds in a contaminated medium. *US Patents* 5,019,175 (28 May 1991); 5,039,350 (13 Aug 1991); 5,064,506 (12 Nov 1991).
12. Kirschner, E. M., Environment, health concerns force shift in use of organic solvents. *C&EN*, 20 June, 1994, pp. 13–20.
13. Grindstaff, T. H., Syntheses without solvents. *C&EN*, 7 Nov, 1994, p. 3.
14. Depree, D. O. and Mattson, G. W., Preparation of sodium alpha-sodioalkanoates by the reaction of sodium alkanoates with sodium metal. *I&EC Product R&D*, **2**, 238–239, 1963.
15. Mendenhall, D., Don't knock solvents. *C&EN*, 6 Feb, 1995, p. 3.
16. Li, C.-J., Organic reactions in aqueous media - with a focus on carbon- carbon bond formation. *Chemical Reviews*, **93**, 2023–2035, 1993.
17. Lubineau, A., Augé, J. and Queneau, Y., Water-promoted organic reactions. *Synthesis*, 741–760, 1994.
18. Breslow, R., Hydrophobic effects on simple organic reactions in water. *Accounts of Chemical Research*, **24**, 159–164, 1991.
19. Pitzer, K. S., Dielectric constant of water at very high temperature and pressure. *Proc. Natl. Acad. Sci. USA*, **80**, 4575–4576, 1983.
20. Siskin, M. and Katritzky, A. R., Reactivity of organic compounds in hot water: geochemical and technological implications. *Science*, **254**, 231–237, 1991.
21. Heylin, M., Bhopal: ten years after. *C&EN*, 19 Dec, 1994, p. 3.
22. Bell, A. T., Manzer, L. E., Chen, N. Y., Weekman, V. W., Hegedus, L. L. and Pereira, C. J., Protecting the environment through catalysis. *Chem. Eng. Progress*, **91**, 26–34, 1995.
23. Pratt, K. C., Contributions of catalytic technology to cleaner production. *Proc. Economic Growth with Clean Production Conf.*, UNIDO / CSIRO, Melbourne, 1994.
24. Haggin, J., Catalysis gains widening role in environmental protection. *C&EN*, 14 Feb, 1994, pp. 22–30.
25. Haggin, J., Innovations in catalysis create environmentally friendly THF process. *C&EN*, 3 Apr, 1995, pp. 20–23.
26. Dixit, R. S. and Murchison, C. B. Crude butadiene to styrene process. *Proc. AIChE Spring National Meeting*, Atlanta, April 1994, pp. 118–133.
27. Yurchenko, E. N., Principal trends in the development of catalysts and processes intended for the protection of the environment. *Russian J. Applied Chem.*. **66**, 1991–1998, 1994, or *Zh. Prikl. Khim.*, **66**, 2641–2650, 1994.
28. Haggin, J., Carbonylation catalyst acts in aqueous media. *C&EN*, 13 Feb, 1995, p. 38.
29. Frost, J., Green chemistry at work. *EPA Journal*, Fall 1994, p. 22.
30. Illman, D. L., Environmentally benign chemistry aims for processes that don't pollute. *C&EN*, 25 Sep, 1994, pp. 22–27.
31. Bennett, B., Microwave chemistry heats up. *Ecos*, **79**, 29–31, 1994.
32. Strauss, C. R. and Trainor, R. W., Developments in microwave-assisted organic chemistry. *Aust. J. Chem.*, in the press.
33. Raner, K. D., Strauss, C. R., Trainor, R. W. and Thorn, J. S., A new microwave reactor for batchwise organic synthesis. *J. Org. Chem.*, **60**, 2456–2460, 1995.
34. Cablewski, T., Faux, A. F. and Strauss, C. R., Development and application of a continuous microwave reactor for organic synthesis. *J. Org. Chem.*, **59**, 3408–3412, 1994.
35. Strauss. C. R. and Trainor, R. W., Improved synthesis of phenylacetamides by Willgerodt reactions with microwave heating. *OPPI Briefs*, **27**, in the press, 1995.

2

RISK MANAGEMENT CONSIDERATIONS FOR COST-EFFECTIVE ENVIRONMENTAL DECISIONMAKING

Marjorie Auyong Gonzalez

University of California
Lawrence Livermore National Laboratory
Livermore, California

BACKGROUND

Scientific publications and media reports continually remind us about the environmental hazards that surround us. We are appraised of the environmental legacies left by chemical industries, the defense complex, and even our local dry cleaning establishments. Governmental regulations have dictated that industry provide detailed listings of their input materials, wastes, and emissions to the public and perform risk assessments to demonstrate compliance with standards. These regulations were designed to make industry more accountable and to give the public information that would allow them to understand risks and either work for change or accept their living conditions. This process would appear to be rational, fair, and acceptable to both industry and the public. However, our inability to reach agreement on questions such as "How Clean is Clean?" or "Is it Safe?" after more than ten years of scientific and public discussions, coupled with the frequency of environmental demonstrations throughout the world, serves as evidence that "acceptable risk" has not yet been defined.

RISK ASSESSMENT DEFINED

Risk assessment is a risk management tool that scientists and government officials use to estimate the increased risk of health problems in people who are exposed to varying amounts of harmful substances. According to the U.S. Environmental Protection Agency, risk assessment is a four step process involving (1) hazard identification, (2) exposure assessment, (3) dose-response assessment, and (4) risk characterization.[1] Hazard identification involves reviewing and analyzing health effects data and studies to determine what health problems are caused by a substance. An exposure assessment estimates the magni-

Chemistry for the Protection of the Environment 2, edited by Pawłowski et al.
Plenum Press, New York, 1996

tude, frequency, duration, and route of exposure, and a dose-response assessment estimates the incidence of health problems at different exposures. The risk characterization uses the dose-response and exposure assessments to describe the type and size of any increased risk expected as a result of exposure to a harmful substance.

RISK ASSESSMENT LIMITATIONS

In recent years, risk estimates derived from risk assessments have been used to set regulatory requirements in the U.S. This has led to complaints from industry because such requirements are projected to cost them $185 billion by the year 2000.[2]

In theory, assessing the risks of activities to determine potential environmental consequences is a good one. However, there are limitations associated with basing decisions strictly on a risk assessment conclusions. These include:

- Absence of human effects data—data from studies of a substance's effect on animals may or may not accurately predict its effect on humans.
- Bias—without concrete scientific evidence, assumptions are created that can overestimate or underestimate risks. Industry representatives argue that regulators have a conservative bias which leads to expensive programs to protect insignificant threats to public health and the environment.
- Difficult comparisons—methods do not allow weighing different health and environmental effects against one another (e.g. dioxin exposure vs. ecosystem degradation vs. cancer death). Furthermore, there are natural (earthquake fatality) or societal (smoking a pack of cigarettes a day) vs. exposure (living next to a hazardous waste storage facility) comparisons that cannot be compared reliably.
- Value Judgments—In the U.S., we have seen many instances where the results of technical risk assessments were disregarded by affected public groups. Most people are able to accept the risks associated with decisions that they personally make. However, they generally want zero risk if they believe another person's actions are creating a risk that they are uncertain about or that they do not welcome. The actions may not be harmful, but an individual's personal fears, may lead to non-acceptance of all risk assessment conclusions other than zero risk. Unfortunately, most of the population only understands two probabilities: *zero,* which means an event will not occur or *one*, which means an event will occur with certainty. They want results expressed simply—safe or unsafe. Coupled with the limitations described above, it is not surprising that the results of risk assessments are often subject to challenge.

COST-EFFECTIVE ENVIRONMENTAL DECISIONMAKING

While technical risk assessments may not lead to conclusive results, they may be useful scientific tools in a broader risk management scheme. Such a scheme would acknowledge that absolute certainty with respect to risk is not a rational outcome and would be qualitative, allowing inclusion of a broader cross section of stakeholders in the decisionmaking process. For example, a value added approach[3] proposed by Vernon Grose, combines risk severity (negligible to catastrophic), risk probability (rare to very frequent), and risk countermeasures cost (exorbitant to inexpensive) with a methodology that collects disparate risks, ranks them for significance, and displays the risks in a format that can be understood by all the ranking process participants. Since people are generally able to accept the risks associated with decisions that they personally make, the costs associated with regulatory burden and court challenges could be averted by employing such approaches.

MIDDLE URALS POLLUTION PREVENTION PRIORITIES ASSESSMENT PROJECT

In 1993, scientists from the Lawrence LIvermore National Laboratory, the Russian Federal Nuclear Center Institute of Technical Physics, and the Institute of Industrial Ecology, Urals Branch established a collaboration to assess pollution priorities in the Middle Urals region of Russia. The project scientists are employing scientific risk assessment tools for evaluating existing data. However, because a broad spectrum of participants from the oblast government, nongovernmental organizations, and scientific institute will participate in the priority setting, a methodology is being developed to allow ranking input by the participants. The urgency of the problems in this region and the limited resources available for remediation are driving priorities toward pollution prevention efforts, stabilization (not clean-up) of contaminated areas, and improvements in food and water supply systems. The risk management approach, which includes science based risk assessment as a tool, should minimize delays and be cost effective.

CONCLUSIONS

Progress in addressing environmental problems has been severely hampered by the inability of stakeholders—government agencies, nongovernmental groups, and industry—to agree on acceptable risk.

Science based conclusions would be a desirable basis for determining acceptable risk, but unfortunately the complex linkage between low levels of harmful substances and cancer or other health effects is not fully understood. The costs involved with further research on the relationship and with examining the behavior and fate of thousands of chemicals is prohibitive. Clearly, we must employ new semi-quantitative or qualitative methodologies that allow us to use existing data and employ interested parties to prioritize actions in a way that maximizes environmental and health quality, but is also cost beneficial.

ACKNOWLEDGMENT

This work was performed under the auspices of the U. S. Department of Energy by the Lawrence Livermore National Laboratory under contract number W-7405-ENG-48.

REFERENCES

1. U.S. Environmental Protection Agency, *Risk Assessment for Toxic Air Pollutants* , EPA 450/3–90–024 (March 1991)
2. Baker, Beth, "Risk Assessment Kills Bills," *Bioscience,* (January 1995), p. 15.
3. Grose, Vernon L. *Managing Risk*, Omega System Group, Arlington, VA (1987)

3

THE GLOBAL PERSPECTIVE—WHY POLLUTION PREVENTION?

CPE X Address

John A. Marchetti[1] and Elizabeth McPherson[2]

[1] United States Department of Energy
[2] McPherson Environmental Resources

Good morning. I certainly appreciate the opportunity to present to an international forum some of the thoughts I have on the importance of a globally unified approach to the environmental defense of this planet. I believe each country represented here will benefit from our combined efforts to use pollution prevention as the spearhead for achieving our economic goals in a new, environmentally safe and protective manner. Why? Because pollution prevention is the right way to go...because pollution prevention preserves natural and economic resources, and jobs.

I am with the Department of Energy's Office of Defense Programs. I want to cite to you today the results of good pollution prevention practices that have been experienced by my organization and U.S. private businesses. The successes do, I think, represent strong initiatives that address barriers and needs common to most members of the international community, and my hopes are that you will find these case studies to have some value and applicability in your country.

It is no coincidence that some of the biggest and most successful companies in America are considered to be among the "Best-in-Class" when it comes to aggressive and effective Pollution Prevention programs. It should be no surprise to hear companies like Du Pont condense their definition of waste to this: "Waste is anything that is not product..." In other words, "Waste is waste."

And waste generation is something we in America have been pretty good at. An article in the April 1991 issue of Scientific American included the finding that, for each dollar of goods sold, American industries produce almost five times more waste than the Japanese, and over twice more than the Germans.

I understand that this audience is made up of individuals who are already interested and, to some degree, educated in the advantages of Pollution Prevention—but let me give you a few numbers. These statistics are based on the successful results of a Department of Energy program that has been ongoing for a little over four years, called the National Industrial Competitiveness through Energy, Environment, and Economics, or NICE3. The

NICE3 program began as a cooperative agreement between the U.S. Environmental Protection Agency and DOE, each of which initially contributed $1 million.

The NICE3 goals are:

1. To promote technology innovations that simultaneously improve energy efficiency, reduce waste production, increase economic competitiveness, improve productivity, preserve and increase jobs, and
2. to improve cooperation at the regional and state levels among DOE, EPA, the States, and the Department of Commerce.

The results: The 17 funded projects now in progress cost DOE $4.4 million and, in 2010, will result in:

- Energy savings of 91 trillion BTUs per year
- Economic savings of 2 billion dollars
- Waste savings of:

64 thousand tons of waste solvents, CFCs, other volatile organics
4.7 million tons of carbon dioxide from reduced use of energy
35 thousand tons of sulfur dioxide from reduced use of energy
17 thousand tons of nitrogen oxides from energy reduction
4 million tons of hazardous wastes
11 million tons of nonhazardous wastes
12 billion tons of wastewater

and finally:

- Job savings. 2500 person years by 2010.

The NICE3 program has commercialized two technologies. NICE3-funded technologies have created an estimated 140 jobs for each $1 million invested.

I can give you inspiring case studies of huge cost and energy savings that can result from easy-to- implement, environmentally sensitive modifications to the way we do business. But I want to talk a little more about jobs, one of the spin-off elements of what is commonly referred to as "sustainable development."

What I'd like to concentrate on for a few minutes is Pollution Prevention awareness, including "culture change," and training and retraining the DOE-directed workforce to understand and participate in the expanding the scope of our mission.

The primary function of DOE's Defense Programs has always been national security. That mission remains the same—but, since the end of the Cold War, the emphasis has changed. We now need to defend the planet against further degradation as well. Strategies to prevent additional pollution should be applied even while we clean up the results of years of neglect. Much more important is that we use this time of relative peace and our considerable resources in the areas of science, engineering and manufacturing to put in place an entire pollution prevention system—a system that will act as a filter to lower the environmental impact of all industrial and research and development activities in the future. This system would also act as a catalyst to give small industry access to DOE's technology resources bringing innovation to economic growth, ensuring a leadership role for our country on the global market and economic security in the decades to come.

If we put this system in place and if we do it correctly, we're talking about high quality jobs—jobs required at every level and in all areas of expertise. There is plenty of work for enlightened and educated technologists, scientists, maintenance personnel, engineers, technicians, machinists, and many others. The key word here is "educated." Our workforce needs to be trained, not only to the awareness level, but to the level required for them to perform, as needed, different job functions—ones that promote the hierarchy of preferred environmental practices stated in the EPA's Pollution Prevention Act of 1990: reduce, recycle, treat, then dispose—and job functions that actively support environmental defense.

Dr. Joseph Romm, Special Assistant to the Deputy Secretary of Energy, in his book Lean-and-Clean, states:

"Lean and clean management saves jobs by replacing capital expenditures and resource consumption with human ingenuity and labor... Lean and clean saves jobs by reducing the non-labor costs of doing business."

He follows up with these words of advice:

- *Use workers' knowledge to cut resource use and costs.*
- *Intensify training of workers and managers.*
- *Reduce energy use everywhere.*

Dr. Romm cites an example of the first option. He tells how a U.S. company, Republic Engineered Steel, kept layoffs to a bare minimum during the recent slowdown in the steel industry by putting in place an employee suggestion system. The employee suggestions, many aimed at recycling and pollution prevention, have already reduced annual costs by $45 million, and will eventually cut another $20 million.

And in the area of intensified training, Dr. Romm states that, when forced to cut production, Honda's Marysville, Ohio, plant increased training time by 5% rather than cutting the work force by the same amount.

In support of the third recommendation regarding energy use, he discusses how a comprehensive efficiency program saved another U.S. company, Southwire Corporation, $40 million in energy costs from 1981 to 1988. During a rough financial period, the savings were almost equal to all of the company's profits and may well have saved Southwire a total of 4000 jobs at ten different plants.

Many of the drivers for pollution prevention are obvious—cost and waste savings, higher quality, safer jobs, and so forth. But if we take a good look at the (spatial and temporal) BIG picture, the trends we are beginning to notice intensify our recognition of the potential benefits. These trends should motivate any country.

The Organization for Economic Cooperation and Development has provided a rough estimate for global environmental spending in 1990 of $200 billion. About one quarter of this spending was attributed to international trade. U.S. environmental export earnings, according to these estimates, were about $8 billion, while Germany earned $11 billion. By the year 2,000, the organization estimates that the world market for environmental goods and services will grow to $300 billion.

Granted, to date many of these environmental dollars have been or are being spent on remediation and end-of-pipe controls. But this will not be the case in the future. New environmental regulations and policies will mandate, and the public will demand, that industry provide cleaner, greener products. Manufacturers will be required to provide these products in an environmentally responsible manner.

From a March 1993 draft document produced by DOE's Office of Domestic and International Energy Policy titled *Economic Growth Through Clean Industry: An Emerging DOE Mission and Opportunity for a National Initiative* comes this statement:

> "Demand for green products creates important new business opportunities, both in manufacturing the products themselves and in developing the infrastructure to support them. But there's

> more to it than simply capitalizing on a new market niche. Failure to recognize the trend towards clean processes, technologies and products could place U.S. industry behind in the race to produce the technologies of the 21st century."

Secretary of Energy O'Leary describes the National Clean Industry Initiative in her May 14, 1993, letter to Vice-President Gore as a proposal that would "provide a customer-driven, decentralized mechanism for delivering a full range of one-stop-shopping services through the states, thus integrating and leveraging current Federal, State and private efforts." She continues, "The result will be to improve industrial efficiency and to better position U.S. businesses in the emerging international market for clean technology."

The "Implementation Plan for the Collaborative Technology Element of the National Clean Industry Initiative" says the mission of the NCII is "to help increase the productivity and competitiveness of U.S. Industry through shared acceleration of the development and deployment of innovative clean manufacturing technologies, processes, and information...This joining and concentration of talent and resources will improve America's overall environment while enhancing U.S. Industry's ability to compete in the global market."

Another federal initiative, the Technology Reinvestment Project, was developed to execute the programs authorized under the Defense Conversion, Reinvestment, and Transition Assistance Act of Fiscal Year 1993 and other legislation. The TRP is chaired by the Advanced Research Projects Agency of the Department of Defense, the Department of Energy/Defense Programs, the Department of Commerce's National Institute of Standards and Technology, the National Science Foundation, and the National Aeronautics and Space Administration. The TRP's programs are structured to expand "high-quality employment opportunities in commercial and dual-use United States industries and demonstrably enhance U.S. competitiveness." One of the specific areas of interest in the TRP's technology development activity is Environmentally Conscious Electronic Systems Manufacturing. The *Program Information Package for the Defense Technology Conversion Investment and Transition Assistance* explains that the electronics and computer industry is the largest manufacturing employer in the world. The document states, "Manufacturing by-products of the electronics industry and the disposition of electronic products are raising increasingly important technical and financial issues, and there is a need to improve processes, materials, and manufacturing equipment to control production of hazardous waste material and to be able to recycle the products."

The messages being sent to American businesses and government agencies and contractors are clear. There is a global need for industries to adopt greener practices including environmentally conscious manufacturing and design for the environment, and the future belongs to those who "seize the day."

Responses to these variously worded appeals have begun. DOE's national laboratories are researching and developing environmentally benign products and processes. Following the R&D phase, other weapons complex installations are poised with operating, fully-equipped production facilities and trained operators to perform full-scale testing of potential advances in manufacturing technologies. Once the technology is proven, these same installations are available to explain the benefits of these new processes and products, and assist U.S. businesses across the country in acquiring and using them.

For instance, DOE's Kansas City Plant has an active technology transfer program that reaches out nationwide to small- and medium-sized businesses. Since the beginning of 1992, the plant's personnel have helped nearly 150 companies analyze their needs and determine cost-effective options in compliance with regulatory requirements. Over half of these activities involve Pollution Prevention and environmentally conscious manufacturing. In addition, the plant sponsored a total of 14 workshops on environmentally safe painting techniques and solvent substitutions attended by nearly 700 individuals for small- and medium-sized manufacturers.

We also have an opportunity to, in a sense, recycle people. If we can cut costs through pollution prevention, we will be able to save jobs. Just as the DOE-supported "Partnership for Environmental Technology Education" or PETE program, is our mechanism for training young people to understand and cope with the environmental challenges of today, we in DOE must likewise retrain our workers to perform, in an environmentally protective manner, jobs that may be fundamentally and functionally different from jobs they previously performed.

The Department of Energy, by reapplying resources to support pollution prevention and environmentally conscious manufacturing, is in a position to assist not only private industry, but also other government departments. In a recent meeting between the Department of Energy and the Department of Labor, it was agreed that training must involve both management and labor in productivity improvement and pollution prevention. DOE and the Department of Labor agreed that labor should consider as a win-win approach the initiative for U.S. industries and federal facilities to become "lean and clean," driven in part by prompting from the White House, because it achieves cost savings without labor reduction.

Many of the initiatives that we have discussed are the results of innovative planning by the individual installations. These initiatives need to be funded and otherwise supported as we undertake what we have called our "expanded mission."

Each of us has the job of promoting the value of our efforts in the field of pollution prevention. It is important that we convince ourselves and we willing to convince others that pollution prevention is, indeed, the best and most effective way to save our dollars, our resources, and our jobs.

I hope I am leaving you with a sense of our, my, commitment to environmental defense through pollution prevention. I believe that the majority of the populace of the United States shares, with the majority of the populations of the countries you represent, a respect for this planet and a sense of responsibility for its continued welfare. I believe Harvard biologist E.O. Wilson, who proposes it is important for us to foster, biophilia or love of life. If this attitude were to be adopted worldwide, I think we could overcome a great number of the barriers we all face. Thank you for your time and attention.

4

ENVIRONMENTAL ENGINEERING IN POLAND*

Availability of Domestic and Foreign Technology

Tomasz Winnicki†

Technical University of Wroclaw
Wroclaw, Poland

INTRODUCTION

Central and Eastern Europe (C&EE) lack effective partnership and financial support for technology transfer and the commercial implementation of practical programmes. Recognition and effective support for the rapidly-growing private sector environmental technology industry is a key to the future well-being of C&EE countries and their sustained technology relationship with the West.

There is sufficient understanding in the West that Poland and its neighbouring countries have a large number of competed and skilled scientists and engineers, capable of understanding and implementing new technologies rapidly. In many instances, scientists from the former Eastern Block countries have, in isolation, developed innovative technology which reflect today's Western state-of-the-art technologies.

However, Poland lacks Western assistance in the techniques and skills needed to commercialise these developments and exploit them in a wider marketplace, which is why joint ventures are so important and potentially have great benefits to both parties.

To understand the current environmental situation in C&EE, it is necessary to provide some background on developments since the Second World War, because the political and economic situation in C&EE after the War greatly influenced the development of environmental technology in the region.

Although a high priority of ecological values was emphasised in all official post-war programmes they were never realised in practice. Either programme costs led to the omission of environmental protection elements, or the choice of technology was inappropriate and enforcement policies were inadequate, finally, it happened quite often, that resources

* Some information and opinions already presented in BMT News (apologies to British Maritime Technology Ltd).

† Wyb.Wyspianskiego 27, 50–370 Wroclaw, Poland. E-Mail: WINNICKI@pwr.wroc.pl

planned for the environmental part of the investment were used for other purposes as a result of delayed realisations and increased costs of the project.

The political climate and financial restrictions have caused the virtual isolation of intellectuals, scientists and technologists from Western information.

SITUATION AT THE POLITICAL BREAKING-POINT OF 1989

Environmental designers were particularly isolated from Western technology and based their designs mainly on the traditional civil engineering "cumulative approach". One outcome of this was the arbitrary collection and mixing of all wastes from municipal, industrial and other services. Concern about the consequences of such compositions came later. The management of hazardous waste production was an almost unknown subject. The prevailing "waste management strategy" for long-term storage of waste mixtures in open lots has invoked secondary effects of self-combustion, dusting and leaching which have resulted in a disastrous "domino effect" of harmful impacts of unprecedented extent.

Although there was a strong emphasis on air pollution control and particulates were captured with quite a high efficiency, gas abatement technology was almost completely lacking. Energy generation based on hard-coal and lignite, not assisted by sufficient acid-gases control, caused a tremendous ecological disaster, widely know now as the "black triangle" or "sulphur triangle" area, which has transferred acid rain up to Scandinavia.

Conventional water and sewage treatment technology had been well taught and both designers and operators of industrial systems are sufficiently prepared. However, also in this domain modern separation technology, based on membrane processes, as well as other means of water and valuable species recovery is still deficient. The discharge of brackish water from coal- and copper-mines became the Number 1 problem in Poland's water economy.

One of the most severe local environmental hazards has been caused by land and ground-water pollution by hydrocarbons and other toxic and harmful products left after military installations. The tremendous extent of the problem was revealed when the post-Soviet army was withdrawn from Poland's western territories. There was no domestic technology to remediate the soil and to clean-up underground waters, hence almost all current attempts have been based on partnership with highly specialised foreign companies or on contracting them.

MAJOR NEEDS FOR NEW TECHNOLOGIES

The extent of environmental degradation now identified in C&EE has highlighted the need for urgent action and created a strategic demand for the introduction of new technologies and implementation techniques. The long, certainly not complete list of technologies required includes:

- separation techniques for solid and liquid waste including physico-chemical and biotechnical methods
- recycling of waste products including recovery of energy and semiprecious materials
- reclamation of polluted land using physical, chemical and biological decontamination and recycling processes
- neutralisation and recovery of acid exhaust components from industrial flues, especially from large power generation plants

- destruction of PCBs, VOC and other toxic organics by reductive methods, avoiding secondary emission of dioxines and furanes
- introduction of a complex brackish water economy programme, including hydrological means as well as salty streams separation & salt recovery processes
- controlled and effective combustion of coal fuels by the implementation of new type furnaces
- introduction of a wide range of new manufacturing technologies to substitute environmentally troublesome products by biodegradable ones (e.g. packing technology)
- implementation of second generation environmental monitoring systems allowing not only the measuring but also the controlling of pollution sources
- introduction of alternative energy resources.

TECHNOLOGY TRANSFER CONDITIONS

Western bankers and management consultants have came in for considerable criticism in C & EE for charging high fees and for proposing business solutions that may work well in London or New York but not, as yet, in Lublin or Minsk.

Effective technology transfer should be based on wider knowledge of local Polish circumstances: (1) intellectual and manpower conditions, (2) state-of-the-art of domestic technology and market to identify the most important gaps to be filled, (3) banking regulations including the current and planned inflation rate and resulting interest rates, as well as cash flow problems, (4) internal structure of privatised companies and existing managerial capabilities (5) the most effective type of links between Polish and foreign companies, including such forms as: (a) trade representation, (b) consulting, (c) licensing, (d) permanent trade & technology cooperation on contract basis and finally (e) the creation of joint venture company partnerships.

It was anticipated that international financial assistance to solve Polish environmental problems would embrace a much wider range and would not concentrate to such an extent on projects realised by foreign contractors. The Polish environmental private sector requires low interest credits not only to compete within international tenders, but also to survive. The long list of domestic environmental tasks to be completed in this and the future decade, inclines us to be optimistic as regards that branch of Polish economy.

5

MIDDLE URALS' POLLUTION PREVENTION PRIORITIES ASSESSMENT PROJECT

Marjorie A. Gonzalez,[1] Ronald L. Ott,[1] Victor Chukanov,[2] Anatoly Kuligin,[2] Vadim Simonenko,[3] Albert Vasyliev,[3] and Nina Barysheva[3]

[1] Lawrence Livermore National Laboratory
Livermore, California
[2] Institute of Industrial Ecology
Ekaterinburg, Russia
[3] Russian Federal Nuclear Center
Snezhinsk, Russia

ABSTRACT

The Middle Urals is an important Russian industrial region. The key industries are also the most environmentally damaging: mining, metallurgical and chemical industries. There are some 600 large-sized and medium-sized enterprises located within the Middle Urals' region. Their annual solid and gaseous chemical releases have led to exceeding some maximum permissible contaminant concentrations by factors of tens and hundreds. The environmental problems of the Middle Urals are of such magnitude, seriousness, and urgency that the limited available resources can be applied only to the problems of the highest priority in the most cost-effective way.

By the combined efforts of scientists from Lawrence Livermore National Laboratory (USA), Institute of Industrial Ecology (Ekaterinburg, Russia) and Russian Federal Nuclear Center (Snezhinsk, Russia) the project on Environmental Priorities Assessment was initiated in 1993. Because the project will cut across a spectrum of Russian environmental, social, and political issues, it has been established as a genuine Russian effort led by Russian principals. Russian participants are the prime movers and decision-makers, and LLNL participants are advisors.

A preliminary project has been completed to gather relevant environmental data and to develop a formal proposal for the full priorities assessment project for submittal to the International Science and Technology Center. The proposed priorities assessment methodology will be described in this paper.

The specific objectives of this project are to develop and to implement a methodology to establish Russian priorities for future pollution prevention efforts in a limited geographic region of the Middle Urals (a part of Chelyabinsk and Sverdlovsk Oblasts). This methodology will be developed on two geographic levels: local (town scale) and regional (region scale). Detailed environmental analysis will be performed on a local scale and extrapolated to the regional scale.

This project will allow Russian federal, regional and local governments to more effectively focus efforts and direct capital investments to reduce the detrimental influence of current pollution sources on the environment and ultimately on human health, at both regional and local levels.

Results of the project and the experience accumulated during its performance will be useful thereafter to solve analogous problems in other regions of the Urals and, possibly, in other parts of Russia and other nations.

INTRODUCTION

The Middle Urals region of Russia, which includes the central and southern part of Sverdlovskaya oblast and the central and northern part of Chelyabinskaya oblast, is an important Russian industrial region. The key industries are also the most environmentally damaging: mining, metallurgical and chemical industries. There are some 600 large- and medium-sized enterprises located within this region.

According to the data presented in "The State Report on Environmental Conditions in Russia in 1992", the Ural region has the largest total atmospheric emissions of gaseous and aerosol wastes in Russia. The main contributors of these discharges are the industrial enterprises and transportation vehicles of Sverdlovskaya and Chelyabinskaya oblasts.

The Middle Urals' region is situated at the junction of three water basins, all of which are seriously contaminated with heavy metals, petroleum products, organic matter, nitrates and other pollutants. The levels and diversity of contaminants make it difficult to solve environmental problems, including those associated with regional food and water supplies.

The territory under discussion is also characterized by intensive agriculture and by intensive consumption of natural resources: mining, water use, State timber purchasing, recreation, wild berries and mushrooms.

The impact on the environment is redoubled by a high population density.

The environmental problems of the Middle Urals are of such magnitude, seriousness, and urgency that the limited available resources can be applied only to the problems of the highest priority in the most cost-effective way.

PROJECT OBJECTIVES

By the combined efforts of scientists from Lawrence Livermore National Laboratory (USA), Institute of Industrial Ecology (Ekaterinburg, Russia) and Russian Federal Nuclear Center (Snezhinsk, Russia) the Middle Urals' Pollution Prevention Priorities Assessment Project was initiated in 1993.

A preliminary project has been completed to gather relevant environmental data and to develop a formal proposal for the full priorities assessment project for submittal to the International Science and Technology Center.

The focus of the priorities assessment project is on immediate pollution problems from current industrial operations, the migration of legacy wastes, and potential accidental situations and their solutions, rather than long-term research related to fundamental environmental investigation. The specific objectives of this project are to develop and to implement a methodology to establish Russian priorities for future pollution prevention efforts in a limited geographic region of the Middle Urals (a part of Chelyabinsk and Sverdlovsk oblasts).

The project will be based on available environmental data obtained by Russian governmental organizations. Existing technical analyses for the assessment of environmental conditions and effects of pollution sources will be used and existing methods for modeling the transport of pollutants in the environment will be the basis for the methodology. The methodology will also include techniques developed at the Russian Federal Nuclear Center by nuclear weapons scientists which can be applied to environmental problems.

Methods for gathering, organizing, and systematically analyzing environmental information, and for formulating and testing priorities assessment models will be created for the proposed project. Diverse opinions, including those from governmental, nongovernmental, and scientific organizations are being integrated into the priority setting process.

This work is being performed in compliance with international standards in environmental protection; legislation and other regulations currently in force in Russia; and original techniques developed by the Russian participants.

The created pollution prevention methodology will be of practical importance at the local and regional levels. It will be useful for governmental planning and for future pollution prevention projects having the greatest impact on ecological improvement.

MIDDLE URALS POLLUTION PREVENTION PROJECT

The project has been divided into the following seven tasks:

1. The initial step will involve the collection of pertinent reference documents and other materials describing methods and approaches for the establishment of environmental priorities, including those adopted by Russian governmental and non-governmental organizations, and those used by environmental specialists in other countries. Based on the analysis of these reference materials, principles and criteria will be developed to define local and regional-level environmental priorities in the Middle Urals.
2. Regional environmental data presented in official reports about sources of direct and potential effects on the environment and the state of environmental objects will be collected, processed, systematically organized, and compiled into data bases. Models for the transport of pollutants in the environment will be reviewed.
3. A methodology (based on a variety of approaches, methods, and physical-mathematical models) for the establishment of pollution prevention priorities on a regional level will be developed. As part of this task, the regional environmental data gathered will be analyzed and summarized to develop a formal model of a regional level priorities assessment to prevent environmental pollution.
   After evaluation, methods will be selected for describing the condition of environmental objects and assessment criteria. Pollution migration models within different

media (water, soil, air) will be considered and applied, with respect to the Middle Urals, based on their usefulness and scientific validity.
Systematic analysis will be performed to relate the influence of a pollution source on the condition of environmental objects on a regional scale. The relevant parameters of this influence will be determined and the object's condition will be characterized.
A priorities assessment model to prevent environmental pollution on a regional level will be formulated and studied.

4. A methodology (based on a variety of approaches, methods, and physical-mathematical models) for establishment of pollution prevention priorities on a local level will be developed.
This task will allow closer examination of pollution sources and effects on a local scale. Two local model territories (one in Sverdlovsk region, another in Chelyabinsk region) typical of regional conditions will be chosen.
Detailed data related to sources of direct and potential influence on the local-scale environment and the state of environmental objects will be gathered, analyzed, and summarized. Modeling of environmental effects of sources not fully described in official reports will be performed, and data obtained will be analyzed and summarized.
After evaluation, methods will be selected for describing the condition of environmental objects and assessment criteria. Pollution migration models within different media (water, soil, air) will be considered and applied, with respect to the Middle Urals, based on their usefulness and scientific validity.
Systematic analysis will be performed to relate the influence of a pollution source on the condition of environmental objects on a local scale. The relevant parameters of this influence will be determined and the object's condition will be characterized.
A priorities assessment model to prevent environmental pollution on a local level will be formulated and examined. If needed, the local-scale model will be improved and then tested.
5. An environmental priorities assessment to prevent pollution in collaboration with local government bodies and non-governmental organizations will be conducted based on this model. Results of this work will be made public.
6. Based on results of the local-scale model, improvement will be made to the regional-scale model of environmental priorities assessment for prevention of pollution.
7. A priorities assessment for prevention of regional environmental pollution will be conducted in collaboration with regional governmental bodies and non-governmental organizations. Results of this work will be made public.

As a result of this project:

- Priorities for future pollution prevention efforts to reduce the negative effects of pollution sources on the environment and, consequently, on the health of people in the Middle Urals will be assessed;
- Local and regional-level methodologies will be created with practical uses which will allow local and oblast-level governing and administrative bodies to make better environmental decisions. This methodology will be developed on two geographic levels: local (town scale) and regional (region scale). Detailed environmental analysis will be performed on a local scale and extrapolated to the regional scale.

Results of the project and the experience accumulated during its performance will be useful thereafter to solve analogous problems in other regions of the Urals and, possibly, in other parts of Russia and other nations.

CURRENT STATUS OF THE PROJECT

A preliminary project to gather relevant environmental data and to develop a formal proposal for the full priorities assessment project for submittal to the International Science and Technology Center has been completed. Russian and LLNL collaborators have worked together to establish a project plan, share perspectives on approaches, and compile and publish regional environmental data.

ACKNOWLEDGMENTS

This work was performed under the auspices of the U. S. Department of Energy by the Lawrence Livermore National Laboratory under contract number W-7405-ENG-48.

REFERENCES

1. N. M. Barysheva, V.N. Chukanov, M.A. Gonzalez, A. P. Kuligin, R. L. Ott, V. A. Simonenko, and A. P. Vasilyev, "Middle Urals' Pollution Prevention Priorities Assessment," International Science and Technology Center International Ecological Workshop, Snezhinsk, Russia, (1995).
2. Chelyabinsk Regional Committee on Ecology and Nature Use, *The State of the Environment in Chelyabinsk Region in 1993*, Complex Report, Chelyabinsk Regional State Ecological Fund, Chelyabinsk, (1994).
3. Commission on Ecological Situation Study in Chelyabinsk Region, "Ecological Characteristic of Chelyabinsk Region," USSR Presidential Order NRP-1283, (1991), p. 157.
4. M. A. Gonzalez and R. L. Ott, "An Environmental Priorities Assessment of the Southern Urals," *Energy and Technology Review* , Lawrence Livermore National Laboratory, (1993), pp. 15–20.
5. Ministry of Environmental Protection and Natural Resources of the Russian Federation, *The State of Environment in Russian Federation in 1993*,, State Report, (1994) p. 237.
6. A.P. Vasilyev, V.J. Zuev, V.I. Legon'kov, V.F. Kuropatenko, V.A. Simonenko, and V. A. Teryohin, *Ecological Monitoring of the Zone with Chemical and Radioactive Contamination in the South Urals,* Institute of Technical Physics, Chelyabinsk, Russia (1992).

CHARACTERIZATIONS OF AIR POLLUTANTS IN THE ATMOSPHERE AT METROPOLITAN AREA IN NORTHERN TAIWAN

Pen-Chi Chiang,[1] Yu-Chun Chiang,[1] Chu-Fang Wang,[2] and Sheng-Chi Chang[3]

[1] Graduate Institute of Environmental Engineering, National Taiwan University
Taipei, Taiwan, R.O.C.
[2] Graduate Institute of Nuclear Science, National Tsing-Hua University
Hsinchu, Taiwan, R.O.C.
[3] Department of Biochemistry, National Yang-Ming University
Taipei, Taiwan, R.O.C.

ABSTRACT

Hazardous air pollutants have been noticed in Taiwan recently, they are known not only to endanger human health but also to diminish the air quality. The primary objective of this investigation is to determine the concentration levels and to explore the characterizations of air pollutants in metropolitan areas of northern Taiwan. Air samples were collected from Oct. 1989 to March 1993 in five sampling stations, located in northern Taiwan. The 24-hour suspended particulate matter samples were obtained by using Hi-vol samplers. The analysis of chemical elemental components was conducted with XRF & ICP-AES; the content of ionic species was analyzed by using IC; the amount of fourteen PAHs was measured with GC/MS; and the mutagenicity levels embraced on the particles was examined from the bioassay. The result of this study reveals that the BaP/BghiPr ratio in downtown stations ranged from 0.1 to 0.3. Examination of the results from EF and FA, suggest that the soil dust, residual oil combustion, traffic, photochemical reactions, oceanic processes, coal-burning, biomass burning, fertilizer factory, waste incinerator, foundry and limestone industry are the predominant pollution sources. Moreover, on the basis of the good relationship existing between the elemental and PAHs data, introduction of PAHs information into the original inorganic data set not only improves the resolvable performance of the modeling, but also establishes better credibility.

1. INTRODUCTION

The sources of pollutants in the atmosphere were classified into natural and anthropogenic sources. The pollutants from anthropogenic sources were primarily caused by the

Chemistry for the Protection of the Environment 2, edited by Pawłowski et al.
Plenum Press, New York, 1996

socio-economical development and industrial technology progression. Among the pollutants emitted from human activities, polycyclic aromatic hydrocarbons (PAHs), certified as hazardous air pollutants, are studied in this investigation.

PAHs, one of the major families of organic compounds in the environment, can be mutagenic and/or carcinogenic. These compounds, mainly associated with small aerosol particles between 0.1 and 1 μm, result especially from the combustion of fossil fuels and biomass burning. Since PAHs are rarely generated from natural sources, they can be considered an excellent anthropogenic tracer (Masclet et al., 1988(1); Jaffrezo et al., 1994(2))

Determination of PAHs profiles and their characteristic ratios are common methods being used to identify the pollution sources of PAHs. For instance, tetracyclic PAH compounds are regarded as the most important emissions coming from domestic heating; while heavy compounds such as coronene (Cor) and benzo(ghi)perylene (BghiPr) are good indicators of automobile traffic. (Masclet et al., 1986(3); Li & Kamens, 1993(4)). Masclet et al. (1986)(3) also suggested unique characteristics of PAHs associated with five specific industry emissions. For example, the Flu/Pyr ratio is 0.5 for a petrol refinery, 1.7 for a fertilizer factory, 65 for the foundry, 3 for a coal-fired power station and 0.15 for a municipal incinerator, which can clearly distinguish those five industries.

Although source and receptor relationships in Taiwan have been under investigation for many years, the source identification was primarily perfomed by using the chemical elemental component data in the particles. Therefore, the objective of this investigation was to examine the source/receptor relationships in northern Taiwan through use of characteristic ratios and factor analysis associated with PAHs.

2. EXPERIMENTAL METHOD

2.1 Sampling

The sampling program was done from Oct. 1989 to March 1993. Five sampling stations, i.e., Tanshui, Fuhsing, Sanchung, Taoyuan and Kuanyin (Figure 1), were located

Figure 1. The location of each sampling stations in this study.

around northern Taiwan. The Tanshui station is on the coast in an area of no significant industrial pollution, thus it could be used as a background station. The Kuanyin station is also near the coast but is in an industrial park. The Fuhsing station is in Taipei city where many construction works are performed in addition to heavy traffic. The Sanchung station is located outside of Taipei city where complicated sources exist. The Taoyuan station is in Taoyuan city in which traffic should be the predominant source. Suspended particulate matter was collected by the Hi-vol sampler for 24 hours at the above five stations.

2.2 Analysis

The collected samples were uniformly divided into four parts for the determination of chemical elemental components, polycyclic aromatic hydrocarbons (PAHs), mutagenicity and ionic species. The XRF & ICP-AES methods were utilized for chemical elemental analysis. PAHs including phenanthrene (PhA), fluoranthene (Flu), pyrene (Pyr), 2,3-benzofluorene (Ben), benzo(a)anthracene (BaA), triphenylene (Tph), chrysene (Chr), benzo(b)fluoranthene (BbF), benzo(e)pyrene (BeP), benzo(a)pyrene (BaP), perylene (Pr), indeno(1,2,3-cd)pyrene (IP), dibenz(a,h)anthracene (DbA), benzo(ghi)perylene (BghiPr) were measured by GC/MS. The mutagenic bioassay was determined by the Ames test. The ionic species, including SO_4^{2-}, NO_3^- and Cl^- were analyzed by IC.

The data sets were utilized to perform enrichment factor analysis (EF) and factor analysis (FA). EF analysis can differentiate other pollution contributors from the background source. In this study, soil dust was considered as the background source by selecting Al as the reference element. The EF value of an element close to 1 indicates this element would be primarily contributed from the background source. An element with EF value higher than 10 can be attributed to non-background sources. However, when the EF value of an element was higher than 1 but below 10, it was not possible to discriminate between the extra-sources and the background source.

FA was performed by the principal component factor analysis. (Hopke, P. K., 1985(5)). The factor axes were VARIMAX rotated and the factors with variances larger than one are retained. In addition, the variables with factor loadings higher than 0.7 are considered as primary variables, while the variables with factor loadings between 0.5 and 0.7 are regarded as secondary variables. Based on the primary and secondary variables, this model could be useful for determining the potential sources qualitatively.

3. RESULTS AND DISCUSSION

3.1 Characteristics of Hazardous Air Pollutants in TSP

The hazardous components in the particles are of serious concern to the EPA regulatory agency in Taiwan. The following discussions concentrate on the lead level and the characteristic ratios of PAHs and the mutagenicity level in TSP, as well as the relationship among them. Figure 2 shows the variations of lead concentration measured in TSP from Oct. 1989 to March 1993, at the Fuhsing, Sanchung and Taoyuan stations which are located in metropolitan areas with a heavy traffic load. It is significant that the lead concentration has been diminishing in each station due to the phase out of lead gasoline.

PAHs adsorbed on the particles measured at the Fuhsing, Sanchung and Taoyuan stations were higher than those at the stations close to the coast, such as Tanshui and Kuanyin stations. However, the fluctuation patterns at each of the stations was very simi-

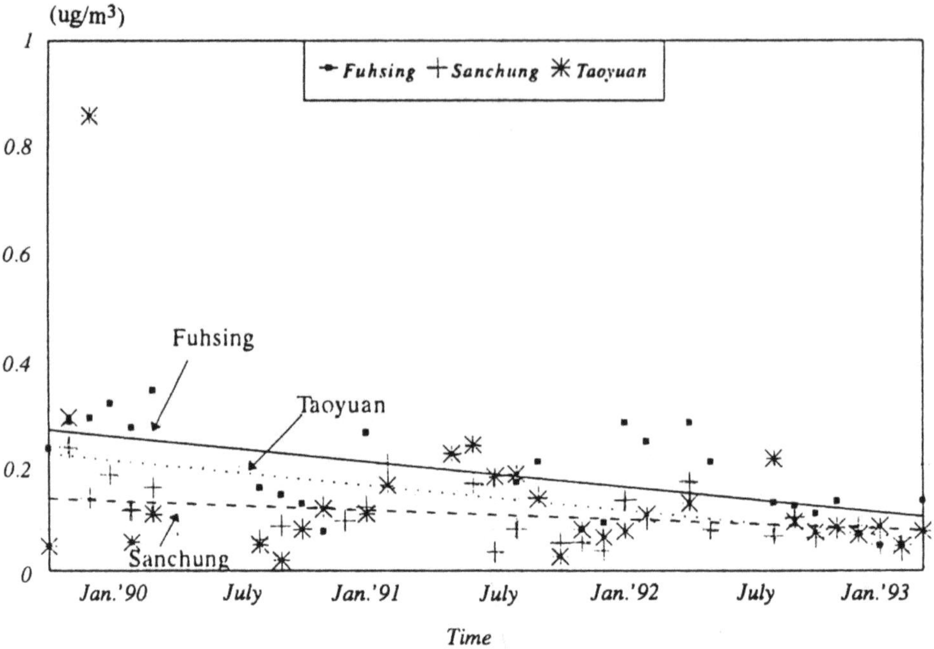

Figure 2. Variations of lead concentration measured in TSP from Oct. 1989 to March 1993 at the Fuhsing, Sanchung and Taoyuan stations.

lar. This implies that the pollution contributors to these five sites could be similar to each other. In addition, the concentration distribution of PAHs measured at the Fuhsing and Tanshui stations shown in Figure 3 presents typical examples for assessment of their characteristic distribution. Figure 3 indicates that the concentration distribution exists a right-skewed tendency. The BghiPr which has been recognized as an indicator of pollution caused from gasoline-powered vehicles, was the most predominant species in each sampling station, especially in the downtown stations.

Table 1 lists the characteristic ratios of PAHs measured at each of the sampling stations from 1989 to 1993. For BaP/BghiPr, this ratio ranged from 0 to 0.6. Moller and Alfheim (1980)(6) reported a ratio of 0.4 for vehicles and higher than 0.8 for heating. Therefore, the BaP and BghiPr measured in TSP most likely results from vehicles in this region. The Pyr/BaP ratio determined at the Tanshui and Kuanyin stations in 1992–1993 exhibited higher values than those at other downtown stations. Since Pyr predominantly from biomass burning, it is most likely attributed to open vegetative burning. Masclet et al. (1986)(3) suggested that Flu/Pyr could discriminately represent five types of industrial pollution, which have been described in our previous discussion. Since this ratio was nearly constant during the 4-year study, it is expected that these five sites have similar pollution sources for each year.

The variations of mutagenicity level exhibited in TSP at each of the stations, presented in Figure 4, indicates that most of the high measurements were at the stations located in metropolitan areas, such as Fuhsing, Sanchung and Taoyuan stations. The results from the correlation analysis show that Fe, Pb and Zn were usually related to mutagenicity level. This suggests that the mutagenicity in TSP should be correlated with vehicle exhaust.

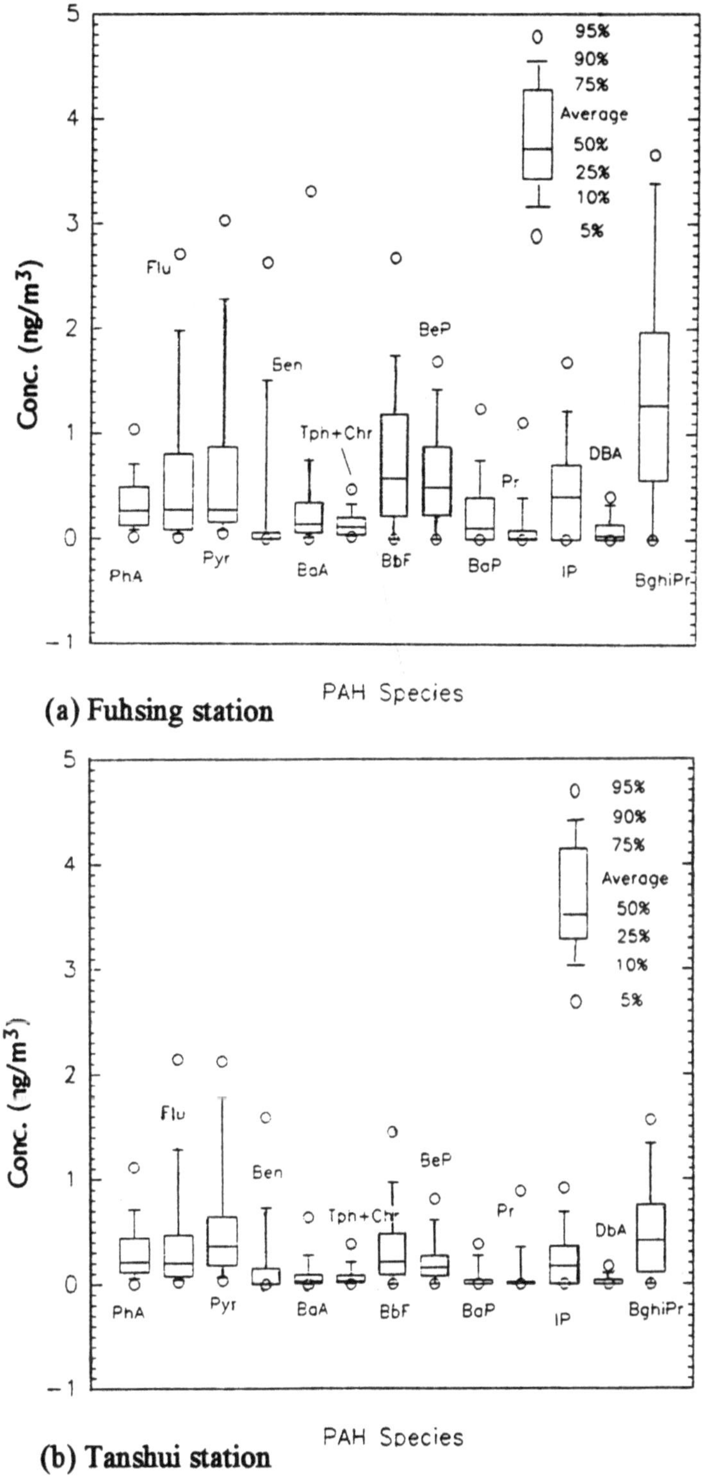

Figure 3. Box plots for demonstrating the concentration distributions of PAHs adsorbed on the particles collected from Oct. 1989 to March 1993. (a) Fuhsing station (b) Tanshui station.

Table 1. Summary of the characteristic ratios of polycyclic aromatic hydrocarbons which were adsorbed on the suspended particulate matter collected in the stations around northern Taiwan during 1989 and 1993

Station	Time	Flu/Pyr	Pyr/BaP	BaA/BaP	BaP/BghiPr	BghiPr/BbF	BghiPr/IP	(Tph+Chr)/BeP
Tanshui	1989-1990	1.6	2.4	1.0	0.2	1.8	3.0	0.5
	1990-1991	2.1	33.3	2.7	0.04	0.7	2.1	0.2
	1991-1992	0.6	1.0	0.3	0.1	2.7	1.9	0.1
	1992-1993	1.1	18.4	2.7	0.1	3.5	1.6	0.3
Fuhsing	1989-1990	1.4	2.6	1.2	0.3	4.3	4.2	1.4
	1990-1991	2.1	12.7	1.2	0.2	1.8	2.9	0.5
	1991-1992	0.9	0.9	0.5	0.2	3.3	2.5	0.1
	1992-1993	1.1	2.6	1.1	0.2	2.4	2.2	0.3
Sanchung	1989-1990	1.0	3.2	1.5	0.3	1.9	3.0	1.2
	1990-1991	1.1	11.4	1.2	0.2	1.4	3.6	0.2
	1991-1992	0.4	3.7	0.6	0.1	2.9	2.3	0.1
	1992-1993	1.0	3.4	1.3	0.1	2.8	2.3	0.3
Taoyuan	1989-1990	1.3	2.3	0.7	0.3	0.7	2.5	0.3
	1990-1991	1.2	15.3	1.1	0.1	0.8	2.1	0.2
	1991-1992	0.6	3.5	0.7	0.2	2.9	1.9	0.1
	1992-1993	1.1	4.1	1.1	0.1	3.5	2.4	0.2
Kuanyin	1989-1990	1.2	6.4	0.5	0.6	1.9	2.8	0.7
	1990-1991	1.6	103.8	2.0	0.0	0.1	1.8	0.4
	1991-1992	0.4	9.3	0.3	0.1	1.9	1.3	0.2
	1992-1993	1.1	21.3	2.5	0.1	1.8	1.4	0.5

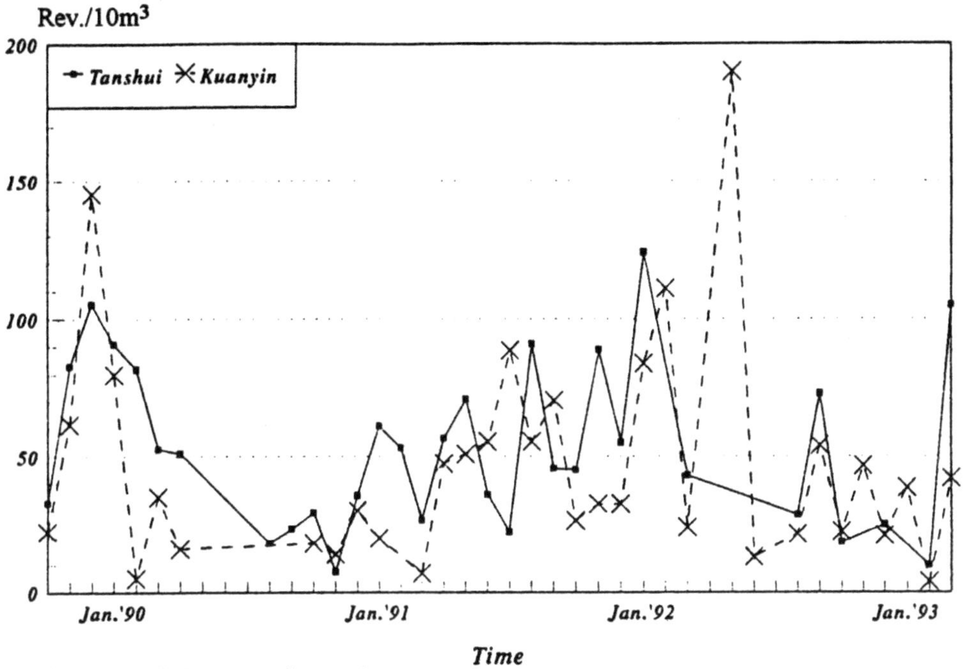

(a) Tanshui & Kuanyin stations

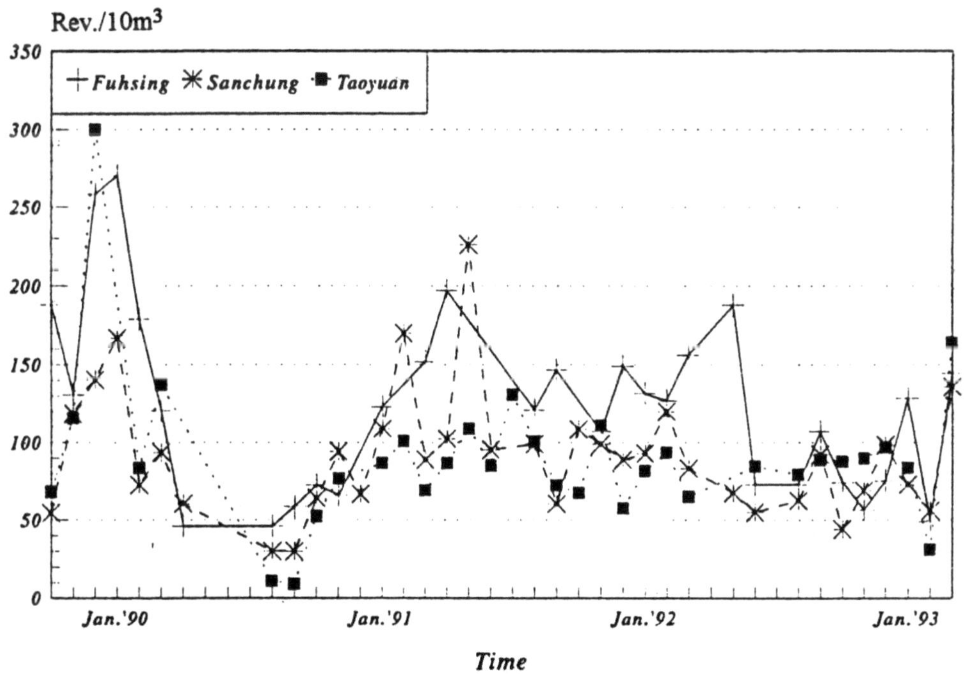

(b) Fuhsing, Sanchung & Taoyuan stations

Figure 4. Variations of potential mutagenicity in TSP from Oct. 1989 to March 1993, which was obtained by using Ames test at (a) Tanshui & Kuanyin stations (b) Fuhsing ,Sanchung & Taoyuan stations.

3.2 Source Identification

Determination of the potential pollution sources is the most critical task for controlling ambient air quality. Therefore, the following discussions would attempt to identify qualitatively the potential sources in the air by using the enrichment factor analysis (EF) and factor analysis (FA) for data sets obtained from 1989 to 1993 at all sampling stations.

3.2.1. in 1989 - 1990. Figure 5 illustrates the EF values for each of the chemical elemental components Na, S, Cl^-, NO_3^- and SO_4^{2-}, with EF values higher than 10, could be generated from non-background sources, thus, oceanic processes and photochemical reactions are suspected. Moreover, the Cl^-/Na ratio is equivalent to 0.56 (1.17 in sea water, Lee et al., 1994(7)), which would imply the effect of photochemical reactions. In addition, since the EF values of Ca, Cr, Fe, K, Mg, Mn, Sr and V were close to 1, these elements would be mostly contributed from soil dust. The other elements, i.e., Ba, Cu, Ni, Pb and Zn, could be related to the diesel vehicles, incinerators, residual oil combustion and gasoline-powered vehicles. Table 2 summarizes the results of factor analysis which indicate that there are four factors to be retained and the percentage of total explainable variance is 68.8%. The first factor was related to Mg, Mn, Ca, Sr and Al; and could be contributed from soil dust. High factor loadings for Cd, V, Ni and Cr were observed in the second factor, which suggests that residual oil combustion is the potential pollution source. The third factor was associated with NO_3^-, Zn, Fe, Pb, SO_4^{2-}, S and K, which indicated contributions from photochemical reactions and gasoline-powered vehicles. The factor loading of Cl^- was predominant in the fourth factor, it could imply oceanic processes as the major pollution source. However, the total explainable

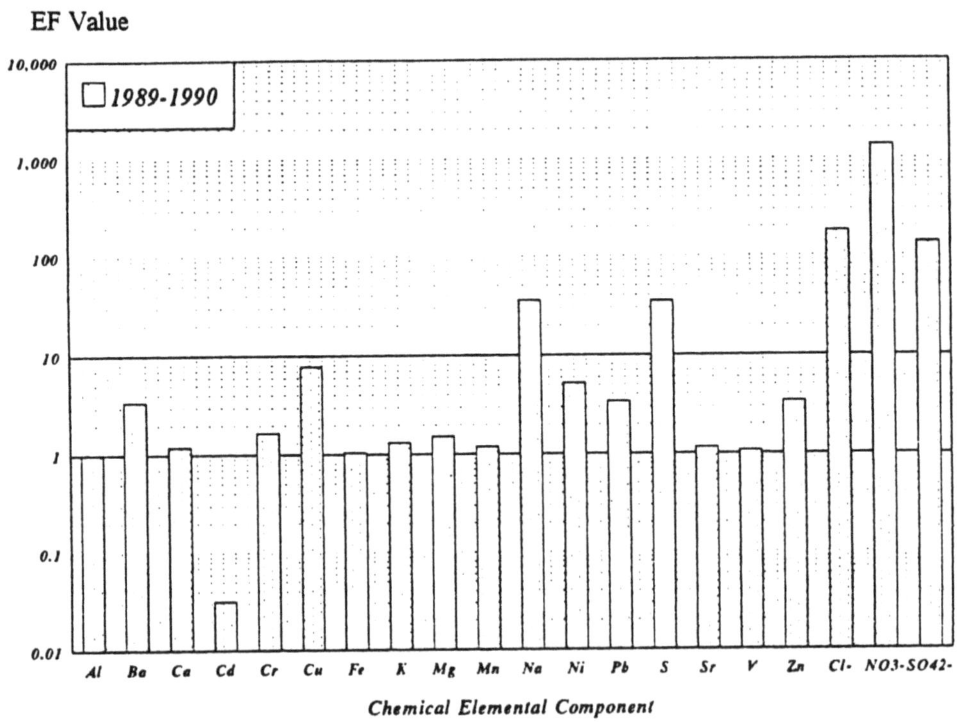

Figure 5. EF values of each chemical elemental components measured in TSP collected from 1989 to 1990.

Table 2. Summary of the results of factor analysis for the pollutants measured in the suspended particulate matters with aerodynamic diameters less than 100 μm which were collected from 1989 to 1990

Data Type	Retained Factor	Factor 1	Factor 2	Factor 3	Factor 4	Factor 5	Factor 6	Factor 7
Element (n=64)	Primary Variable	Mg(0.91) Mn(0.75) Ca(0.72) Sr(0.72) Al(0.72)	Cd(0.89) V(0.84) Ni(0.83) Cr(0.80)	NO_3^-(0.74)	—	—	—	—
	Secondary Variable	K(0.66) S(0.58) Fe(0.57) Na(0.53)	Zn(0.61) Pb(0.56)	Zn(0.69) Fe(0.54) Pb(0.53) SO_4^{2-}(0.53) S(0.52) K(0.51)	Cl⁻(0.57)	—	—	—
	Variance (Total=68.8%)	25.1%	22.0%	15.3%	5%	—	—	—
	Potential Sources	Soil dust	Residual oil combustion	Gasoline vehicle	Oceanic processes	—	—	—
Element & PAHs (n=64)	Primary Variable	V(0.94) Cd(0.89) Ni(0.87) Zn(0.82) Cr(0.75) Pb(0.72)	Mg(0.85) Ca(0.79) K(0.76)	BghiPr(0.79) BaP(0.76)	Tph+Chr(0.74)	Flu(0.77) Pyr(0.75)	—	—
	Secondary Variable	Fe(0.53)	Sr(0.69) Fe(0.68) Mn(0.66) Na(0.62) S(0.54) Al(0.51)	BbF(0.67) BeP(0.65) IP+DbA(0.63)	BaA(0.69) Pr(0.64)	Ben(0.55) PhA(0.54)	Cl⁻(0.57) SO_4^{2-}(0.50)	Al(0.62)
	Variance (Total=67.0%)	18.8%	17.0%	10.3%	6.9%	6.6%	3.7%	3.7%
	Potential Sources	Traffic	Soil dust & Oceanic processes	Gasoline vehicle	Fertilizer factory	Diesel vehicle	Oceanic processes	Soil dust

※The number in the parentheses represents factor loading.

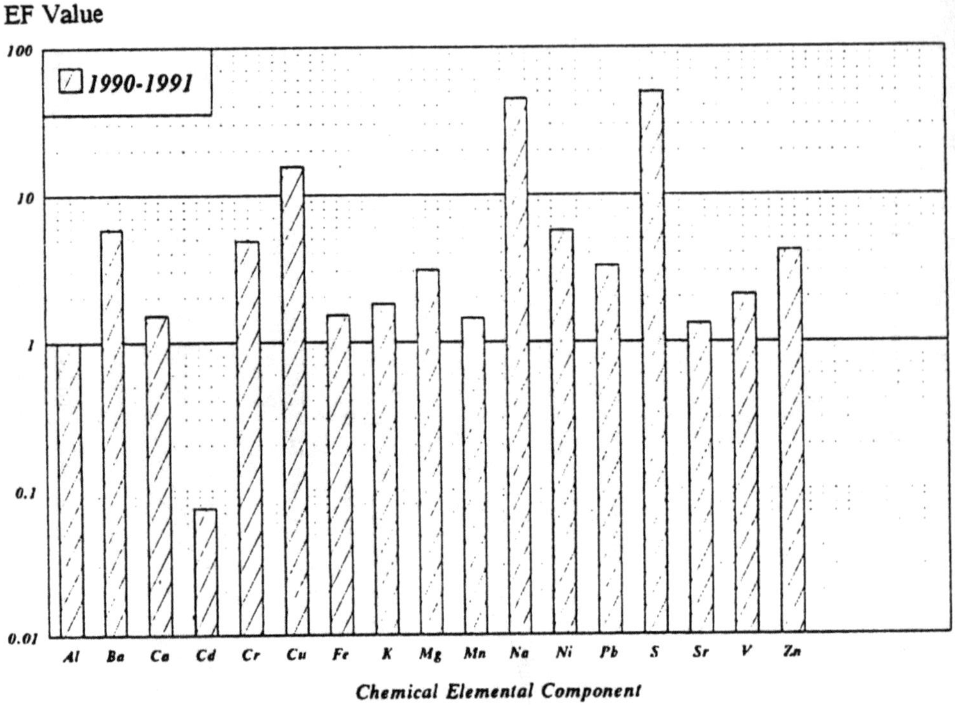

Figure 6. EF values of each chemical elemental components measured in TSP collected from 1990 to 1991.

variance was not raised after introducing PAH data into the original elemental data, although the retained factors were as high as seven. A possible explanation is that there was no significant correlation among the chemical elements and PAHs data in 1989–1990.

3.2.2. in 1990 - 1991. The results of EF analysis are shown in Figure 6. The EF values for Cu, Na and S were greater than 10, which suggests the pollution sources to be metal industry, oceanic processes and fuel burning. Subsequently, Ca, Fe, K, Mn and Sr could primarily be contributed from soil dust because their EF values were close to 1. Finally, Ba, Cr, Mg, Ni, Pb, V and Zn with EF values much higher than 1 but lower than 10, the diesel vehicle, residual oil combustion and gasoline-powered vehicles can be considered as the pollution sources. Table 3 summarizes the results of FA, which clearly demonstrates that three factors were resolved with a percentage of total explainable variance approaching 50.0%. The first factor was significantly associated with K, Al, Fe, Sr, Ca and Mn, it can be interpreted soil dust pollution. High factor loadings of Zn, Ni, Pb, V and S were obtained in the second factor, which accounts for the contribution of traffic. The third factor showed high Na and Mg loadings, which indicates marine pollution. Compare this result with that from the EF analysis, the modeling of FA does not seem to resolve well. However, it can be improved by introducing PAHs data, shown in Table 3. There were three more factors to be resolved and the total explainable variance was raised up to 65.1%. High-cyclic PAHs were predominant in the first factor, which resulted from foundry industries and the gasoline-powered vehicle. The second factor could be considered as soil dust in response to our previous discussions. Ni, V and Zn were related to the third factor indicating the pollution of residual oil combustion. The fourth factor is associated with PhA, Cd and Flu, a strong relationship exhibited among these tracers suggests a source contribution from the waste incinerator. Ba

Table 3. Summary of the results of factor analysis for the pollutants measured in the suspended particulate matters with aerodynamic diameters less than 100 μm which were collected from 1990 to 1991

Data Type	Retained Factor	Factor 1	Factor 2	Factor 3	Factor 4	Factor 5	Factor 6
Element (n=57)	Primary Variable	K(0.93) Al(0.90) Fe(0.90) Sr(0.76) Ca(0.71) Mn(0.71)	Zn(0.77) Ni(0.71)	—	—	—	—
	Secondary Variable	—	Pb(0.68) V(0.58) S(0.55)	Na(0.57) Mg(0.54)	—	—	—
	Variance (Total=50.1%)	27.4%	16.3%	6.4%	—	—	—
	Potential Sources	Soil dust	Traffic	Oceanic Processes	—	—	—
Element & PAHs (n=57)	Primary Variable	BeP(0.93) Pr(0.92) BbF(0.92) BaP(0.91) IP(0.90) BghiP(0.83) BaA(0.81) DbA(0.72) Tph+Chr(0.71)	K(0.97) Fe(0.95) Al(0.94) Mn(0.82) Sr(0.76) Ca(0.74)	Ni(0.72) V(0.71)	PhA(0.71)	—	—
	Secondary Variable	Ben(0.51)	S(0.50)	Zn(0.69)	Cd(0.67) Flu(0.64)	Ba(0.63) Ben(0.58)	Na(0.60)
	Variance (Total=65.1%)	23.9%	18.7%	7.4%	6.5%	5.0%	3.6%
	Potential Sources	Gasoline vehicle	Soil dust	Residual oil combustion	Waste incinerator	Diesel vehicle	Oceanic processes

※The number in the parentheses represents factor loading.

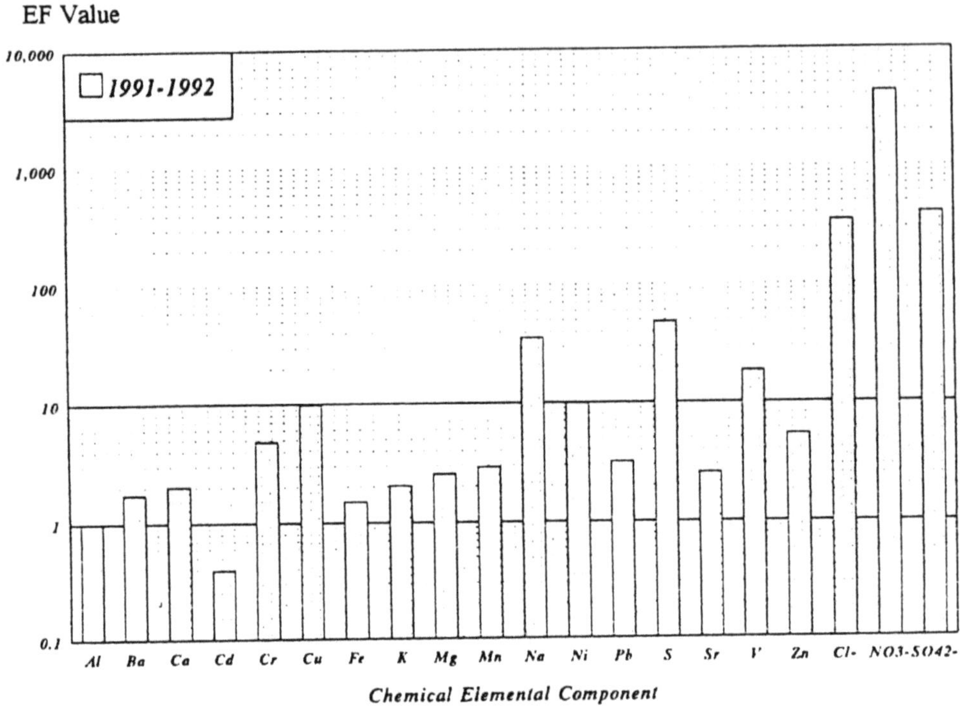

Figure 7. EF values of each chemical elemental components measured in TSP collected from 1991 to 1992.

and Ben were predominant in the fifth factor, and reflect pollution from the diesel vehicle. The last factor with high Na loading reflects the source of oceanic processes. The above discussions provide the evidence demonstrating the advantages of introducing PAH data into the original elemental data in order to determine the pollution sources.

3.2.3. in 1991 - 1992. Figure 7 plots the results of EF analysis for chemical components measured in TSP. With the exception of soil dust, it implies that oceanic processes, residual oil combustion, photochemical reactions, coal-burning, metal industry and the gasoline-powered vehicle were the primary pollution sources. Table 4 summarizes the results of the FA by using the chemical element data only and the combination of chemical elements and PAHs. There were six factors to be retained by adding PAH data. And the total explainable variance was improved from 61.6% to 68.8%, compared to that without PAHs data set. The gasoline-powered vehicle, foundry industry, soil dust, coal-burning, oceanic processes and diesel vehicle can be qualitatively determined as the potential sources.

3.2.4. in 1992 - 1993. Figure 8 represents the results of EF analysis for chemical elemental components measured in TSP. It implies oceanic processes, fuel burning, incineration, coal-burning industry, the gasoline-powered vehicle and soil dust as the predominant sources. Table 5 shows the results of FA, from which there are six factors to be retained and the percentage of total explainable variance is 70.6% by using data sets containing chemical elements and PAHs. This resolution is much better than that obtained from elemental data only which resolves five factors with a total explainable variance of 65.1%. By examining the primary variables in each factor, the soil dust, foundry industry, gasoline-powered vehi-

Table 4. Summary of the results of factor analysis for the pollutants measured in the suspended particulate matters with aerodynamic diameters less than 100 μm which were collected from 1991 to 1992

Data Type	Retained Factor	Factor 1	Factor 2	Factor 3	Factor 4	Factor 5	Factor 6
Element (n=66)	Primary Variable	Fe(0.95) Al(0.92) Ca(0.86) K(0.86) S(0.71)	Ni(0.97) Cr(0.96) Na(0.87) Ba(0.73)	Zn(0.75)	—	—	—
	Secondary Variable	Mg(0.68) Ba(0.54)	—	Pb(0.67) SO_4^{2-}(0.67) S(0.56) NO_3^-(0.52)	Cl^-(0.57) Mg(0.56)	—	—
	Variance (Total=61.6%)	24.2%	18.7%	12.0%	6.6%	—	—
	Potential Sources	Soil dust	Coal-burning	Gasoline vehicle	Oceanic processes	—	—
Element & PAHs (n=66)	Primary Variable	BeP(0.98) BghiPr(0.98) BbF(0.97) IP(0.96) BaA(0.92) BaP(0.92) DbA(0.85) Pr(0.85) Tph+Chr(0.85)	Fe(0.95) Al(0.93) Ca(0.87) K(0.85)	Ni(0.95) Cr(0.95) Na(0.84) Pyr(0.78) Ba(0.71)	Zn(0.72)	—	—
	Secondary Variable	PhA(0.67)	S(0.69) Mg(0.67) Ba(0.55)	—	SO_4^{2-}(0.67) Pb(0.66) S(0.59)	Mg(0.59) Cl^-(0.58)	Flu(0.60)
	Variance (Total=68.8%)	25.9%	15.0%	13.2%	7.2%	4.2%	3.3%
	Potential Sources	Gasoline & Foundry	Soil dust	Coal-burning	Gasoline vehicle	Oceanic processes	Diesel vehicle

※The number in the parentheses represents factor loading.

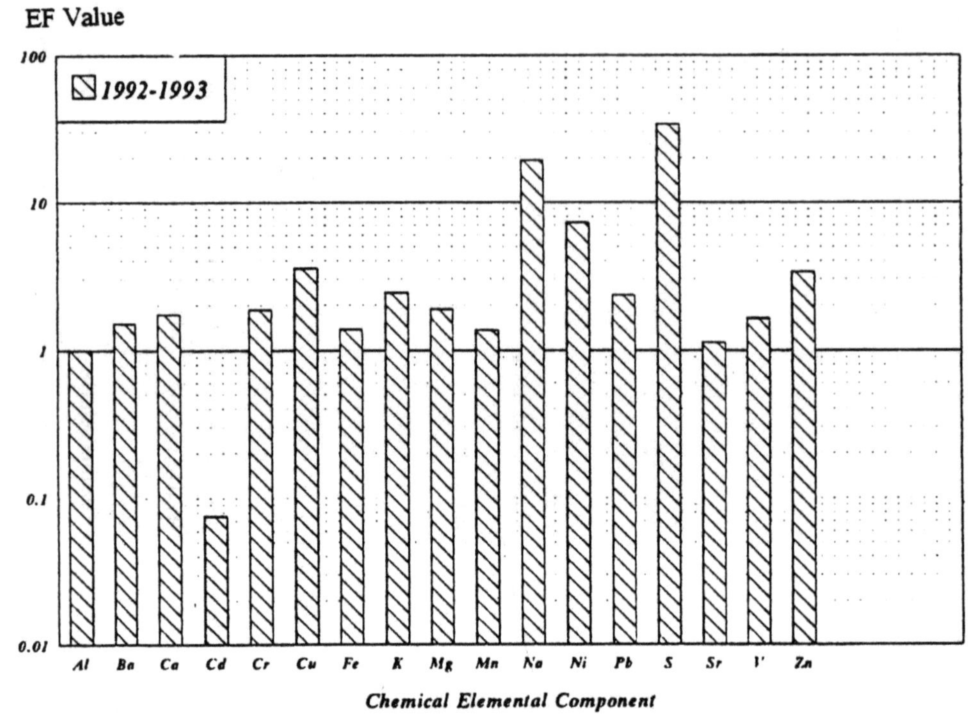

Figure 8. EF values of each chemical elemental components measured in TSP collected from 1992 to 1993.

cles, diesel vehicles, biomass burning, the waste incinerator and coal-burning industry could be suspected. The results with only elemental data could resolve the extra factor, i.e, the fourth factor—limestone, based on the factor loadings of Mg and Ca, but missed some other contributors. Therefore, introducing PAH data still exhibits a valuable performance for source resolution.

In summary, after introducing PAH information into the inorganic data set, the predominant potential pollution sources resolved by the FA modeling were presented in Table 6. It is clearly found that the transformation of predominant sources is associated with recent development of industry in northern Taiwan during these four years.

4. CONCLUSIONS AND RECOMMENDATIONS

In this research, it was clearly demonstrated that the receptors were highly correlated with the pollution sources. From the examination of characteristic ratios of PAHs and results of enrichment factor analysis and factor analysis for the organic and inorganic components measured in TSP, the potential primary sources in northern Taiwan were revealed. They were soil dust, residual oil combustion, traffic, photochemical reactions, oceanic processes, coal-burning, biomass burning, fertilizer factories, waste incinerators, foundries and the limestone industry. The contribution from incomplete combustion of carbonaceous materials can be certified in detail through the PAH profiles. Therefore, introducing PAH data into the original chemical elemental data sets can not only improve the resolvable performance of the modeling but also establish better credibility. However,

Table 5. Summary of the results of factor analysis for the pollutants measured in the suspended particulate matters with aerodynamic diameters less than 100 μm which were collected from 1992 to 1993

Data Type	Retained Factor	Factor 1	Factor 2	Factor 3	Factor 4	Factor 5	Factor 6
Element (n=72)	Primary Variable	Al(0.88) Fe(0.81) Sr(0.73)	Mn(0.86) Zn(0.84) Ba(0.73)	Ni(0.83) Cr(0.80)	Mg(0.72)	—	—
	Secondary Variable	V(0.59) As(0.57)	—	—	Ca(0.50)	Cd(0.56)	—
	Variance (Total=65.1%)	22.2%	18.1%	9.0%	8.1%	7.7%	—
	Potential Sources	Soil dust & Oil burning	Diesel vehicle	Coal-burning	Limestone	Waste incinerator	—
Element & PAHs (n=72)	Primary Variable	Al(0.92) Sr(0.91) Fe(0.85) V(0.84) As(0.82)	BaA(0.89) BaP(0.87) Tph+Chr(0.83) Pr(0.80) BbF(0.79) BeP(0.72)	Flu(0.87) Pyr(0.80)	DbA(0.70)	Mn(0.77) Zn(0.73)	Cr(0.78) Ni(0.78)
	Secondary Variable	Cd(0.67) Ba(0.60) Pb(0.59) Ca(0.57) Mg(0.56) Mn(0.50) Se(0.50)	BghiPr(0.67) IP(0.66) PhA(0.57)	PhA(0.68)	Cd(0.52)	Ba(0.64)	—
	Variance (Total=70.6%)	21.9%	20.4%	9.3%	6.9%	6.8%	5.2%
	Potential Sources	Soil dust	Foundry & Gasoline vehicle	Diesel & Biomass burning	Waste incinerator	Diesel vehicle	Coal-burning

※The number in the parentheses represents factor loading.

Table 6. Summary of the predominant air pollution sources resolved by using FA modeling in sequential four years from 1989 to 1993 with inorganic chemical data and introducing PAHs information into the former data set

Period	1989-1990		1990-1991		1991-1992		1992-1993	
Item	Element & Ion	Element, Ion & PAHs	Element	Element & PAHs	Element & Ion	Element, Ion & PAHs	Element	Element & PAHs
Soil dust	√	√	√	√	√	√	√	√
Oceanic processes	√	√	√	√	√	√	-	-
Fertilizer factory	-	√	-	-	-	-	-	-
Photochemical reactions	√	√	-	-	-	-	-	-
Gasoline vehicles	√	√	√	√	√	√	-	√
Diesel vehicles	-	√	√	√	-	√	√	√
Residual oil combustion	√	√	-	√	-	-	√	-
Foundry industry	-	-	-	√	-	√	-	√
Waste incinerator	-	-	-	√	-	-	√	√
Coal-burning	-	-	-	-	√	√	√	√
Biomass burning	-	-	-	-	-	-	-	√
Limestone	-	-	-	-	-	-	√	-
Total explainable variance(%)	68.8%	67.0%	50.1%	65.1%	61.6%	68.8%	65.1%	70.6%

this conclusion must be made on the basis of a good relationship between the elemental and PAH data. In this study, the BaP/BghiPr ratio at the downtown stations, which were primarily influenced by traffic, ranged from 0.1 to 0.3 which was consistent with the study reported by Masclet et al. (1986) (3). The potential pollution sources in this region had not varied significantly for the past four years, however, the predominant sources were gradually changed year by year, which was confirmed by observation of the characteristic ratios of PAHs.

5. REFERENCES

1. Masclet, P., Pistikopoulos, P., Beyne, S. and Mouvier, G. (1988) Long Range Transport and Gas/Particle Distribution of Polycyclic Aromatic Hydrocarbons at a Remote Site in The Mediterranean Sea, Atmospheric Environment, 22:639–650.
2. Jaffrezo, J. L., Clain, M. P. and Masclet, P. (1994) Polycyclic Aromatic Hydrocarbons in The Polar Ice of Greenland. Geochemical Use of These Atmospheric Tracers, Atmospheric Environment, 28:1139–1145.
3. Masclet, P., Mouvier, G. and Nikolaou, K. (1986) Relative Decay Index and Sources of Polycyclic Aromatic Hydrocarbons, Atmospheric Environment, 20:439–446.
4. Li, C. K. and Kamens, R. M. (1993) The Use of Polycyclic Aromatic Hydrocarbons as Source Signatures in Receptor Modeling, Atmospheric Environment, 27:523–532.
5. Kopke, P. K. (1985) Receptor Modeling in Environmental Chemistry, JOHN WILEY & SONS.
6. Moller, M. and Alfheim, I. (1980) Mutagenicity and PAH-analysis of Airborne Particulate Matter, Atmospheric Environment, 14:83–88.
7. Lee, D. S., Garland, J. A. and Fox, A. A. (1994) Atmospheric Concentrations of Trace Elements in Urban Areas of the United Kingdom, Atmospheric Environment, 28:2691–2713.

7

AIR POLLUTION CONTROL WITH FIBROUS ION EXCHANGERS

V. S. Soldatov,[1] I. S. Elinson,[2] A. A. Shunkevich,[2] L. Pawlowski,[1] and H. Wasag[1]

[1] Technical University of Lublin
40, 20–618, Lublin, Poland
[2] Institute of Physical Organic Chemistry of the Belarus National Academy of Sciences
13 Surganov Str., Minsk 220072, Republic of Belarus

1. INTRODUCTION

Ion exchange treatment of gases can be, in principle, as vast of a field for application of ion exchangers as liquid ion exchange technologies. Investigation of ion exchangers as sorbents from gaseous media began in the fifties, when the first patents, papers and monographs appeared [1–6]. Since then this research has developed in a number of countries, especially in USSR, Japan, Germany, USA and Poland [7–11].

In spite of a large number of studies on optimization of gas treatment processes, granular ion exchangers did not find significant application in this field.

One of the most important cases of these processes is air purification from acid, base and the other chemically active substances. The experience gained in the previous research shows that ion exchangers satisfy from a chemical point of view all necessary requirements, but the granular form is not suitable for creation of efficient technologies of air purification. The reasons why granular ion exchangers have not found large scale application in this process have to be analyzed in order to choose a strategy for future work in this direction.

In technological schemes granular ion exchangers can be used in a fixed or boiling bed. Both methods can provide any degree of the gas flows purification with correct choice of the ion exchanger and the process conditions: filtration rate, the filtering bed length, gas humidity etc. Nevertheless, there are some negative factors of a principle nature enhancing when the process is scaled up and their application becomes unprofitable.

Slow diffusion of the sorbing species in ion exchangers requires application of either small particles or(and) long filtering layers. At the same time, gas flow resistance must be low enough not to exceed the pressure on the filtering plant 1000 N/m^2. This requirement is critical since this is the highest possible pressure drop in ventilation systems. It is also

Chemistry for the Protection of the Environment 2, edited by Pawłowski et al.
Plenum Press, New York, 1996

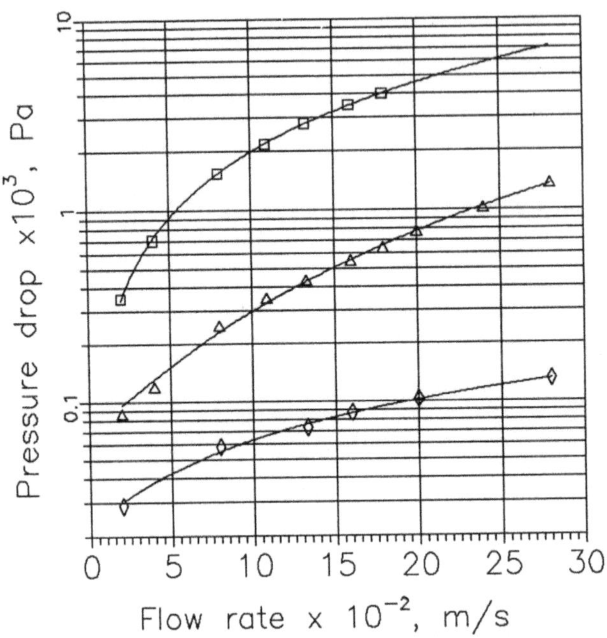

Figure 1. Pressure drop as a function of the air flow rate. 1—granular resin Vofatit SBF, the filtering layer thickness 0.300 m.; 2—the same, 0.030 m thickness; 3—fibrous cation exchanger VION KN-1, 0.010 m layer thickness.

necessary to take into account that the smallest air filtering apparatuses should have a working capacity of 500 m^3/h and be desk top size in order to be convenient for home or small industry use. The apparatuses for industrial exhaust gas treatment of medium and large size (galvanic, food, glass, pharmaceutics, chemical, metal processing, etc.) should have a capacity of 5–50 thousands m^3/h the apparatuses having their own volume of few cubic meters.

Attempts to compromise these requirements were unsuccessful. This is illustrated by Figs. 1, 2 in which the pressure drop is plotted as a function of air flow rate and radius of granules.

The flow rates 0.2 m/s can be considered as minimal and maximal acceptable for practical apparatuses with granular sorbents. Under this condition the height of the filtering layer may not be less than 0.3 m to provide a sufficient degree of purification. This means that the pressure drop must be at least

3.5 kPa which is much above the critical value. Attempts to improve the aerodynamic characteristics by increasing the particle size are doomed to fail since this leads to drop of the sorption rate and thus to the working capacity of the column. It is seen from the figure that the fibrous ion exchanger filtering layer with a thickness of 10 mm, sufficient for air flow purification under practical conditions, has a satisfactory pressure drop up to flow rate 0.3 m/s. This is unattainable with granules with a diameter below 1.5 mm. Summarizing we conclude that the set of parameters required for gaseous ion exchange technologies is incompatible for granular ion exchangers, but sufficient for ion exchange fibers.

The second critical point in the fixed bed gaseous ion exchange technologies is granules volume change in sorption/desorption cycles. This causes two undesirable effects: changing the fixed bed volume accompanied by a change in its aerodynamic resistance; mechanical destruction of the granules caused by osmotic shocks in swelling/shrinking in sorption/regeneration cycles. Fig. 3 shows that the sorption layer volume change is signifi-

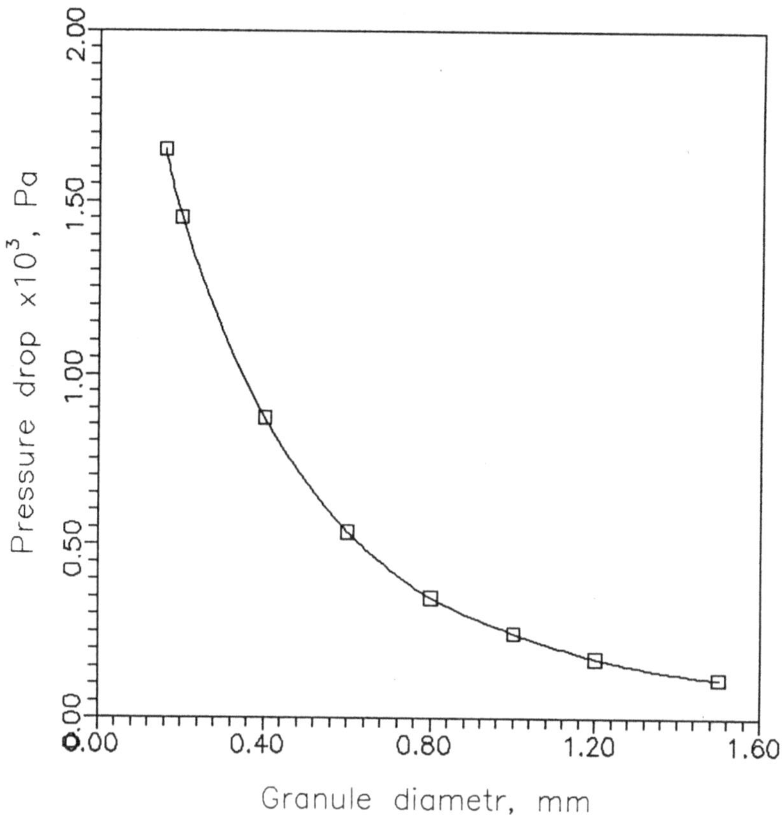

Figure 2. Pressure drop as a function of granules diameter for sulfonic type cation exchanger KU-2 at the filtering layer thickness 0.030 m and flow rate 0.20 m/s.

cant even for macroporous ion exchangers. Compensation for both unfavorable phenomena causes heavy technological complications and can not be applied in practical systems.

There are a number of other unfavorable peculiarities of ion exchange systems for gas treatment preventing their practical applications.

Nevertheless, a number of attempts to develop and use ion exchange gas-filtering plants have been made.

In the first communication from 1963 [12] the workings of one of the Ural factories filtering plants on exhaust gases purification from sulfur dioxide have been described. Its capacity was about 1000 m^3/h. The exhaust gases, preliminary purified from dust and cooled to 30–40°C, containing 0.06 volume percent SO_2 was purified by weak base ion exchanger AN-2F in a column with diameter 0.4 and height 4.6 meters. The plant provided complete purification of the exhaust gases from SO_2.

In Ref [8] is described sorption of more than 30 inorganic and organic gases on pilot plants of the Swiss company Geigy.

In [13] were reported the applications of ion exchange resins in air purification after soda gas washing of selenium dioxide and acidic components of air in regeneration chambers of electrolytic baths of Kyshtym and Moscow copper-electrolyte factories.

The apparatus for air purification from NO_2 and H_2SO_4 aerosol with capacity 1400 m^3/h was described in [11]. It was a shelf type device with column diameter 1 m and height

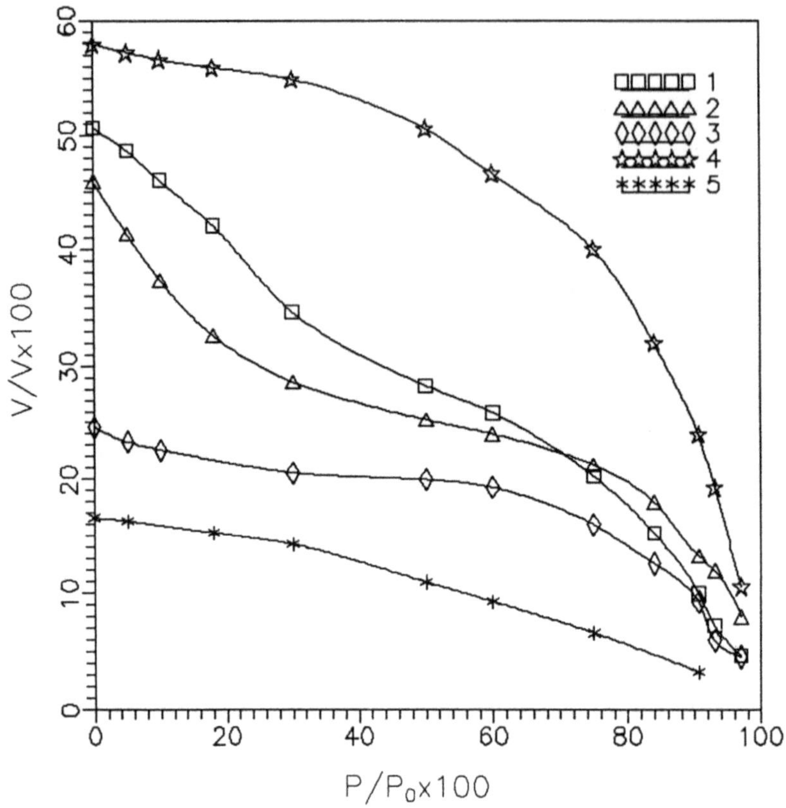

Figure 3. Relative volume change of the filtering layer of granular ion exchangers as a function of relative humidity of air in equilibrium. The ion exchangers: 1—AV-17X8 (CO_3^{2-}), 2—AV-17X12 (CO_3^{2-}), 3—AN-211 (OH^-), 4—AN-31g (OH^-), 5—AN–511 (OH^-).

3 m. The strong base ion exchanger in the column was continuously sprayed with regeneration solution of alkali whose consumption was 12.7 m^3/h. The filtration cross-section in the apparatus was 0.66 m^2/m^3.

The technological parameters of the described plants were close to the best possible and yet too low for practical large scale applications.

The apparatuses with boiling beds did not find wide application either. Only in Ref [13] a continuos process of exhaust gases of sulfuric acid production from SO_2 was described. It was stated that application of the boiling bed apparatuses is sensible for purification of concentrated gas flows containing 1–10 g/m^3 of the impurity.

The fact that an insignificantly small number of ion exchange gas filtering plants have been used in industries during more than a 30 year period since the beginning of the related research shows that this method is a dead-end.

We believe that application of fibrous ion exchange materials can be a way out and that technologies of gas purification utilizing specific properties of fibrous ion exchangers can satisfy modern requirements.

Two practical important examples are taken to illustrate properties and possibilities of fibrous ion exchangers in air purification processes from acidic and basic impurities: hydrogen fluoride removal by fibrous aminocarboxylic ion exchange FIBAN AK-22 and ammonia removal by carboxylic ion exchanger FIBAN K-4 produced by Ecofil-Deco Ltd. in Minsk (Address: P.O. BOX 47, Minsk 220050, Rep. of Belarus. Fax (0172) 352543).

2. HYDROGEN FLUORIDE REMOVAL

Hydrogen fluoride is one of the most dangerous air pollutants in glass, chemical fertilizers, non-ferrous metals and petrochemical industries. Its MPC level in breathing air is 0.5 mg/m^3. The usual way of air purification from HF is water or alkali scrubbers, packed columns or foam absorbers. These apparatuses are expensive, are large in size and not always efficient at HF concentration below 500 mg/m^3. In these cases two step purification technology with fibrous ion exchange filter as the second step seems to be efficient.

An ion exchange technology for HF removal from air in glass industry was developed and 8 plants with total working capacity 80,000 m3/h were constructed and installed in two factories in Belarus and Russia.

For development of the technology and in order to determine the apparatuses' influence on the main factors on the sorption process the following have been investigated: the concentration of HF in the air, air flow rate and air humidity. Conditions of the ion exchanger regeneration have also been investigated.

The characteristics of FIBAN AK-22 are as follows: the functional groups are carboxylic (1.5 mmole/g) and the amido aminogroup $R\text{-}CONH(CH_2CH_2NH)_2H$ (3.5 mmole/g). Since the latter contains two protonizable N atoms, the total anion exchange capacity is 7.0 meq/g. The water uptake is 0.69 gH_2O/g of H^+/OH^- form. The filament diameter was 25±5 μ.

Initially it was found that 10 mm thickness of the filtering layer is sufficient for efficient removal of HF from the air and it was used in all further experiments as well as in the industrial filtering plants.

Investigation of dependence of the equilibrium HF sorption as a function of its concentration in the gas phase have shown (Fig. 4) that about 80% of the theoretical value 126 mg HF/g (0.8 mole HF per mole of functional groups) is reached at negligibly small HF concentrations (below 17 mg/m^3). This corresponds to HF sorption due to addition reaction

$$R\text{-}NH_2 + HF = RNH_3^+F^-$$

The further increase of HF concentration leads to proportional sorption rise up to 3.5 HF molecules per functional group.

This phenomenon was observed in our experiments with the other acids and can be used in the technology of their removal from the air, as will be shown below. It can be interpreted in the following way: there are two reasons for superequivalent HF sorption. The first, universal effect, is connected with association of the HF molecules in the ion exchanger phase leading to formation of a series of associates

$$R\text{-}NH_3^+F^- + iHF = R\text{-}NH_3^+F(HF)_i^-,$$

where i = 1, 2, 3, ...

The second effect is specific for the FIBAN AK-22 ion exchanger, whose functional group contains two protonizing nitrogen atoms. The protonization constant of the second step is about three orders of magnitude lower than the first one [14]. The reaction of addition of HF to the second nitrogen atom can have a free energy comparable to the associa-

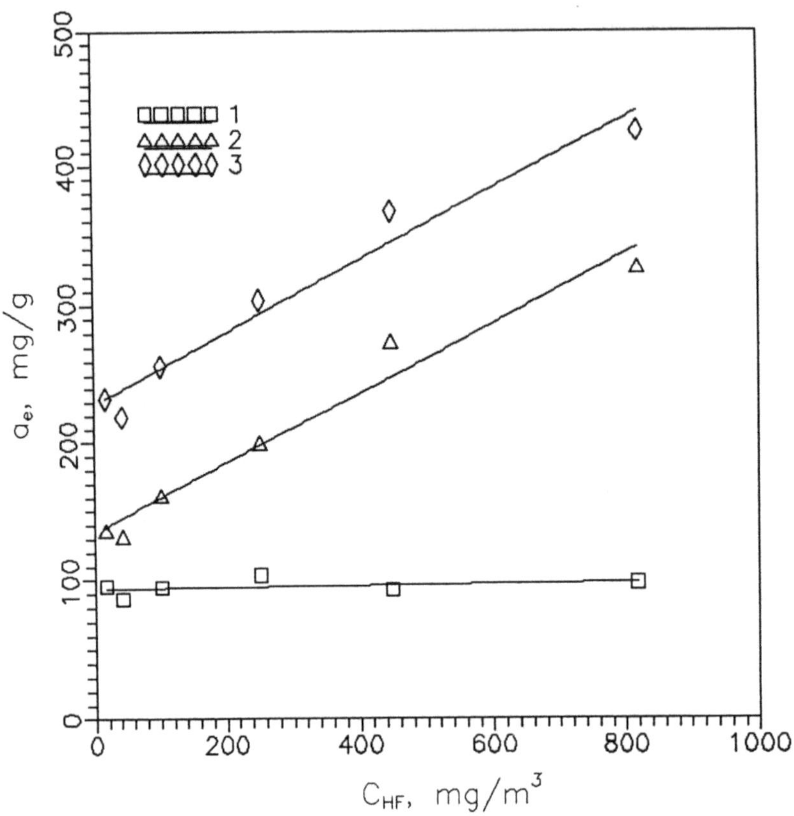

Figure 4. Sorption of HF by FIBAN AK-22. 1—ion exchange sorption, 2—non-exchange sorption, 3—the total.

tion process and the HF molecules in these different states are indistinguishable from each other according to energy of sorption.

The superequivalent HF can be washed up by water while the equivalent HF requires desorption with soda or alkali. In Fig. 4 these two forms of the HF sorbed are presented separately. The particularly important conclusion from these data is that alkaline regeneration is necessary only if trace amounts of HF are removed from the air or in the cases when the air contains other accompanying acid impurities such as SO_2 and H_2S0_4, as often happens. In the majority of practical cases where the level of air contamination is 10–100 mg/m^3 water regeneration can be used providing the working capacity 130–330 mg HF per gram of the fiber. This allows us to avoid the alkali regeneration step.

The process of HF sorption exhibits an interesting specific feature absent in the other cases of gas sorption: in a wide range of air humidity (P/P_0 = <0.1 -1) the sorption does not depend on humidity (Fig. 5). Only at extremely low humidity the sorption efficiency drops down. The physical reason for this phenomenon is high solvation ability of HF molecules, similar to water. This means that the HF sorbed by an ion exchanger can play, similar to water, the role of diffusion medium and the critical air humidity [15] is very low.

This phenomenon gives an additional advantage for the HF removal technology since the stage of the flow humidisation is not necessary (in the other cases it might be needed for purification of dry gas flows).

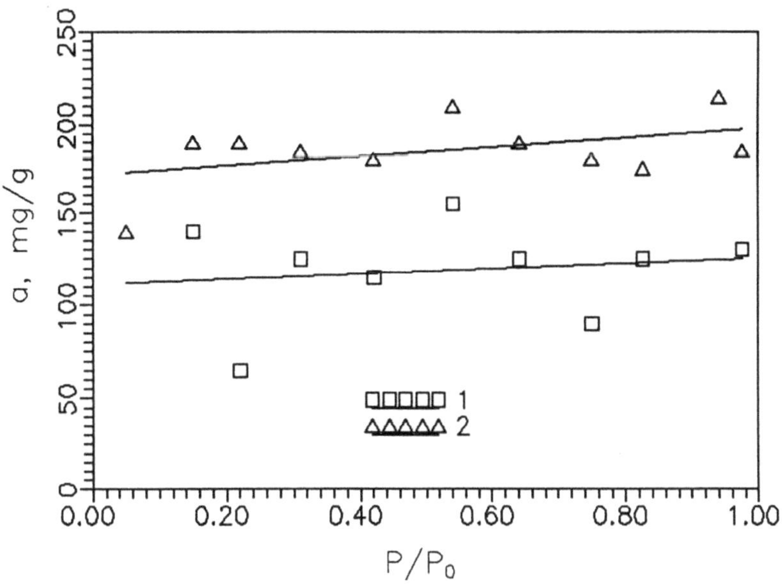

Figure 5. Dynamic (1) and equilibrium (2) activities of FIBAN AK-22 in HF sorption as a function of the air humidity. Conditions: V =9.8 m/min, C = 52 mg/m^3, layer 9 mm.

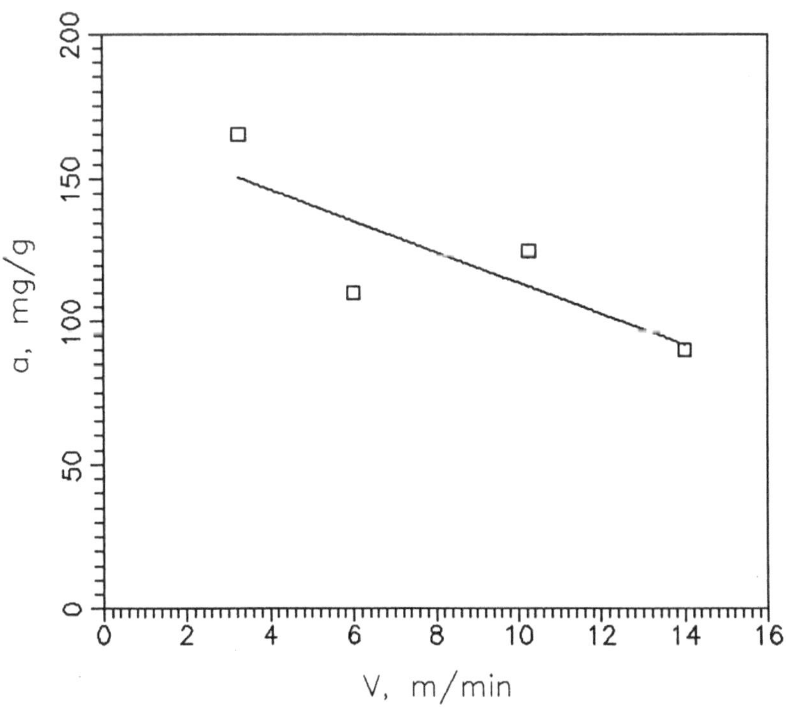

Figure 6. Dynamic activity as a function of the flow rate. HF on FIBAN AK-22. Conditions: C = 52 mg/m^3, P/P$_0$) = 0.65, layer 9 mm.

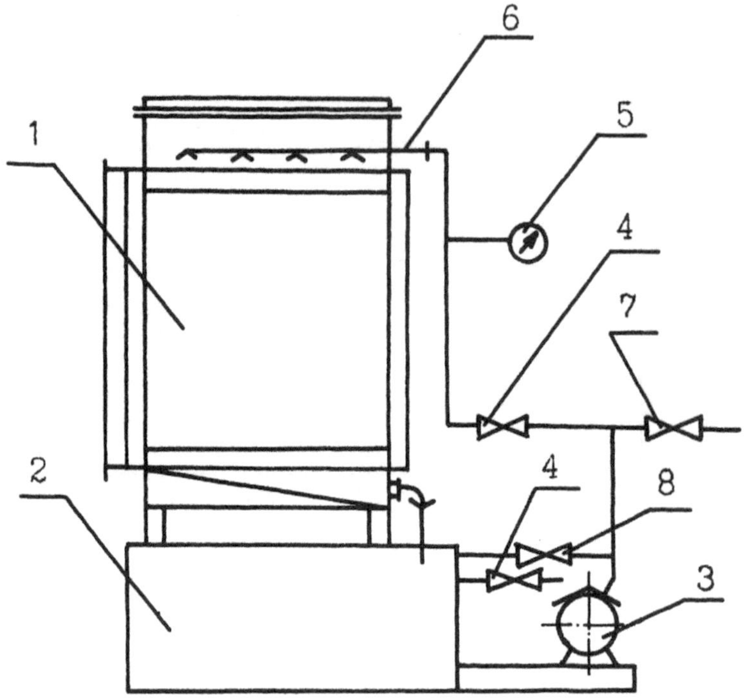

Figure 7. Scheme of the industrial filtering plant FIF-10 applied at NEMAN glass factory (Belarus) for air purification from HF. I—filter with FIBAN AK-22g, 2—regeneration solution tank, 3—pump, 4,7—valves, 5—manometers, 6—distribution collector.

In Fig. 6 dependence of HF sorption on the flow rate is presented. It is seen that at linear gas flow velocity 3–7 m/min, usually applied in filtering plants, the HF sorption is in the range 110–170 mg/g at purification degree level 95%, which means extremely high efficiency.

The scheme of the industrial filtering plant with nominal working capacity 10,000 m3/h applied in the crystal chemical polishing workshop of the glass factory NEMAN in Belarus is presented in Fig. 7. Six units of this type were installed in the factory in 1990 and are in use at present without changing of the ion exchange non-woven filtering fabric. Concentration of the HF in the inlet ranges between 3 and 300 mg/m^3 (depending on the production stage). Purification degree of the air is 92–100% depending on the inlet concentration. Regeneration is performed by washing the filter with water or 3–5% soda solutions down flows along the layers of the filtering materials in film regime. The regeneration time is 2 hours. The sorption chamber volume is 1.3 m^3 the area of the plant is 1.7 m^2. Comparing these parameters with those reported in the cited above works it can be noted that the efficiency of the ion exchange fiber plants exceeds those for the filtering plants with ion exchange resins by at least a power of magnitude.

The other important field of application of the described process is respirators for human's lungs protection from HF. The fibrous form of ion exchanger is extremely convenient for that purpose and allows for the use of ordinary dust respirators in which the dust filter material is substituted by the FIBAN AK-22 non-woven fabric with thickness 5 mm. Such respirators have passed official medical sanitary examination which shows that the respirator containing 12 g of the ion exchanger provides complete protection for humans from HF during 16 hours in the atmosphere with a ten fold HF MPC contamination level. Production of this type of respirator for protection from acidic contaminants including HF has started (Acacia-V type).

3. AMMONIA REMOVAL

Ammonia is one of the most wide-spread air pollutants. Massive ammonia air contamination takes place in agriculture (cattle and chicken farms), chemical, galvanic and food industries and hydrometallurgy. The NH_3 MPC level is 10 mg/m^3. Scrubber air purification with water or acids is efficient and economic at large ammonia concentrations(>500 mg/m^3) and at large volumes of purifying air. The scrubbers are large in size and are inconvenient and expensive for small scale processes. For instance, a scrubber with the working capacity 10,000 m^3/h has a height of 9 m and a diameter of at least 1.5 m. This makes construction of local purifying systems extremely difficult.

Application ion exchange filters of the FIF type provide efficient and convenient air purification from NH_3. At present, an industrial experimental plant with working capacity 10,000 m^3/h is functioning in a prolonged official experiment at the Minsk watch factory.

Sorption of ammonia as a function of its concentration, flow rate and air humidity have been studied for development of the apparatus and air purification technology.

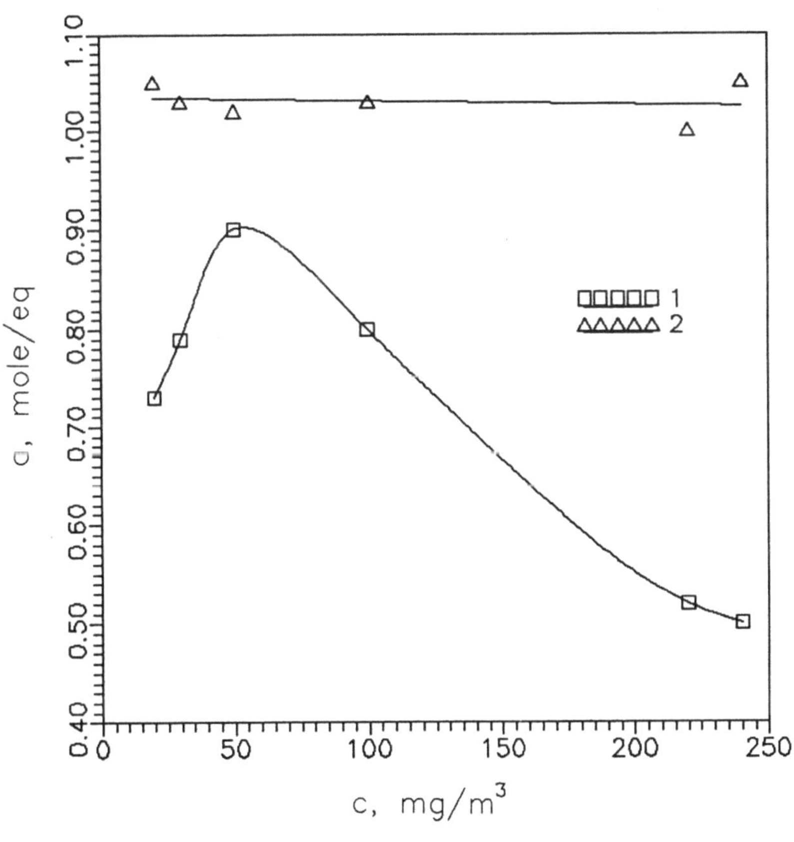

Figure 8. Dynamic (1) and equilibrium (2) activities as a function of concentration of NH_3 at sorption by FIBAN K-4. Conditions: V = 7.0 m/min, P/P_0) = 0.82, filtering layer 10 mm.

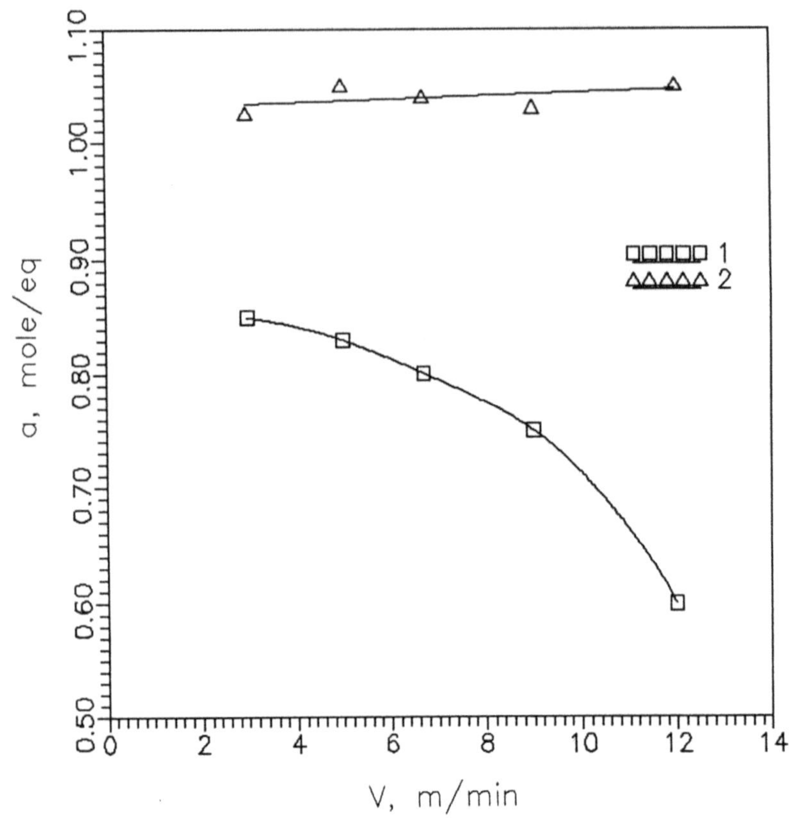

Figure 9. Dynamic (1) and equilibrium (2) activities as a function of flow rate by FIBAN K-4. Conditions: C = 107 mg/m^3, P/P$_0$) = 0.82, filtering layer 10 mm.

Carboxylic cation exchange fiber FIBAN K-4 has been chosen as a sorbent. This ion exchanger is a product of radioactive grafting of acrylic acid onto polypropylene fibers. Its exchange capacity is 7 meq/g, swelling is 1.2 g H_2O/g. The filaments diameter is 40±5μ, the filtering layer thickness in the experiments and the filtering plant is 10 mm.

In Fig. 8 the sorption isotherm is presented. It is sharply convex with the saturation level corresponding to equimolecular quantity of the NH_3 sorbed to the ion exchanger capacity according to reaction

$$\text{R-COOH} + NH_3 = \text{R-COO}^-NH_4^+$$

As follows from the data in Fig. 9 equilibrium capacity does not depend on the air flow rate and in the all range corresponds to the maximum sorption equal to 11.9% mass. The dynamic capacity in the usually applied flow rate range 3–7 m/min is also almost constant and is about 75–80% of the equilibrium value. Further increase in the flow rate causes its decrease. In the range of NH_3 concentrations 15–55 mg/m^3 the protection time of the layer 10 mm is 8–25 hours which can be considered high efficiency.

Ammonia is related to gases whose sorption is strongly dependent on the air humidity. In the range P/P_0 = 0–0.4 the dynamic capacity of the FIBAN K-4 is practically zero (Fig. 10). A sharp increase up to maximum of the sorption is observed at air relative humidity 0.4–0.75. At that humidity the water content in the ion exchanger increases from 0.1 to 1 ware molecule

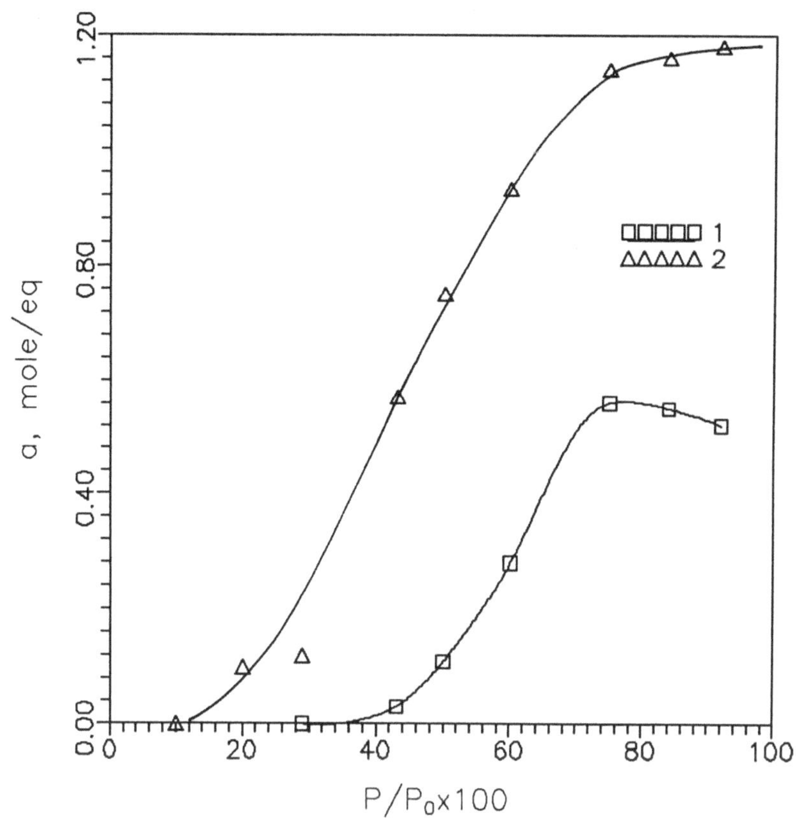

Figure 10. Dynamic (1) and equilibrium (2) activities as a function of air humidity by FIBAN K-4. Conditions: C = 110 mg/m^3, V = 7.0 m/min, layer 10 mm.

per functional group. Further humidity increase (P/P_0 = 0.75–1) leads to the water content increase in the sorbent from 1 to 10 mole H_2O per functional group and some increase of NH_3 sorption (O.25 molc NH_3/cq) duc to NH_3 dissolution in the intrinsic water. Some drop in the dynamic activity at high humidity is caused by increasing length of the diffusion path of the sorbing molecules in the ion exchanger. The interval of maximum sorption P/P_0=0.5–1, fortunately, corresponds to most practical cases of air purification.

The technical parameters of the filtering plant for ammonia removal from air are similar to those of the HF-plant described above. In this case regeneration is performed with 3–5% sulfuric acid solution which can be used practically until its complete neutralization (pH = 3–4). The six month period of industrial examination has shown that the plant provides at least 95% removal of NH_3 from the air of the galvanic workshop in which its concentration ranged between 20 and 90 mg/m^3. The working capacity in real exploitation was 7,000 m^3/h, amount of the absorbed NH_3 5 kg per cycle, the time before regeneration 2 -3 days, regeneration time—2 hours.

4. CONCLUSION

The two considered examples have shown that fibrous ion exchangers can be efficiently used in practical air purification both from acidic and basic impurities. The technical parameters of frame type filtering plants with short (10 mm) filtration layers of fibrous ion exchangers is much higher than those of plants with granular ion exchangers. The current experience with exploitation of that type of filter in Russia, the Ukraine and Belarus (about 50 plants for removal of different acid and alkali aerosols and vapours, SO_2, HCl, a.o. with total working capacity about 0.5 million m^3/h) confirms efficiency of these materials and technologies in air purification.

REFERENCES

1. Watanabe. *Jap. Pat.* 3480, 30. 10. 50. C.A., 1952, **46**, 9812.
2. R. W. Zane, F.B. Larson, G.W. Pankey. *Ind. Eng. Chem.*, 1955, **47**, 47.
3. F. Runge, F. Wolf, R. Griessbach. *Pat. DDR* No 7758, 26.07.54.
4. A. Richter. *Pat. DDR* No 9657, 20.04.55.
5. R.C. Smart, W.S. Derrick. *Anesthesiology*, 1957, **18**, 216.
6. R. Kunin. *Ion Exchange Resins*. N-Y., J. Wiley, 1958, p.113.
7. I. Hashida. *Bull. Osaka Res. Inst.*, 1974. **50**, 1.
8. S. Turkolmez. *Wasser, Luft und Betr.*, 1965, **9**, 737, 812.
9. A.I. Vulih, V.L. Bogatyriov, A.A. Aloviajnikov. *Z. Vsesojuznogo Khomicheskogo Obschestva imeni Mendeleeva*. 1970, **15**, 425.
10. A.I. Vulih, M.K. Zagorskaya, A.A. Aloviajnikov , *Nauchnie Trudy Gintsvetmeta*, 1970, **31**, 120.
11. M.A. Gostomczyk. *Prace Nauk Inst. Inz.. Ocr. Srodow.*, 1974, No. 22, 3.
12. V.P. Plutenko. *Gorodskoe Hoziaistvo Goroda Moskvi*. 1963, No. 11, 37.
13. A.I. Vulih, A.A. Aloviajnikov. *Nauchnie Trudy Gintsvetmeta*, 1979, **47**, 194.
14. V.S. Soldatov a. o. *J. Appl. Chem.* (USSR). 1988, **61**, 46.
15. V.S. Soldatov, I.S. Elinson, A.A. Shunkevich. In: *Chemistry for Protection of the Environment*, Eds L. Pawlowski, G. Alaerts, W.J. Lacy, Elsevier, Amsterdam–Oxford–N.-Y.–London, 1986. P. 369.

SOLVENT SUBLATION

A Method Suitable for Protection of the Atmosphere against Organic Pollutants

Krystyna Medrzycka and Sebastian Pastewski

Faculty of Chemistry
Technical University of Gdansk
Narutowicza 11/12, 80–952 Gdansk, Poland

ABSTRACT

A solvent sublation technique has been applied to the removal of mesitylene from aqueous emulsions. The aim of the investigation was to select an organic solvent which will protect the atmosphere against pollution from mesitylene vapours which could be released from the airstream coming out of a flotation column. Dodecane appeared to be an appropriate solvent for this purpose. Retention of mesitylene in the dodecane layer was extremely good.

INTRODUCTION

Protection of the environment, especially of the aqueous environment, is mostly realized by waste water treatment before its discharge to a receiver. The proper technique for the removal of the pollutants should be applied. For waste waters containing emulsified oils (e.g. hydrocarbons), flotation was found to be an effective method if realized in a proper manner. Our earlier investigations on the flotation of hydrocarbons from aqueous emulsions gave satisfactory results[1]. However, in some cases the purification of waste water results in contamination of the atmosphere. This may happen when the removed substance is volatile and slightly soluble in water. In such a case, simultaneously with the interceptive mechanism for their removal, evaporation into air bubbles takes place[2]. After leaving the aqueous phase the bubbles break and release the vapours to the atmosphere. In our investigations this was observed for mesitylene[2,3]. In order to avoid secondary pollution of the environment, namely the pollution of the atmosphere, the flotation process could be modified by a solvent sublation technique.

Up till now, from the practical point of view, the solvent sublation technique has been considered as a method useful for the removal of certain classes of water pollutants,

Chemistry for the Protection of the Environment 2, edited by Pawłowski et al.
Plenum Press, New York, 1996

mainly refractory organics, which are toxic and persistent in the environment[4–7]. These substances are transported from the aqueous phase to the overlying immiscible layer of non-volatile organic liquid by gas bubbles rising through the solvent sublation column (within or on the surface of the bubble). The mechanism of sublation depends on the sublate properties, and it may be based on the surface adsorption of the solute at the air-water interface (surface active substances), on evaporation of the solute into the interior of the bubble (volatile compounds), on the formation of air bubble–particle aggregates (dispersed substances, e.g., oil droplets), or on the adsorption of complexes formed with surfactants (metals, dyes). As sublation proceeds, transport of the solute is also possible due to diffusion between the phases. In the initial stages this transport is driven from the aqueous to the solvent phase, and as the solute concentration builds up in the organic solvent, the reverse transfer becomes important[8]. Another possibility is solute transport with the thin film of water dragged into the solvent phase with the bubbles. The returning water droplets may be depleted in solute concentration[8].

In the present work the solvent sublation technique has been applied to remove semivolatile hydrocarbons (e.g., mesitylene) from their emulsions. There are two main mechanisms in the transport of mesitylene from water: interception of its droplets by rising bubbles and evaporation of its molecules into the bubble. Transport due to diffusion between the aqueous and organic phases probably also takes place; at this moment we will

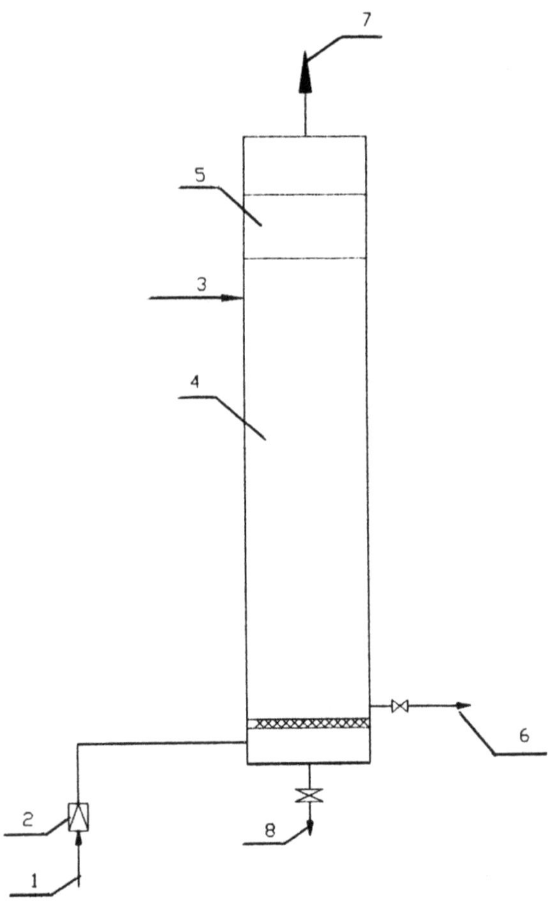

Figure 1. Solvent sublation column. 1—gas inlet; 2—rotameter; 3—mesitylene emulsion inlet; 4—aqueous layer; 5—mesitylene emulsion inlet; 6—raffinate outlet; 7—gas outlet; 8—outlet valve.

not consider the problem of the contribution of particulate mechanisms to the removal of mesitylene.

The aim of our investigations was to study the usefulness of this technique for waste water treatment and protection of the atmosphere at the same time. Thus, we planned to select an organic solvent which ensures the protection of the atmosphere against volatile organics which could be released from the airstream coming out of flotation units. Simultaneously, a high efficiency of sublate removal from water was desirable.

EXPERIMENTAL

Solvent sublation experiments were carried out in a glass column 120 cm long and 4.5 cm in diameter (Fig.1). The column was filled with the experimental mesitylene emulsions (1000 or 1300 ml) and the organic liquid (50 or 100 ml) was poured on top. Dodecane or n-octanol were used as the organic solvent. The sintered glass sparger, fitted at the lower end of the column, produces air bubbles which rise through the aqueous and organic solvent layers. The gas flow rate was equal to 4 dm^3/h. Runs were conducted in a batch system.

The concentration of mesitylene in the organic and aqueous phases was determined by GLC. The efficiency of the process was evaluated on the basis of the removal of mesitylene from water and on the mesitylene retention in the organic phase.

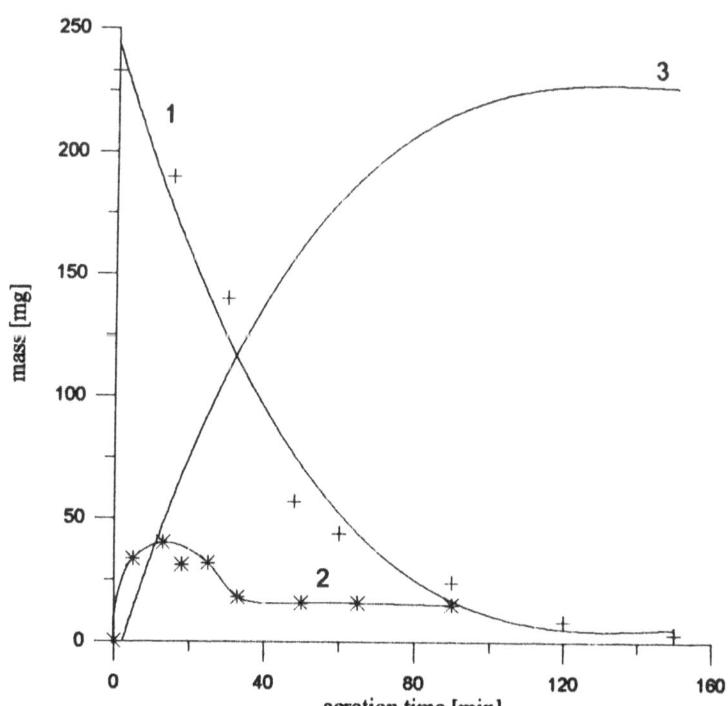

Figure 2. Solvent sublation of mesitylene into n-octanol. 1—content of mesitylene in the aqueous phase; 2—content of mesitylene in the organic phase; 3—"theoretical" content of mesitylene in the organic phase.

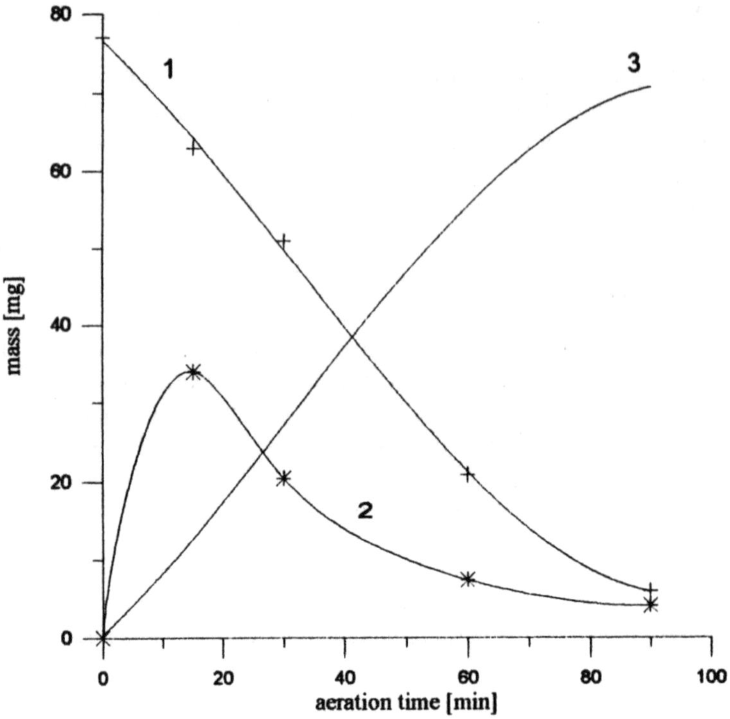

Figure 3. Solvent sublation of mesitylene into n-octanol. Curves description as in Fig. 2.

RESULTS

Figures 2 and 3 show the sublation results when octanol was used as a solvent phase. Curve 1 presents the mesitylene concentration changes in the aqueous phase during the batch runs. Curve 2 presents the changes of mesitylene concentration in the organic phase (octanol). Curve 3 corresponds to the expected mesitylene concentration in octanol (calculated on the basis of the mass balance relating to curve 1).

Figures 4 and 5 show the mesitylene sublation results in cases when dodecane was used as the organic phase.

As can be seen from the Figures, in cases when octanol was used the removal of mesitylene is faster compared to the processes with the use of dodecane. About 90% of mesitylene removal was achieved after 1.5 h in the case of octanol, but 3 h was required in the case of dodecane. Probably, the difference in the initial mesitylene concentration in the emulsion could affect the curves (in case of dodecane the mesitylene concentration was much higher than that in the case of octanol).

The most important difference is the retention of mesitylene in the organic solvents. In octanol the mesitylene content increases, reaching 0.05%, but after 15 min. the mesitylene concentration decreases to 0.007% after 2 h. The mesitylene concentration never reaches its "theoretical" value. For example, after 50 min. from the beginning of the run the content of mesitylene in the octanol phase was 19 mg instead of 176 mg, as predicted from curve 3 (Fig.2). Analysing Figures 4 and 5 one can state that the agreement between the "theoretical" (curves 3) and experimental content of mesitylene in dodecane (curves 2) is really good.

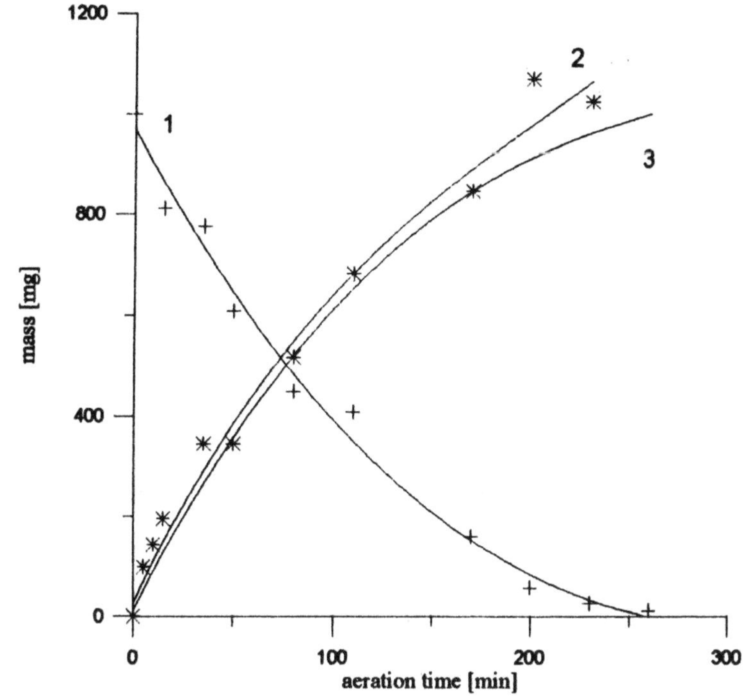

Figure 4. Solvent sublation of mesitylene into dodecane. Curves description as in Fig. 2.

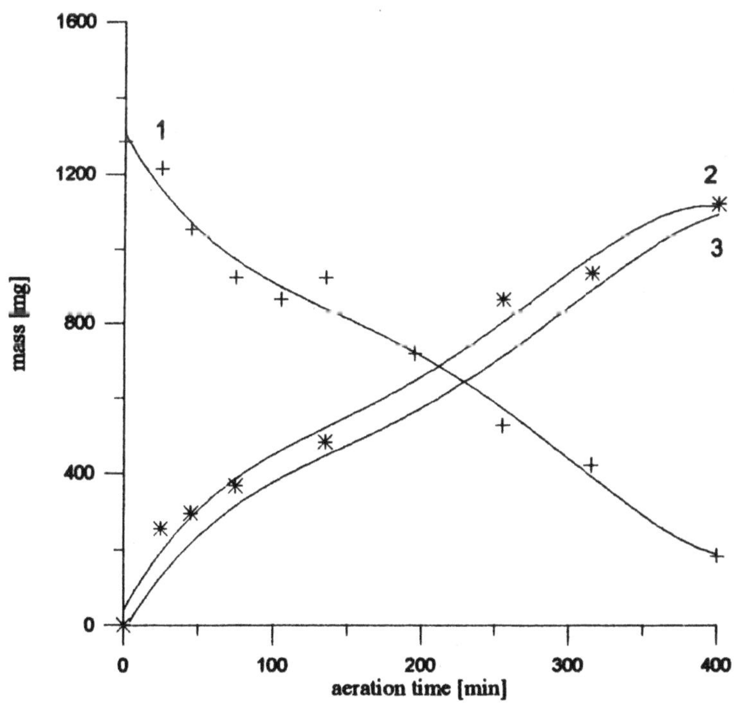

Figure 5. Solvent sublation of mesitylene into dodecane. Curves description as in Fig. 2.

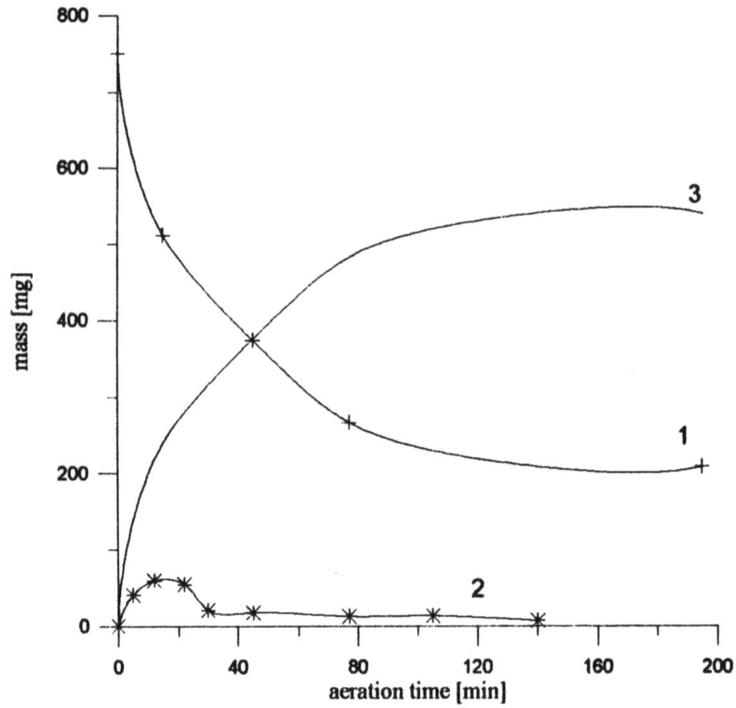

Figure 6. Solvent sublation of mesitylene into n-octanol. Emulsions of mesitylene stabilized with NaDBS. Curves description as in Fig. 2.

On the basis of the presented results one can state that dodecane is an appropriate organic solvent for mesitylene removal from water by solvent sublation. Simultaneously, this solvent protects the atmosphere against pollution by the removed, volatile hydrocarbons. In contrast n-octanol does not protect the atmosphere because mesitylene is very easily transferred from n-octanol to the atmosphere by rising bubbles.

Surfactants, usually anionic ones, are very often present in oily wastewaters. It is known that surfactants influence the efficiency of oil removal by flotation. Hence we investigated their effect on the sublation of hydrocarbons. Mesitylene emulsions, stabilized with the anionic surfactant sodium dodecylbenzenesulfonate (NaDBS) were used in the experiments. Dodecanol or dodecane were used as an organic solvent. The concentration of NaDBS was equal to 10^{-5} mol/l. Figures 6 and 7 present the results of mesitylene sublation to n-octanol and to dodecane.

It can be seen that the removal of mesitylene from its emulsions stabilized by NaDBS is at the beginning of the run faster than the removal from nonstabilized emulsions; however, after about 45 min of the run, the rate of removal decreases and the final concentration of mesitylene (e.g., after 3 h) is higher than the concentration achieved in the absence of surfactant. Besides, it was found that there is no effect of NaDBS presence on mesitylene retention in the organic phase. Curves 2 are similar to those obtained in processes without surfactant.

Summarizing, one can state that it is possible to apply the solvent sublation technique to flotation for the removal of semivolatile hydrocarbons from wastewaters. The atmosphere can be protected against pollution by these compounds if the organic solvent is

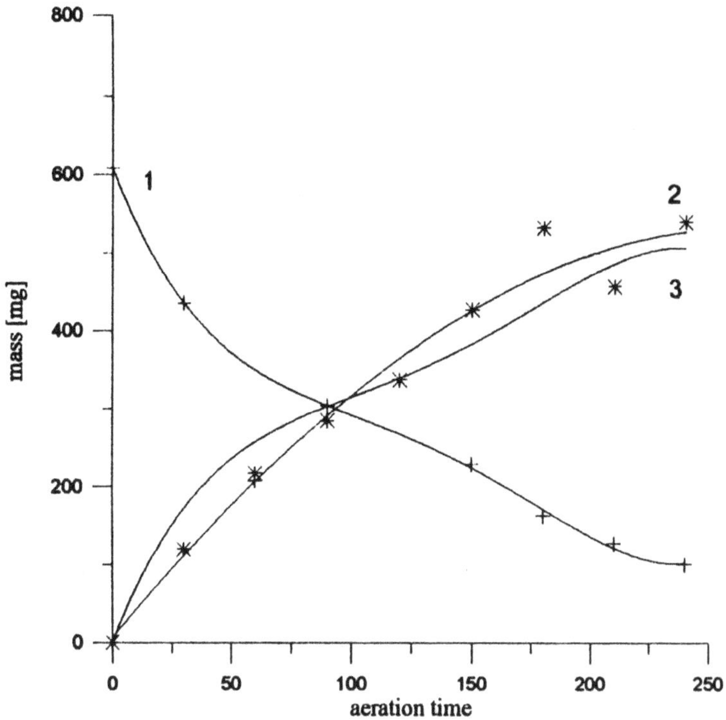

Figure 7. Solvent sublation of mesitylene into dodecane. Emulsions of mesitylene stabilized with NaDBS. Curves description as in Fig. 2.

properly chosen. For mesitylene removal dodecane appeared to be a good solvent. Generally, for hydrocarbons removal by the solvent sublation method, the best solvent is a hydrocarbon e.g., paraffin oil, dodecane, tetradecane.

Some problems arise when the aqueous phase contains an anionic surfactant. In such a case the organic upper phase can be emulsified in water. In the literature we have found some reports in which redissolution of the organic solvent has been observed[7–9]. Valsaraj[7] and Grieves[9] have found that redissolution of the organic solvent in the aqueous phase is smaller than that observed in conventional solvent extraction. For example, the methyl isobutylketone concentration after a single-stage semibatch solvent sublation is equal to 0.135 mol/l, while after extraction it is 0.194 mol/l [7].

Two advantages of solvent sublation over solvent extraction are pointed out:

- lower solubility of the solvent in the aqueous phase (owing to nonequilibrium conditions);
- the use of a phase volume ratio lower than that in extraction[9].

The significance of the solvent redissolution effect on the sublate removal efficiency has been investigated. The same problem exists when the solvent extraction technique is applied. However, there are no reports on the emulsification of the organic solvent in the aqueous phase during solvent sublation processes, which has been observed in our investigations.

In order to avoid this secondary pollution of water, further investigations are necessary. Our effort focuses on finding ways of preventing the emulsification of the organic solvent in the aqueous phase.

REFERENCES

1. K. B. Medrzycka, Sep. Sci. Technol., 25 (1990), 825.
2. K. B. Medrzycka, Sep. Sci. Technol., 23 (1988), 565.
3. K. B. Medrzycka, Sep. Sci. Technol., 27 (1992), 1077.
4. K. T. Valsaraj, D. J. Wilson, Colloids Surf., 8 (1983), 203.
5. L. K. Foltz, K. N. Carter, D. J. Wilson, Sep. Sci. Technol., 21 (1986), 57.
6. K. T. Valsaraj, C. Springer, Sep. Sci. Technol., 21 (1986), 789.
7. K. T. Valsaraj, Flotation Sci. Eng., 13 (1995), 365.
8. K. T. Valsaraj, L. J. Thibodeaux, Sep. Sci. Technol., 26 (1991), 37.
9. R. B. Grieves, W. Charewicz, S. M. Brien, Anal. Chim. Acta, 13 (1974), 293.

9

A POSSIBILITY TO REDUCE METHANE EMISSION FROM LANDFILLS BY ITS OXIDATION IN THE SOIL COVER

Witold Stępniewski[1,2] and Małgorzata Pawłowska[1]

[1] Technical University of Lublin
40, Nadbystrzycka Str., 20-618 Lublin, Poland
[2] Institute of Agrophisics PAN
4, Doświadczalna Str., 20-280 Lublin, Poland

ABSTRACT

Landfills of municipal wastes are considered to be a source of total methane emission to the atmosphere estimated at 20 to 70 Tg annually which is an important contribution (6–20%) to total anthropogenic emission (about 360 Tg). Atmospheric concentration of this very radiatively active gas (about 30 times more than CO_2) is increasing at a rate of approximately 1% per year. One of the possibilities to reduce the ongoing methane concentration growth in the atmosphere is the reduction of anthropogenic emission by about 10%, which corresponds to the oxidation of 3/4 of methane emitted from landfills. According to some authors the cheapest and the most feasible way of decreasing methane emission from municipal wastes is increasing the degree of methane oxidation in the landfill cover soil.

This paper presents a general discussion of the problem and analyses the effects of different factors determining methane oxidation efficiency in landfill cover soil. Optimization of the methane oxidation process in landfill cover might be achieved based on such parameters as: type and thickness of the soil cover, pH and gas diffusion coefficient.

The possibility of a substantial increase in the degree of methane oxidation in landfill soil cover has been confirmed. It has been demonstrated that methane oxidation in soil obeys Michaelis–Menten kinetics with respect to both oxygen and methane. The methanotrophic activity in soil increases with time in the presence of elevated methane concentrations and stabilizes after about one month. The optimum pH for methane oxidation ranges from 6 to 8.

1. INTRODUCTION

Total annual emission of methane from natural and anthropogenic sources is estimated to be 470–650 Tg[27]. The predominant source of methane is bacterial decay of or-

Table 1. Estimated present-day sources and sinks of methane (from Lelieveld and Crutzen[27])

	Emission ($Tg\ yr^{-1}$)
Source	
Total	560 ± 90
Natural wetlands	125 ± 70
Rice fields	70 ± 50
Enteric fermentation (mainly ruminants)	80 ± 20
Landfills	40 ± 25
Biomass burning	30 ± 15
Animal wastes	25 ± 10
Domestic sewage	25 ± 10
Coal mining	30 ± 30
Oceans	10 ± 5
Freshwater	5 ± 5
CH_4 hydrate destabilization	5 ± 5
Sink	
Total	530 ± 85
Reaction with OH in the troposphere	455 ± 50
Removal by soils	30 ± 25
Reaction with OH, Cl in the stratosphere	45 ± 10

ganic matter under anaerobic conditions. The main natural sources are: wetlands, enteric fermentation in the stomachs of wild ruminants, oceans and freshwaters. The emission from anthropogenic source, i.e., rice fields, domestic ruminants, agricultural biomass burning in the tropics, organic waste in sewage systems and landfills, fossil fuel production, transport and utilization, constitute about 70% of total annual methane emission (329–455 Tg).

According to Bouwman[6] methane is the second most important green-house gas with a chemical lifetime in the atmosphere of about 10 years. It contributes already now approximately 20% to global warming. It is about 30 times more active than carbon dioxide in absorbing infrared radiation. Its concentration has increased over the past century from 0.75 ppm to 1.7 ppm, mainly as the result increasing intensity of emission from the anthropogenic sources. This growth during the last few years is about 1% per year. However, the reason for this large annual increment is not only the growth of emission derived from industry and agriculture. It is possible that CH_4 sinks have decreased recently. The methane sinks are: (1) the reaction with OH radicals, (2) transport to the stratosphere and (3) microbial oxidation in terrestrial environments[18]. Soil investigations[13, 15, 22, 23, 39] show that soils might be very effective oxidants of methane under oxygenated conditions. The confirmation of the possibility of methane uptake in the soil is very important in formation of municipal landfill covering layers.

2. METHANE PRODUCTION IN LANDFILLS

One significant anthropogenic methane source is landfills of municipal waste. They are considered to be a source of total methane emission to the atmosphere estimated at 40 Tg of methane annually [27, 43]. It should be noted that landfills are a source of methane for several decades and that already existing landfills will still be producing methane for many years.

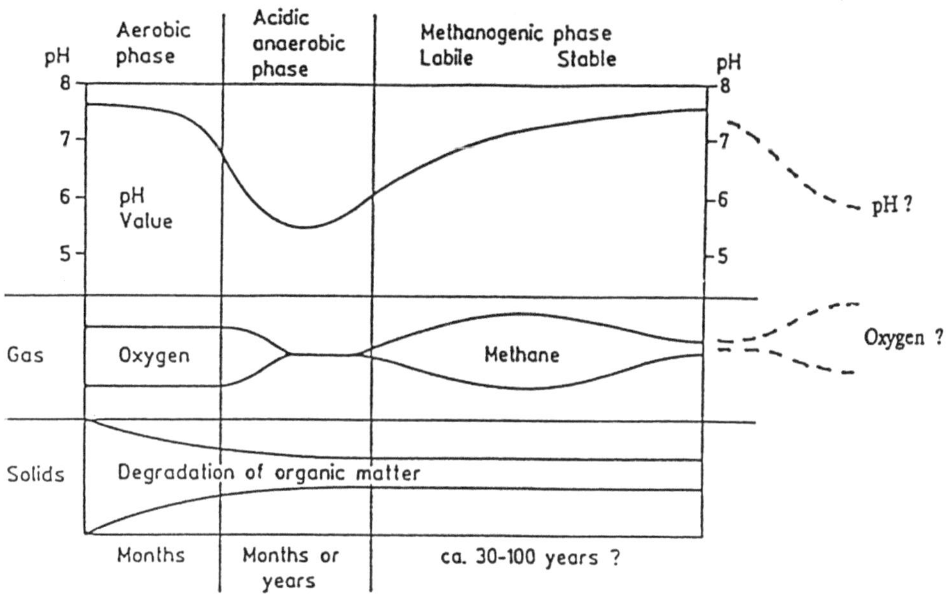

Figure 1. Scheme of chemical evolution of municipal solid waste landfills (from Förstner et al.[11]).

Methane production in landfills is connected with the anaerobic phase of its decomposition which is presented in Figure 1.

Potential production of methane from landfill waste ranges from 35–100 kg methane per Mg of refuse[12 42], which corresponds to 50–150 m^3 methane per Mg of refuse. The total volume of landfill gas containing approximate equimolar quantities of CO_2 and CH_4 is twice as much. Hoeks[15] assumes that the average gas production (CO_2+ CH_4) rate in waste tips for a period of 5–15 years is about 6 m^3 of biogas, i.e., $3m^3$ (2 kg) CH_4 Mg^{-1} refuse $year^{-1}$. The gas production rate is dependent on many factors such as quantity of particular refuse fraction (more or less biodegradable or non-biodegradable), moisture content, pH level, temperature, presence and distribution of methanogenic microorganisms. Total gas production rates and long-term total gas production are demonstrated in Figure 2.

3. POSSIBILITY OF REDUCTION OF METHANE EMISSION FROM LANDFILLS

The most efficient method of methane emission reduction is its practical exploitation as a source of energy. The greatest methane emission rate from waste tips is noted in the period 0–5 years after deposition of the waste (about 45% of total gas production). Usually a gas extraction system is installed after termination of dumping activities and thus the whole emission of the first 5 years is lost. The recovery of methane is not always economically justified especially in the case of the already existing small and closed landfills requiring recultivation. In this case covering the landfill with the soil layer is the only feasible method to reduce methane emission to the atmosphere.

According to Aitchison[2] and Gaudioso et al.[12] the increasing of the degree of methane oxidation in the soil cover may be the cheapest and the most feasible way of decreasing of methane emission from landfills. In different models the assumed oxidation degree varies from 10% to 40% which is a result of little scientific recognition of that problem.

One possibility for reducing the ongoing methane concentration growth in the atmosphere is the reduction of anthropogenic emission by about 10%, which corresponds to the oxidation of 3/4 of methane emitted from landfills. Limitation of methane emission from landfills seems to be necessary and international regulation in that area as well as international control systems should be introduced [12,47]. Based on such parameters as: type and thickness of cover, pH and gas diffusion coefficient and eventually the kind of plant covering, the optimization of the methane oxidation process in landfill cover might be reached. Investigations in this subject were conducted by Hoeks[16]. According to him the methane oxidation capacity in the soil cover of landfills is quite large. He has calculated that about 5–25 m^3 CH_4 can be oxidized per square metre per year, depending on soil type, thickness of soil cover, physical and chemical conditions and methanotroph microorgan-

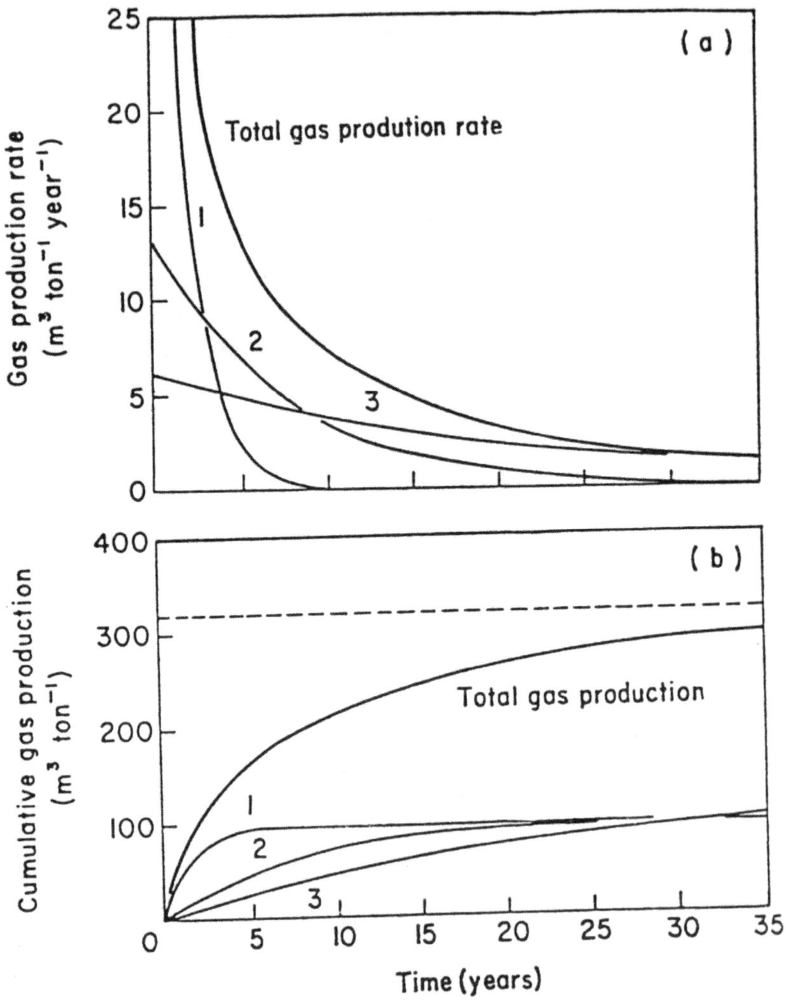

Figure 2. Calculated gas (CO_2+ CH_4) production rate (a) and cumulative gas production (b) in municipal solid waste with the following composition: (1) 30% of readily degradable fraction: half-life = 1 year; (2) 30% of moderately degradable fraction: half-life = 5 years; (3) 40% of slowly degradable fraction: half-life = 15 years (from Hoeks[15]).

isms evolution. The maximum methanotrophic capacity reported in the literature[23] is 10.4 mole of CH_4 $m^{-2}day^{-1}$ which corresponds to about 85 m^3 of CH_4 m^{-2} $year^{-1}$ which is much higher than the data reported by Hoeks[16].

According to Kightley and Nedwell[22] the typical rate of methane production derived from a 20m deep layer of landfill is on the order of 18 mole of CH_4 $m^{-2}day^{-1}$ that is about 150 m^3 of CH_4 $m^{-2}year^{-1}$. The calculation based on the data of Hoeks[16] for the emission of 6 m^3 biogas Mg^{-1} of the refuse $year^{-1}$ at its 20 m deep layer and the bulk density of 0.8 Mg m^{-3}, gives the value of about 50 m^3 CH_4 per year which is about 1/3 of the value given by Kightley and Nedwell[22]. The comparison of these values indicates that there is a great chance to obtain substantial or even complete oxidation of methane emitted from landfills in the soil cover. The question arises how to increase the capacity of the soil cover of landfills to oxidize methane. To answer that question a detailed recognition of factors affecting this process is discussed further.

4. POSSIBILITY OF ANAEROBIC CH_4 OXIDATION

The possibility of methane oxidation under anaerobic conditions has been shown by many authors in oceanographic and limnological studies and association with sulphate or ferric oxide reduction has been suggested [9, 19, 26, 30, 36, 37, 48]. The capability to perform anaerobic oxidation of methane has been demonstrated for *Desulfovibrio desulfuricans*[9]. The existence of methane oxidation under anaerobic conditions in rice paddy soil has been reported by Murase and Kimura[32, 33, 34] who stated that the methane content in the leachate from anaerobically incubated soil was substantially reduced in the presence of 1% amendment with the oxidants (Figure 3). It should be emphasised that this process is possible due to the presence of such compounds as nitrates, manganic and ferric oxides as well as sulphates playing the role of electron acceptors during methane oxidation.

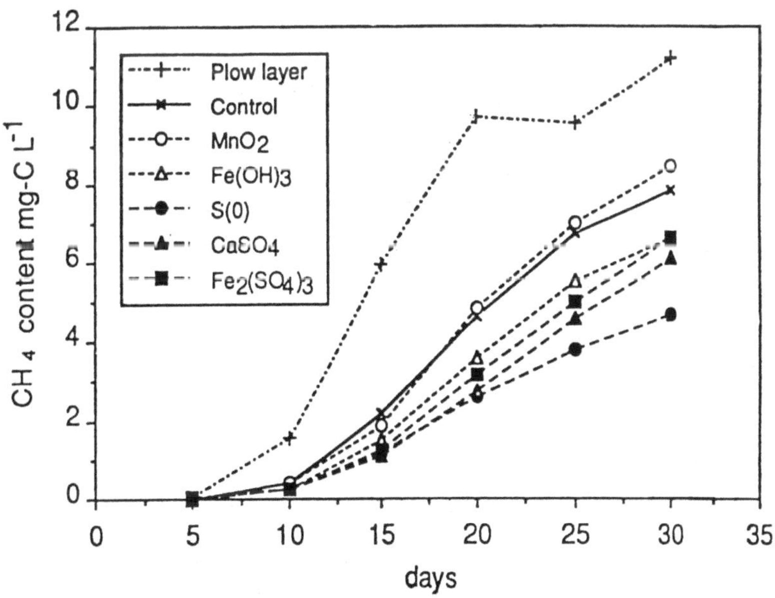

Figure 3. Effect of 1% amendment of oxidants on methane content in the leachate from anaerobically incubated plow layer soil of a paddy field. The control consisted of the soil material amended with the basal fertilizers and rice straw (from Murase and Kimura[34]).

Table 2. Equivalents of the probable methane oxidants under anoxic conditions in soil according to reactions 1–5 (expressed in moles per mole, in kg per m^3 and in Mg per Mg of methane oxidized)

Oxidation of CH_4	Amount of oxidant				
	KNO_3[a]	KNO_3[b]	MnO_2	Fe_2O_3	$CaSO_4\,2H_2O$ gypsum
1 mol	1 mol	2 mol	4 mol	4 mol	1 mol
1 m^3	4.22 kg	8.44 kg	14.58 kg	26.72 kg	7.18 kg
1 Mg	6.31 Mg	12.63 Mg	21.75 Mg	40.00 Mg	10.75 Mg

[a]Reduction to NH_4^+ (reaction 1).
[b]Denitryfication to N_2O (reaction 2).

The possibility of anoxic methane oxidation occuring in landfills is connected with the presence of the electron acceptors which would substitute for the function of oxygen.

The hypothetical reactions would be as follows:

$$CH_4 + NO_3^- + H^+ = HCO_3^- + NH_4^+ \tag{1}$$

$$CH_4 + 2NO_3^- + H^+ = HCO_3^- + N_2O + 2H_2O \tag{2}$$

$$CH_4 + 4MnO_2 + 7H^+ = HCO_3^- + 4Mn^{++} + 5H_2O \tag{3}$$

$$CH_4 + 4Fe_2O_3 + 15H^+ = HCO_3^- + 8Fe^{++} + 9H_2O \tag{4}$$

$$CH_4 + SO_4^{--} = HCO_3^- + HS^- + H_2O \; (G = -25 kJ\,mol^{-1}) \tag{5}$$

The stoichiometric calculations for the above reactions are presented in Table 2. It can be noticed that the oxidation of 1m^3 of methane requires 7.18 kg of gypsum and the amounts of manganic and ferric oxides are much higher. More than 10 Mg of gypsum is needed for oxidation of 1 Mg of methane. Nitrate is the most efficient of the oxidants in the methane oxidation process but it does not seem to have practical importance under landfill conditions. Of the oxidants presented above only gypsum may have a practical application for reduction of methane flux from landfills to the atmosphere as it can be applied at least in the regions where it is easily available for covering the landfills or for layering them. If the total methane produced per one square meter of landfill surface during the entire period of its methane activity is on the order of 0.6 to 1.6 Mg (assuming bulk density of refuse 0.8 Mg m^{-3} and a landfill depth of 20 m and the methane production 35 - 100 kg CH_4 Mg^{-1} of refuse) then covering of the landfill with 1Mg gypsum layer per 1 square meter (layer of about 0.5 m thickness) would reduce total methane flux by 93 kg m^{-2}. This amount constitutes 6–16% of the above mentioned values of the total methane emission and can play a significant role in the initial period of intensive methane production when the landfill is not fully recultivated or when the oxidation in the soil cover is not sufficient.

5. AEROBIC METHANE OXIDATION

5.1. Methanotrophic Microorganisms

Aerobic oxidation of methane is performed by methanotrophic microorganisms. To this group belong bacteria such as *Bacillus methanicus*[41]; *Methanomonas methanooxidans*; *Methylococcus capsulatus* (after Hoeks[15]); *Methylomonas methanica, Pseudomonas methanica*[10]; *Methylosinus sp., Methylococcus sp., Methylobacter sp., Paracoccus denitrificans*[29]; *Methanomonas margaritae, Methylomonas albus*[8]; chemo-autotrophic NH_4 oxidizers *Nitrosomonas europaea and N. oceanus*[20].

According to Adamse et al[1] the optimum temperature for growth of *Pseudomonas methanica* and *Methylosinus trichosporium* is about 30°C. The optimum pH value is 7.0, while growth is possible within a pH-range of 5.5–8.0. It is possible even at pH of 4.0 although the microbial activity was low under these conditions[15].

Moreover, methane oxidizing mixotrophic fungi were isolated from dune sand soil[1].

5.2. Kinetics of Methane Oxidation in Soil

Methane oxidation reaction can be written as follows:

$$CH_4 + 2O_2 \longrightarrow CO_2 + 2H_2O \qquad (6)$$

The sequence of the oxidation reactions is presented after Hoeks[15] below:

$$CH_4 \xrightarrow{-26.12} CH_3OH \xrightarrow{-44.81} HCOH \xrightarrow{-57.15} HCOOH \xrightarrow{-58.25} CO_2 \qquad (7)$$

Ribbons et al.[38] calculated the change in free energy (kcal $mole^{-1}$) for each oxidation step (presented above).

As it can be noted from Eq. 6 the theoretical ratio of oxygen consumed to oxidized methane is equal to 2 (on molar basis). It has been found that in soil the ratio is lower than theoretically predicted as the oxidation process is brought about by microbes using part of the methane carbon for the synthesis of the cell material. Literature review of Hoeks[15] showed that the measurements with suspensions of methane oxidizing bacteria yielded divergent values for the ratios of the gases involved in the reaction. Per unit volume of oxidized methane the oxygen consumption ranged from 0.2 to 1.8 (instead of theoretical value 2) and the production of carbon dioxide from 0.2 to 0.9 volume units (instead of theoretical value 1). The relation for methane oxidation in dune sand was experimentally determined by Hoeks[15] and is presented in Figure 4.

Methane oxidation in soil medium undergoes kinetics of Michaelis - Menten being intermediate between the first order and zero order reaction according to the equation:

$$V = V_{max} \frac{1}{\left(1 + \frac{K_M}{C}\right)} \qquad (8)$$

where:

V- is the reaction rate,

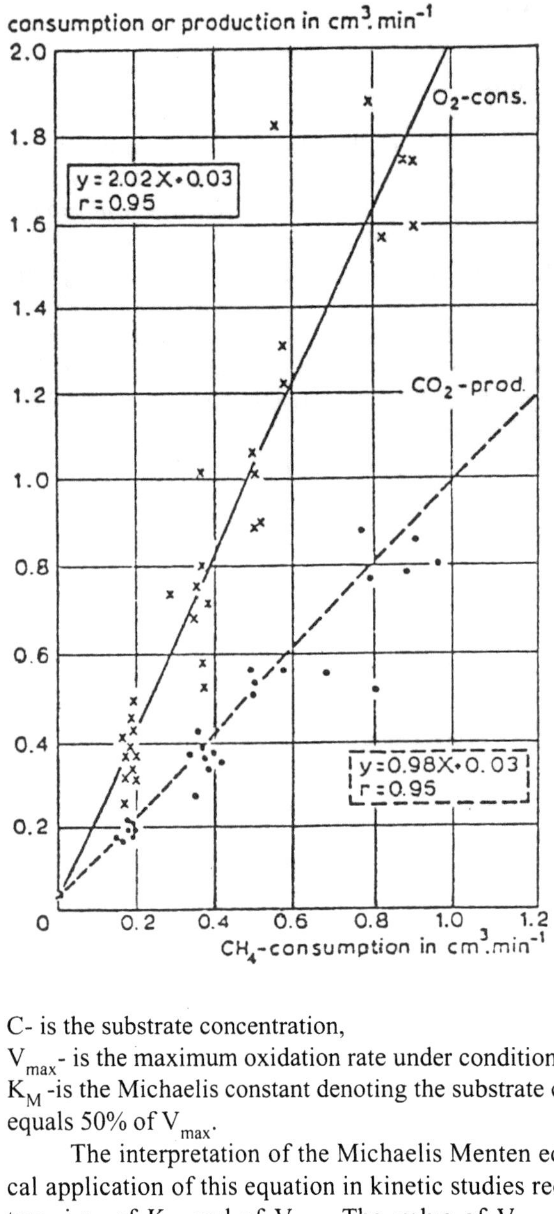

Figure 4. Relation between the O_2-consumption rate, the CO_2- production rate, and the CH_4-consumption rate in dune sand soil at about 20°C (from Hoeks[16]).

C- is the substrate concentration,
V_{max}- is the maximum oxidation rate under conditions of saturation with the substrate,
K_M -is the Michaelis constant denoting the substrate concentration at which the reaction rate equals 50% of V_{max}.

The interpretation of the Michaelis Menten equation is presented in Figure 5. Practical application of this equation in kinetic studies requires determination of its two parameters, i.e., of K_M and of V_{max}. The value of V_{max} is interpreted as the potential activity being a measure of the oxidation capacity of the reaction medium, while the K_M value is a measure of the affinity (defined as the reciprocal of K_M) of the reaction medium for a given substrate. It should be noted that high capacity is not always connected with high affinity and sometimes the relation may be just the opposite.

Bender and Conrad[4] detected two types of populations of soil methanotrophs: (1) high capacity/low affinity population exposed to elevated CH_4 concentrations and (2) low capacity/high affinity population exposed to atmospheric CH_4 level. Investigation of Kightley et al.[23] confirmed the existence of two types of soil communities in landfill cover soils. Methanotrophic microbial community with a high capacity (V_{max} = 998 nmole g of soil^{-1}h^{-1}) and low affinity for CH_4 (K_M =31.7 μM) developed in CH_4-purged microcosms.

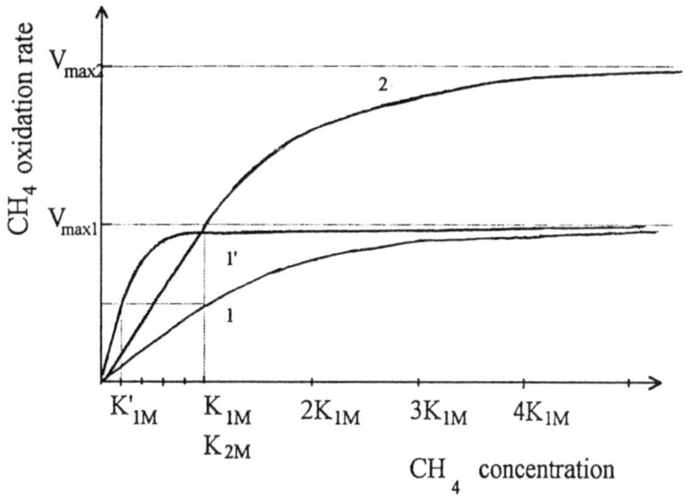

Figure 5. Graphical interpretation of the Michaelis- Menten equation for three situations: (1) - low capacity (V_{1max}) and low affinity (K_{1M})for methane; (1') - the same low capacity (V_{1max}) and 5 times higher affinity ($K'_{1M}=0.2\ K_{1M}$), and (2) - twice as much capacity ($V_{2max}=2V_{1max}$) as in the case of (1) and the same affinity ($K_{2M}=K_{1M}$).

In the case when the soil remained for 6 months without CH_4, microorganisms with a low methanotrophic capacity (V_{max} of 258 nmol g of soil^{-1}h^{-1}) but relatively high affinity (K_M of 1.6 μM) were found.

The reaction of methane oxidation in soil is probably related by Michaelis -Menten equation to both substrates, i.e., to methane and oxygen. The equation in this case should be as follows:

$$V = V_{max} \frac{1}{\left(1+\frac{K_{M_1}}{C_1}\right)} \frac{1}{\left(1+\frac{K_{M_2}}{C_2}\right)} \quad (9)$$

where.
V- is the reaction rate,
V_{max}- is the maximum oxidation rate under the conditions of saturation with both the substrates,
C_1 - is the methane concentration,
C_2 - is the oxygen concentration,
K_{M1} -is Michaelis constant for methane.
K_{M2} -is Michaelis constant for oxygen.

It should be noticed that the time of stabilization of the methane oxidation rate in soil seems to be on the order of one month as shown in Figure 6.

5.3. Effect of Chemical Parameters on CH_4 Uptake in Soils

5.3.1. The Effect of Oxygen Concentration on Aerobic Methane Oxidation in Soil. The effect of oxygen concentration on methane oxidation rate is presented in Figure 7.

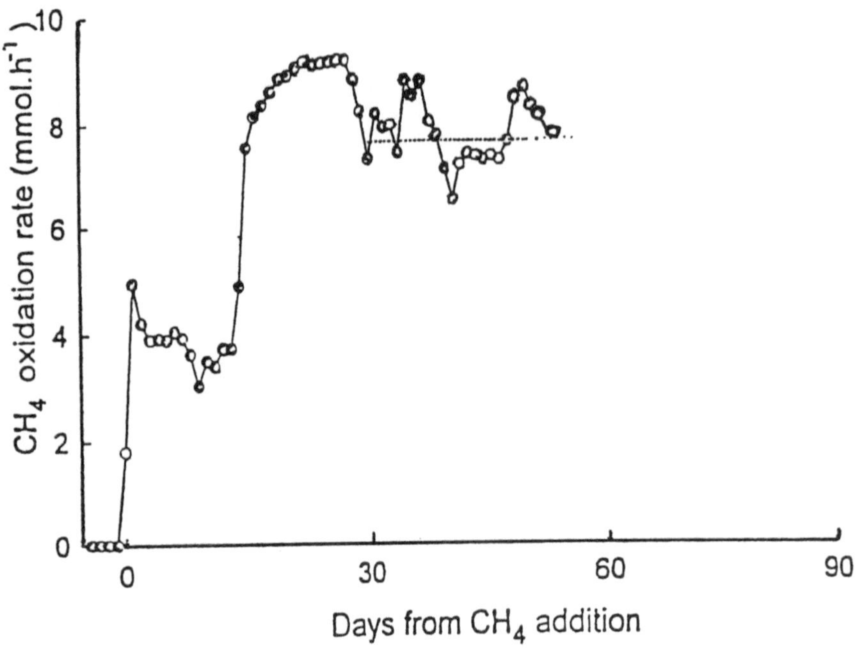

Figure 6. Methane oxidation rate for a coarse sand soil purged with this gas from the base. Points represent mean daily rates. After Kightley et al.[23].

This figure confirms the Michaelis -Menten relation in the entire temperature range studied. As can be assessed from this Figure, the K_M value for oxygen is between 0.2% (at 13.5°C) and 1.2% (at 20.5°).

5.3.2. the Effect of Methane Concentration on Aerobic Methane Oxidation Rate in Soil. According to King[25], the threshold value of CH_4 concentration for methanotrophs, below which no uptake occurs, can be a critical determinant of overall CH_4 consumption. Hütsch et al.[17], maintain that this value for soils is generally below the atmospheric methane level (1.7 μl CH_4 l^{-1}). However, Yavitt et al.[46] have observed threshold values of about 10 μl CH_4 l^{-1} for a forest soil that apparently did not consume atmospheric CH_4. Hütsch et al.[17] explain these elevated threshold values by the inhibiting properties of some chemicals such as, ammonia (see further). Whalen et al.[44] observed higher CH_4 oxidation rates with increasing CH_4 concentration in the headspace over the range of 1.7 to > 10,000 μl CH_4 l^{-1}.

Above this threshold value the relation of methane oxidation to its concentration obeys the above mentioned Michaelis - Menten equation.

Hoeks[15] did not find this effect of methane concentration in the interval 2.5–5% of methane on its oxidation rate in sand which suggests that the methane saturation was already reached at this level. In other words, the K_M constant must have been much below the 2.5% value. The data reported by other authors also give the K_M values in the range of 0.12- 2.41% of CH_4 which are dependent on the type of amendments (Table 3).

5.3.3. Effect of Soil pH on Methane Oxidation. King[24] demonstrated that CH_4 uptake in soil slurries from Danish wetland sediment was dependent on pH values. In the pH range between 6 - 8 the CH_4 oxidation was relatively stable but it significantly decreased at pH>9. Hütsch et al.[18] showed that methane consumption in soil of the Broadbalk Wheat Experiment

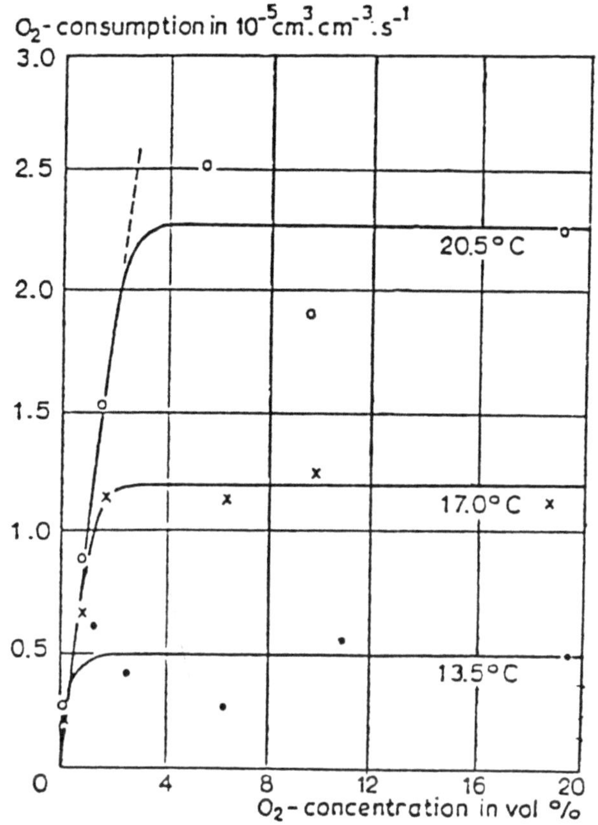

Figure 7. O_2- consumption rate as a function of oxygen concentration at three different temperatures in a dune sand soil through which a natural gas-air mixture (CH_4-concentration about 2.5 vol.%) was leaked (Hoeks [15]).

decreased doubly with decreasing pH values from 6.3 to 5.6 and declined to zero between pH 5.6 - 5.1.

5.3.4. Effect of Fertilization on Methane Oxidation in Soil. $-NH_4^+$- fertilization. The addition of NH_4^+ to the soil reduces considerably CH_4 consumption. This statement has been confirmed by many authors. Bronson and Mosier[7] reported that the inhibiting effect of NH_4Cl-N or NH_4NO_3-N in fine sandy loam and sandy clay loam had been very strong (Table 4). Hütsch et al.[18] showed that mineral N-applied annually as $(NH_4)_2SO_4$ at a dose 96 or 144

Table 3. Kinetic parameters for methane oxidation in a coarse sand landfill cover soil after a steady-state for CH_4 oxidation had been reached (from Kightley et al.[22])

Soil treatment	V_{max} (µmol kg of soil^{-1} h^{-1})	K_M (% of CH_4)	K_M Conc. of CH_4 (µM)
Unamended	2,342 ± 165	0.56	8.1
CH_4-purged	998 ± 58	2.41	31.7
N_2-purged	258 ± 12	0.12	1.6
NH_4NO_3 (2g of N kg^{-1})	390 ± 146	1.00	14.4
K_2HPO_4 (100mg of P kg^{-1})	470 ± 160	1.28	18.6
Sewage sludge (equiv. to 100 mg of N + 100 mg of P kg^{-1} of soil)	719 ± 222	1.66	24.0

Table 4. Inhibition of CH_4 oxidation by inorganic N and K compounds (doses in mg of pure N or K) as well as by acetylene and nitrapirin (nitrification inhibitors) in a Vona fine sandy loam after 6 days of incubation at 2 µmol mol^{-1} methane concentration (from Bronson and Mosier[7])

Treatment	Concentration (mg kg^{-1} soil)	Inhibition (%) of CH_4 oxidation after 6 days
NH_4Cl-N	25	89
NH_4NO_3-N	25	73
KNO_3-N	25	46
$Ca(NO_3)_2$-N	25	44
K_2SO_4-K	300	39
$KHCO_3$-K	300	35
Nitrapyrin	5	89
Acetylene	5 (mol mol^{-1})	100

kg N ha^{-1} for 130 years, completely held up CH_4 oxidation and withholding of fertilization for 3 years caused no significant recovery of CH_4-oxidizing activity. According to Kightley et al.[23] addition of NH_4NO_3 reduced the CH_4 oxidation capacity by 64%.

Recent investigations have shown that the application of NH_4^+-N fertilizer is associated with a decrease in CH_4 consumption rate in soil, probably because NH_4^+ (and/or NH_3) can competitively inhibit CH_4 oxidation by methanotrophs. The CH_4 monooxygenase enzyme and the NH_4^+ monooxygenase enzyme in chemo-autotrophic nitrifiers are apparently similar[3].

The rate of CH_4 uptake was negatively correlated with the nitrification rate. An inverse relationship between CH_4 uptake and N_2O emission, as a product of nitrification[40] is illustrated in Figure 8.

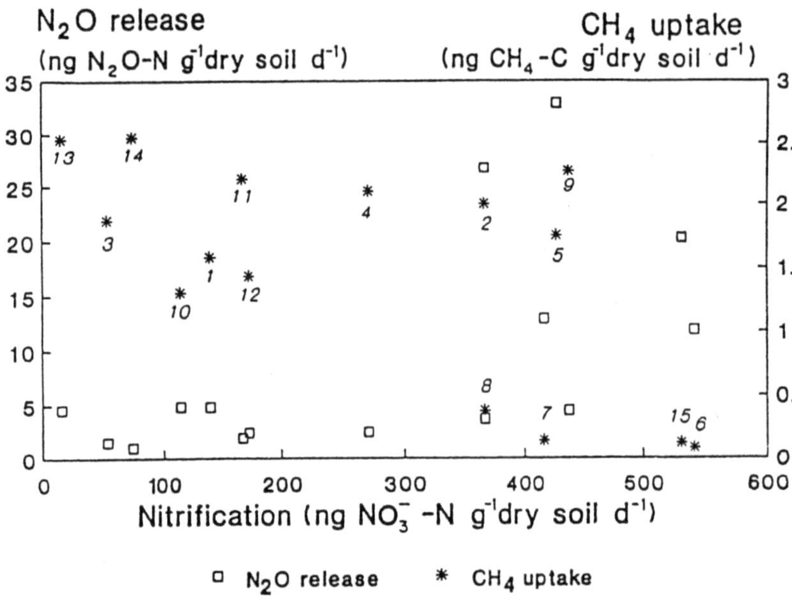

Figure 8. Relationship between nitrification rate (expressed as N_2O release) and CH_4 uptake in soil). The samples were taken at random within a 100 m^2 forest stand. Both gas fluxes were significantly correlated with the nitrification rate (P<0.01). From Situala and Bakken.[40]

As it was mentioned above, King[25] suggested that NH_4 application can elevate the threshold of methane concentration below which no uptake occurs. Nesbit and Breitenbeck[35] demonstrated that methane uptake could be re-established by exposure of the soil to a concentration of methane >1000 μmol mol^{-1} for several days. It suggests that the re-establishment of methanotrophic activity in the landfill cover is possible in this way.

Kightley et al.[23] have found that amendment of NH_4NO_3-N in the amount equivalent to 2 g N kg^{-1}soil decreased the maximum methane oxidation rate (V_{max}) about 6 times and increased the K_M value for methane almost twice which denotes a decrease of the affinity. Similar effects were caused by application of sewage sludge in a dose corresponding to 100 mg N kg^{-1}soil (cf. Table 3).

-NO_3^- - Fertilization. This kind of fertilization is probably not involved in competitive inhibition of CH_4 oxidation. Even long-term application of mineral N as $NaNO_3$ (96 kg N ha^{-1} $year^{-1}$) did not cause a significant inhibition effect[23]. However, studies conducted by Bronson and Mosier[7] have shown that 46% and 44% inhibition of CH_4 consumption was obtained with the addition of 25 μg KNO_3-N g^{-1} soil, and 25 μg $Ca(NO_3)_2$-N g^{-1} soil, respectively (Table 4). Their observations are in contrast to the results reported by Nesbit and Breitenbeck[35], who found that CH_4 oxidation was inhibited only at high concentrations of KNO_3 (490 mg kg^{-1} soil).

5.3.5. Effect of Other Chemical Factors. Kightley et al.[23] have compared K_2HPO_4 (100 mg P kg^{-1} soil) amended and unamended control soil. They found a 5 fold decrease of V_{max} and more than a 2 fold increase of K_M (see Table 3).

The effects of SO_4^{2-} and HCO_3^- were assumed to be negligible.

Bronson and Mosier[7] conducted investigations regarding the effects of nitrification inhibitors: nitrapyrin [2-chloro-6-(trichloromethyl) pyridine] and acetylene (Table 4 and Figure 9) on CH_4 consumption in a fine sandy loam and a sandy clay loam. They showed that both nitrification inhibitors have been strong inhibitors of CH_4 oxidation.

5.4. Effect of Physical Parameters on CH_4 Uptake in Soils

5.4.1. The Effect of Temperature on Aerobic Methane Oxidation in Soil. The effect of temperature on the rate of methane oxidation in soil is presented in Figure 7.

As it can be seen the effect of temperature is very strong. At lower temperatures the microbial activity decreases. Normally, the rate of biological reaction increases 2–3 fold with the increase in temperature of 10°C. In this case the increase of temperature from 13.5°C to 17°C is accompanied by a 2.4 fold increase of the reaction rate. Further increase of the temperature to 20.5°C caused 4.5 fold increase of the rate of methane oxidation reaction as compared to 13.5°C.

According to Hoeks[15] the effect of temperature on methane oxidation rate was approximated by application of Arrhenius equation to the experimental data for the temperature interval from 13.5–20.5°C:

$$\alpha_m = \frac{k \cdot e^{-\mu}}{2.3RT} \tag{10}$$

where:

α_m - is methane oxidation rate, [m^3 m^{-3} s^{-1}]

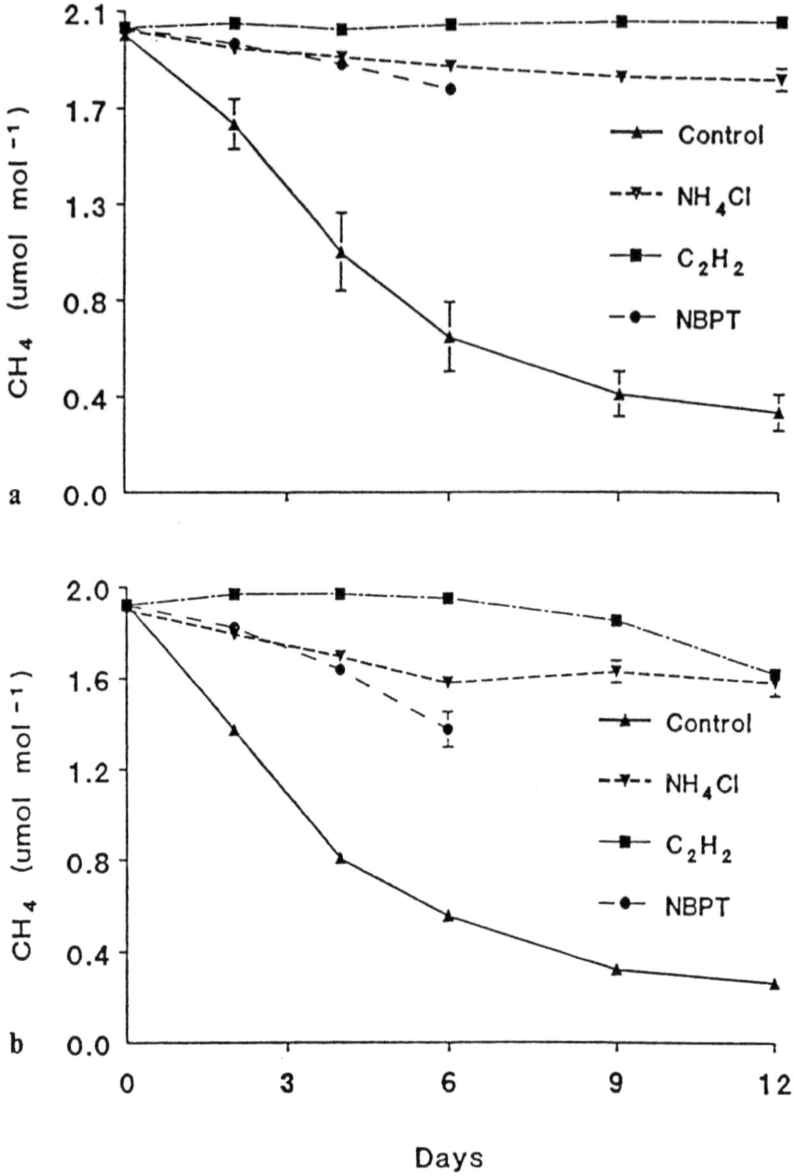

Figure 9. The effect of NH_4Cl and acetylene on methane concentration in Vona fine sandy loam (a) and Albinas sandy clay loam (b) treated with 25 µg NH_4Cl-N g^{-1} soil and 5 µmol C_2H_2 mol^{-1} in soil headspace, or untreated at initial headspace methane concentration equal to 2 µmol mol^{-1}. The bars represent standard deviations > 0.03 µmol mol^{-1} (Bronson and Mosier[7]).

k - is the reaction rate constant,
μ - is the activation energy [3.41×10^5 $Jmol^{-1}$]
R - is the universal gas constant,
T - is the absolute temperature in kelvins,

No other data related to methane oxidation rate versus soil temperature were found by us.

Table 5. Steady-state CH_4 oxidation rate measured in soil cores of a landfill cover soil (from Kightley et al.[22])

Soil type	Methane oxidation rate (mol of CH_4 $m^{-2}day^{-1}$)	% Oxidation
Coarse sand	10.4 ± 0.1	61
Clay topsoil	6.8 ± 0.3	40
Fine sand	6.9 ± 0.3	41

5.4.2. Effect of Soil Moisture Content and Soil Compaction on CH_4 Oxidation. The effect of soil moisture content on methane uptake should show a relationship typical for most of the biological processes in soil with a maximum in the range of intermediate moisture contents[13]. The low moisture content range limitation is connected with the deficiency of water to microbial life while the high moisture content range limitation is connected with the limitation of the reaction rate by gas diffusion. The diffusion of the gases involved in the process of methane oxidation in soil, i.e., of oxygen and of methane in water, is about 10^4 times slower as compared to the gas phase. Thus the replacement of air in the soil pores connected with increasing of the moisture content should be accompanied by the reduction of the oxidation rate.

5.4.3. Effect of Soil Type. The effect of soil type on methane oxidation rate is presented in Table 5. Of the three materials studied, the porous, coarse sand soil developed the highest methanotrophic capacity (10.4 mol of CH_4 $m^{-2}day^{-1}$), the greatest value reported as yet in the pertinent literature. It seems to be connected with the highest gas diffusion rate in this case as the experiment was performed in columns with methane purged from the bottom and the oxygen diffusing downwards from the atmosphere. Hoeks[15] also found very high methane oxidation rates in a sandy soil after prolonged contact with methane leaking from natural gas pipes. This was 50–100 times higher than compared to normal soils and ranged up to 65–95 ml of CH_4 ml of $soil^{-1}$ s^{-1}.

This effect has been confirmed for methane oxidation under field conditions for moisture content range 30–45% (v/v) by Lessard et al.[28]. The investigation of Boeckx and Van Cleemput[5] showed that methane oxidation rate for a landfill cover soil was maximal in the moisture content interval 15–25% (w/w) which is presented in Figure 10. According to Whalen et al.[44] soil covering the landfill site showed the highest methane consumption rate at a moisture content of 11% (w/w). In turn Hütsch et al.[17] showed that methane uptake was higher at a moisture content of 16–17% as compared to that of 19–20% in arable soil. The differences in these values are connected with the differentiation of the material used by different authors. Obviously the high moisture range limitation is connected with the gas diffusion coefficient of a given soil material and not with its water content. However, this parameter was not included in the studies quoted above. The limitation in the range of low water contents is connected with the decrease of water availability for the microorganisms. This is determined y the matrix potential of soil water and not by its content expressed by weight or on volumetric basis. Yet no data relating methane oxidation rate to water potential or to gas diffusion coefficients have been found by us. As concerns the soil compaction it was shown[5] that it adversally affected the methane oxidation rate as is presented in Figure 10.

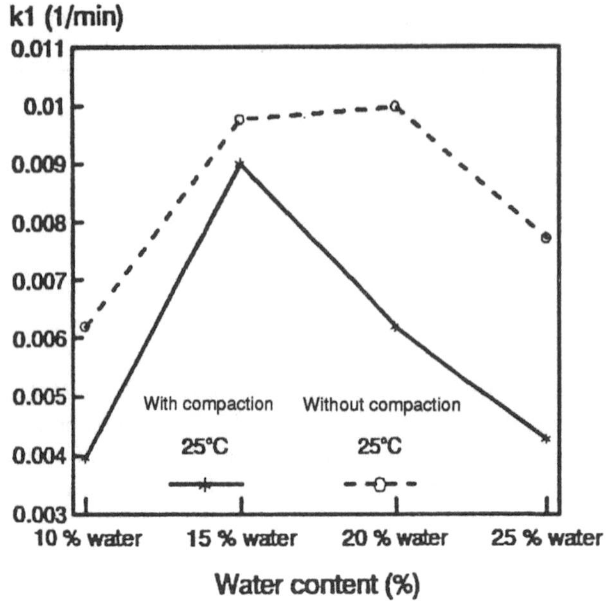

Figure 10. The water content and compaction influence on the methane oxidation rate constant (k1). From Boeckx and Van Cleemput[5].

5.4.4. Effect of Land Use on CH_4 Oxidation. Methane oxidation in soil is also affected by land use. It has been confirmed by Hütsch et al.[18] They have observed the highest methane oxidation rate in a woodland soil (Table 6). Keller et al.[21] showed that arable land in the tropics has about 4 times lower CH_4 uptake rate than tropical forest soil.

CONCLUSIONS

The above review confirms the possibility of asubstantial increase in the degree of methane oxidation in landfill soil cover but the practical recommendations require further

Table 6. Effect of land use on CH_4 oxidation rate at its concentration of 10 $\mu l\ l^{-1}$. Undisturbed soil cores in plastic tubes 6.4 cm i.d. and 12 cm deep were incubated in 1 litre Kilner jars (from Hütsch et al.[17])

The land use	CH_4 concentration decrease rate in headspace (nl $l^{-1}h^{-1}$)
Mixed forest	80
Arable plot receiving FYM[a]	13
Arable plot receiving N-mineral fertilizer[b]	3
Unamended arable plot	7
Grass-land	17.3

[a]FYM- farmyard manure (N addition equivalent to 240 kg $ha^{-1}a^{-1}$)
[b]NH_4NO_3 (N-addition equivalent to 144 kg $ha^{-1}a^{-1}$)

research in this direction. It has been demonstrated that methane oxidation in soil obeys Michaelis - Menten kinetics. The metanotrophic activity in soil increases in the presence of elevated methane concentrations and stabilizes after about one month. The optimum pH for methane oxidation ranges from 6 to 8.

REFERENCES

1. Adamse et al., Hoeks J., de Bont J.A.M.: Microbial activities in soil near natural gas leaks, Antonie van Leeuwenhoek, 37, 251–252, 1971.
2. Aitchison E.: Options for controlling methane emissions from landfill sites. IPCC Workshop "Methane and Nitrous oxide", Amersfoort, 221–230, 1993.
3. Bedard C., Knowles R.: Physiology, biochemistry, and specifics inhibitors of CH_4, NH_4^+, and CO oxidation by methanotrophs and nitrifiers, Microbiol. Rev., 53,68–84, 1989.
4. Bender M., Conrad R.: Kinetics of methane oxidation in oxic soils. Chemosphere 26, 687–696, 1993.
5. Boeckx P., Van Cleemput O.: Methane Oxidation in a Landfill Cover Soil. A contribution to subproject BIATEX, The Proceedings of EUROTRAC Symposium '94 (ed. P.M. Borrell et al.), 521–524, 1994.
6. Bouwman A. F.: Introduction. In "Soils and Greenhouse Effects" (A.F. Bouwman, Ed.), 25–32. Wiley, Chichester, UK, 1990.
7. Bronson K.F., Mosier A.R.: Suppression of methane oxidation in aerobic soil by nitrogen fertilization, nirtification inhibitors, and urease inhibitors, Biol. Fertil. Soils, 17, 263–268,1994.
8. Buchholz L.A., Klump V.J., Collins M.L.P., Brantner C.A., Remsen C.C.: Activity of methanotrophic bacteria in Green Bay sediments, FEMS Microbiology Ecology, 16, 1–8, 1995.
9. Davis J.B., Yarbrough H.F.: Anaerobic oxidation of hydrocarbons by *Desufovibrio desulfuricans*. Chem. Ecol., 1, 137–144, 1966.
10. Ferenci T., Strom T., Quale J.R.: Oxidation of carbon monoxide and methane by *Pseudomonas methanica*, J. Gen. Microbiol., 91, 79–91, 1975.
11. Förstner V., Calmano W., Kienz W.: Assessment of long-term metal mobility in heat-processing wastes, Water Air Soil Pollut., 57–58, 319–328, 1991.
12. Gaudioso D.,Trozzi C., Vaccaro R.: Methane emissions from landfills in Italy. IPCC Workshop "Methane and Nitrous oxide", Amersfoort, 405–415, 1993.
13. Gliñski J., Stêpniewski W.: Soil Aeration and its Role for Plants. CRC Press Inc., Boca Raton Florida, 1985.
14. Ham R.K., Hekinian K.K., Katten S.L., Lockman W.J., Lofy R.J., McFaddin D.E., Daley E.J.: *Recovery*, processing and utilization of gas from sanitary landfills, EPA-600/2–79–001. U.S Enviromnental Protection Agency (Solid and Hazardous Waste Res. Div.) Cincinnatti, USA, 1979.
15. Hoeks J.: Effect of Leaking Natural Gas on Soil and Vegetation in Urban Areas. Agricultural Research Reports 778, 120. Centre for Agricultural Publishing and Documentation (Pudoc), Wageningen, The Netherlands, 1972.
16. Hoeks J.: Significance of biogas production in waste tips. Waste Management & Research. 1, 323–335, 1983.
17. Hütsch B.W., Webster C.P., Powlson D.S.: Long-term effects of nitrogen fertilization on methane oxidation in soil of the Broadbalk Wheat Experiment, Soil Biology and Biochemistry 25, 1307–1315, 1993.
18. Hütsch B.W., Webster C.P., Powlson D.S.: Methane oxidation in soil as affected by land use, soil pH and N fertilization, Soil Biology &. Biochemistry., Vol. 26, No.12,1613–1622, 1994.
19. Iverson N., Oremland R.S., Klug M.J.: Big Soda Lake (Nevada). 3. Pelagic methanogenesis and anaerobic methane oxidation, Limnol. Oceanogr., 32, 804–814, 1987.
20. Jones R.D., Morita R.Y.: Methane oxidation by *Nitrosococcus oceanus* and *Nitrosomonas europaea*, Appl. Environ. Microbiol., 45, 401–410, 1983.
21. Keller M., Mitre M.E., Stallard R.F.: Consumption of atmospheric methane in soils of Central Panama: effects of agricultural development, Global Biogeochemistry Cycles, 4, 21–27, 1990.
22. Kightley D., Nedwell D.B.: Methane oxidation in landfill cover soils: can bacteria solve the emission problem? Environ. Manage. J., 3, 24–26, 1994.
23. Kightley D., Nedwell D.B., Cooper M.: Capacity for Methane Oxidation in Landfill Cover Soil Measured in Laboratory-Scale Soil Microcosms, Appl. and Environ. Microbiol., 61(2), 592–601, 1995.
24. King G.M.: Dynamics and control of methane oxidation in a Danish wetland sediments, FEMS Microbiology Ecology 74, 309–324, 1990.
25. King G.M.: Ecological aspects of methane consumption, a key determinant of global methane dynamics, Advences in Microbial Ecology, 12, 431–468, 1992

26. Kosiur D.R., Warford A.L.: Methane production and oxidation in Santa Barbara sediments, Estuarine Coastal Mar. Sci. 8, 379–385, 1979.
27. Lelieveld J., Crutzen P.J.: Methane emissions into the atmosphere an overview, IPCC Workshop "Methane and Nitrous oxide", Amersfoort, 17–25, 1993.
28. Lessard R., Rochette P., Topp E., Pattey E., Desjardins R.L.,Beaumont G.: Methane and carbon dioxide fluxes from poorly drained adjacent cultivated and forest sites, Canadian Journal of Soil Science, 74, 2, 1994.
29. Lidstrom L.E., Stirling D.I.: Methylotrophs: Genetics and commercial application. Annu Rev. Microbiol. 44, 27–58, 1990.
30. Miura Y., Watanabe A., Murase J., Kimura M.: Methane Production and Its Fate in Paddy Fields. II. Oxidation of Methane and Its Coupled Ferric Oxide Reduction in Subsoil, Soil Sci. Plant Nutr., 38(4), 673–679, 1992.
31. Mosier A.R., Schimel D.S.: Influence of agricultural nitrogen on atmospheric methane and nitrous oxide, Chemistry and Industry 23, 874–877, 1991.
32. Murase J., Kimura M.: Methane Production and Its Fate in Paddy Fields. IV. Sources of Microorganisms and Substrates Responsible for Anaerobic Methane Oxidation in Subsoil, Soil Sci. Plant Nutr., 40(1), 57–61, 1994.
33. Murase J., Kimura M.: Methane Production and Its Fate in Paddy Fields. VI. Anaerobic oxidation of methane in plow layer soil, Soil Sci. Plant Nutr., 40(3), 505–514, 1994.
34. Murase J., Kimura M.: Methane Production and Its Fate in Paddy Fields. VII. Electron Acceptors Responsible for Anaerobic Methane Oxidation. Soil Sci. Plant Nutr., 40(4), 647–654, 1994.
35. Nesbit S.P., Breitenbeck G.A.: A laboratory study of factors influencing methane uptake by soil, Agriculture, Ecosystems and Envronment, 421, 39–54, 1992.
36. Panganiban A.T. Patt T.E. Hart W., Hanson R.S.: Oxidation of methane in the absence of oxygen in lake water samples, Appl. Environ. Microbiol. 37, 303–309, 1979.
37. Reeburgh R.S., Heggie D.T.: Microbial methane consumption reactions and their effect on methane distribution in freshwater and marine environments, Limnol. Oceanogr., 22, 1–9, 1977.
38. Ribbons D.W., Harrison J.E., Wadzinski A.M.: Metabolism of single carbon compounds, A. Rev. Microbiol., 24, 135–158, 1970.
39. Seiler C.W., Scharffe D.: Field studies of methane emission from termite nests into the atmosphere and measurements of methane uptake by tropical soil. Journal of Atmospheric Chemistry 1, 171–186, 1984.
40. Sitaula B.K., Bakken L.R.: Nitrous oxide release from spruce forest soil: relationships with nitrification, methane uptake, temperature, moisture and fertilization, Soil Biology & Biochem., 25 (10), 1415–1421, 1993.
41. Söhngen N.L.: Ueber Bakterien, welche Methan als Kohlenstoffennahrung Energiequelle gebrauchen. Zentralb. Bakteriol. 15, 513–517, 1906.
42. Stegmann R.: Beschreibung eines Verfahrens zur Untersuchung anaerober Umsentzungprozesse von festen Abfallstoffen im Labormassstab (Laboratory research on anaerobic degradation processes in solid waste), Müll und Abfall 2/81, 35–39, 1981.
43. Thorneloe S.A.: Methane emissions from landfills and open dumps. IPCC Workshop "Methane and Nitrous oxide", Amersfoort, 93–113, 1993.
44. Whalen S.C., Reeburgh WS., Sandbeck K.A.: Rapid methane oxidation in a landfill cover soil, Applied and Environmental Microbiology 56, 3405–3411, 1990.
45. Yavitt J.B., Downey D.M., Lancaster E., Lang G.E.: Methane consumption in decomposing Sphagnum- derived peat, Soil Biology and Biochemistry, 22, 441–447, 1990.
46. Yavitt J.B., Downey D.M., Lang G.E., Sexstone A.J.: Methane consumption in two temperate forest soils, Biogeochemistry 9, 39–52, 1990.
47. Yavitt J.B., Fahey J.: Production of methane and nitrous oxide by organic soils within a northern hardwood ecosystem, In "Biogeochemistry of global change - radiatively active trace gases", 10th Int. Symp. Environ. Biogeochemistry, San Francisco, 261–277, 1991.
48. Zehnder A.J.B., Brock T.D.: Anaerobic methane oxidation: Occurence and ecology, Appl. Environ. Microbiol., 39, 194–204, 1980.

CATALYTIC HYDRODEHALOGENATION OF CFC'S (FREONS)

P. Kirszensztejn,[1] L. Wachowski,[1] B. Czajka,[1] R. Łopatka,[1] and T. N. Bell[2]

[1] Faculty of Chemistry
A. Mickiewicz University
Grunwaldzka 6, 60-780 Poznań, Poland
[2] Department of Chemistry
Simon Fraser University
Burnaby, B.C. V5A 1S6 Canada

INTRODUCTION

The photolysis of fully halogenated chlorofluorocarbons (CFC's) in the stratosphere, and their resultant role in ozone depletion is of great global concern[1]. In addition to the role played by CFC's in the destruction of ozone, they also contribute to the "green-house effect" and hence to climatic changes[2]. The replacement of halogen by hydrogen makes the halo-carbon susceptible to destructive free radical attack in the troposphere and considerably reduces the depletion of ozone[3–6]. The following reaction may serve as a pertinent example :

$$CCl_2F_2 \xrightarrow[\text{catalyst}]{H_2} CH_2F_2 + HCl\ (HF)$$

The starting materials (halocarbons) as well as by-products (HCl or /and HF) formed during the reaction, may interact with the catalyst metal, and also with the catalyst support, leading to formation quite new properties of catalyst surface in comparison to originating one[7,8]. For example, it is reasonable to aspect distinctive changes in acidic-basic behaviour of surface and in consequence it implication on mechanism of conversion of CFC's[9]. Understanding of its role may have crucial importance in the catalytical system applied in hydrogenolysis of CFC's.

EXPERIMENTAL

The surface acidic properties of γ-Al_2O_3 as well as Al_2O_3-SnO_2 system used in the catalytic conversion of CF_2Cl_2/H_2 (as a model reaction) have been investigated by adsorption of molecular probe trimethylphosphine (TMP) and using the solid state high resolution CP/MAS ^{31}P NMR. Some of these results have been confirmed by FT-IR spectroscopy with pyridine as a molecular probe.

NMR spectroscopy - solid state NMR experiments were carried out using a 3.5 T instrument on which ^{31}P resonates at 60.457n MHz ^{31}P chemical shifts are refered to 85% H_3PO_4. A proton decoupling field strenght of 45 kHz was used.

Adsorption of trimethylphosphine (TMP) - A standard pretreatment for the catalyst after reaction involved vacuum activation for 2 hr at 450°C at a base presurre of about 10^{-5} torr. Adsorption of TMP on the samples was carried out at room temperature, and then the sample was sealed off from the vacuum line for NMR measurement. The sealed sample tubes were spun at the magic angle at a speed of about 2 kHz.

FT-IR Spectroscopy - were carried out in the range 4000–900 cm^{-1} on a Bruker IFS88 FTIR spectrometer. The spectra (resolution 1 cm^{-1}; 250 scans) were recorded in the transmittance mode. The adsorption of the probe molecule, pyridine, was carried out on a particular sample as follows. Pyridine vapour was allowed to be adsorbed at room temperature for 5 min. This procedure allows for the complete saturation of the alumina surface as a large excess of the probe molecules is admitted into the IR cell. The adsorption step was followed by outgassing for 1h at 150°C (with controled heating 10°C/min).

Catalytic experiment in order to introduce Cl- and/or F- ions into the samples, alumina and tin-modified alumina were subjected to the catalytic reaction involving CFC-12/H_2 at 300°C for 1 or 6 hrs. The catalytic reaction was carried out in a flow system with a contact time of ≅ 13 sek. Samples of catalyst after catalytical reaction were subject of NMR investigation. In the case of IR investigation catalytic reaction has been carry on *in situ* in IR cell.

RESULTS AND DISCUSSION

Surface acidity of the support plays an important role in the conversion of hydrocarbons on metallic catalysts[10]. In the case of Pt/γ-Al_2O_3 catalysts, it has been commonly accepted that acidic alumina catalyses the izomerization of alkenes. Primet[11] discovered that chlorinated alumina is an effective alkane isomerization catalyst even at relative mild temperatures. Understanding the role of acicity could thus be very important, especially in the case where such a system is used as a catalyst support for the conversion of CFC's where HCl and HF are involved in the catylitic reaction.

Fig.1 shows the spectra using different pulse programs for the γ-Al_2O_3 sample after reaction with CFC-12/H_2 for 6 hrs. The peak at -47 ppm is due to Lewis sites on γ-Al_2O_3 and the at -3 ppm to Brönsted sites. The latter peak does not occur with pure γ-Al_2O_3 samples before reaction with CFC-12/H_2. The -3ppm peak became bigger under CP excitation (Fig1.b), and disappeared under delayed decoupling (Fig.1,d). These confirm the assignments made and show that the species at -3ppm are firmly bonded to surface.

Two γ-Al_2O_3 samples were examined, one was reacted with CFC-12/H_2 for 1 hr and the other for 6 hr. Lewis acid lines centered at -43.5 ppm became broader and shifted to lower field for these two samples (Fig.2, b-d), as compare with pure γ-Al_2O_3 not reacted with CFC-12/H_2 (Fig.2,a). The reason for the broadening and the slight shift could be that some Lewis acid sites on γ-Al_2O_3 had reacted with HCl; or HF, and created new Lewis acid sites which have different chemical shift.

No significant numbers of Brönsted acid centers were observed on the pure γ-Al_2O_3 sample, however after its reaction with CFC-12/H_2, some Brönsted acid sites were ob-

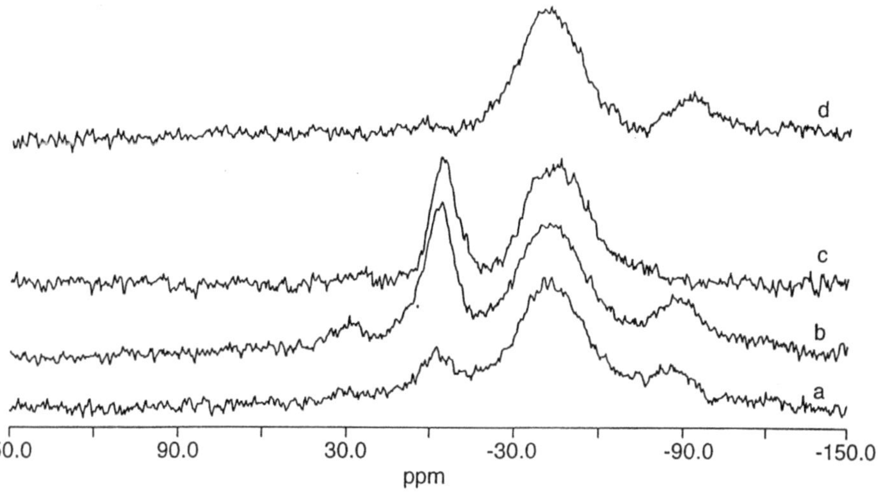

Figure 1. ^{31}P MAS NMR spectra of trimethylphosphine (TMP) adsorbed on γ-Al_2O_3. a) pure, b) after reaction with CFC-12/H_2 for 6 hrs., c) CP (cross-polarization), d) under delayed decoupling.

served through peaks at -3 ppm as shown in Fig.2, b-e. This assignment was confirmed by the delayed decouple program under which this peak disappeared.

Fig.2 also shows the efect of different reaction times on the amount and the rigidity of the species bonded to the Brönsted acid sites. It was found that the number of Brönsted acid sites on the two samples are almost the same. This conclusion is interferred from the cross polarization excitation program, where the species strongly bonded on the surface cause an increased spectral intensity.

It was found that on the γ-Al_2O_3-SnO_2 catalysts (see Fig.3.) there are three types of Lewis acid sites giving lines at -12, -22 and -47 ppm. No Brönsted acid sites were found. The peak at -12 and -22 ppm are attributed to the Lewis acid sites from tin oxides, and the

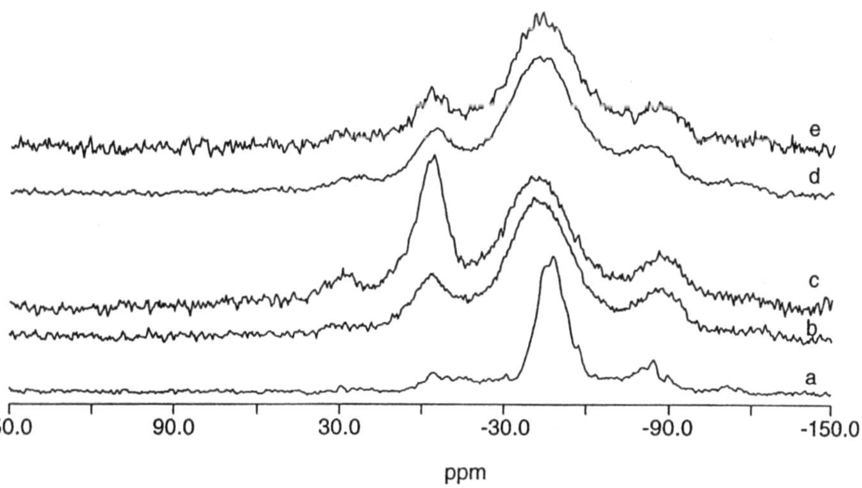

Figure 2. ^{31}P MAS NMR spectra of trimethylphosphine (TMP) on γ-Al_2O_3. a) pure, (b)-(c) after reaction with CFC-12/H_2 for 1 hr and adsorbed TMP, (d)-(e) after reaction with CFC-12/H_2 for 6 hrs and adsorbed TMP.

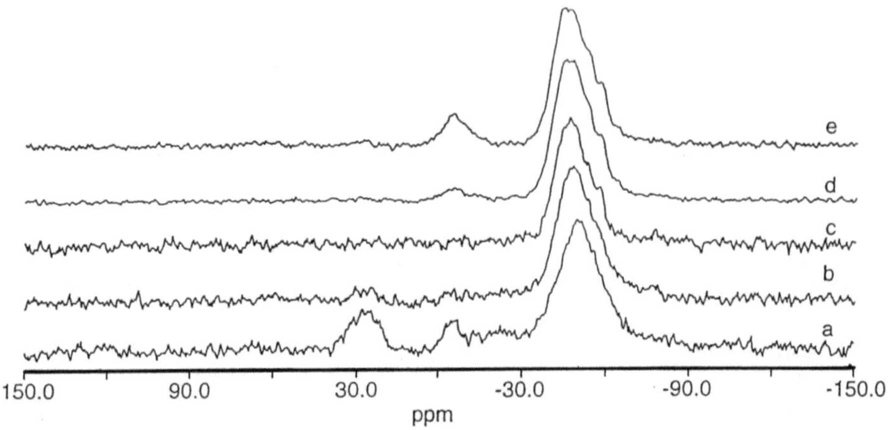

Figure 3. ^{31}P MAS NMR spectra of trimethylphosphine (TMP) adsorbed on (a), (d) and (e) γ-Al_2O_3-SnO_2 with mole ratio of Al_2O_3:SnO_2 equal 1:1, 1:0.05 and 1:0.01 respectively, (b), (c) SnO_2 and γ-Al_2O_3 after reaction with CFC-12/H_2 respectively.

peak at -47 ppm is from Lewis sites on γ-Al_2O_3. After reaction with CFC-12/H_2, the samples with different amounts of SnO_2 demonstrate different behavior towards the formation of acid sites. For the sample with the highest SnO_2 content, the two peaks at -12 and -22 ppm disappeared and new resonance at +27.1 and -4.8 ppm appeared (Fig.3,a). The peak at -4.8 ppm has been assigned from our previous study to Brönsted acid site. We now assign the peak at -4.8 ppm to Brönsted acid sites on the SnO_2 surface created by reaction between SnO_2 and CFC-12/H_2. The peak at -3.2 ppm created by reaction between γ-Al_2O_3 and CFC-12/H_2 for pure γ-Al_2O_3 samples, and also for Sn-0.05 and Sn-0.01 samples (Fig.3, d and e) is assigned to Brönsted acid sites on the γ-Al_2O_3 surface. It should be noted that the peak at -3.2 ppm in sample Sn-0.01 is distinguished from the peak at -4.8

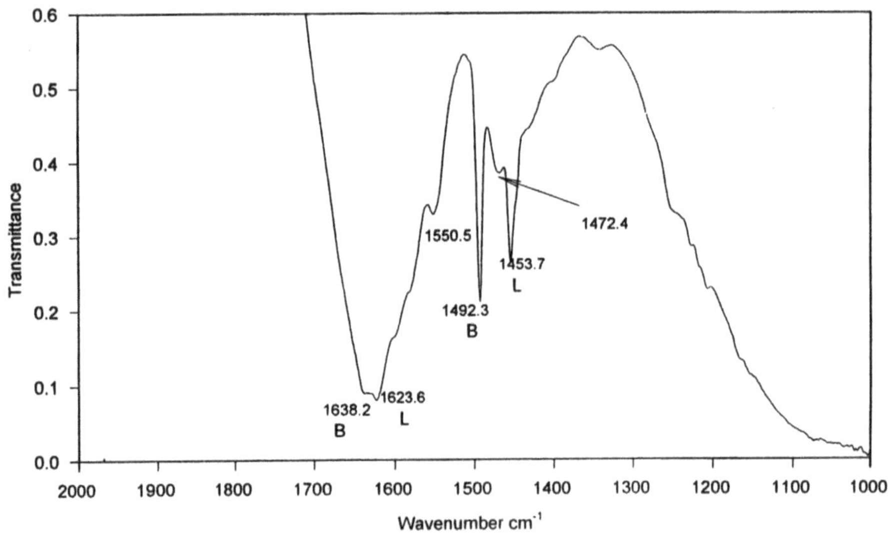

Figure 4. IR spectra of pyridine adsorbed on γ-Al_2O_3 after reaction with CFC-12/H_2.

ppm in the Sn-0.5 sample. The former originates from γ-Al_2O_3, the later from SnO_2. It is quite possible that there is a special structural configuration available on the γ-Al_2O_3 surface which can react with CFC-12/H_2 to form Brönsted acid sites. This configuration can be covered by SnO_2 leading to Brönsted acid sites on the SnO_2 surface after reaction with CFC-12/H_2. NMR results has bccn confirmcd with FT-IR spcctroscopy with adsorption pyridine as sond probe on the γ-Al_2O_3 (Fig.4). After reaction with CFC-12/H_2 is clear to see appearance a new bands at 1492 and 1638.2 cm^{-1} assigned to Brönsted acid sites.

CONCLUSIONS

Brönsted acid sites are formed on pure γ-Al_2O_3 and γ-Al_2O_3-SnO_2 system after reaction with CFC-12/H_2.

Lewis acid sites on the SnO_2 component disappeared after the reaction with CFC-12/H_2.

Indirect evidence is to the effect that the sites on the γ-Al_2O_3 surface which form the Brönsted acid, are the same sites that are covered up by the SnO_2 component.

REFERENCES

1. M.J. Molina and F.S. Rowland, Nature, 249 (1974) 810.
2. Climate Change 1992 (The Suplementary Report to the IPCC Scientific Assesment) Cambridge University Press 1992, ISBN 0 521 43829 2
3. J.D. Farman, B.G. Gardiner and J.D. Shanklin, Nature, 315 (1985) 207.
4. J.D. Farman, New Scientist, 116 (1987) 50.
5. S. Biniak, J. Siedlewski and J. Gacek, Przem. Chem., 72 (1993) 89.
6. T.J.Wallington, W.F.Schneider, D.R.Worsnop, O.J.Nielsen, J.Schested, W.J.Debruyn and J.A.Shorter, Environ. Sci. Technol., 28 (1994) 320 A.
7. L.E. Mazner and V.N.M. Rao, Adv. Catal., 39 (1993) 336.
8. T.N.Bell, P.Kirszensztejn and B.Czajka, Catal. Lett., 30 (1995) 305.
9. T.C.Sheng, P.Kirszensztejn, T.N.Bell and I.D.Gay, Cat. Lett., 23 (1994) 119.
10. J.R.Baker, R.L.Mc Cornik and H.W.Haynes Jr., Ind. Eng. Chem. Res., 26 (1987) 1885.
11. P.M.Bernard and M.Primet, J. Chem. Soc. Faraday Trans., 86 (1990) 567.

11

SOLUBLE ORGANIC POLYMERS IN WATER AND WASTEWATER TREATMENT

B. A. Bolto, D. R. Dixon, and S. R. Gray

CSIRO Chemicals and Polymers
Private Bag 10, Rosebank MDC, Clayton, Victoria 3169, Australia

ABSTRACT

Organic polymeric flocculants have been used in water purification processes for several decades as coagulant aids or floc builders, after the addition of inorganic coagulants like alum, iron salts or lime. The application of polymers as primary coagulants in cleaning up low quality natural waters for drinking water production by direct filtration and in weighted flocculation is discussed. For industrial wastewaters new information is presented on primary coagulation where separation is based on flotation, the burden of solids to be separated now being much reduced relative to systems utilising inorganic coagulants. As well, the dosage of chemicals overall is reduced, and there is less final sludge. The major advantages in polyelectrolyte use generally are faster processing and much greater operational stabilty and reliability.

1. INTRODUCTION

The relatively recent use of polyelectrolytes as primary coagulants replacing all or some of the inorganic salts has a number of inherent advantages.[1] Among them are

- an increase in the solid-liquid separation rate
- only 10% w/w of the dose needed, compared with an inorganic salt
- a much smaller sludge volume is produced
- extended filter runs are obtained in direct filtration
- the process is less pH dependent
- there is a lowered usage of other chemicals
- the treated water has a lower dissolved salt level.

There are disadvantages of course, the main one being that operating costs may be higher; this must be balanced against the benefits. The situation for a number of purification processes will be outlined.

Chemistry for the Protection of the Environment 2, edited by Pawłowski et al.
Plenum Press, New York, 1996

In drinking water production by direct filtration a mixture of inorganic and organic polymer additives is usually employed. The involvement of polymers in new heterocoagulation techniques for a variety of raw waters where use is made of weighted particles to accelerate sedimentation is now well established . Polymers have been used as primary coagulants for industrial wastewater treatment where the separation step is sedimentation. But for many industrial wastewaters, especially those containing suspended or emulsified oil and grease, as occurs in the food industry and in the processing of agricultural products, the organic phase prefers to float rather than settle. A new method of advanced primary treatment of sewage which includes phosphorus removal via pre-precipitation, but utilises polymers extensively, is now in use. The load of metal salt normally needed for phosphorus removal can be halved, with resultant benefits to the efficiency of operation of the subsequent secondary biological and sludge digestion stages because of the reduced presence of inorganic solids. A further new approach is the recovery of valuable or toxic metals from waste streams which is based on binding the metal ions to polymeric sequestering agents. The resultant macromolecular complex can be separated on a more open membrane requiring a lower operating pressure than would otherwise be necessary.

2. DRINKING WATER PRODUCTION

2.1 Direct Filtration

In the production of drinking water from natural waters of low turbidity and moderate to high colour, a highly charged cationic polyelectrolyte like poly(diallyldimethylammonium chloride), or PDADMAC, can be used as the primary coagulant instead of a metal salt. The best performance occurs when the mobility or zeta potential of the impurity particles is close to zero.[2] Such waters are ideal candidates for polymer coagulation. Avoiding the use of an inorganic coagulant has the advantages that a smaller amount of sludge is produced because no metal hydroxide precipitate is formed, the system is less pH sensitive, and there is less dissolved salt in the product water.[3, 4]

Settleable flocs are not usually formed, but the solids can be readily removed by filtration, which requires strong porous flocs. The settling stage normally employed in conventional sedimentation plants is kinetically inefficient when there is a low level of suspended matter. In direct filtration, where some time is allowed for floc growth to achieve optimum floc size, there is no settling stage, which means a capital cost saving.

The effect of polymer charge type and molecular weight in direct filtration has been reviewed.[5] Charge neutralisation is the main mechanism, with anionic polymers being the least effective, although they can be preferred if calcium or magnesium cations are present. Polymers give the advantages of extended filter runs and the production of a deformable floc. The MW of the polymer is not a major variable.

A substantial polymer dose is needed for waters high in organics, but compared to alum usage the filter run can be more than double in length.[2] If it is found that direct filtration is not economic because of the high polymer dose required, a smaller amount of conventional metal ion coagulant than normally employed can be used to minimise the amount of metal hydroxide precipitated; this can then be followed by PDADMAC.

Pretreating the filter media rather than dosing polymer to the full feedwater flow has been explored, with the polymer being added to the backwash water for a portion of the backflow time. The binding of polymers to sand is ten times more intense for cationic and nonionic polymers than for anionic ones.[5]

A 1–10 mg/L dose of cationic polymer can coagulate freshwater algae.[6] The dose must be more than that necessary to overcome the effects of exocellular matter released by the algae, when good cell removal results.[7] When an inorganic coagulant is used for this purpose the cell membrane is usually disrupted and compounds are released which give rise to taste and odour in the product water.[8] With blue green algae the toxins produced can be a health hazard. The compounds can be removed by oxidation or by adsorption on activated carbon, but these are expensive treatment options. It has been shown that an amphoteric polymer made by reacting polyacrylonitrile with dicyandiamide can aggregate algal cells without killing them.[9]

2.2 Weighted Flocculation

Sedimentation may be greatly enhanced by binding impurities to a weighting agent. One weighted system is based on sand which is recovered from the final sludge by cycloning and reused.[10, 11] A full scale plant is now in operation,[12] and application of the technique to stormwater treatment is reported.[13]

In a more radical example of this technique a recyclable magnetic coagulant is employed which has its settling rate further accelerated by applying a magnetic field, such that the size of the clarifier needed is less than a third of that of conventional processes. Magnetite, Fe_3O_4, can be used as a primary coagulant, and with a particle size of 1–10 μm the reaction rate is four times greater than with metal salt coagulants, using a slurry of up to 1% w/w. The process is assisted by the addition of a cationic polyelectrolyte, at levels of up to 1 mg/L. The impurity-laden magnetite is capable of very rapid sedimentation after it is passed through a simple magnetic field, when intense magnetic flocculation occurs.[14] Magnetic particle technologies have been utilised also for many wastewater applications, including the recovery of useful metal ions, when an anionic polymer assists the heteroflocculation step as the pH is raised to precipitate the metal hydroxide.[15]

Most recently, the technology has provided a completely new approach to sewage treatment, when magnetite is used as a 0.5% w/w slurry and alum is dosed at 8 mg/L to bind the suspended material to the magnetite at pH 6, without the addition of polymer. Typical performance is 87% suspended solids (SS), 90% oil and grease and 60% COD removal. The biochemical oxygen demand (BOD) is about halved, as soluble material is not removed. Bacteria are 99% removed, phosphate 89% and heavy metals 74–89%. These are all achieved within 15 minutes, so very little land space is required.[16]

3. TREATMENT OF INDUSTRIAL WASTEWATERS

3.1 Sedimentation Processes

Industrial effluents can be cleaned up by coagulation with organic polymers, followed by sedimentation if the flocs formed are dense enough. Those most successfully treated are from the food industry, where chitosan can give a 90% reduction in the turbidity of bean processing wastewaters, allowing the reuse of some wash waters.[17] The treatment of wastes from many food processing industries and the recovery of useful solids is possible with chitosan.[18] Polyethyleneimine[19] and the polymer made from polyacrylonitrile and dicyandiamide are capable of removing colour from paper mill effluents; the latter is claimed to eliminate colour from dyehouse wastes.[9] Partial colour removal from dyehouse wastes is possible with a highly charged cationic polymer, but at very high doses.[20] In other examples, applying a cationic polymer with clay, which acts as both an adsorbent and a weighting agent, takes out some colour, but is less effective than alum.[21] In fact, inorganic coagulants are much used in industrial effluent treatment, often in conjunction with a polymeric flocculant.

3.2 Flotation Processes

In the clean up of a chromium laden tannery effluent, which also contained fats and protein, the customary treatment by dissolved air flotation required the addition of 500 mg/L of poly(aluminium chloride), or PAC, and 7.5 mg/L of a high charge density (CD), medium molecular weight (MW) anionic polymer; the pH of 8.9–9.3 was not adjusted. The waste dosed in this manner could on occasion not be satisfactorily treated. Large flocs were seen in the underflow and in the sludge, so inadequate flocculation did not seem to be the cause. However, at times there was no increased concentration of the solids in the froth, with both the underflow and the froth having similar composition. Non-attachment of the floc to air bubbles could be one reason, perhaps because the flocs were not sufficiently hydrophobic. Laboratory scale flotation experiments were carried out to test the effect of an anionic surfactant, sodium dodecyl sulfate (SDS), as well as the normal additives.

It was noticed that treatment performance improved when there was a pre-settling stage, suggesting that the solids loading was too high. As the large dose of inorganic coagulant adds to the solids load, it seemed that it would be beneficial to utilise a cationic organic polymer instead. Flotation with a 10 minute float time, decreasing amounts of PAC, varying amounts of a medium charge density and very high MW cationic polymer, and some SDS. The data in Table 1 show that under these conditions the surfactant influence is detrimental, and that the major parameter is the cationic polymer, which effectively replaces PAC. Hence the dominant feature of the system is one of overloading of the solids, which are added to by the presence of inorganic coagulant.

A series of experiments was then done in the absence of PAC and SDS, with fresh effluent at pH 9, a 10 minute float time, and varying types of cationic polymer. Table 2 shows that the best performance is obtained with a polymer of very high cationic charge, and that the MW is not crucial. It is notable that it is possible to overdose the system, as indicated by the first two sets of results; this may account for the variation in the behaviour with the high MW polymer. It is important that the subnatant volume remains substantial, because it represents a higher concentration of solids in the sludge. This is very much the case with the most highly charged polymer, which is consequently the preferred additive for this particular effluent.

Table 1. The replacement of PAC with cationic polymer, with and without added surfactant

PAC mg/L	Polymer mg/L	SDS mg/L	Turbidity NTU
500	7.5	100	12,100
500	50	100	7,680
0	100	0	6,400
0	50	0	4,680
0	50	20	6,020

In the treatment of industrial effluents that are loaded with emulsified oil and grease as occurs in steel mill waste or in wool scouring, high separation rates can be achieved by first coalescing the organics and then using foam flotation.[22] Large air bubbles are formed which result in faster rise rates and lower capital costs than are achieved with the small bubble systems prevailing in the competing technology of dissolved air flotation. Such emulsions can be successfully treated with only a medium CD, high MW cationic polyacrylamide, which adequately neutralises the negatively charged organic phase. The lack of a dependence on MW which was observed indicates that bridging of particles by the polymer does not occur to any significant degree, and that charge neutralisation to give coagulation and flocculation. Typical results for steel mill and wool scouring wastes are given in Table 3.

In applying foam flotation to wool scouring waste, performance is dependent on pH.[26] The reductions shown in Table 3 were achieved in pilot-scale work at pH 4.1 by adding 130 mg/L of polymer. At the natural pH level of 7.3 there was greatly reduced performance, and the dose required was then of the order of 50 mg/L of the standard polymer, plus 500 mg/L of a high CD cationic polymer of medium MW made from epichlorohydrin and dimethylamine.

Some recent research on the influence of polymer structure on the efficiency of emulsion breaking and foam flotation has been carried out with a more hydrophobic polyelectrolyte.[23] Laboratory studies were carried out at pH 7 on a synthetic waste comprising 1.6 g/L of mineral oil dispersed in 0.01 *M* KCl with the aid of 4.5% w/w of sodium oleate. To achieve a more than 95% removal of turbidity a dose of at least 30 mg/L of a cationic polyacrylamide of low CD was required. The polymer has the cationic site linked to an acrylate unit by an ethylene bridge. A cationic polyacrylamide of the same CD was tested which had the charge linked to a methacrylamide unit via a propylene bridge; it required only 15 mg/L for the same duty. The beneficial effects of hydrophobically modified polymers in sludge dewatering have been demonstrated.[24, 25]

The concept of primary coagulation of industrial wastewaters with polymers, followed by flotation, is applicable as well to wastes from the cosmetic, detergent, plastics, dyehouse, paper, food processing and brewing industries.[26] In some cases the addition as well of an inorganic coagulant is unavoidable, as with highly coloured effluents; in others, an anionic surfactant is also needed to facilitate flotation.

4. PHOSPHORUS REMOVAL FROM SEWAGE

Partial phosphate elimination from sewage by precipitation with metal salts has long been in use. A recent variation of the method has been the replacement of some of the

Table 2. Variation of the cationic polymer. The raw material had a turbidity of 18,080 NTU

Polymer dose, mg/L	Charge density	MW	Final turbidity, NTU	Subnatant volume, %
50	medium	v high	575	55
100	medium	high	3,400	80
50	high	high	277	70
50	medium	high	174	55
50	low	high	198	45
50	v high	medium	67	83

Table 3. Treatment of wastes containing oil and grease

Parameter	Level in raw waste	Amount removed, %
Steel mill waste, pH 6.5 and 30 mg/L of polymer:		
Turbidity	6,000 NTU	95
COD	4 g/L	80
Grease	3.5 g/L	95
Wool scouring waste, pH 4.1 and 130 mg/L of polymer:		
Turbidity	79,100 NTU	99
COD	118 g/L	87
Grease	28 g/L	97
Wool scouring waste, pH 7.3, 50 mg/L of standard polymer plus 500 mg/L of a high CD, medium MW polymer:*		
Turbidity	13,200 NTU	91 (97)
COD	31 g/L	64 (73)
Grease	7.6 g/L	84 (96)

*The data in parentheses are for the same experiment without the second polymer, but at pH 4.0.

metal salt with a cationic polymer without loss of efficiency of phosphate removal, indicating that the polymer is involved in adsorption of anionic phosphorus species.[27] In a full-scale plant the normal dose of 23 mg/L of ferrous sulfate was reduced to 12 mg/L plus 3 mg/L of a high CD cationic polymer, along with 0.2 mg/L of a high MW anionic polymer to act as a flocculant. The removal of BOD was increased from 30 to 55% while the phosphorus uptake was left unchanged. The sludge produced after the activated sludge step was then more readily biodegraded because of the lower inorganics content.

More recent work makes use of permethylated polyethyleneimine in a hybrid membrane approach.[28] The original polymer before reaction with dimethyl sulfate had a MW of 30–40,000. A 1% solution was employed in treating a 10 mg/L phosphate solution at pH 7. The membrane had a MW cut off of 10,000 and was operated at 300 kPa. A 90% removal of phosphate was claimed to be feasible in a three-step procedure, with both the polymer and the phosphate being recycled by variation of the pH.

5. METAL ION RECOVERY

A further opportunity for polyelectrolytes is the recovery of valuable or toxic metals from waste streams, based on binding the metal ions to soluble polymeric sequestering agents. After attachment of the metal to the polymer, separation is readily achieved if precipitation occurs when the metal ions crosslink the polymer chains. This happens for example when cellulose xanthate is used to recover Cd^{2+} from plating wastewater.[29] Analogous systems are Cu^{2+} or Cr^{3+} precipitation with a copolymer of maleic and acrylic acids.[30] Another variation is the extraction of Pb^{2+} from contaminated soil by sequestering it with polyethyleneimine containing carboxylic acid groups, after which the polymer is regenerated by precipitation with acid to leave the lead in solution.[31] A polymer of opposite charge may be the means of precipitation, as when polyethyleneimine is reacted with poly(acrylic acid) bound Cu^{2+}. Regeneration with acid releases the metal, leaving the insoluble metal-free polymer complex behind.[32]

An alternative approach is to concentrate the soluble metal-loaded polymer using membrane technology. Thus Cu^{2+} has been bound to sodium polystyrenesulfonate, and an essentially metal-free filtrate obtained by ultrafiltration.[33] Similar work has been done on

Cu^{2+} abstraction with carboxymethyl cellulose, which was superior to polystyrenesulfonate,[34] and with alginic acid.[35] Poly(acrylic acid) has been used for Cu^{2+} and Zn^{2+} removal.[36] An extensive study of metals removal from plating effluents and other metal finishing wastewaters with low MW poly(acrylic acid) and polyethyleneimine has obtained data for Cu^{2+} and Zn^{2+}, and proposed electrolysis or acid treatment and a second ultrafiltration step to regain the original polymer for further use.[38]

6. CONCLUSIONS

Organic polymeric flocculants can been used as primary coagulants, instead of inorganic salts, in many treatment situations. The benefits include faster and more stable and reliable processing, a lower content of insoluble solids to handle, whether by direct filtration, sedimentation, flotation or even in biological conversions, and a much smaller sludge volume. The use of alum can likewise be avoided in breaking emulsified oil and grease wastes from a range of industries. Polyelectrolytes are used to advantage in a number of new processes, such as in weighted systems, in the removal of phosphate, and in hybrid membrane systems.

7. REFERENCES

1. Bolto, B. A., Soluble polymers in water purification. *Progress in Polymer Science,* 1995, accepted for publication.
2. Edzwald, J. K., Conventional water treatment and direct filtration: treatment and removal of total organic carbon and trihalomethane precursors. *Organic Carcinogens In Drinking Water* (N. M. Ram, E. J. Calabrese and R. F. Christman Ed.), p. 208, Wiley, New York, 1986.
3. Schlauch, R. M., Coagulation for gravity type clarification and thickening. *Polyelectrolytes for Water and Wastewater Treatment* (W. L. K. Schwoyer Ed.), p. 91, CRC Press, Boca Raton, 1981.
4. Vik, E. A. and Eikebrokk, B., Coagulation process for removal of humic substances from drinking water. *Aquatic Humic Substances* (I. H. Suffet and P. MacCarthy Ed.), p. 385, Am. Chem. Soc., Washington, 1989.
5. Jackson, G. E., Granular media filtration in water and wastewater treatment: Part 2. *CRC Critical Reviews in Environmental Control* **11**, 1–36, 1980.
6. Bilanovich, B., Shelef, G. and Sukenik, A., Flocculation of microalgae with cationic polymers - effects of medium salinity. *Biomass* **17**, 65–76, 1988.
7. Haarhoff, J. and Cleasby, J. L., Direct filtration of chlorella with cationic polymer. *J. Environ. Eng.* **115**, 348–366, 1989.
8. Himberg, K., Keijola, A. M., Hiisvirta, L., Pyysalo, H. and Sivonen, K., The effect of water treatment processes on the removal of hepatotoxins from *Microcystis* and *Oscillatoria* cyanobacteria: a laboratory study. *Water Research* **23**, 979–984, 1989.
9. Philipp, B., Jaeger, W., Gohlke, U., Wandrey, C., Hahn, M., Dietrich, K., Linow, K-J., Dautzenberg, H. and Kötz, J., Ergebnisse Der Polymer-Forschung im Zusammenhang mit der Wasserreinigung. *Paperi ja Puu* **68**, 419–430, 1986.
10. Demeter, L., Galgoczi, B., Bozzay, E. and Zagyvai, I., Weighted flocculation. *US Pat.* 3,350,302 (1967) assigned to Nikex Nehézipari Külkereskedelmi Vallalat, Budapest.
11. Cailleaux, C., Pujol, E., De Dianous, F. and Druoton, J. C., Study of weighted flocculation in view of a new type of clarifier. *Aqua* **41**, 18–27, 1992.
12. Pujol, E., Vullierme, M., Druoton, J. C. and Sibony, J., Performance of the Actiflow settling process in the Neuilly-sur-Marne plant. *Tech., Sci., Methodes: Genie Urbane-Genie Rural* **11**, 557–562, 1993.
13. Guibelin, E., Delsalle, F. and Binot, P., High speed settling of stormwater with microsand. *Chem. Water Wastewater Treat. III; Proc. Gothenburg Symp., 6th* (R. Klute and H. H. Hahn Ed.), p. 307, Springer, Berlin, 1994.

14. Dixon, D. R. and Kolarik, L. O., The role of colloid and surface chemistry in the SIROFLOC process. *Surface and Colloid Chemistry in Natural Waters and Water Treatment* (R. Beckett Ed), pp. 103–118, Plenum, New York, 1990.
15. Bolto, B. A., Magnetic particle technologies for wastewater treatment. *Waste Management,* **10**, 11–21, 1990.
16. Akyel, G., Booker, N., Cooney, E. and Priestley, A., Rapid sewage clarification using magnetite particles. *Proc. AWWA Fed. Conv.,* Australian Water and Wastewater Association, Gold Coast, Vol. 1, pp. 137–144, 1993.
17. Moore, K. G., Johnson, M. G. and Sistrunk, W. A., Effect of polyelectrolyte treatments on waste strength of snap and dry bean wastewaters. *J. Food Sci.* **52**, 491–492, 1987.
18. Pariser, E. R. and Lombardi, D. P., A guide to the research literature. *Chitin Sourcebook,* p. II: L - 1, Wiley, New York, 1989.
19. Kisla, T. C. and McKelvey, R. D., Color removal from softwood kraft pulp bleach plant effluent by polyamines. *Environ. Sci. Technol.* **12**, 207–211, 1978.
20. Michelsen, D. L., Fulk, L. L., Woodby, R. M. and Boardman, G. D., Adsorptive and chemical pretreatment of reactive dye discharges. *Emerging Technologies In Hazardous Waste Management III* (D. W. Tedder and F. G. Pohland Ed.), p. 119, Am. Chem. Soc., Washington, 1993.
21. Türkman, A., Polymer application examples in industrial wastewater treatment. *New Developments in Industrial Wastewater Treatment* (A. Türkman and O. Uslu Ed.), p. 93, Kluwer Academic, Dortrecht, 1991.
22. Beeby, J. P., Dixon, D. R., Gray, S. R. and Smitham, J. B., Foam flotation in rolling mill effluent treatment. *15th Fed. Conv. Aust. Water & Wastewater Assoc.* Vol. 2, pp. 548–552, Gold Coast, Australia, 1993.
23. Bolto, B. A., Dixon, D. R., Gray, S. R., Ha, C., Harbour, P. J., Ngoc, L. and Ware, A. J., The use of soluble organic polymers in wastes treatment. *Proc. 18th Biennial International Conference,* International Assocn. on Water Quality, Singapore, 1996.
24. Burkert, H., Hartmann, J., Jung, D., Heide, W., Hartmann, H. and Spoor, H., Flocculants based on polymers containing water-soluble dialkylaminoneopentyl (meth)acrylate. *US Pat.* 4,465,599 (1984), assigned to BASF A.-G.
25. Sato, S., Cationic polymer coagulant for sludge dewatering. *Jpn. Kokai Tokkyo Koho* 92 07,100 (1992), assigned to Kurita Water Industries Ltd.
26. Dixon, D. R., The development of improved methods for the purification of wool scouring effluents to minimise the total cost of treatment and discharge to the sewer. *Final Report for Project CPO2*, Australian Wool Research and Promotion Organisation, Melbourne, 1994.
27. Sörensen, J. and Larsson, S. G., Particle separation in wastewater treatment. *Chem. Water Wastewater Treat. II, Proc. Gothenburg Symp., 5th* (R. Klute and H. H. Hahn Ed.), pp. 181–190, Springer, Berlin, 1992.
28. Palmer, V., Zhou, R., Geckeler, K. E. and Bayer, E., Removal of phosphate in aqueous solution by permethylated poly(ethyleneimine). *Acta Hydrochim. Hydrobiol.* **22**, 231–236, 1994.
29. G. Tiravanti, G. Macchi, D. Marani, M. Pagano, M. Santori and R.Passino, *Heavy Met. Environ., Int. Conf., 5th* (T. D. Lekkas Ed), P. 631, Cep Consult., Edinburgh (1985).
30. L. Campanella, V. Crescenzi, M. Dentini, C. Fabiani, F. Mazzei and A. I. N. Scheffino, *Metals Speciation, Separation, And Recovery* (J. W. Patterson and R. Passino Ed), Vol. 2, P. 359, Lewis, Chelsea (1990).
31. B. L. Duran, D. S. Ehler and N. N. Sauer, *Waste Treat. Technol. News* **9**(11), 286 (1994).
32. H. H. G. Jellinek and M. D. Luh, *J. Polym. Sci. A-1* **7**, 2445–2449 (1969).
33. K. J. Sasaki, S. L. Burnett, S. D. Christian, E. E. Tucker and J. F. Scamehorn, *Langmuir* **5**, 363–369 (1989).
34. S. D. Mundkur and J. C. Watters, *Sep. Sci. Technol.* **28,** 1157–1168 (1992).
35. A. G. Fane, A. R. Awang, M. Bolko, R. Macoun, R. Schofield, Y. R. Shen and F. Zha, *Water Sci. Technol.* **25**, 5–18 (1992).
36. R-S. Juang and J-F. Liang, *J. Mem. Sci.* **82**, 175–183 (1993).
37. K. Volchek, E. Krentsel, Y. Zhilin, G. Shtereva and Y. Dytnersky, *J. Mem. Sci.* **79**, 253–272 (1993).

12

PROSPECTS OF FIBROUS ION EXCHANGERS IN WATER POLLUTION CONTROL (CHROMATES SORPTION BY AMINOCARBOXYLIC FIBERS EXAMPLE)

V. Soldatov,[1] L. Pawłowski,[1] H. Wasag,[1] I. Elinson,[2] and A. Shunkevich[2]

[1] Technical University of Lublin
20-618, Lublin, Poland
[2] Inst. Phys. Organic Chem. of the Belarus Acad. Sci.
13 Surganov Str., Minsk 220072, Belarus

1. INTRODUCTION

Application of granular ion exchangers for treatment of large volumes of polluted water with low concentrations of pollutants meets serious technological and economic problems. One of them is large energy and time consumption for pumping of the purified water through the column; the other one is destruction of the granules under mechanical and osmotic stress in prolonged exploitation.

These principle difficulties can be overcome by using fibrous ion exchangers which provide possibilities for increasing the filtering cross-section, having high rate of ion exchange processes in combination with low hydrodynamic resistance, high osmotic and mechanic stability.

Starting in the sixties extensive works on development of synthetic ion exchange fibrous materials have been done by scientists of different countries, especially USSR and Japan where their industrial or semi-industrial production has been organized. At present fibrous ion exchangers of all main types are produced as staple and non-woven fabric in Belarus (FIBAN) and in Russia (VION, KOPAN); and short-cut fibers in Japan (IONEX) [1–5].

A large variety of potentially available fibrous ion exchange materials (cotton, threads, woven and non-woven cloths, belts, mats etc.) opens new possibilities for development of non-traditional ion exchange technologies in addition to or alternative to the existing ion exchange industrial processes. A useful peculiarity of ion exchange fibers is the small diameter of their mono-filaments 5–50 μ in combination with its narrow size distribution (not more than ±10%), which predetermines high sorption rate and excellent dynamic characteristics of the filtering layers. Experimental research on kinetics of ion exchange on strong and weak base materials FIBAN A-1 and FIBAN AK-22 (Cl^--SO_4^{2-}) [6] and strong acid ion exchange

IONEX TIN 110 (Na^+-Ca^{2+}) [7] has shown that the half process period of ion exchange in this case can be as short as 3–7 s which is 1–2 powers of magnitude shorter than on conventional ion exchange resins with the same functional groups. The high sorption rate allows to use these materials in the form of short filtering layers (10–20 mm).

Fibrous ion exchange materials have found significant application in Russia, the Ukraine and Belarus for purification of industrial exhaust gases and in gas-dust respirators. There are several hundred of industrial gas purification plants in operation in these countries [8,9].

Experience in application of fibrous ion exchangers for water treatment is restricted at present with rather small scale devices for individual potable water preparation (FIBAN and VION in Aquapor A1 and Barrier filters, Russia) and high degree water desalination processes (IONEX, Japan) [3]. Large scale water treatment technologies with fibrous ion exchangers are still not developed.

In this paper an attempt to illustrate the applicability of fibrous ion exchangers exemplified by one of the most promising materials, aminicarboxylic ion exchanger FIBAN AK-22 in the process of Chromates CrO_4^{2-}($Cr_2O_7^{2-}$) sorption, will be made. This model system is chosen in view of its own practical importance and good prospects for development of the industrial process based on the FIBAN AK-22 material and the laboratory model of a filtering plant described below. At the same time there is no doubt that with small modifications the method suggested can be applied to water purification for other impurities.

2. AMINOCARBOXYLIC FIBROUS ION EXCHANGERS

2.1. Preparation

Initial material for aminocarboxylic fibrous ion exchangers is industrial fiber NITRON, copolymer of acrylonitrile (92.5%), metacrylate (6.3%) and itaconic acid (1.2%). Amino and carboxylic groups are formed in this polymer in a two step process: reaction of nitrile groups with different polyethylene polyamines to reach anion exchange capacity 1–6 meq/g; hydrolysis of the residual nitryle groups with formation of carboxylic groups to reach the cation exchange capacity 1–7 meq/g. Finally, the material contains in controlled ratios functional groups $R\text{-}CONH(CH_2CH_2NH)_{n+1}H$ and
R-COOH, $n = 0, 1, 2$. An example of these ion exchangers is given in Table 1.

Mechanical properties of the sorbents obtained are satisfactory for textile reprocessing. Their significant drop comes when the exchange capacity reaches 6 m-mole/g. Above this value the fibers become rigid and brittle in a dry state which makes their textile reprocessing and use in the form of non-woven materials difficult. If the requirements of textile properties are not very high (for instance short cut fibers, sponges, filters etc., then the capacity can be increased up to 10 mmole/g.

2.2. Properties

Potentiometric titration curves exhibit only one sufficiently expressed inflection point which is connected with the presence in the structure of the ion exchangers large variety of amino and carboxylic type groups without a sharp difference in their acido-basic strength (Figs 1, 2).

Swelling of the ion exchangers as a function of the equilibrium solution pH is illustrated by Fig. 3. It is seen that the ion exchanger with a high carboxylic groups content has

Table 1. Exchange capacities of amino carboxylic ion exchangers

Sample no	Capacity amino-groups, meq/g	Capacity carboxy-lic groups, meq/g	Summary capacity, meq/g
1	1.05	7.20	8.25
2	1.75	5.30	7.05
3	2.00	5.80	7.80
4	2.38	5.40	7.78
5	2.10	5.50	7.60
6	3.50	2.75	6.25
7	3.50	3.35	6.85
8	3.4	3.5	6.9
9	3.2	3.3	6.5
10	3.2	3.7	6.9
11	4.0	3.0	7.0
12	3.1	3.9	7.0
13	6.0	0.8	6.8

a typical polyampholyte type curve with a sharp minimum at the isoelectric point. On the other hand, if the content of the carboxylic groups is sufficiently low, only a small increase in the swelling with increasing pH is observed. Almost constant swelling is a great technological advantage of this ion exchanger since it predetermines invariable with pH hydrodynamic filtering layer resistance and accessibility of the functional groups to the

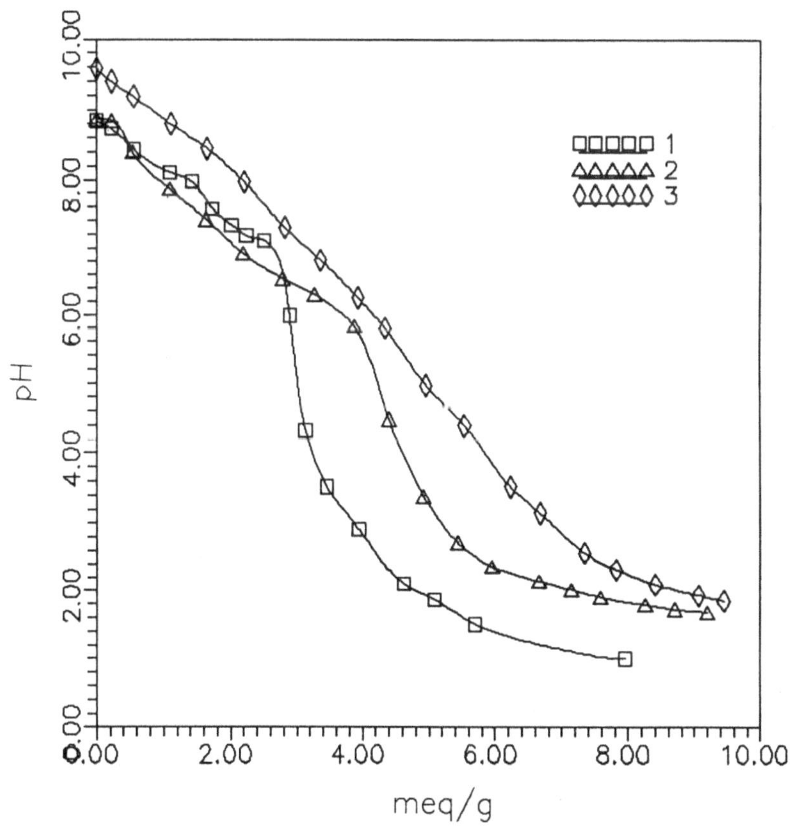

Figure 1. Potentiometric titration curves of FIBAN AK-22 ion exchangers with different structure of the amido-amino groups: 1- n = 0, 2 - n = 1, 3 - n = 2. Background electrolyte is 1 m NaCl.

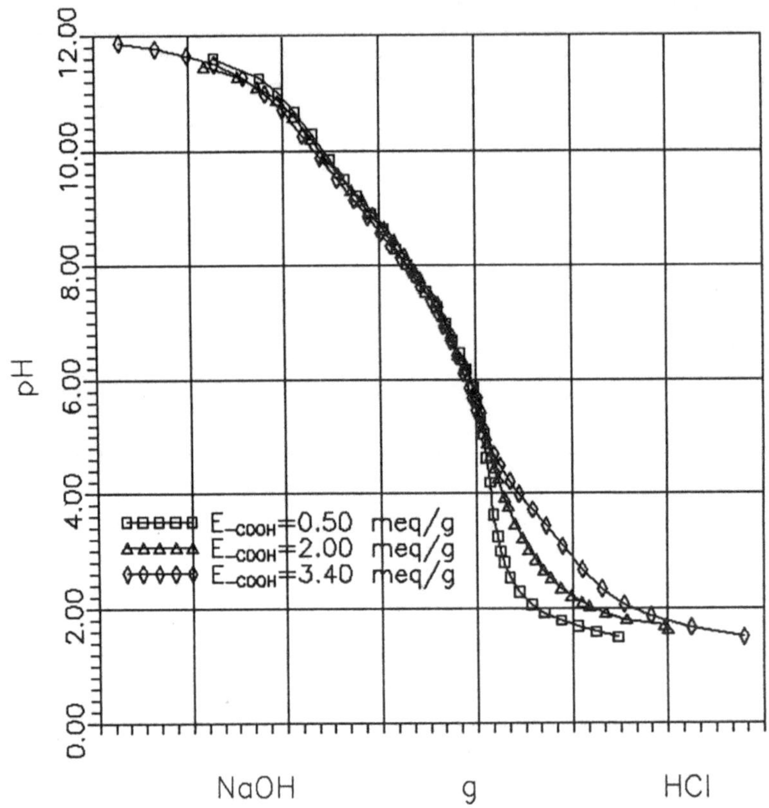

Figure 2. Potentiometric titration curves of FIBAN AK-22 ion exchangers with different contents of carboxylic groups, n =1.

ions as well as favors osmotic stability. Because of this property it is reasonable for technological processes to take ion exchangers with moderate contents of carboxylic groups as was done in the present work, where the ion exchanger with anion- and cation exchange capacity 3.75 and 1.00 meq/g respectively has been chosen.

2.3. Sorption and Ion Exchange

Due to their amphoteric nature aminocarboxylic sorbents can act as anion or cation exchangers depending on the solution pH. In addition, both functional groups can act as ligands for complex forming cations which can be sorbed by non-exchange mechanisms. It has been reported elsewhere [10, 11] that FIBAN AK-22 can be used for sorption of anions, including metal containing anionic species type $E_xO_y^{z-}$ (E= V, Mo, W, Cr) or EAx^{z-} (E= Pd, Pt, etc.) and a large variety of cations: Cu^{2+}, Pb^{2+}, Hg^{2+}, Co^{2+}, Fe^{3+}, Ag^{+} and others. Nevertheless, the sorption processes have not been investigated in detail sufficient to make a conclusion of their applicability for practical application. In this paper such a study was performed with one of the most important water pollutants - Cr(VI) anions.

An important feature of FIBAN aminocarboxylic ion exchangers is their high resistivity to oxidation by Chromates in alkali media. Keeping them in 1 m Na_2CrO_4 solution at pH=13 during 100 days did not cause reduction in their sorption capacity according to

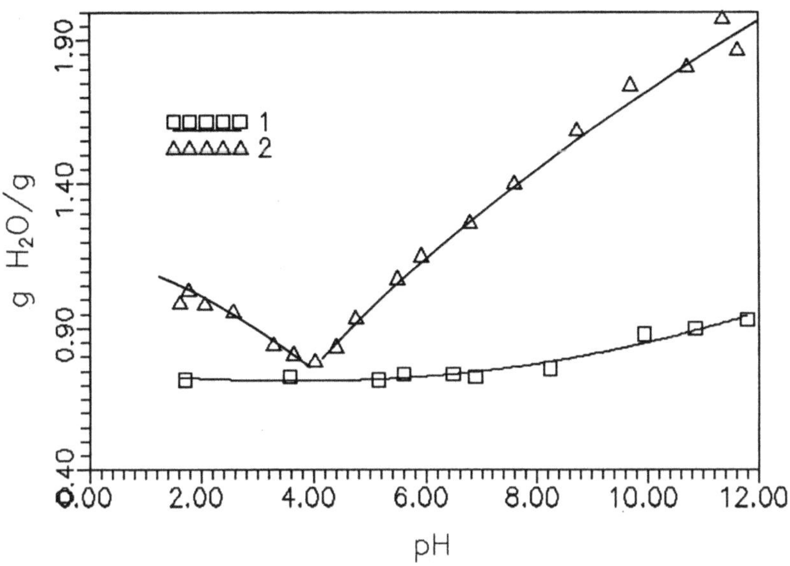

Figure 3. FIBAN AK-22 swelling , n = 1, as a function of the equilibrium solution pH at the contents of carboxylic groups 1.5 (1) and 3.6 meq/g (2).

chromate ions. This makes it possible to apply FIBAN aminocarboxylic ion exchanger in a multi-cycle sorption/desorption processes.

3. CHROMATE SORPTION PROCESS INVESTIGATION

Basic information required for evaluation of the possibility of development of technology for chromium (VI) removal from waste water should include data on the sorption equilibrium, kinetics and dynamics under different conditions: chromium (VI) concentration, solution pH, type of accompanying ions, solution flow rate. The appropriate data have been obtained and are reported further.

3.1. Equilibrium

The usually applied "pH profile" method for characterization of ionic sorbents efficiency has a serious drawback. It does not allow us to keep ionic concentration constant with changing pH and vice versa. The "pH-profile" is always related to non-constant ion concentration and their theoretical consideration is extremely difficult.

We have applied a dynamic method for studying dependencies sorption - metal concentration - pH.

The initial solutions with desired pH and Cr(VI) concentrations have been pumped through a cell with the ion exchanger until the equilibrium was reached. The cell contained 0.3–0.5 g of ion exchanger and was a cylinder shape with the height of 10 mm and diameter of 15 mm. With this experimental technique it was possible to keep constant either pH or the metal concentration in the solution and obtain cuts of the sorption surface.

Cr(VI) sorption isotherms at 20°C at different pH are presented in Figs 4a, 4b. Dependence of Cr(VI) sorption at its constant concentrations has a maximum at pH≃3 (Fig 5), showing that both high and low solution pH can be used for the sorbent regeneration.

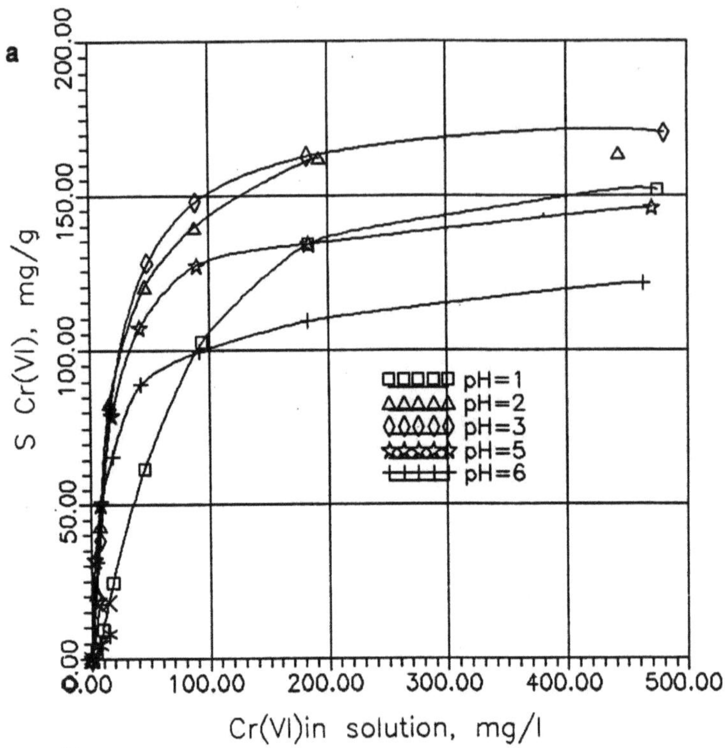

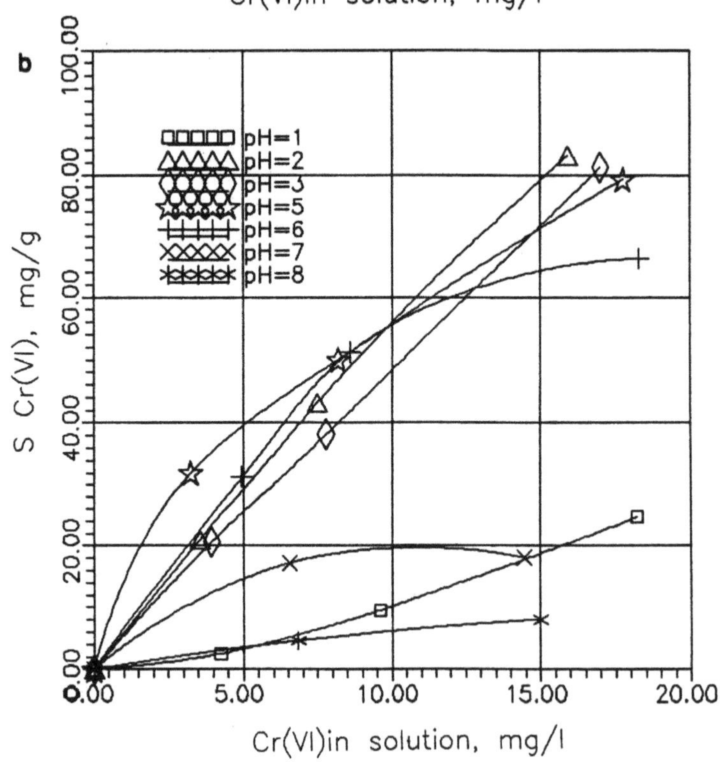

Figure 4. Chromate sorption isotherms.

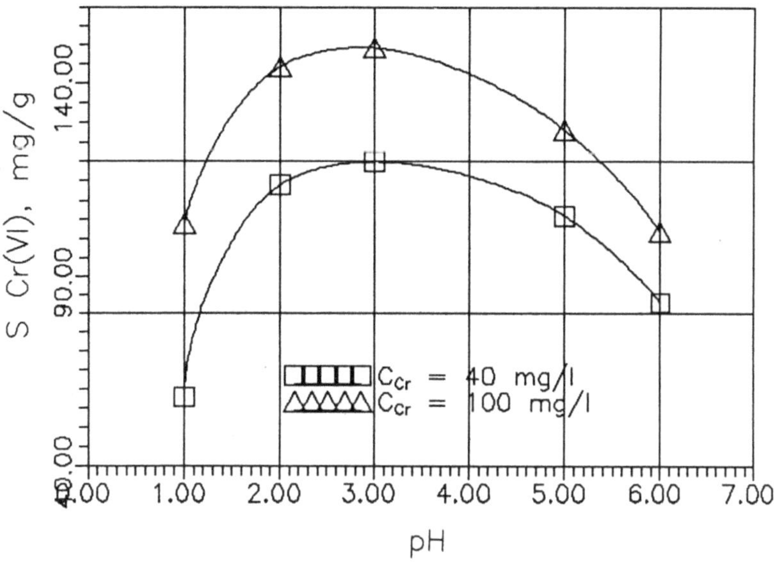

Figure 5. Cr (VI) anions sorption as a function of the equilibrium solution pH.

3.2. Kinetics

A modification of the shallow bed method has been used for the kinetics studies. It was necessary due to an extremely high process rate.

Kinetic curves of Cr(VI) sorption at fixed solution pH=2 and Cr(VI) concentration 1–200 mg/l are presented in Fig. 6. For comparison the same curve was obtained for a granular ion exchanged EDE-10P (Fig. 7). It is seen that the rate of sorption is drastically slower in the case of granular resin.

Influence of pH on the sorption rate is illustrated by Fig. 8 indicating rather weak influence of this factor. On the other hand, substitution of HCL with H_2SO_4 leads to a great decrease of the process rate (Fig. 9).

Analysis of the experimental kinetic curves of the Cr(VI)- Cl^- and - SO_4^{2-}has shown that they satisfactorily coincide with the theoretic ones with the assumption that the rate controlling process step is intrinsic diffusion. This allowed us to calculate the diffusion coefficients of Cr(VI) anions in the fibers (D_f) and granular resin EDE-10P (D_r) from equations

$$D_f = 0.065R_f^2/t_{1/2}$$

$$D_r = 0.030R_r^2/t_{1/2}$$

where t=1/2 is a half exchange time, R_f and R_r are the fiber and granules radius.

The diffusion coefficients are presented in Table 2.

The data obtained can be summarized as follows.

High sorption rate of Cr(VI) on fibrous ion exchangers is due exclusively to short diffusion path compared to the granular ion exchangers. Diffusion itself is slower in the fibers (D is four times lower than on the granules at C_{Cr}=5 mg/g).

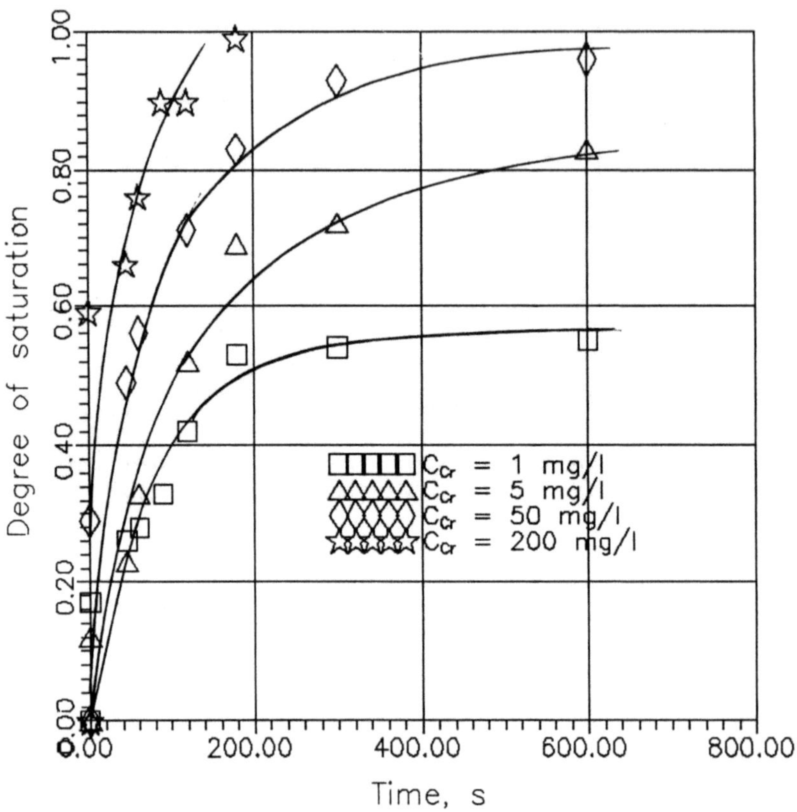

Figure 6. Kinetic curves of Cr(VI) sorption at different concentrations and pH = 2.0. FIBAN AK-22.

The diffusion coefficient drops by a power of magnitude when sulfate is used instead of chloride as a background electrolyte. This indicates the importance of the chemical interaction step in the overall ion transport process.

Increase of Cr(VI) concentration leads to a significant rise of the diffusion coefficient while pH effect is of little importance in the range pH= 2–5.

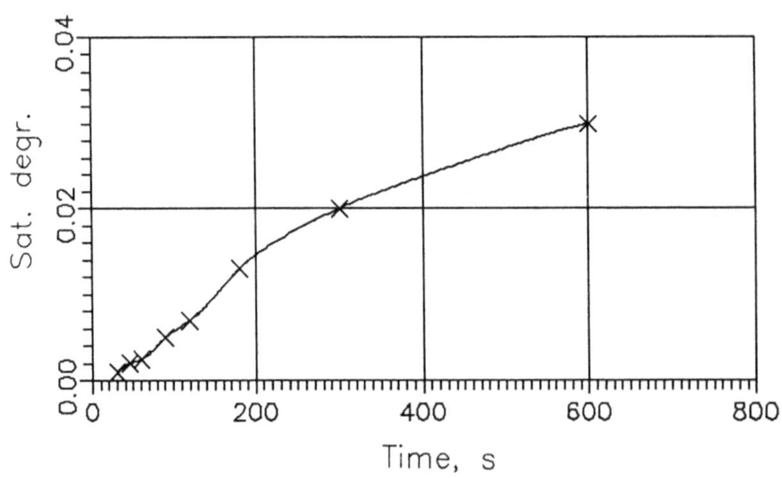

Figure 7. Kinetic curves of Cr(VI) sorption by industrial ion exchanger EDE-10P. C_{Cr} = 5 mg/l, pH = 2.0.

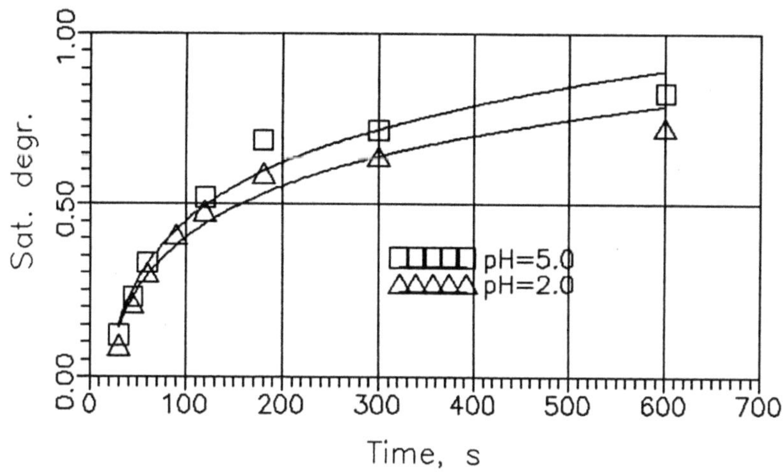

Figure 8. Effect of pH onto Cr(VI) sorption at C_{Cr} = 5 mg/L for FIBAN AK-22.

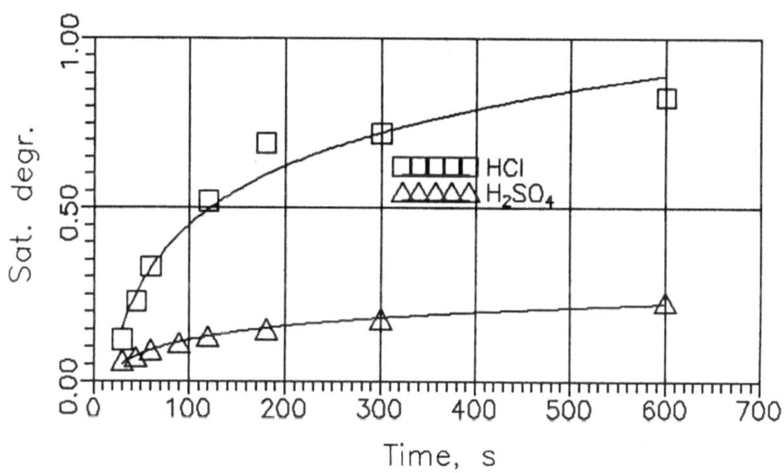

Figure 9. Kinetic curves of Cr(VI) sorption from hydrochloric and sulfuric acids . FIBAN AK-22, C_{Cr} = 5 mg/l, pH = 2.0.

Table 2. Kinetic characteristics of Chromate ions sorption

Sorbent	Backgr. acid	pH	Cr(VI) conc., mg/l	Half-proc. time, s	D, m^2/s
Fiban	HCL	2	1	173	$5.75 \cdot 10^{-14}$
Fiban	HCL	2	5	114	$8.89 \cdot 10^{-14}$
Fiban	HCL	2	50	50	$2.05 \cdot 10^{-13}$
Fiban	HCL	2	200	20	$5.12 \cdot 10^{-13}$
Fiban	HCL	5	5	126	$8.10 \cdot 10^{-14}$
Fiban	H_2SO_4	2	5	2245	$4.54 \cdot 10^{-15}$
EDE-10P	HCL	2	5	6765	$4.00 \cdot 10^{-13}$

3.3. Dynamics

It is almost obvious that application of fibrous sorbents in traditional ion exchange columns is irrational. It is difficult technologically to obtain high package density for fibrous materials and void volume in this case will be 80–90%. The filtering layer forms elastic system compressing in high rate liquid flows with increasing resistance. This means that in order to get use from excellent selective and kinetic properties of fibrous ion exchangers special technologies and apparatuses should be developed taking into account properties of fibrous materials.

One of the possible solutions to this problem is suggested in the present work.

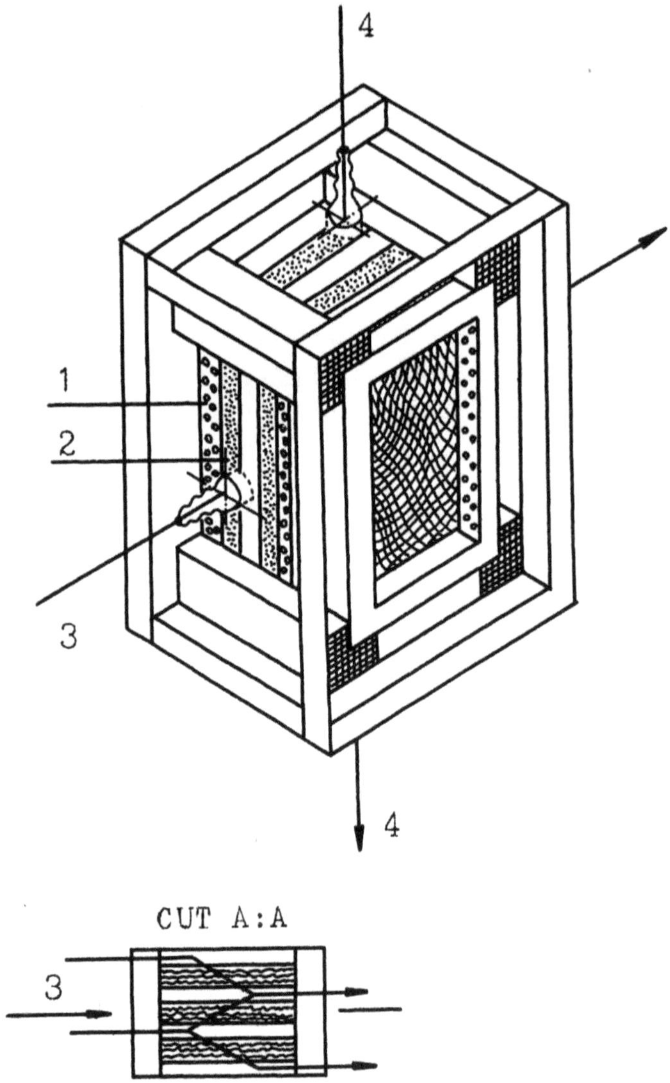

Figure 10. Scheme of the laboratory module for dynamic sorption experiments. 1 - spacer, 2 - fibrous ion exchanger, 3 - solution flow direction, 4 - regeneration flow direction.

The absorber is a rectangular chamber with layers of the filtering ion exchange material situated as shown in Fig. 10. The fibrous ion exchanger is a form of needle-punctured non-woven material with density of package 0.2 kg/dm^3 and has a filtering layer thickness of 7 mm. The laboratory apparatus contained 3 filtering separators with volume of the working chamber 0.4 l and contained 264 cm^2 filtering area (33.8 g sorbent FIBAN AK-22). In the dynamic experiments the flow rate varied in a range 0.092–1.48 m/h. This corresponds to 1.5 -24 l of water solution per hour passed through the device.

In Fig. 11 the breakthrough curves are presented for a model system with Cr(VI) concentration 20 mg/l and pH=2.5. The background electrolyte was HCl. It is seen that raising the flow rate 16 times does not have a significant effect and high flow rates can be used due to high dynamic capacity of the fibrous sorbent. Table 3 summarizes dynamic activity of the sorbent showing that breakthrough dynamic capacity (C/Co=0.05) is 25–42 mg/g. This corresponds to the degree of saturation of the sorbent at the breakthrough point 42–56%.

In solutions with sulfuric acid as a background dynamic, activity of the sorbent decreases which is caused by slower diffusion described above. The breakthrough curves presented in Fig. 12 show that a 7-fold increase in the flow rate decreases the dynamic activity four times. At the same time, equilibrium capacity does not depend on the flow rate (Table3).

It can be noted that values of dynamic activities obtained are 2–3 times higher than those for the fibrous ion exchanger VION AN-1 measured in a column with thickness of filtering layer 100 mm [12].

The regeneration was carried out by 1 m NaOH solution with the following neutralization of the NaOH with HCl or H_2SO_4.

Simple calculations show that scaling of the described apparatuses up to 1 m^3 operational chamber volume would allow room for 30–60 m^2 filtering layer of 10

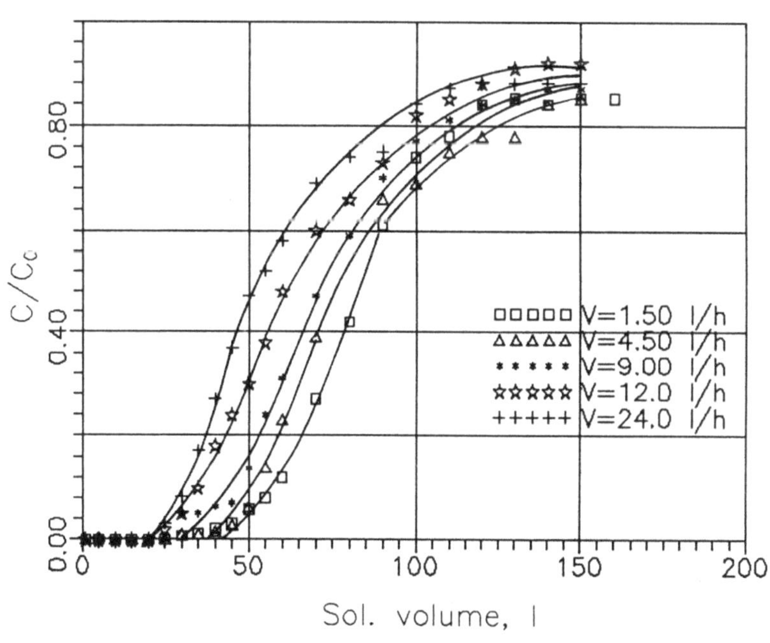

Figure 11. Breakthrough curves of Cr(VI) at pH = 2.5 at different flow rates.

Table 3. Dynamic characteristics of Cr(VI) sorption from solution with C_{cr}=20 mg/l and pH=2.5

Flow rate, l/h	Breakthrough dyn. capacity, mg/g	Equilibrium capacity, mg/g	Degree of saturation	Background acid
1.5	42.14	78.40	0.54	HCl
4.5	42.80	76.40	0.50	HCl
9.0	30.43	68.10	0.45	HCl
12.0	28.40	62.30	0.46	HCl
24.0	25.08	59.12	0.42	HCl
1.5	6.99	19.90	0.35	H_2SO_4
12.0	1.75	21.00	0.08	H_2SO_4

mm thickness. This would provide an apparatus working capacity above 20 m^3/hour, which is substantially higher than the usual values for column absorbers of the same volume (about 1 m^3/hour).

4. APPLICATION OF AMINOCARBOXYLIC FIBROUS ION EXCHANGER TO AIR PURIFICATION FROM CHROMATE AEROSOLS

High sorption ability and easy regeneration of Fiban AK-22 material in cyclic processes of chromate sorption made it possible to apply this material for air purification from aerosols of chromate solutions formed in galvanic operations for watch manufacturing at Minsk Watch Factory stock company. The scale of production is large: 11,000,000

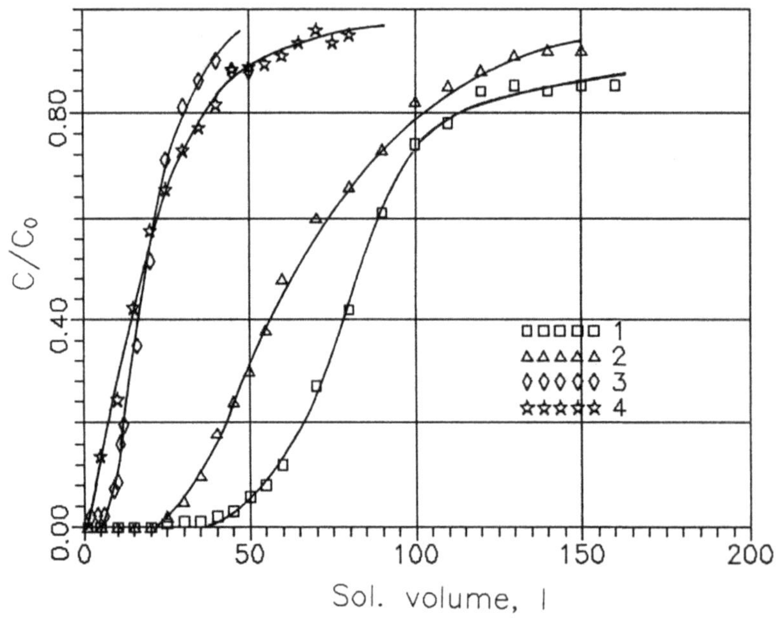

Figure 12. Breakthrough curves of Cr(VI). flow rate: 1, 3 - 1.5 l/h; 2, 4 - 12 l/h. 1, 2- HCl; 3, 4 - H_2SO_4.

watches per year, 8,000 workers. The amount of chromium aerosol with Cr(VI) concentration 0.2 - 5.5 mg/m^3 is approximately 35,000 m^3 per hour.

Two filtering plants with capacity each 20,000 m^3 air/hour have been constructed by Ecofil-Deco Ltd. and installed at the Minsk Watch Factory . The official testing was performed by a special commission including representatives of the Factory, Ecofil-Deco Ltd., the Institute of Physical Organic Chemistry and the Minsk City Council Ecological Department. The protocol of the examination is amended. It shows that at mean concentrations of chromic anhydride 1.4–1.8 mg/m^3 purification degree in the prolonged industrial experiment (14 months) was 93–99%. Regeneration was done by water (hydrolysis), 15 kg of water per 1 m^2 of the ion exchanger. The period between regeneration was 440 hours, duration of regeneration was 3 hours.

The filtering plants have been recommended for industrial exploitation.

5. CONCLUSION

FIBAN type aminocarboxylic fibrous ion exchangers have a set of properties sufficient for their application in water purification technologies. The chromium (VI) example shows that sorption apparatuses of a flat frame type with a short filtering layer provide high efficiency of the sorption/regeneration process utilizing advantages of fibrous ion exchange materials.

Application of fibrous ion exchangers in water treatment processes is promising for creation of a new type of ion intensive exchange technologies.

6. ACKNOWLEDGMENT

This work was supported by the International Association for the promotion of cooperation with Scientists from the independent states of the former Soviet Union (INTAS), project INTAS-93–1193.

REFERENCES

1. M.P. Zverev. *Chemosorption Fibers* (Russ.), Moscow, Khimia, 1981.
2. *Fibers With Special Properties* (Russ.), ed. L.A. Volf, Moscow, Khimia 1980.
3. M. Shimamura, T. Teramoto, T. Yoshioka. In: *Handbook of Fiber Science and Technology* V.3, part B, ed. M. Levin, J. Preston. Marcel Dekker Inc., N-Y and Basel, 1989, P. 209.
4. V.S. Soldatov, A.A. Shunkevich, G.I. Sergeev. *Reactive Polymers*. 1988, **7**, 159.
5. G.V. Miasoedova, S.B. Savin. *Chelate forming sorbents* (Russ.). Moscow, Nauka, 1984.
6. D. Petruzzelli, G. Tiravanti, L. Liberti a.o.. In: *Ion Exchange Processes: Advances and Applications*, Royal Soc. Chem., Cambridge, UK, 1993, P. 167.
7. T. Yoshioka and M. Shimamura. *Bull. Chem. Soc. Jap.*, 1983, **56**, 3726.
8. A.I. Vulih, A.A. Aloviajnikov, G.A. Nikandrov. In: *Ion Exchange* (Russ.), ed. M.M. Seniavin, Moscow, Nauka, 1981, P. 214.
9. A.A. Shunkevich, G.I. Sergeev, I.S. Elinson, *Z. Vsesojuznogo Khimicheskogo Obshestva imeni Mendeleeva*, 1990, **35**, 64.
10. V.S. Soldatov, R.V. Matrzinkevich, G.I. Sergeev, A.I. Pokrovskaya, In: *Theory and Practice of Ion Exchange Processes* (Russ.) Voronez, 1991, **21**, 154.
11. V.S. Soldatov, G.I. Sergeev, *Z. Vsesojuznogo Khimicheskogo Obshestva imeni Mendeleeva*, 1990, **35**, 101.
12. E.P. Doskina, T.I. Miagkaya, A.I. Starovatyh, In: *Prospective Methods for Purification of Natural and Rinsing Waters* (Russ.), Kuibyshev, 1981, P.69.

13

BEHAVIOR OF MACRORETICULAR CHELATING RESINS HAVING OXY ACIDS OF PHOSPHORUS IN ADSORPTION AND ELUTION OF LEAD ION

Akinori Jyo,[1] Kazunori Yamabe,[1] and Hiroaki Egawa[2]

[1] Department of Applied Chemistry
Faculty of Engineering
Kumamoto University
Kumamoto 860, Japan
[2] Kumamoto Institute of Technology
Kumamoto 860, Japan

ABSTRACT

Behavior of two macroreticular resins having oxy acids of phosphorus in adsorption and elution of lead ion was studied. One is a phosphonic acid resin based on styrene-divinylbenzene copolymer, and the other is a phosphoric acid resin based on glycidyl methacrylate-divinylbenzene copolymer. Both the resins exhibit the highest selectivity toward lead ion among common divalent metal ions and have high breakthrough capacities comparable to those of sulfonic acid resins in the dynamic uptake of lead ion. In addition, both the presented resins are much superior to sulfonic acid resins in dynamic elution of lead ion.

1. INTRODUCTION

As well known, strong-acid cation exchange resins containing sulfonic acid groups exhibit high selectivity toward lead ion. However, problems encountered are difficulty in smooth elution of adsorbed lead ion with acid solutions, since their selectivity toward lead ion over hydrogen ion is high. On the other hand, with weak-acid cation exchange resins having carboxylic groups the situation is quite reversed; these resins in the hydrogen ion form are not suitable for uptake of lead ion from neutral and weakly acidic solutions, even though the elution with acid solutions can be smoothly achieved. Thus, of interest is the behavior of resins having oxy acid groups of phosphorus in adsorption and elution of lead

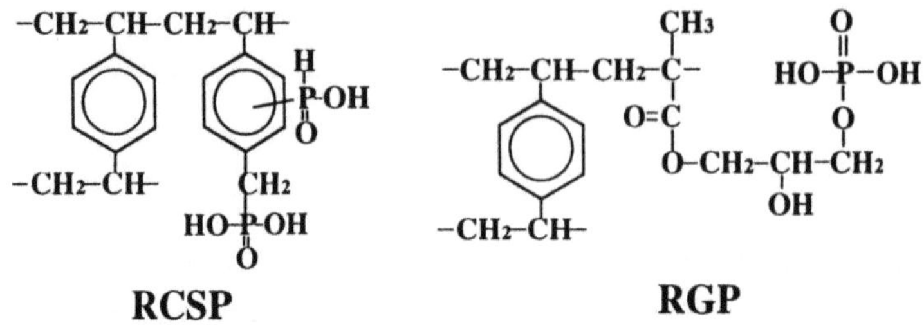

Scheme 1. Structures of RCSP and RGP

ion, since these groups exhibit medium acidities between sulfonic and carboxylic acid groups. However, little is known about the behavior of these kinds of resins toward lead ion. In this connection, we have been studying the preparation and properties of resins containing phosphonic acid groups (RCSP) as well as resins containing phosphoric acid groups (RGP).[1–4] The present work was planned in order to clarify the behavior of RCSP and RGP toward lead ion.

2. EXPERIMENTAL

2.1 Material

The resin RCSP was prepared according to the reported method.[1] The alternative resin RGP was prepared by a modified method of the previously reported one;[2] namely, glycidyl methacrylate-divinylbenzene copolymer beads (RG) were prepared by the suspension polymerization of a monomeric mixture (divinylbenzene 10 mol %) in the presence of isobutyl acetate as porogen (volume ratio of the porogen to the monomer mixture =1.4), and the resulting copolymer was heated in phosphoric acid (85 %) at 140 °C for 2 h. Table 1 summarizes properties of RCSP and RGP. In the column-mode study, commercial resins Diaion PK216 (strong-acid resin) and Diaion WK11 (weak-acid resin) were also used for the sake of comparison. All resins were used in the hydrogen ion form.

2.2 Batchwise Study

A capacity for uptake of lead ion was determined by equilibrating RCSP or RGP (0.125g) with a lead nitrate solution (0.01 M, 50 mL) of a given pH under mechanical

Table 1. Properties of RCSP and RGP

Resin[a]	Phosphorus content (mmol/g)	Cation exchange capacity (meq/g)	Salt splitting capacity (meq/g)	Volume of dry resin (mL/g)	Volume of wet resin (mL/g)	Specific surface area (m^2/g)
RCSP[b]	4.0–4.3	7.2–7.4	2.0–2.1	1.8–2.0	2.6–2.9	19–25
RGP[c]	3.4–3.7	6.9–7.0	2.7–2.9	1.8–2.1	3.2–3.3	17–28

[a]Degree of crosslinking is 10 nominal mol % of divinylbenzene.
[b]32–60 mesh(batchwise study) 60–80 mesh (column-mode study).
[c]32–60 mesh.

shaking at 30 °C for 24 h. From a decrease in the lead ion concentration in the aqueous phase, the capacity was calculated. Alternatively, distribution behavior of several metal ions between resin and aqueous phases was evaluated by equilibrating a resin (0.04g) with a metal ion solution (0.0001 M, 25 mL). A distribution ratio (D), which is designated by a ratio of an amount of metal ion in the resin phase (mmol/g) to that in the aqueous phase (mmol/mL), was calculated from a concentration decrease of the metal ion in the aqueous phase. Nitric acid, acetic acid, and sodium acetate were used in the pH adjustment.

2.3 Column-Mode Study

A given volume of wet settled resin was packed into a fritted glass column. In adsorption of lead ion, a given volume of feed solution containing lead nitrate was supplied to the column at an hourly rate of 10 bed volumes. After washing with 10 bed volumes of water, lead ion adsorbed on the column was eluted with aqueous nitric acid at an hourly rate of 3 bed volumes. All column effluents including washings were collected on a fraction collector, and the metal ion composition in each fraction was determined. In evaluation of the effect of a foreign cation on dynamic uptake of lead ion, a solution containing lead nitrate as well as a nitrate salt of an appropriate metal ion was used as feed. Other detailed experimental conditions will be given in later sections with results.

3. RESULTS AND DISCUSSION

3.1 Batchwise Study

Figure 1 shows capacities of RCSP and RGP for lead ion as a function of pH. Although the phosphorus content and the cation exchange capacity of RGP are less than those of RCSP (Table 1), both the resins exhibit almost the same capacities for lead ion in the tested pH region. This means that the phosphoric acid group in RGP more effectively takes up lead ion than does the phosphonic acid group in RCSP as in the case of salt split-

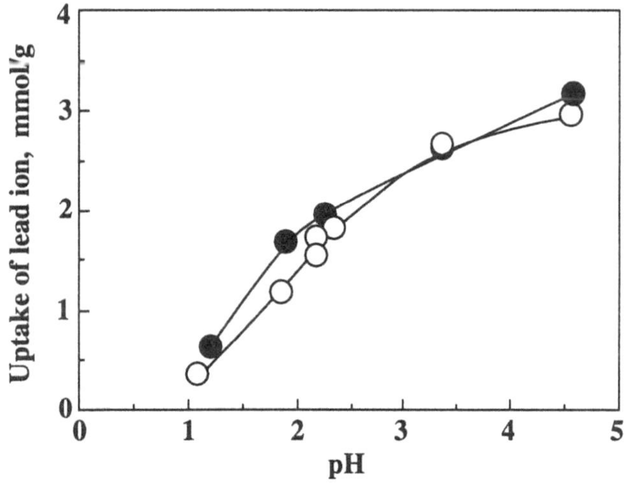

Figure 1. Dependence of uptake of lead ion on pH. ○ RCSP; ● RGP.

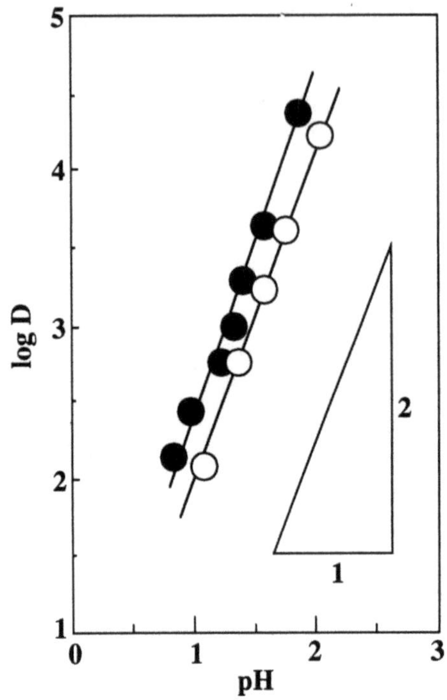

Figure 2. Distribution ratios of lead ion as a function of pH. O RCSP; ● RGP.

ting capacities, which are designated by an amount of hydrogen ion replaced with sodium ion. At pH 4.6, both the resins can adsorb lead ion as high as ca. 3 mmol/g dry resin. The lead ion uptake at pH 4.6 is less than the phosphorus content of each resin, indicating that the second acid dissociation steps of the functional groups in each resin do not completely contribute to uptake of lead ion in the tested pH region.

Figure 2 shows the relationship between distribution coefficients of lead ion (log D) and equilibrium pH. Slopes of plots of log D vs. pH are nearly equal to 2. Thus, uptake of 1 mol of lead ion is accompanied by dissociation of 2 mol of hydrogen ion. Consequently, the following two mechanisms can be proposed for uptake of lead ion with RCSP and RGP of hydrogen ion form.

$$2RH_2 + Pb^{2+} \xrightleftharpoons{K_{21}} (RH^-)_2Pb^{2+} + 2H^+ \quad (1)$$

$$RH_2 + Pb^{2+} \xrightleftharpoons{K_{11}} R^{2-}Pb^{2+} + 2H^+ \quad (2)$$

Here, RH_2 represents RCSP or RGP in the hydrogen ion form. From Eqs. (1) and (2), the following relation can be obtained.

$$\log D = 2pH + \log(K_{21}[RH_2]^2 + K_{11}[RH_2]) \quad (3)$$

In the evaluation of distribution ratios, the amount of functional groups adopted (0.14 - 0.17 mmol) is much greater than that of lead ion (0.0025 mmol). Thus, the value of the term $(K_{21}[RH_2]^2 + K_{11}[RH_2])$ can be approximated to a constant so long as an aqueous

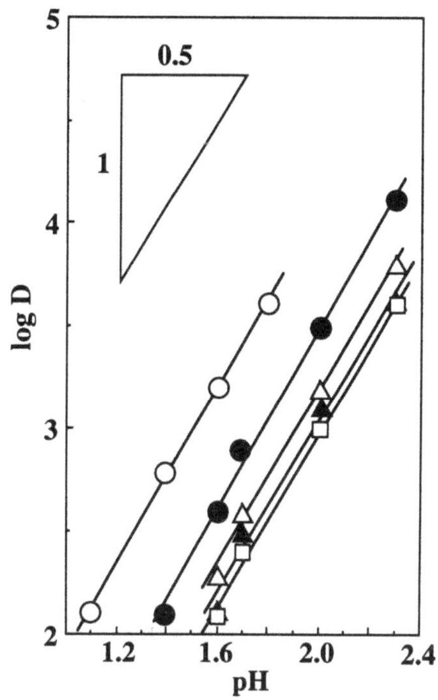

Figure 3. Distributions of several metal ions between RCSP and nitric acid solutions. ○ Pb^{2+}; ● Cd^{2+}; △ Ca^{2+}; ▲ Mg^{2+}; □ Ni^{2+}.

phase contains only hydrogen and lead ions as counter ion. Under these approximations, we have the relation (4).

$$\log D = \text{Constant} + 2\text{pH} \tag{4}$$

Eq. (4) semi-quantitatively explains the results shown in Fig. 2, although the detailed mechanisms are not yet clear.

Next, we briefly refer to the metal ion-selectivity of RCSP and RGP. Figure 3 shows distribution ratios (log D) of several metal ions as a function of pH and leads to the selectivity sequence: $Pb^{2+} >> Cd^{2+} >> Ca^{2+} >> Mg^{2+}$, Ni^{2+}. This sequence is somewhat different from that for strong-acid cation exchange resins: $Pb^{2+} >> Ca^{2+} >> Ni^{2+} >> Cd^{2+} >> Mg^{2+}$.[5)] It is also clear that the selectivity sequence of RGP toward divalent metal ions is similar to that of RCSP.

3.2 Column Mode Adsorption

Breakthrough experiments of lead ion were conducted by using RCSP, RGP, Diaion PK216, and Diaion WK11 columns. Figure 4 shows the results. Diaion PK216 and RGP give almost the same breakthrough capacities for lead ion (ca. 0.7 mmol/mL-wet resin), but RCSP gives a lower one (ca. 0.5 mmol/mL-wet resin). The weakly acidic Diaion WK11 does not have a meaningful breakthrough capacity. Among RCSP, RGP and Diaion PK216 columns, significant differences can be seen in the shapes of the breakthrough curves after each breakthrough point. In the case of the Diaion PK216 column, the con-

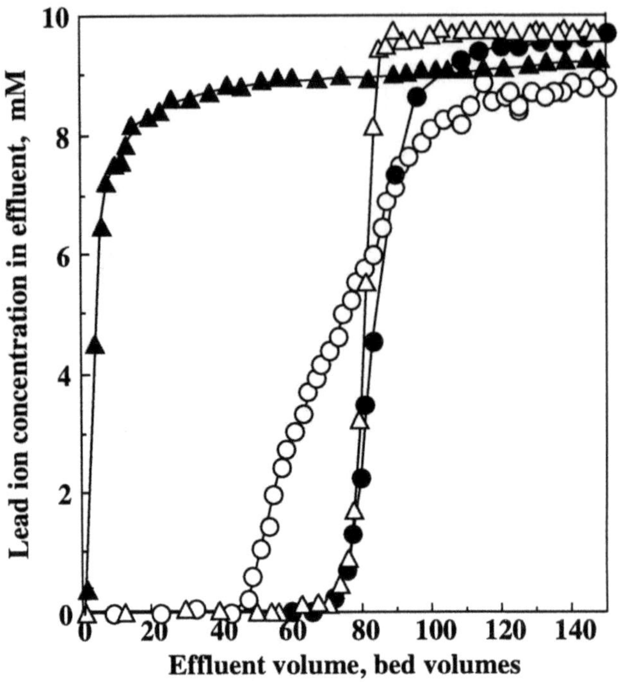

Figure 4. Breakthrough curves of lead ion. *Column:* 2 mL of wet resin (○ RCSP; △ Diaion PK216; ▲ Diaion WK11), 5 mL of wet resin (● RGP). *Feed:* 0.01 M lead nitrate in water.

centration of lead ion in the effluent after the breakthrough point rapidly attains that of the feed. However, such normal behavior was not observed in the RCSP and RGP columns, and this tendency is more marked in the RCSP column than in the RGP column. These differences among the three resins can be explained by their differences in acidity. In the strong-acid resin (Diaion PK216) hydrogen ion as counter ion can be easily replaced by the more preferred lead ion, leading to ideal breakthrough behavior. On the other hand, hydrogen ion in RGP and RCSP is not so completely displaced by lead ion, since the acidity of RCSP and RGP is much lower than that of Diaion PK216. Egawa et al. have reported that pKa_1 and pKa_2 of RCSP are 2.9 and 8.4, respectively[1)] and those of RGP 2.5 and 6.5.[2)] Then, the acidity of RCSP is lower than that of RGP. This may be the reason that RCSP gives a rather strange breakthrough behavior.

3.3 Column Mode Elution

Figure 5 shows elution of lead ion from RCSP, RGP, Diaion PK216, and Diaion WK11 columns with nitric acid. In the case of Diaion PK216, prolonged tailing was observed and ca. 20 bed volumes of 1 M nitric acid were required for the quantitative elution. On the other hand, much more smooth elution was accomplished in the case of the RCSP, RGP and Diaion WK11, and quantitative elution was achieved within 10 bed volumes of 1 M or 0.5 M nitric acid. For RGP, the effect of the nitric acid concentration on the elution of lead ion was examined, and the results are shown in Fig. 6; the higher the nitric acid concentration the higher the elution efficiency. Thus, it can be concluded that

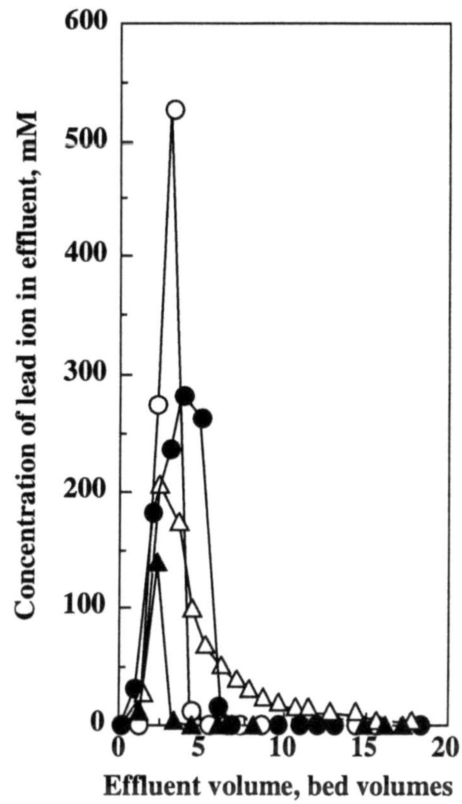

Figure 5. Elution of lead ion with nitric acid. *Column:* ○ RCSP; ● RGP; △ Diaion PK216; ▲ Diaion WK11. *Eluent:* 1 M nitric acid for RCSP, Diaion PK216 and Diaion WK11; 0.5 M nitric acid for RGP. For loading of lead ion, see Fig. 4.

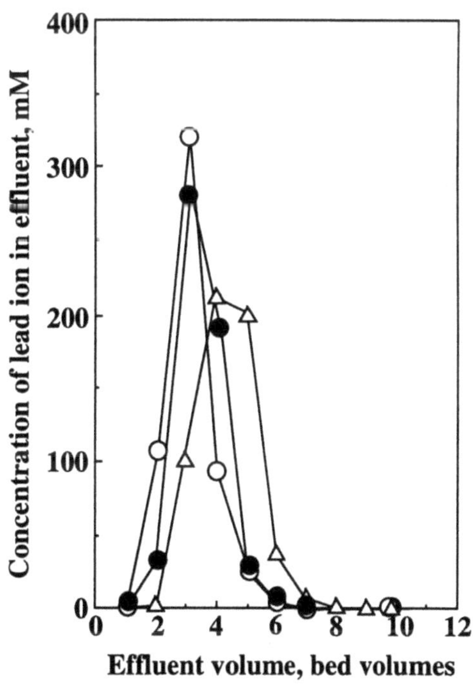

Figure 6. Effect of nitric acid concentration on elution of lead ion adsorbed on RGP. *Column:* 2 mL of wet RGP. *Concentration of nitric acid:* ○ 2 M; ● 1 M; △ 0.5 M. *Amount of loaded lead ion:* 1.00 mmol.

RGP and RCSP are superior to strong-acid resins in the efficiency of dynamic elution of lead ion with hydrogen ion.

3.4 Effect of Foreign Cations

Figure 7 shows the effect of calcium ion on the dynamic adsorption of lead ion with the RCSP. An equimolar calcium ion does not significantly interfere with the adsorption of lead ion, and the concentration of calcium ion in the effluent from 15 to 70 bed volumes markedly exceeds that of the feed, indicating that RCSP prefers lead ion to calcium ion. Even in the presence of calcium ion in excess (molar ratio Ca/Pb = 10), RCSP still remains an appreciable breakthrough capacity for lead ion. The effect of sodium ion was also examined up to 10 of molar ratio Na/Pb, and it is clear that sodium ion hardly interferes with the uptake of lead ion. RGP gave similar results.

4. CONCLUSIONS

The resins RCSP and RGP gave breakthrough capacities comparable to those of commercially available strong-acid resins in dynamic uptake of lead ion from neutral pH solutions, and the lead ion adsorbed was smoothly and completely eluted within ten bed volumes of 0.5 - 1 M nitric acid, indicating their excellent performances in the adsorption as well as in the elution of lead ion. In addition, RCSP and RGP exhibit the highest selectivity toward lead ion among common divalent metal ions. During this work, RCSP and RGP columns were repeatedly used by cycling adsorption-elution experiments. No deterioration in the column performances was observed.

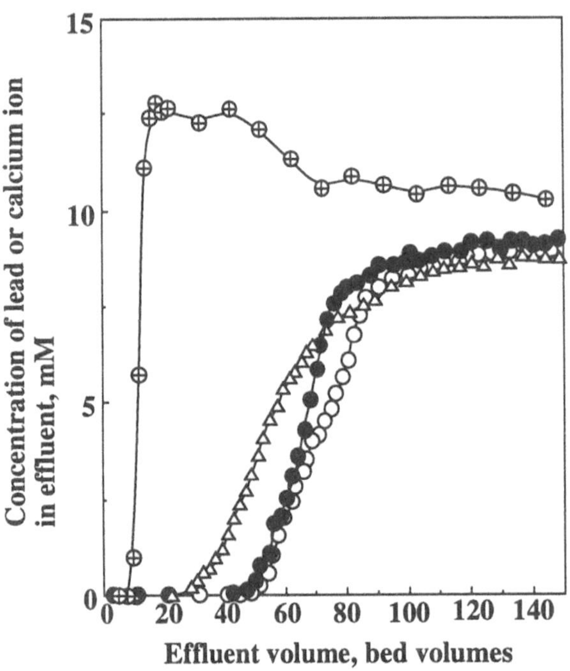

Figure 7. Effect of calcium ion on breakthrough of lead ion. *Column:* 2 mL of wet RCSP. *Feed:* 0.01 M lead nitrate solution containing calcium ion at the molar ratio Ca/Pb =0, Ca/Pb=1 or Ca/Pb=10. *Breakthrough curves of lead ion:* ○ Ca/Pb=0; ● Ca/Pb=1; △ Ca/Pb=10. Here, breakthrough curve of calcium ion is given by the symbol (⊕) for the case of Ca/Pb = 1.

5. REFERENCES

1. H. Egawa, T. Nonaka, and M. Ikari, *J. Appl. Polym. Sci.*, **29**, 2045(1984).
2. H. Egawa, T. Nonaka, and H. Maeda, *J. Appl. Polym. Sci.*, **30**, 3239(1985).
3. H. Egawa, K. Yamabe, and A. Jyo, *J. Appl. Polym. Sci.*, **52**, 1153(1994).
4. A. Jyo, K. Yamabe, and H. Egawa, *Sep. Sci. Tech.*, in press.
5. F. Helfferich, "Ion Exchange", McGraw-Hill, New York, 1962.

KINETICS OF H^+/CU^{2+} ION EXCHANGE ON KPS AND KS-10 WOFATITS

Anna Jóźwiak and Zbigniew Gorzka

Institute of General and Ecological Chemistry
Technical University of Łódź

ABSTRACT

Ion exchange allows the recovery of water and heavy metals from exhausted electroplating baths and wastewaters. The rate of ion exchange depends on experimental conditions and is determined by diffusion processes.

Investigations of the kinetics of H^+/Cu^{2+} ion exchange were carried out on strongly acid cation exchangers: gel Wofatit KPS and macroporous Wofatit KS-10. Dependence of rate of H^+/Cu^{2+} ion exchange on temperature in the range from 303 to 363 K was determined at initial concentration of Cu(II) from 0.0125 up to 0.05 mole/dm^3 and at acidity of solution up to 0.5 mole H^+/dm^3. The data obtained illustrate the effect of increase in H^+ on H^+/Cu^{2+} exchange. This has no influence on the half-life time of ion exchange in Wofatit KS-10, and in the case of Wofatit KPS only at H^+ concentrations lower than 0.25 mole/dm^3. However, the concentration of H^+ ions has a significant influence on the amount of Cu^{2+} ions exchanged.

A simplified kinetic equation, according to equation of a first order chemical reaction, is given. This equation was used to determine rate constants of H^+/Cu^{2+} ion exchange; the average values proved the diffusion character of the process.

INTRODUCTION

Ion-exchange is used for water and heavy metal recovery from exhausted electroplating baths and wastewaters [1–5]. Its effectiveness is higher than the effectiveness of other classical methods of treatment of wastewaters from superficial metalworking. It enables heavy metals to be removed from wastewaters toxic for water environment. Closed loop washery effluent circulation in electroplating plants is associated with a decrease in production costs.

The possibility of ion-exchanger application for copper recovery from washery effluents depends among other things on the kinetics of H^+/Cu^{2+} ion exchange. In the kinet-

Chemistry for the Protection of the Environment 2, edited by Pawłowski et al.
Plenum Press, New York, 1996

ics of ion exchange, three quasi - phases can be distinguished in the ion exchanger - solution system: [6,7]:

1. solution inside the ion exchanger. There are active ions which compensate the charge of ionogenic functional groups,
2. static film layer,
3. external solution.

During ion exchange, active ions are displaced from the ion exchanger phase to solution. Ions from the external solution take their place in the ion exchanger. Ion transfer proceeds in the ion exchanger as well as in the film around the resin particles. Stirring influences neither the interior of the ion exchanger phase nor ion transfer through the film, where transfer proceeds by diffusion. The ion exchange rate depends on experimental conditions and is determined by diffusion processes.

If diffusion in the exchanger particle is the main parameter which determines the rate of ion exchange, then differences in concentrations in the film layer immediately decay. The concentration gradient is only in the particle.

If diffusion in the film of solution around the resin is the main parameter which determines ion exchange rate, then differences in concentrations inside the ion exchanger phase are quickly compensated. The concentration gradient is observed only in the film.

The process of ion exchange is controlled by diffusion in the film layer of a dilute solution without efficient stirring if ion exchangers are characterized by high capacity, low cross-linking degree and small particle size.

EXPERIMENTAL

In order to determine the course of H^+/Cu^{2+} ion exchange and parameters which influence its rate, experiments were carried out in a glass thermostated reactor fitted with a stirrer.

In experiments, two types of Wofatit resins produced by Chemiecombinat - Bitterfeld: gel KPS and macroporous KS-10 [4] were used. The weighed amounts of dry ion exchangers with known humidity were applied. They were allowed to swell in water for 24 hours. This procedure eliminated differences in sorptive capacity and the rate of attainment of equilibrium. The copper concentration during experiments was determined according to a known analytical procedure [8].

The effects of the initial concentration of copper ions, H^+ ions and the temperature on the ion exchange rate were determined.

RESULTS AND DISCUSSION

The dependence of the attainment of the degree of equilibrium (U) for H^+/Cu^{2+} ion exchange on time for different initial copper concentrations in the range from 0.0125 to 0.050 mole/dm^3 is presented in Table 1 and 2.

Attainment degrees of equilibrium state were calculated from the following equation:

$$U = (C_0 - C_\tau)/(C_0 - C_r)$$

Table 1. Degree of equilibrium attained versus τ for the kinetics of H^+/Cu^{2+} ion exchange in Wofatit KS-10 for different initial concentrations of Cu(II) ions. T = 298 K

Lp	τ[min] U	C_0 = 0.0125 mole/dm^3 m_j = 0.5394 g	C_0 = 0.025 mole/dm^3 m_j = 1.0789 g	C_0 = 0.0375 mole/dm^3 m_j = 1.6185 g	C_0 = 0.050 mole/dm^3 m_j = 2.1578 g
1	0	—	—	—	—
2	1	0.421	0.432	0.468	0.366
3	3	0.633	0.731	0.598	0.664
4	5	0.744	0.785	0.695	0.794
5	7	0.786	—	—	—
6	10	0.841	0.823	0.864	0.886
7	20	0.942	0.966	0.943	0.970
8	60	0.983	0.994	0.981	1.0
9	120	1.0	1.0	1.0	1.0

where:
C_0 - initial concentration of Cu(II) in a solution,
C_τ - Cu(II) concentration in a solution after contact for time τ with ion exchanger,
C_r - Cu(II) concentration in a solution in equilibrium state.

Attainment degrees of equilibrium state depend on initial copper(II) concentration in solution in the case of both tested ion exchangers. However, this dependence is only apparent for KS-10 Wofatit in the higher range.

The effect of H^+ counter-ion on the course of ion exchange in the range of H_2SO_4 concentrations from 0.0 to 0.5 mole/dm^3 is presented in Table 3 and 4. The increase in H^+ ion causes shortening the time of equilibrium attainment in the case of both ion exchangers. The concentration of H^+ ions has a significant effect, decreasing the amount of exchanged Cu(II) ions.

An increase in temperature causes an increase in rate in the case of both ion exchangers (Table 5 and 6). Temperature changes have no effect on the half-life time in exchange reaction in Wofatit KS-10. In the case of Wofatit KPS, the increase in temperature above 313 K causes a decrease in the half-life time. This confirms a change in reaction mechanism from film diffusion to diffusion inside the ion exchanger particle.

Table 2. Degree of equilibrium attained versus τ for the kinetics of H^+/Cu^{2+} ion exchange in Wofatit KPS for different initial concentrations of Cu(II) ions. T=298 K

Lp	τ[min] U	C_0 = 0.0125 mole/dm^3 m_j = 0.6052 g	C_0 = 0.025 mole/dm^3 m_j = 1.2104 g	C_0 = 0.0375 mole/dm^3 m_j = 1.8155 g	C_0 = 0.050 mole/dm^3 m_j = 2.4207 g
1	0	—	—	—	—
2	1	0.177	0.255	0.330	0.209
3	3	0.444	0.469	0.560	0.669
4	5	0.705	0.582	0.671	0.828
5	10	0.902	0.746	0.783	0.935
6	20	0.911	0.877	0.896	0.964
7	60	0.957	0.962	1.0	1.0
8	120	1.0	1.0	1.0	1.0

Table 3. Kinetics of H^+/Cu^{2+} ion exchange in Wofatit KS-10 for different concentrations of H_2SO_4 , showing degree attainment of equilibrium as a function of τ. T = 298 K, m_j = 2.1578g, C_0 = 0.05 mole Cu(II)/dm^3

Lp	τ[min] U	$C_{H_2SO_4}$ 0.05 mole/dm^3	0.25 mole/dm^3	0.50 mole/dm^3
1	0	—	—	—
2	1	0.371	0.422	0.175
3	3	0.651	0.645	0.480
4	5	0.768	0.773	0.673
5	10	0.846	0.935	0.938
6	20	0.948	0.988	0.990
7	60	1.0	1.0	1.0
8	120	1.0	1.0	1.0

Table 4. Kinetics of H^+/Cu^{2+} ion exchange in Wofatit KPS for different concentrations of H_2SO_4, showing degree attainment of equilibrium as a function of τ. T = 298 K, m_j = 2.4207g, C_0 = 0.05 mole Cu(II)/dm^3

Lp	τ[min] U	$C_{H_2SO_4}$ 0.05 mole/dm^3	0.25 mole/dm^3	0.50 mole/dm^3
1	0	—	—	—
2	1	0.460	0.627	0.532
3	3	0.677	0.872	0.726
4	5	0.849	0.947	0.768
5	10	0.946	0.962	0.823
6	20	0.975	0.980	0.871
7	60	1.0	1.0	0.957
8	120	1.0	1.0	1.0

Table 5. Kinetics of H^+/Cu^{2+} ion exchange in Wofatit KS-10 at different temperatures, showing degree attainment of equilibrium as a function of τ. C_0 = 0.05 mole Cu(II)/dm^3, m_j=2.1578g

Lp	τ [min] U	T = 303 K	T = 313 K	T = 323 K	T = 363 K
1	0	—	—	—	—
2	1	0.266	0.254	0.260	0.457
3	3	0.614	0.595	0.716	0.750
4	5	0.754	0.762	0.802	0.823
5	10	0.890	0.873	0.895	0.925
6	20	0.948	0.967	0.934	0.963
7	60	0.984	1.0	1.0	0.989
8	120	1.0	1.0	1.0	1.0

Table 6. Kinetics of H^+/Cu^{2+} ion exchange in Wofatit KPS at different temperatures, showing degree attainment of equilibrium as a function of τ. $C_0 = 0.05$ mole Cu(II)/dm^3, $m_j = 2.1578$g

Lp	τ[min] U	T = 303 K	T = 313 K	T = 323 K	T = 363 K
1	0	—	—	—	—
2	1	0.276	0.157	0.159	0.337
3	3	0.648	0.609	0.704	0.602
4	5	0.907	0.776	0.806	0.781
5	10	0.983	0.922	0.953	0.920
6	20	0.987	0.968	0.974	0.972
7	60	1.0	1.0	0.996	1.0
8	120	1.0	1.0	1.0	1.0

The dependence between U and τ can be presented in the form of the following equation:

$$U = 1 - e^{-k\tau}$$

This simplified kinetic equation is satisfied for conversion degrees in the range from 0.44 to 0.99. Certain deviations observed at the beginning of the process are caused by changes in ion exchange mechanism.

Determination of rate constants for the reaction (Table 7 and 8) at different temperatures enables one to calculate the apparent activation energy of ion exchange.

Its average values are as follows:

for Wofatit KS-10 - 17.8 kJ/mole,
for Wofatit KPS - 5.1 kJ/mole.

Table 7. Rate constants of exchange reaction (H^+/Cu^{2+}) in Wofatit KS-10. σ - standard devation, k_0 - correlation coefficient

Parameter	k[min^{-1}]	k_0	$\sigma * 10^3$
C_0 Cu(II) =			
0.0125 mole/dm^3	0.105	0.997	0.80
0.0250 "	0.123	0.978	0.96
0.0375 "	0.114	0.988	0.88
0.050 "	0.137	0.966	1.05
C_{H+} =			
0.00 mole/dm^3	0.043	0.995	1.03
0.05 "	0.110	0.995	0.84
0.25 "	0.199	0.995	1.51
0.50 "	0.234	0.990	1.79
T =			
303 K	0.114	0.968	0,89
313 K	0.142	0.995	1.08
323 K	0.082	0.959	0.65
363 K	0.111	0.974	0.80

Table 8. Rate constants of exchange reaction (H^+/Cu^{2+}) in Wofatit KPS σ - standard devation, k_0 - correlation coefficient

Parameter	k[min^{-1}]	k_0	σ * 10^3
C_0 Cu(II) =			
0.0125 mole/dm^3	0.137	0.996	1.04
0.0250 "	0.108	0.995	0.82
0.0375 "	0.082	0.994	0.62
0.050 "	0.122	0.940	0.98
C_{H+} =			
0.00 mole/dm^3	0.122	0.940	0.99
0.05 "	0.141	0.949	0.13
0.25 "	0.120	0.950	0.95
0.50 "	0.043	0.983	0.33
T =			
303 K	0.173	0.857	1.55
311 K	0.142	0.970	1.11
323 K	0.141	0.937	1.11
363 K	0.150	0.978	1.17

These values prove the diffusion character of the process.

CONCLUSIONS

In the case of macroporous Wofatit KS-10, the process of ion exchange is controlled by internal diffusion. Wofatit KPS as a gel ion exchanger shows greater changes of volume with the change of ionic form.

Therefore, the rate process depends to different degrees on internal diffusion and diffusion in the surrounding film. Both exchangers can be used to adsorb Cu(II) from solutions such as washery effluents.

In solutions with low pH values, Wofatit KS-10 has more favourable features due to its higher rate of equilibrium state attainment. In solutions with low concentration of H^+ ions, Wofatit KPS has more favourable features.

REFERENCES

1. Hayashi T., Nojisi K.; Regeneration of chromic plating bath, IP 7930950 (1977)
2. Świderska-Bróż M.; Usuwanie metali ciężkich w procesach fizyczno-chemicznych stosowanych w ochronie wody, GWTS, 55, 185–188, (1981)
3. Lipsztajn M.; Electrochemical removal of chromium from chlorate solutions, EP 267704 (1988)
4. Gorzka Z., Jóźwiak A., Jasińska K., Socha A.; Regeneration of exhausted chromium baths, Environment Protection Engineering, 3–4, 16, (1990)
5. Jóźwiak A., Gorzka Z.; The use of the ion exchangers for the recovery of nickel from washery effluents after the process of nickel plating, Conf. Proc. 8th International Conference, Chemistry For Protection of the Environment, Lublin 1991
6. Winnicki T.; Polimery czynne w inżynierii ochrony środowiska, Arkady, W-wa, 1978
7. Pawłowski L., Hefty J.M.; Wodociągi i kanalizacja, Arkady, W-wa 1978, 81–154
8. Hermanowicz W. i in.; Fizykochemiczne badania wody i ścieków, Arkady, W-wa, 1976

DEOXYGENATION AND DESALINATION OF WATER ON DIFUNCTIONAL STRONG BASIC ANION EXCHANGER

R. Tosik

Institute of General and Ecological Chemistry
Technical University of Łódź
36 Żwirki Street, 90–924 Łódź, Poland

ABSTRACT

Corrosivity of natural water is controlled by concentration of soluble oxygen and amounts of mineral salts. In order to decrease the corrosivity of water, deoxygenation and demineralization methods are applied. This article presents investigations on oxygen and mineral salts removal by ion exchangers. A special kind of bed which contained two ionic forms: sulphite and hydroxyl was investigated for simultaneous removal of oxygen and anions from water. Natural water from a deep well supplied the bed after its initial decationization on a strong acid cation exchanger. The model solutions were used with hydrochloric and sulphuric acids in amounts equal to decationization of water from the well.

The removal efficiency of oxygen and anions was very high. The bed may be used as a difunctional ionic column for the simultaneous removal of oxygen and anions if a demineralization process is designed by the ion exchanger's method. The corrosivity of water can be eliminated and technological procedures should be simplified using such a bed.

1. INTRODUCTION

In oxygenated natural water, corrosivity depends on electrochemical processes resulting from soluble oxygen and mineral salts. Oxygen causes a cathodic reaction which has a potential much higher than hydrogen electrodes and is a more strongly depolarized agent than hydrogen ions [1]. Oxygen controls corrosion also by production of hydroxyl ions—eq 1—that take a role in multistage anodic reactions of iron corrosion, particularly in the slowest step which is markedly sped up by hydroxyl ions eq 2 [2, 3]:

Chemistry for the Protection of the Environment 2, edited by Pawłowski et al.
Plenum Press, New York, 1996

Table 1. The requirements of quality for feeding water

Source	Oxygen mg/L	Conductance μS/cm	pH	Hardness meq/L	Carbon dioxide, mg/L
Steam generator feedwater water treatment handbook [5]	< 0.005	3–13	9–9.2		
Instruction-feeding water quality polish producer of boilers pressure 20–42 bar [6]	< 0.02		7–9.5	< 0.00357	< 1

$$O_2 + 2\,H_2O + 4\,e^- = 4\,OH^- \tag{1}$$

$$(FeOH)ads + OH^- = FeO + H_2O \tag{2}$$

It is known that chloride and sulphate ions are also very aggressive in causing individual specific effects as well as increasing water electrical conductivity [4]. When oxygen is present in the water that effect is much greater.

Deoxygenation and demineralization of the natural water are often required for many industrial purposes. The treated water, depending on working parameters, must not exceed the maximum indices values which are in a criterion published by producers of water equipment [5,6]. In it the concentrations of oxygen and mineral salts are the main indices which are limited in the treated water. The examples of the requirements are presented in Table 1.

2. PROBLEM

Removal of the oxygen and the mineral salts causes elimination or the decrease of water corrosivity . Industrial processes of water demineralization are well known such that there are no problems from a technological view point. However, removal of oxygen from the water is still being investigated and modified [7]. One can observe improvements in the processes of oxygen reduction using advanced reduction catalysts [8]. These methods are applied as separate steps of oxygen reduction from water that was treated initially. If demineralization is made by the ion exchange processes, for instance, the technological installation consists of two steps of the treatment, namely cation and anion removal. Thus, the process of the water treatment together with the deoxygenation process includes three steps. Sometimes, a three step system is not neccessary and then it is difficult to execute. It might be used, for example, for an installation with a small capacity or a system where thermal process cannot be applied.

Thist article presents the results of investigations upon a new method of oxygen removal with simultaneous demineralization of water. As the basic process of the water treatment we assumed an on exchange in order to obtain the elimination of its corrosivity as the result of salts and oxygen removal. A strong basic anion exchanger was used in two layers—upper and lower—in sulphite and hydroxyl forms, respectively. The column was fed with the water containing chloride and sulphate acids in the amount equal to values of the decationization process of natural water by strong cation exchanger. The water was saturated with oxygen. The removal efficiency of oxygen and anions in the two layers bed and some operational parameters were determined. Some reactions are presented in the context of explanation of redox and ion exchange processes in the bed.

3. METHODS AND MATERIALS

Investigations were conducted in laboratory scale equipment detailed in the earlier work [9]. The apparatus consisted of the water supply system with flow meter and the ion exchanger's column with special construction, and it was equipied with the following meters: pH, the concentration of oxygen and the conductivity. The measurements were made according to the Polish Standards [10]. At the same time, concentration of residual oxygen in the effluent was measured by colorimetric indigo method, as well [11]. The later was used for determination up to value 100 μg/L O_2 in the period before and after breakpoint.

Experiments were made during continuous flow of water feeding according to the ion exchange standard technique for determining breakthrough curves.

The strong basic anion exchanger II type Wofatit SBK 0.3 L made in Germany was used in the column with a perforated bottom, of 28 mm diameter and 1 m high. Due to special construction of the column, regeneration of the bed by two the same volumes 0.15 L ionic forms, upper—sulphite and lower—hydroxyl, was possible. The bed was regenerated by the solutions of 4 % natrium sulphite and 4 % natrium hydroxide in divided standarized flow operations through each layer during 45 minutes and after that rinsing was done.

Cycles of ion exchange were made for each of the investigated waters and then the effluent was determined with the assumed program.

As the supply waters were used the HCl solutions of conc. 1.1 meq/L and HCl + H_2SO_4 3.7 meq/L (1.1 + 2.6 meq/L, respectively), and the natural water from the deep well after decationization on strong acid cation exchanger, also. The specific velocity of the water was 20 BV/h of water or flow rate 10 m/h. That water had 3.5 meq/L of the mineral acidity after decationization on the strong cation exchanger in the hydrogen form Wofatit KPS, made in Germany.

The breakpoints of the bed both oxygen and salinity were determined and all breakthrough curves were obtained.

Full characteristics of water deoxygenation and demineralization were made with measurement of indices required for evaluation of both processes running in two ionic form layers of the bed. The amounts of oxygen and anions which were removed by each ionic form of the bed—sulphite and hydroxyl—were estimated. The values of utilization degree of total capacity were calculated from experimental findings. Among others they were applied basic indices of the process such a volume of deoxygenation and demineralization of water. These values are presented in relation to the ionic form as well as to total capacity of the bed, the later named "general".

4. RESULTS

The results are presented in Table 2 and Table3. They characterize both processes in the bed, deoxygenation and demineralization.

As we may see from Table 2 the efficiency of oxygen removal was very high because residual concentration of oxygen was zero and 0.025 mg/L for the model solutions with 1.1 meq/L HCl and 3.7 meq / L HCl plus H_2SO_4, respectively. A very good effect of deoxygenation for water from the well was obtained. The trace amount of oxygen 5 ppb in the effluent has been stated, see Table 3. These findings, in comparison with high degrees of acids removal which were obtained for all supplied waters, pointed out the very good

Table 2. The results of oxygen and acids removal from the model solutions

		HCl solution 1,1 meq/L		$HCl + H_2SO_4$ solution 3.7 meq/L (1.1 + 2,6)	
Indices	Unit	Influent	Effluent	Influent	Effluent
Oxygen concentration	mg / L	7.68	0.000	7.36	0.025
Mineral acidity	meq / L	1.1	0.00	3.7	0.00
Conductance	µS / cm	500	4–6	1920	4.5–6.1
pH		2.95	7.5–8.5	2.4	8.1–8.3
Volume of treated water	Bed volume B V cu m / cu m	300			100
Utilization degree of total capacity in relation to form	%				
sulphite		60.4		19.3	
hydroxyl ion		62.9		70.5	
general		61.7		46.0	

acting of the two layer bed in the simultaneous treatment process. Volumes of treated water are different and they depend on the amount of acids contained in the water. It is obvious that at higher amounts of acids volumes of treated water were lower.

Operation capacity was estimated in relation to each of the ionic forms of the bed. We can see, that the sulphite form removed only oxygen, and gave a decrease of its percentage of utilization degree of capacity with an increase of acid amounts in the influent. For higher values of acids it was only ca 20 %. However, the same values for the hydroxyl form were very high, better than 60 % or even above 70 and 80 % for the highest amount of acids. Lastly, the general indices of utilization degree in relation to total capacity of all volumes of ion exchanger had average values in the range from ca 46 % to ca 61 % as amounts of acid decreased.

5. DISCUSSION

Difunctional anion exchanger in the sulphite and hydroxyl ion forms is capable of removal of oxygen and anions from waters that contain acid ions. The sulphite form basi-

Table 3. The results of oxygen and acids removal from the well water after decationization and aeration

Indices	Unit	Influent	Effluent
Oxygen concentration	mg/L	7.36	0.005
Mineral acidivity	meq/L	3.5	0.00
Conductance	µS/cm	2100	2–6.2
pH		2.4	7.1–7.9
Volume of treated water	Bed volume B V cu m/cu m		140
Utilization degree of total capacity in relation to form	%		
sulphite		60.4	
hydroxyl ion		62.9	
general		61.7	

caly removed soluble oxygen but that very high efficiency was probably caused by its combination with hydroxyl form. Anion exchange of chloride and sulphate ions was in hydroxyl form of the bed and a good effect of water deanionization was obtained. That effect was the same as in conventional demineralization processes of water by ion exchangers.

The results obtained for the model solutions of acids were in good rclation with the findings for natural water from the well. This good agreement corroborates the presumption that mechanism of process is with similar.

The lower values of utilization degree of total capacity of the ion exchanger obtained in the experiments in comparision with other operation of water demineralization may be explained as a result of "no opitimization" of reaction conditions in the applied bed. Optimization has to be done in the future, as well as elucidation funtions of the ionic forms of the bed in chemical mechanisms. At this time we may only roughly illustrate processes in the difunctional bed.

If quoted with decationization processes of the natural water on the strong acid cation exchanger, the summarized reactions are as follows:
on strong cation exchanger in hydrogen form

$$CaSO_4 + 2\,Kt\,H \Rightarrow Kt_2Ca + H_2SO_4 \tag{3}$$

$$CaCl_2 + 2\,Kt\,H \Rightarrow Kt_2Ca + 2\,HCl \tag{4}$$

$$Ca(HCO_3)_2 + 2\,Kt\,H \Rightarrow Kt_2Ca + 2\,CO_2 + 2\,H_2O \tag{5}$$

on difunctional bed

$$An_2SO_3 + 1/2\,O_2 \Rightarrow An_2SO_4 \tag{6}$$

$$AnOH + HCl \Rightarrow AnCl + H_2O \tag{7}$$

$$2\,AnOH + H_2SO_4 \Rightarrow An_2SO_4 + H_2O \tag{8}$$

Decationization reactions eq 3–5, also occur for other cations contained in the natural water such as Na, K, Mg and the effluent contains the same mineral acids and soluble carbon dioxide. Desorption of carbon dioxide is made by means of aeration process and after that the water is saturated with oxygen.

It must be mentioned that equations 6–8 illustrate only the final reactions in the bed, but effective processes are more complicated. These assumptions were useful for the calculation of ion and mass balances in order to estimate the characteristic process indices. Such assumptions were correct, because the calculations were made for all volumes of the bed. Obviously the precise mechanisms of the processes in the difunctional bed should be explained and it is neccessary for optimizing utilization's total capacity of the ion exchanger.

6. CONCLUSIONS

The difunctional bed with sulphite and hydroxyl forms gave a high efficiency of oxygen and anions removal from the acid solutions. It may be used in the process of water demineralization by the ion exchanger methods.

Such bed, in conjunction with decationizing of water on a strong acid cation exchanger gave high deoxygenation and demineralization of the natural water from the deep well.

Degrees of demineralation and deoxygenation of water are similar as they may be obtained in the standard processes with the water treatment in higher parameters of its application (for instance as boiler supply water).

Application of the difunctional bed will cause simplification of the water treatment systems using ion exchangers, because in the anion removal step by strong basic anion exchanger, oxygen may be also removed.

The difunctional bed may be very useful when the water must be deoxygenated without thermic processes, after which "cold process" must be applied.

The efficiency of water demineralization with the bed is the same as one can obtain using the conventional ion exchanger's method. Simultaneous removal of oxygen from water will cause the elimination of its corrosivity properties.

It may be concluded that the proposed method of water treatment might be characterized with good operation parameters if the mechanisms of the processes were better known.

The author patented the method of simultaneous demineralization and deoxygenation of natural water using this bed [12].

7. REFERENCES

1. Water Treatment Handbook, ed. Degremont 1991 p.422
2. J.O.M. Bockris, A.K.N. Reddy: Modern Electrochemistry Plenum Rossete Edition, New York 1970 p. 1091
3. B. Andrzejaczek, Ochrona przed korozją XXVII 2, 25 1984
4. R.A. Legut, S. Mori and H.P. Leckie Corrosion vol. 26 No 7, 1970
5. Water Treatment Handbook ed. Degremont 1991 p. 57
6. Instruction -Feeding Water Quality, Sosnowieckie Zakłady Budowy Kotłów, Polish Producer of Boilers
7. F.B. Martinola, P. Tomas, Saving Energy by Catalytic Reduction of Oxygen in Feedwater, International Water Conference, Pittsburgh, Pensylwania, October 20–22, 1980
8. W.S. Miller, Oxygen Removal by Catalyzed Carbon Beds, EPRI Condensate Polishing Workshop, October 29–91, 1985 Richmond Virginia
9. R. Tosik, "Badanie procesu jednoczesnego odtleniania i demineralizacji wody metodą jonitową " Report Institute of General and Ecological Chemistry, Technical University of Łódź, Poland, 1994
10. Polish Standards, Determination of Water Quality
11. L. Buchoff, Colorimetric determination of low concentration of dissolved oxygen in water Anal. Chem. 27, 1401, 1955
12. R. Tosik "Sposób jednoczesnego demineralizowania i odtleniania wody" PL 164523 B1, 31.08.1994 WUP 08/94, IntCl5: C02F 1/42

16

DECOLORING MECHANISMS OF DYE STUFF BY CHEMICAL OXIDATION

Cheng-Nan Chang,[1] Jih-Gaw Lin,[2] Allen C. Chao,[3] and Chu-Sung Liu[1]

[1] Graduate Institute of Envi. Science, Tunghai University
Taichung, Taiwan 40704, ROC
[2] Graduate Institute of Envi. Engineering, Chiao-Tung University
Hsin Chu, Taiwan, ROC
[3] Department of Civil Engineering, North Carolina State University
Raleigh, North Carolina

ABSTRACT

This study utilizes sodium hypochlorite (NaOCl) to reduce the color of synthetic dye solutions. Five dyes including methyl red (MR), methyl violet (MV), methyl blue (MB), malachite green (MG) and methyl orange (MO) were studied. The color was measured using an automatic ADMI (American Dye Manufacturer Institute) measurement system. A visible spectrophotometer is connected to a personal computer so that samples can be scanned and the results are recorded automatically. During the batch decoloring study, several factors: ADMI, ORP, temperature and pH were also monitored continuously using the computer and makes it possible to have an automated decoloring process.

A modified Nernst equation was used to delineate the relationship between the degree of decoloring reaction and the solution ORP value. A "decoloring coefficient" (S) was proposed to represent the oxidative potential (or decoloring property) of the dyes studied. In this study, the oxidative or decoloring potential can be shown in the following order: MG>MV>MO> MR>MB. Using the on-line measurements of ORP, color and pH, the oxidation process can be automated to achieve a pre-selected percentage of color removal.

INTRODUCTION

Many industries discharge visible color that causes severe environmental pollution problems. Normally, organic matter in the effluent can be destroyed or removed using secondary biological treatment processes, but the color must be reduced by use of physicochemical methods [1]. Although color in many industry effluents is traditionally considered

Chemistry for the Protection of the Environment 2, edited by Pawłowski et al.
Plenum Press, New York, 1996

harmless and is usually not removed from the final effluent discharge, it is aesthetically unacceptable and also interferes with the biological process in receiving bodies. With recently more stringent effluent color standards to be implemented in Taiwan, the discharge of industrial effluent with visible color will not be accepted in the future, and the same is expected for many other Asian countries.

Among the many physicochemical methods such as activated carbon adsorption, oxidation, coagulation [2,3], etc. proposed for color removal from wastewater, chemical oxidation is considered to be the most cost-effective alternative. The chemical oxidants which is to destroy the chemical bonds that cause color in dye stuff by Perkin *et al.* [4] include : ozone, chlorine and sodium hypochlorite.

The treatment cost of all chemical oxidation methods depends primarily on the quantity of the oxidant consumed. Generally speaking, more chemical addition will result in a better effluent quality but the treatment cost may greatly increase. Thus, the most cost-effective operation of the oxidation process is after exceeding a threshold concentration, to apply the minimum chemical dosage to reduce the effluent color to an acceptable level. This requires that the chemical oxidation property of the dye stuff be understood and that a system to adjust the chemical dosage in response to changes of the chemical reaction be available. In the past, it was almost impossible to assemble a device to achieve such an automatic dosage control mechanism. With the advent of less expensive and more powerful personal computers, it is feasible to implement an on-line color measurement and feed-back device to automatically control the chemical dosage.

ON-LINE COLOR MEASUREMENT

The studies to be presented in this paper concentrate on on-line monitoring of the parameters relevant to color removal and using these parameters for achieving automation of the decoloring process. For color measurement, there are several methods available for determining the color in water samples. According to the method included in "Standard Methods" [5], the sample color is compared to a series of dilutions of a standard Pt-Co solution. This method has been traditionally used for measuring the color in drinking water and the results are expressed in unit or equivalent concentration of the Pt-Co solution. It is used in the Philippines for the effluent color limitations (100 Pt-Co unit). For many south-eastern Asian countries, however, the Lovibond (LV) method developed in England is commonly used.

In the field application, the visibility of a sample can be obtained by viewing a special cross sign through the water sample. The maximum column height (cm) through which the cross can be clearly seen is inversely proportional to the color being measured. In Taiwan, this "visibility" method is used and the effluent color limitation is currently set at 15 cm.

The American Dye Manufacturers Institute (ADMI) has developed the ADMI method based on Nikerson's chromatic formula for a more scientific measurement of the color in textile wastewaters [6]. Both the laboratory procedures and data calculations of this method are considered very tedious, if performed manually. It has not been applied to field practice except in laboratory studies. Using a personal computer, however, the ADMI method can be automated [7]. The measurements can be quickly obtained and the complicated calculations can be instantaneously performed by use of the computer. Thus on-line measurement of the sample color is made possible with this method. There is a linear relationship between the ADMI color and the visibility color value. The visibility value of 15 cm currently enforced in Taiwan corresponds to an ADMI value of 567 units.

On-line measurements of color are carried out by pumping the liquid with an Eyela MP-3 micro-tube pump into a 1-cm quartz cell which is placed inside a Hitachi U-2000 Spectrophotometer. The range of scanning wavelengths is set from 400 nm to 700 nm with a 10-nm interval. A total of 31 sets of data can be obtained for each color measurement and the signal is transmitted to a personal computer for recording and processing the measured ADMI data. Using the computerized data acquisition system, automatic on-line monitoring of the color variation of the water sample being treated can be accomplished. With appropriate arrangements, the computerized monitoring system can be used as a feed-back control of the chemical dosing of the oxidant to reach a selected standard of color removal.

THE OXIDATION/REDUCTION POTENTIAL

The oxidation process is known to be caused by transfer of electrons from the oxidant to the substance being oxidized. Nernst Equation can be used to relate the Oxidation/Reduction Potential (ORP) of the system to the concentration ratio of the oxidant and the substance being oxidized as [8] :

$$E=E^0 + (RT/nF) \ln ([Oxid]/[Red]) \quad (1)$$

Where: R = gas constant (1.98 cal/mol K).
T = the absolute temperature (K).
n = the number of electrochemical gram equivalent per gram mole exchanged during the redox reaction (mol).
F = Faraday's constant (96,500 coulombs/mol).
E^0 = the standard electrode potential (mV).
[Oxid] = the molar concentration of oxidant (mole/L).
[Red] = the molar concentration of reductant (mole/L).

According to Eq.(1), the system Oxidation Reduction Potential (ORP or E) is indicative of the ratio of the substance that is being oxidized to its oxidized form. The ORP measurement can be applied for process control of both chemical and biological oxidation/reduction reactions. Charpentier *et al.* [9] used the ORP measurement as a control parameter to optimize the aeration of a pilot-scale plant. They reported a correlation between the ORP value of the aeration tank and the effluent TKN and NO_3^-. In addition to biological systems, the ORP measurement is also applicable to controlling the chemical oxidation/reduction wastewater process. It has been proposed for the breakpoint chlorination to remove cyanide and ammonia nitrogen as well as to treat chromium sludge [10] and metal finishing wastewater. Chang *et al.* [11] applied a modified Nernst equation to simulate the decoloring reaction of secondary textile effluent and obtained satisfactory results. However, due to a lack of both theoretical bases and automatic ORP measurements in the past, the ORP technique has not been widely accepted for wastewater treatment.

Using on-line measurements of ORP, color and other parameters, a feed-back control of the decoloring process has been pioneered and successfully carried out in our laboratory. Variations of the system ORP and its relationship to other system parameters under different experimental conditions will be studied with the ultimate goal of applying the

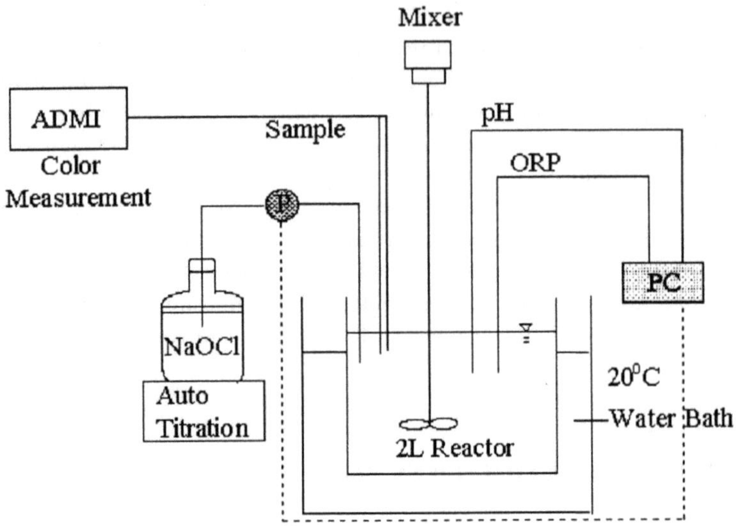

Figure 1. Automatic NaOCl titrator and monitor system of decolor apparatus.

ORP technique as the on-line monitoring and control parameter of the dye decoloring process. In this paper, comparison of the oxidative or decoloring potentials of five representative textile dyes and use of Nernst equation for process control are presented.

EXPERIMENTAL PROCEDURES

Five synthetic basic dye stuffs: methyl red (MR), malachite green (MG), methyl violet (MV), methylene blue (MB) and methyl orange (MO) purchased from Merck Inc. are used to prepare the synthetic dye wastewater. The solid dye is dissolved in distilled water in a 2-L reactor with an initial concentration of 10 mg/l.

Figure 1 shows the schematic drawing of the experimental apparatus. The initial solution pH is adjusted to different levels (e.g. 5, 8 and 10) by addition of NaOH or H_2SO_4, without using buffer solution and the reactor temperature is controlled at 20^0C with a water bath. For on-line monitoring and control of the chemical process, the ORP probe (AgCl reference) and pH electrode are connected to a personal computer which is linked to a Schott Titronic T200 automatic titrator. Both ORP and pH signals of the reactor content are continuously recorded and calculated. Meanwhile a control signal is sent to the automatic titrator to control addition of sodium hypochlorite with a rate of 0.024 mg/l until a pre-selected ORP value of 900mV is reached.

RESULTS AND DISCUSSION

Influence of pH and NaOCl Dosages on ORP

Fig.2 shows typical variations of the pH, ORP and color of the dye stuff solution undergoing decoloring oxidation studies with addition of sodium hypochlorite. During the initial reaction period, all the measured parameters show rapid changes. An increase of the

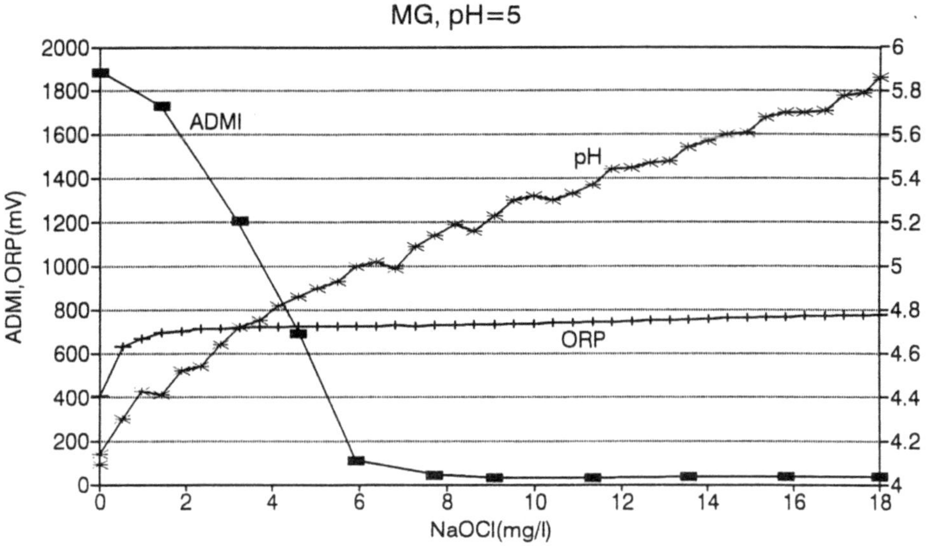

Figure 2. Variations of the ADMI unit and the ORP value of the dye stuff sample with NaOCl.

sample pH and ORP was observed while the measured color was noted to drop. After the initial active reaction period, both color (expressed as the ADMI value) and ORP level off and change slightly until the end of the test period. This initial vigorous oxidation may be caused by the reaction of free chlorine.

ORP vs. pH and NaOCl

The ORP value is related to the NaOCl dosage and solution pH by a linear equation shown as:

$$ORP = a + b\ pH + c\ \ln[NaOCl] \quad (2)$$

Where: a = intercept
b = the characteristic parameter of pH
c = the characteristic parameter of oxidant

In Eq.(2), the sign and value of "b" and "c" indicate how solution pH and NaOCl dosage affect the system ORP value. Addition of NaOCl will generally cause the system ORP value to increase. Thus, the "c" value in Eq.(2) is always positive, except for the MG dye. Because of a negative "c" value in Eq.(2) for MG dye, during the initial reaction period of sodium hypochlorite oxidation, its ORP was observed to decrease. The influence of solution pH on the system ORP is indicated by the value of "b" in Eq.(2). If b is negative, a lower pH will result in a higher ORP value than a higher pH value, and vice versa. The values of "a", "b" and "c" for all dyes as obtained using regressional analyses on laboratory results are listed in Table 1.

Table 1. Calibrated equations relating ORP to other parameters for different dyes

Dye	pH	Model equation	R^2
Red (MR)	5	ORP=1161-64pH+71ln[NaOCl]	0.909
	8	ORP=1413-96pH+86ln[NaOCl]	0.996
	10	ORP=-6633+707pH+16ln[NaOCl]	0.831
Green (MG)	5	ORP=315+81pH-17ln[NaOCl]	0.971
	8	ORP=752-22pH+97ln[NaOCl]	0.983
	10	ORP=1757-140pH+125ln[NaOCl]	0.995
Violet (MV)	5	ORP=683+21pH+28ln[NaOCl]	0.992
	8	ORP=1916-168pH+74ln[NaOCl]	0.878
	10	ORP=12134-1198pH+46ln[NaOCl]	0.723
Orange (MO)	5	ORP=-29+141pH+81ln[NaOCl]	0.886
	8	ORP=5447-568pH+147ln[NaOCl]	0.961
	10	ORP=-1.16+1199pH+291ln[NaOCl]	0.829
Blue (MB)	5	ORP=1174-67pH+60ln[NaOCl]	0.993
	8	ORP=3673-373pH+152ln[NaOCl]	0.982
	10	ORP=1757-140pH+125ln[NaOCl]	0.995

Modified Nernst Equation

The Nernst equation is used to generalize the oxidation/reduction reaction for on-line monitoring and control of the decoloring process. If the ORP value of the wastewater sample prior to addition of NaOCl is taken as the initial electrochemical potential value (E^0) as shown in Eq.(1), then E in the same equation represents the ORP value of the sample after addition of chemical oxidant. The ADMI value of the sample after the oxidation reaction is used to indicate the concentration of the dye stuff in its reduced form, or the [Red] term in Eq(1), while the sodium hypochlorite concentration represents the oxidant form or the [Oxid] term in the same equation. Hence, the Nernst equation or Eq.(1) can be modified as:

$$ORP_0\text{-}ORP = -(RT/nF)\ \ln\ (ADMI_t/NaOCl) \quad (3)$$

Where: ORP = sample ORP value before addition of NaOCl (mV)
ORP_0 = sample ORP value after addition of NaOCl (mV)
$ADMI_t$ = sample color after addition of NaOCl (ADMI)
NaOCl = concentration of NaOCl (mg/L)

The above equation can be re-written in the following form:

$$ORP\text{-}ORP_0 = S\ \ln\ (ADMI_t/NaOCl),\ \text{and}\ S = RT/nF \quad (4)$$

The parameter, "S", is an index representing the oxidation characteristics of the dye being studied. A dye with a higher S value can be oxidized by the oxidant at a faster rate and vice versa. It corresponds to the parameter "K" proposed in a previous paper [11]. The difference is that "S" is obtained by using $ADMI_t$, which is the color remaining at time t, instead of $ADMI_0$ as the oxidant part of Nernst equation. Using $ADMI_t$ and [NaOCl] terms will result in a better simulation of the overall oxidation/reduction reaction. The magnitude of S values is dependent on the type of dyes and the initial solution pH. Fig. 3 shows variation of all S values for the five dyes studied vs. sodium hypochlorite dosages applied.

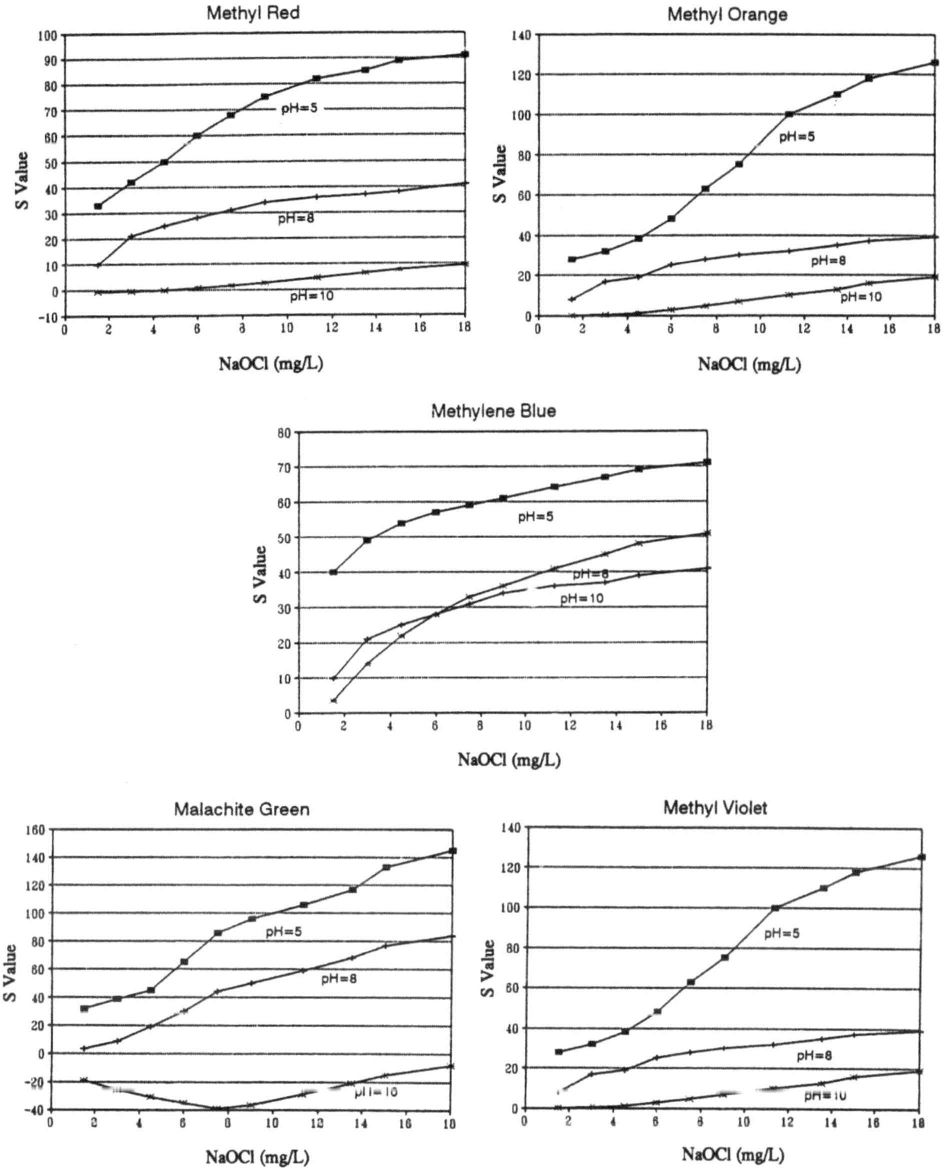

Figure 3. The S values of 5 dyes under various pH conditions.

Table 2 lists the equation of S for different dyes and solution pH values. These equations are calibrated using the laboratory results.

For lower initial pH (=5), all five dyes show higher S values than at higher pH of 8 and 10. In the low pH range, hydrogen ions combine with the hydroxyl ions of dyes and the decoloring reaction is enhanced. Where the type of dyes is concerned, MG has the highest S values of 18.13:10.15:-1 (for pH=5, 8, 10, respectively), while MB has the lowest ratio range with 1.39: 0.8: 1 (for pH=5, 8, 10, respectively). Using the S value for pH 5 as a basis for comparison, the decoloring coefficients (S) are found to have the following order of magnitude: MG(145), MV(126), MO(126), MR(91) and MB(71). If the S value of

Table 2. Calibrated equations relating S to pH and oxidant dosages for different dyes

Dye	pH	Model equation	R^2
Red (MR)	5	S=-11+13.6pH+15.7ln[NaOCl]	0.980
	8	S=239-23pH+25ln[NaOCl]	0.978
	10	S=-1292+129pH+2.1ln[NaOCl]	0.781
Green (MG)	5	S=-482+110pH-13ln[NaOCl]	0.961
	8	S=-527+75pH+18ln[NaOCl]	0.984
	10	S=680-64pH+20ln[NaOCl]	0.995
Violet (MV)	5	S=-437+101pH-7ln[NaOCl]	0.940
	8	S=-478+67pH+9.4ln[NaOCl]	0.989
	10	S=3458-349pH+12.9ln[NaOCl]	0.658
Orange (MO)	5	S=-395+88pH-2.13ln[NaOCl]	0.960
	8	S=-59+10pH+11ln[NaOCl]	0.995
	10	S=-1841+185pH+4.36ln[NaOCl]	0.797
Blue (MB)	5	S=45+2.6pH+11ln[NaOCl]	0.997
	8	S=203-20.9pH+16.8ln[NaOCl]	0.996
	10	S=680-64pH+20ln[NaOCl]	0.995

MB is assumed to be 1, the decoloring capabilities (or S value) of MG, MV, MO, MR and MB are of 2.04: 1.78: 1.78: 1.28: 1.00, respectively.

The Process Control Parameter

Values of the parameters ORP, ORP_0 and S are all related to the dosage of NaOCl and can be combined into a single parameter P as:

$$P=(ORP-ORP_0)/S \tag{5}$$

Combining Eq.(4) and Eq.(5), P can be expressed as :

$$P=\ln(ADMI_t/NaOCl) \tag{6}$$

Rearrange the term in Eq.(6), the color remaining at any time t ($ADMI_t$) can be related to the NaOCl dosage by :

$$ADMIt=(NaOCl)e^P \tag{7}$$

The parameter P which is a function of the initial solution pH and the NaOCl dosage applied, relates the $ADMI_t$ value of the treated sample to oxidant dosage. Thus, it is a practical parameter for control of the degree of the oxidation reaction to achieve a certain level of color remaining in the treated sample. Table 3 lists the equation of P with calibrated coefficience using laboratory results for different dyes and solution pH's.

Laboratory studies to verify the validity of the proposed model equations have been carried out. In these verifying studies, dosages of NaOCl for reducing the color to different pre-selected ADMI values and the expected ORP values can be calculated using the proposed model equations. These predicted values are then compared to the values actually obtained with laboratory oxidation studies. The results show that for all the oxidation studies, the predicted NaOCl dosage requirement and the ORP values of the decolored

Table 3. Calibrated equations relating P to pH and oxidant dosages for different dyes

Dye	pH	Model equation	R^2
Red (MR)	5	P=3.1+0.28pH-1.97ln[NaOCl]	0.994
	8	P=59-6.3pH-0.21ln[NaOCl]	0.985
	10	P=75-6.7pH-1.21ln[NaOCl]	0.993
Green (MG)	5	P=-10+2.8pH-4.5ln[NaOCl]	0.935
	8	P=-3.5+1.1pH-2.7ln[NaOCl]	0.926
	10	P=75-7pH-2.2ln[NaOCl]	0.948
Violet (MV)	5	P=20-2.8pH-1.51ln[NaOCl]	0.960
	8	P=-120+12.7pH-2.7ln[NaOCl]	0.921
	10	P=54-6pH-1.4ln[NaOCl]	0.972
Orange (MO)	5	P=6-0.11pH-2.57ln[NaOCl]	0.953
	8	P=9.5-0.35pH-1.61ln[NaOCl]	0.956
	10	P=21.7-1.4pH-1.1ln[NaOCl]	0.999
Blue (MB)	5	P=8.8-0.26pH-0.99ln[NaOCl]	0.999
	8	P=7.9-0.05pH-1.02ln[NaOCl]	1
	10	P=2.6+0.48pH-1ln[NaOCl]	0.999

Note:*P=ORP/S

sample are within 5% accuracy of the measured values, but the former are consistently lower than the latter. The lower predicted values may be related to the sensitivity of the ORP probe used in this experiment or other unknown causes. However, the percentage of error can be practically reflected. Using MB as an example of the comparison, the observed S value varies linearly with the measure S value for the solution with an initial pH of 10 (Fig.4).

Additionally, the measured ADMI and ORP values show a high degree of repeatability indicating that the results obtained with the proposed model equation are reliable. The automatic measurements of color and ORP as well as the proposed model are dependable tools to implement on-line monitoring and control of the oxidation process for reducing dye color.

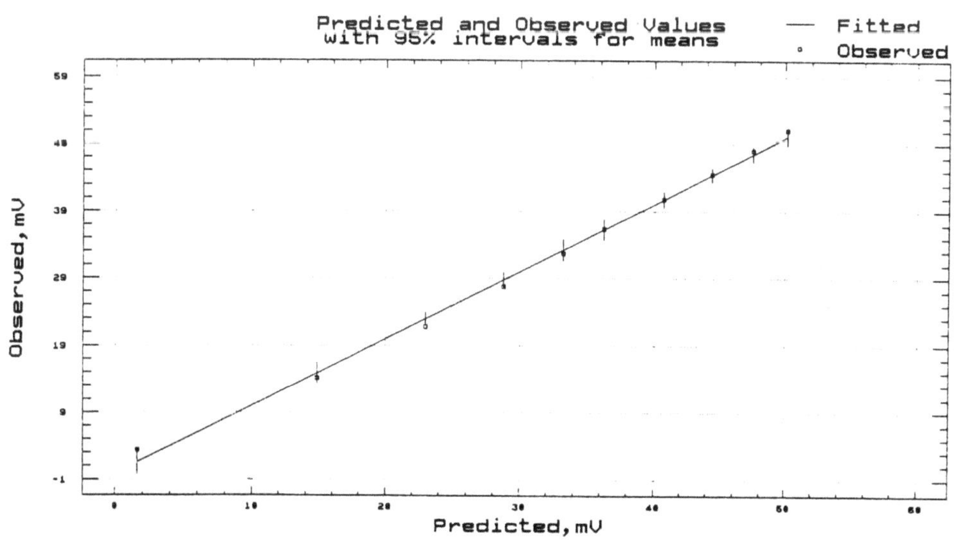

Figure 4. The model predicted S and observed value of MB at pH = 10.

The proposed model may have several potential applications for the design and control of the chemical oxidation process to reduce color to a desirable level. With known influent ADMI and ORP value, the initial solution pH at which the reaction is intended, and the desirable effluent ADMI value, the dosage of NaOCl can be calculated. Once the chemical dosage is known, the sample ORP value can be determined for control of the oxidation process. Further, the time required to complete the desirable oxidation reaction for a given flow rate and the reactor volume can be estimated.

CONCLUSION

1. On-line monitoring of color and automation of the oxidation process for color removal is made possible by use of appropriate probes connected to a personal computer.
2. Results of the oxidation decoloring studies results are modeled using a modified Nernst equation to relate the oxidant dosage requirement for reducing the dye color to a pre-selected ADMI value to the working solution pH, the initial NaOCl concentration, the initial ORP and the final ORP of the dye solution being treated.
3. The oxidation characteristics of the various dye solutions being studied can be represented by their oxidation potential or "S" values. The order of the decoloring potentials are : MG> MV> MO> MR> MB and with numerical ratios of 2.04: 1.78: 1.78: 1.28: 1.00, respectively.
4. For field applications, the proposed model equations can be used to predict the dosage requirement, the end-point of the oxidation reaction and the color removal efficiency of the oxidation process. If a desirable color removal or color remaining in the treated sample is known, the model can be used to estimate the dosage of NaOCl needed and the final ORP value for on-line control of the oxidation process.

ACKNOWLEDGMENT

This study was financially supported by National Science Council, ROC, under grant NSC 84–2211-E029–001.

REFERENCES

1. Weeter, D. W. and A. G. Hodgson, (1977), " Alternative for Biological Waste Treatment of Dye Wastewater", *American Dyestuff Report,* Vol. 8, pp. 32~40.
2. Kreye, W. C., (1974), "Polymer Acided Alum Coagulation of Textile Dyeing and Finishing Wastes", *Proceeding the 29th Industrial Waste Conference,* Purdue, pp.174 ~182.
3. Shelley, M. C., (1976), "Evaluation of Chemical-Biological and Chemical-Physical Treatment for Textile Dyeing and Fishing Waste", *Journal WPCF* , Vol. 48, No. 4, pp.35 ~ 41.
4. Perkins, W. S., J. F. Judkins and W. D. Perry, (1980), "Removation of Dyebath Water by Chlorination or Ozonation", Auburn University.
5. APHA, AWWA and WPCF, (1992), "Standard Method for the Examination of Wastewater", 18th Ed., Method 2120, pp.2–1~2–8.

6. Allen, W., W. B. Prescott, R. E. Derby Jr., C. E. Garland, J. M. Peret and Saltzman M., (1973), "Determination of Color of Water and Wastewater by Means of ADMI Color Values", *Proceedings of the 28th Purdue Industrial Waste Conference*, Ann Arbor Science Publisher, Ann Arbor, MI., pp.661~670.
7. Chang, C. N., W. L. Shang, S. J. Yeh and Y. S. Huang, (1992), "The Development of Auto- matic Water Color Measurement Technique", *8th ASPAC-IWSA Regional Water Supply Conference & Exhibition*, Malaysia, pp.6a2 1~6a2 12.
8. Weber, J. W., (1972), "Redox Control in Breakpoint Chlorination of Amine Complexes", *Wat. Res.*, Vol. 18, No. 1, pp. 21~24.
9. Charpentier, J. H., G. M. Godart and Y. Mogno, (1989), "Oxidation-Reduction Potential (ORP) Regulation as a Way to Optimize Aeration and C, N and P Removal: Experimental Basis and Various Full-Scale Examples", *Wat. Sci. & Tech.*, Vol. 21, Brighton, pp.1209~ 1223.
10. Eilbeck, W. J., (1984), "Redox Control in Breakpoint Chlorination of Amine Complexes", *Wat. Res.*, Vol. 18, No. 1, pp.21~24.
11. Chang, C. N., R. F. Yu, A. C. Chao and Seishu Tojo, (1994), "On-line Monitoring and Control of the Textile Wastewater Color Removal Process", *Wat. Sci. & Tech.*, Vol. 30, pp.264~274.

17

SEPARATION OF OIL-CONTAINING PARTICLES FROM WATER IN A CYCLONE FLOTATION MACHINE

R. Bokotko,[1] J. Hupka,[1] D. Leliński,[2] and J. D. Miller[2]

[1] Department of Chemical Technology
Technical University of Gdańsk, 80–952 Gdańsk, Poland
[2] Department of Metallurgical Engineering
University of Utah, Salt Lake City, Utah 84112

ABSTRACT

Bubble size is an important factor in the flotation of oil and other contaminants from industrial wastewater. A novel cyclone flotation technique—the air-sparged hydrocyclone (ASH) technology—offers fine bubble size distribution required for the efficient removal of dispersed oil and has a specific capacity 100 times greater than conventional flotation equipment. In this investigation several systems containing dispersed oil were examined for treatment by ASH flotation. The influence of surfactant and flocculant concentrations, air and water flow rates and the overflow/underflow split were examined. The results of the testwork indicate that the cyclone flotation machine can be effectively used to remove oily particles if appropriate surface chemistry conditions are guaranteed.

1. INTRODUCTION

The air-sparged hydrocyclone (ASH) which imposes a centrifugal force field during the flotation separation has already been shown to be adequate for the fast flotation of dispersed oil from oily water streams with a corresponding low retention time of a few seconds or even less [1,2]. In coal flotation kerosene and heavier oils are added to the slurry as collectors [3] which may indicate the feasibility and potential of the ASH for other systems containing dispersed hydrophobic particles. Some years ago, our preliminary investigation with water deoiling in the ASH had indicated that an overflow stream of up to six times the oil concentration in the feed could be produced [4]. In classical hydrocyclones, liquids to be separated must differ in their density. In wastewater treatment, oil/solid agglomerates and droplets of very heavy oils have densities close to the density of the aqueous phase. Such suspensions, unsuitable for treatment in conventional hydrocyclones, should be handled effectively in the ASH. The principles of operation of the air-sparged hydrocyclone have been given in previous papers[5,6].

Chemistry for the Protection of the Environment 2, edited by Pawłowski et al.
Plenum Press, New York, 1996

Table 1. Characteristics of oil phase used in experiments

Properties		Turpentine	Waste printing ink	Tar sand bitumen*
Viscosity (20°C),	mPa·s	1.6	>106	1000**
Density (20°C),	g/cm^3	0.85	>1.2	0.96
Sauter mean droplet diameter d_{32},	μm	14	15	150
Zeta potential/pH (20°C),	mV	−40/6.5	−35/7.0	−55/7.2
Dispersion stability***,	s^{-1}	10^{-7}	10^{-4}	10^{-3}

* - diluted with 15% kerosene

** - at 50°C

*** - rate constant in first order kinetics equation

In the present investigation three dispersed systems containing oil were submitted for separation by ASH flotation. In the first system, involving turpentine dispersed in steam condensate, the oil phase consisted of many organic substances, however, it had typical appearance of oil. In the second system, a multicomponent system—comprising printing ink suspended in process water, the hydrocarbons made-up only a portion of the hydrophobic particles. The third dispersed system contained slightly diluted tar sand bitumen and solid particles, as received during hot water processing of tar sands. It should be emphasized that the feed in each system constituted an actual (not model) process stream.

2. EXPERIMENTAL PROCEDURE

In the ASH unit, shown schematically in Fig. 1, the feed enters tangentially through a conventional cyclone header into a porous tube to develop a swirl flow of a certain thickness in the radial direction. The flow is then discharged through the annu-

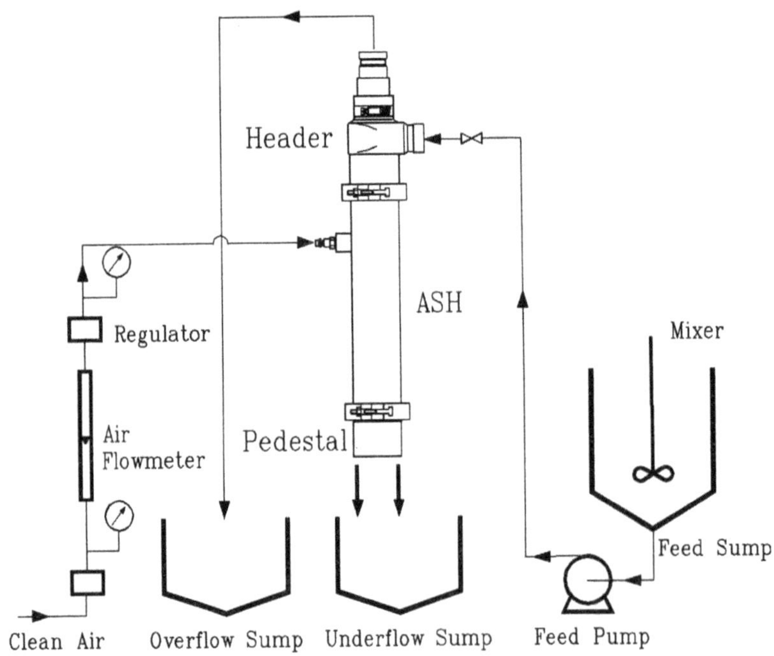

Figure 1. Schematic representation of experimental set-up.

lar opening created between the tube wall and a froth pedestal which is located on the cylindrical axis at the bottom of the ASH. Gas/air is sparged through the jacketed porous tube wall and is sheared into numerous small bubbles by the high-velocity swirl flow of the aqueous phase. Hydrophobic particles in the slurry collide with these bubbles, and, after attachment, are significantly reduced in their tangential velocity and transported radially into a froth phase which forms on the cylindrical axis. The froth phase is supported and constrained by the froth pedestal and thus moves towards the vortex finder of the cyclone header, being discharged as an overflow product. Hydrophilic particles generally remain in the slurry phase and are discharged as an underflow product through the annulus between the porous tube wall and the froth pedestal. For low density particles it is advised that only fines smaller than 150 micrometers should be used for the 50 mm dia. and 50 cm long ASH flotation unit due to the strong centrifugal force developed. This top particle size limit can be increased to about 0.7 mm when a 150 mm ASH system is utilized [3].

All data were collected using the experimental setup presented in Fig. 1. The ASH flotation system consisted of either a single 50 mm ASH unit (capacity 75 dm^3/min.) or a battery of three 50 mm ASH (total capacity 225 dm^3/min.). The feed dispersion in all experiments was an actual wastewater stream fed to the ASH flotation system either from a 200 dm^3 barrel in the laboratory or fed directly at the plant site in the case of the waste printing ink.

2.1. Turpentine

A single ASH unit was used for the removal of turpentine from wastewater. Flotation was conducted in closed cycle in the laboratory. Samples of feed, underflow and overflow were collected during a constant period of time. Turbidity analysis was used to determine the concentration of turpentine in the products and the effect of air and froth flow rate on the efficiency of separation was investigated.

2.2. Ink Flotation

The ASH battery was used with a nominal capacity of 225 dm^3/min. The air flow was constant throughout the whole process and corresponded to the volume flow of the aqueous phase. The separation process was conducted in such a way as to obtain as little froth as possible. In this case the overflow was only 2 dm^3/min., less than 1 % of the feed. Special effort was made to develop an appropriate technique to determine ink particle concentration in the process water [7].

2.3. Tar Sand Processing and Bitumen Scavenger Flotation

Bitumen separation by ASH flotation was preceded by tar sand digestion using an experimental procedure described previously [8,9]. After digestion, the slurry was discharged to a container, diluted with warm water to 55 wt% sand content, and the bitumen concentrate was skimmed-off. Next, the slurry was transferred to a 525 dm^3 agitated tank and diluted with warm water to 6 wt% solids. See schematic of the experimental set-up shown in Figure 1. The tar sand suspension was continuously agitated to maintain the sand in suspension (with limited success, however). The suspension was subsequently discharged to a 175 dm^3 tank from which a submerged centrifugal pump fed the hydrocyclone. All undigested tar sand particles larger than 2 mm were screened and removed from suspension on a screen located between the 525 dm^3 tank and the 175 dm^3 tank. A single ASH unit was used for these tests. The air flow rate was 90 dm^3/min. The Whiterocks (North-West location, 6 wt% bitumen) and the Asphalt Ridge (county open-pit mine, 13 wt% bitumen) tar sands from deposits in Uinta Basin Utah were tested. In each run suspension from the 525 dm^3 tank was continuously fed to the ASH system until exhausted. The overflow and the underflow streams were collected in 280 dm^3 containers, and the bitumen content in the froth and in the sediment product was analyzed by Dean Stark extraction.

3. RESULTS AND DISCUSSION

3.1. Turpentine

The emulsion stability has a profound effect on the process efficiency [2]. In this regard the stability of the turpentine emulsion was considered in more detail. Samples of the emulsion were taken from the container right after dilution. Fig. 2 shows relation between turpentine concentration and sedimentation time. This relationship is linear which means that sedimentation follows first order kinetics. The calculated sedimentation coefficient is $19*10^{-7}$ s^{-1}. Such a low value means that the emulsion of turpentine in water can be considered to be very stable.

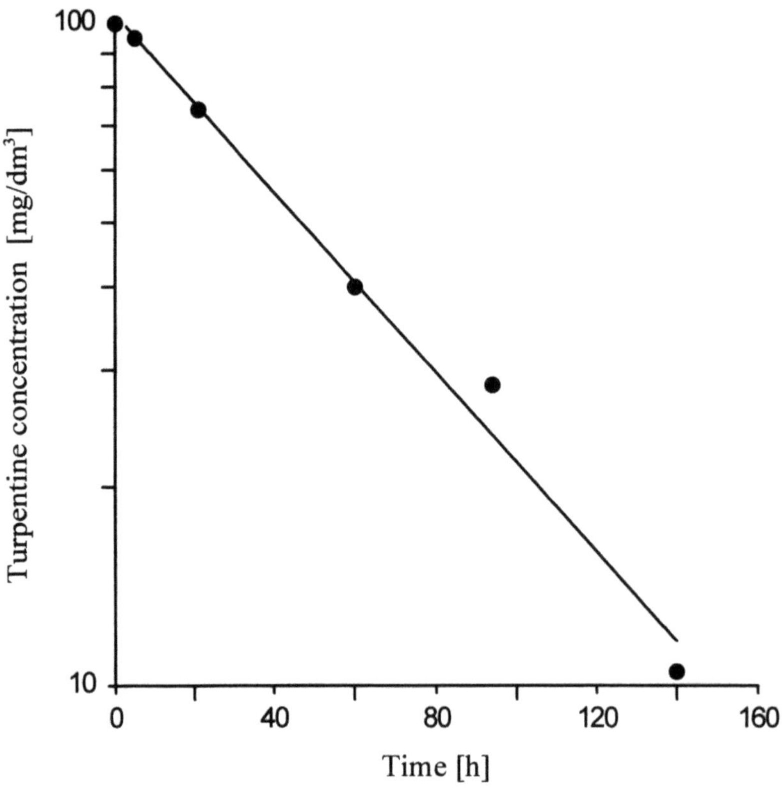

Figure 2. Sedimentation curve of turpentine emulsion.

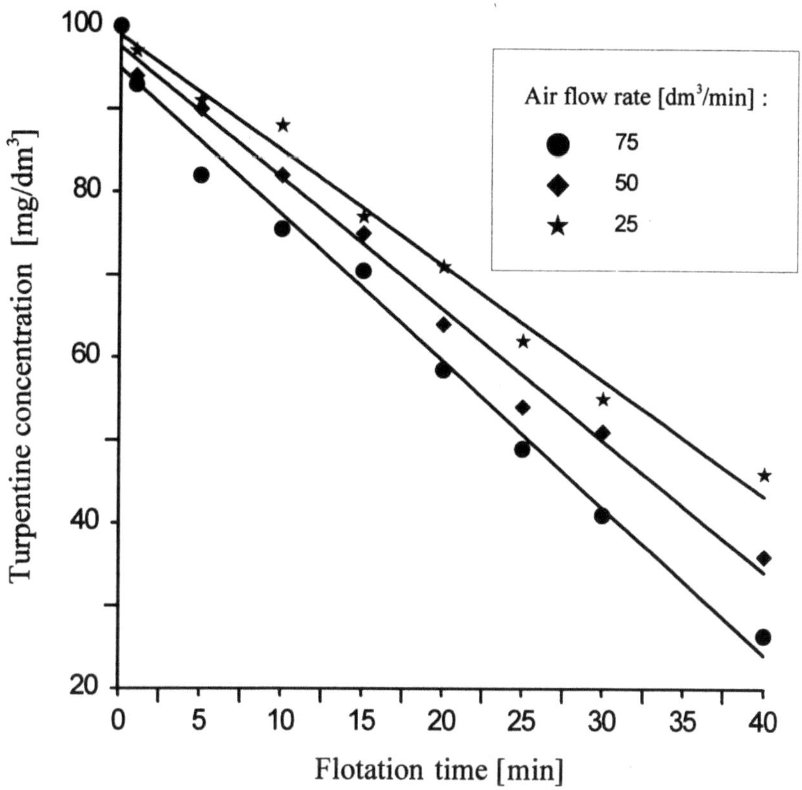

Figure 3. Effect of air flow rate on the efficiency of turpentine separation.

Fig. 3 shows the change of turpentine concentration in the underflow as a function of processing time for different levels of air flow. With an increase in the air flow the deoiling efficiency increases proportionally. Increased concentration of the gas phase apparently results in increased collision between drops and bubbles. Turpentine, which is volatile is also presumably stripped by the air flow.

The influence of overflow flow rate on flotation efficiency is given in Fig. 4. By increasing the split to the overflow from 1.5% to 6% the turpentine recovery to the overflow increases 15% after 40 min. flotation.

As anticipated from other papers [1,2] the addition of polyelectrolyte to the emulsion, stabilized with SDS surfactant, improved the extent of separation. The best results were obtained for 2 mg/dm^3 PE Magnafloc 592 (Allied Colloids). At such a PE dosage the zeta potential of turpentine droplets was close to zero, see Fig. 5. The reduction in oil concentration after a single run trough the ASH amounted 25%, see Fig. 6. Recycling of the emulsion resulted in a subsequent drop in turpentine content; however the removal efficiency decreased with each recycle.

According to Beey & Nicol [1] more concentrated emulsions were more susceptible for separation in the ASH. In the case of 440 mg/dm^3 of oil dispersed in water, 96% of the oil was removed for a very unfavorable 1:1 split, overflow/underflow. The same emulsion for 4% of the stream directed to the overflow resulted in 87% removal of the oil. Similar observations were made for the flotation of bitumens [2]. Emulsions containing 57 mg/dm^3 of oil with 18% of the volume directed to the overflow resulted in 38% removal [1].

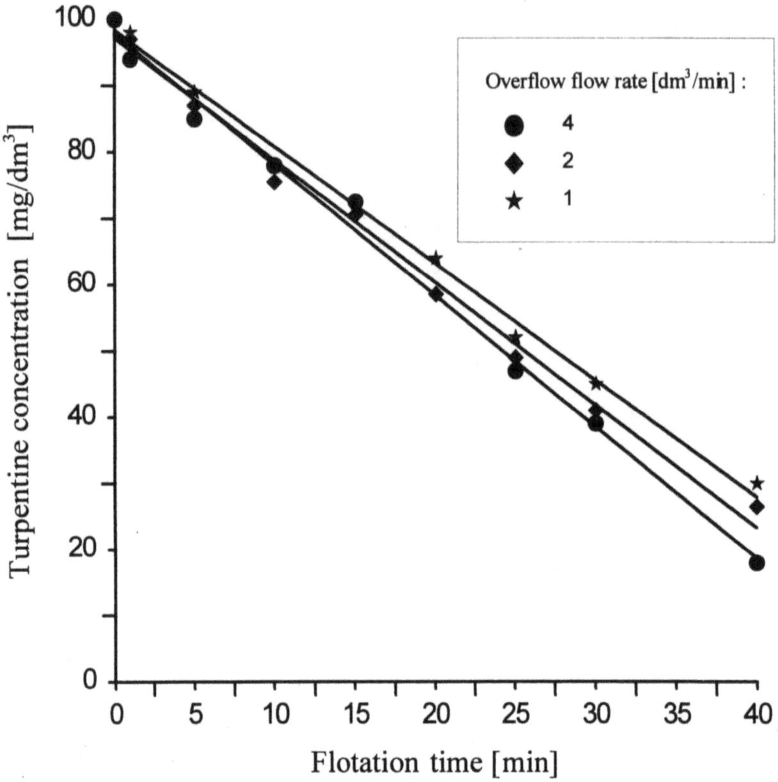

Figure 4. Effect of the overflow rate (split) on the efficiency of turpentine separation.

3.2. Flotational Removing of Ink Particles from Process Water

3.2.1. Process Water. The composition of the aqueous phase and the dispersed particles determines the effectiveness of the flotation separation. The waste paper process water, besides printing ink particles, contained surface active substances, sodium silicate, sodium hydroxide, cellulose fibers (~1 g/dm^3), filler particles (mainly kaolin) and colloid substances (glues ~1 g/dm^3). More than 90% of ink particles were less than 10 μm in size. Particles of diameter of more than 10 μm were mainly agglomerates of smaller particles, very often of ink of different colors. Microscopic examination revealed that color ink constituted 6 weight per cent of the total ink.

3.2.2. Results. Results given in Table 2 show that during testing the amount of the suspended matter and concentration of the printing ink in water have similar values. The 30% reduction of ink in the water can be determined based on the ink concentration in feed water and in underflow. Ink concentration in the froth was in the range of 0.21% to 0.68% depending on the process conditions and the amount of the overflow stream. Ink concentration ratio in the overflow varied from 8 to 26. Cellulose concentration in the froth was approx. 0.06%, which is 5% with respect to the suspended matter. It is thus seen that only a small amount of the fibre reports to the froth phase. The mean size of ink particles is smaller for the underflow when compared to the feed size, which indicates better floatability of larger particles (Table 3).

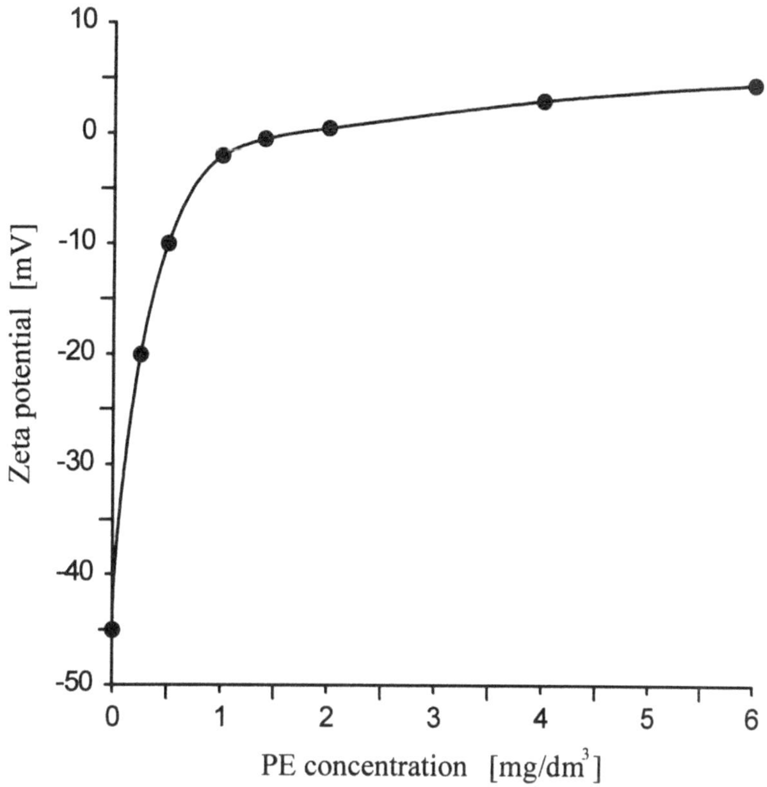

Figure 5. Relationship between zeta potential of turpentine droplets and PE concentration for SDS concentration 30 mg/dm^3.

3.3. Tar Sands Processing

The high capacity air-sparged hydrocyclone should be useful for scavenger flotation of tar sand tailings. Aeration under turbulent shearing conditions, as being developed in

Table 2. Concentration of printing ink and suspended matter in ASH streams

N°	Stream	Suspended matter [%]	Printing ink [%]
1			
	Process water	0.34	0.026
	Overflow	1.29	0.680
	Underflow	0.33	0.019
2			
	Process water	0.43	0.028
	Overflow	1.04	0.280
	Underflow	0.41	0.020
3			
	Process water	0.45	0.027
	Overflow	0.54	0.210
	Underflow	0.44	0.021

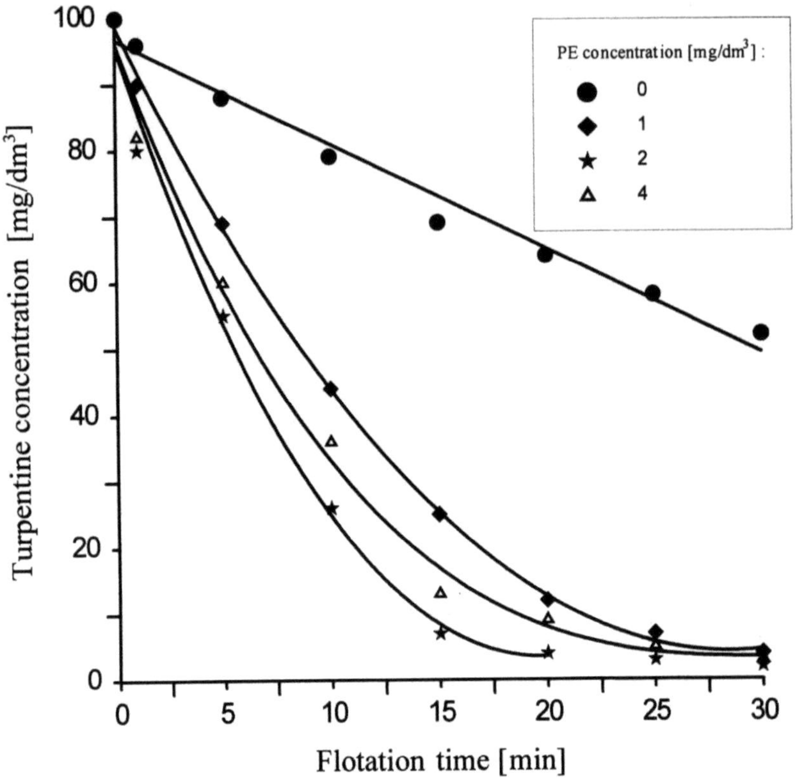

Figure 6. Changes in turpentine concentration in the underflow during flotation. Water flow rate and air was equal to 75 dm^3/min.

the ASH, should benefit the separation process by disrupting relatively week mineral-bitumen interactions, according to experimental data and model of Hall and Tollefson [10,11].

Addition of frothers, other than intrinsic surfactants released to the aqueous phase during digestion, is not practiced in the hot water process. In fact, excessive foaming was encountered during conventional flotation (38 dm^3, Denver type cell) of tar sand samples from certain locations within several deposits, e.g. northern part of the Whiterocks deposit. It was expected that ASH flotation may benefit from such a substantial presence of natural frothers. Also it was expected that the ASH flotation would be influenced by the extent of bitumen displacement from sand and by the ability of bitumen to encapsulate air bubbles.

It has been established that encapsulation of air bubbles in bitumen is controlled by air bubble entrapment mechanism [12] or by attachment and filming [13]. The encapsulation of

Table 3. Characteristic features of process streams for ink flotation

Stream	Cellulose concentration [%]	Kaolin concentration [%]	Mean printing ink particle diameter [μm]
Process water	0.11	0.20	15
Overflow	0.06	0.56	18
Underflow	0.10	0.22	7

air bubbles by the bitumen provides exceptional strength for the air-bitumen-sand particles and should allow their flotation in the highly turbulent flow in the ASH. Our own data indicate [14] that the necessary surface condition for oil spreading at the air-water interface can be satisfied for the tar sand processing strategy developed at the University of Utah [15]. The coalescence of bitumen droplets and the bitumen-air attachment steps are considered to be "fast" [16] which is a necessary condition for effective ASH flotation.

3.4. Performance of the ASH

It was already noted that the size of particles floated in the 50 mm ASH should not exceed 150 μm, but 85 % of the sand grains in the Asphalt Ridge tar sand, and 70 % of the sand grains in the Whiterocks tar sand are larger than 150 μm. The particle size limit can be increased to about 700 μm for the 150 mm ASH. Unfortunately, employment of the 150 mm ASH requires treatment of large flow rates of the tar sand suspension which is difficult under laboratory conditions. Our ASH flotation tests were carried out with the above constraints in mind.

The immediate consequence of the large size of mineral particles was their rapid settling in the 525 dm^3 storage tank, and in the 175 dm^3 distribution tank—despite vigorous stirring. About 85 to 90 % of mineral matter composing the tar sand was screened-off or settled out, and did not enter the ASH. Thus the suspension fed to the ASH contained only 1% solids compared to the recommended 6% solids. Due to a considerable dilution of the tailings the concentration of natural frothers was low and did not result in a dense froth. In an actual tar sand processing plant the process water would be recycled and therefore would contain a sufficient surfactants concentration to form a stable froth [17].

On standing samples from both ASH streams developed both a froth layer and a sediment. Solids (aggregated with bitumen) which reported to the overflow apparently were no longer supported by air bubbles (when the original froth collapsed) and formed the sediment. The froth layer in the underflow collecting tank was formed from sand-bitumen aggregates which were dragged to the underflow due to strong centrifugal forces acting in the hydrocyclone. As expected, most of the free (i.e. not associated with sand grains) bitumen droplets reported to the overflow. The receiving stream for the bitumen-sand aggregates depended on the bitumen/sand weight ratio in this aggregates, and the amount of air entrapped in the bitumen (as revealed by a microscopic examination). For these reasons the mean bitumen recovery to the overflow was not more than twice that in the underflow.

Natural hydrophobicity of tar sand bitumen (enhanced by air bubble filming) and the density difference between bitumen and sand should enhance bitumen recovery by ASH flotation. However, the separation results become unpredictable when bitumen firmly adheres to sand grains, such as in insufficiently digested tar sand slurry.

A thorough testing in a pilot plant utilizing recycled process water is indispensable for a more detailed evaluation of the ASH flotation technology in tar sands processing. In order to float oil-sand aggregates a larger-diameter ASH is recommended in which smaller centrifugal forces counteract the bubble buoyancy forces.

4. CONCLUSIONS

Results from the testwork have indicated that the cyclone flotation machine can be effectively used to remove oily particles if appropriate surface chemistry conditions are provided.

The experimental results lead to the conclusion that the flotation of a very stable emulsion such as the turpentine in steam condensate does not allow for efficient separation during one pass of the feed through the ASH equipment. The reduction of oil content was approx. 5% when no coagulants were used and reached 25% for 2 mg/dm^3 of polyelectrolyte. In such a case it is indispensable to have a multi-level process. Increased separation efficiency may be obtained only by destabilizing the emulsion.

The reduction of ink content in the wastepaper water process water was significantly higher and reached 30% in a single run. Nevertheless two stage separation strategy will be necessary for industrial application.

A high separation efficiency was achieved with respect to bitumen droplets, while bitumen-sand aggregates were distributed between the overflow and the underflow streams due to strong centrifugal force acting on particles.

5. REFERENCES

1. Beeby J.P., Nicol S.K., Concentration of Oil-Water Emulsion Using the Air-Sparged Hydrocyclone, Filtr.Sep., **30**(3), 141–146 (1993).
2. Leliński D., Miller J.D., Hupka J., The Potential of Air-Sparged Hydrocyclone Flotation in the Removal of Oil from Oil-in-Water Emulsion, First Word Congress on Emulsions, Paris-France, 19–22 October 1993.
3. Goplakrishnan S., Ye Y., Miller J.D., Dimensionless Analysis of Process Variables in Air-Sparged Hydrocyclone (ASH) Flotation of Fine Coal, Coal Preparation: An International Journal, **9**, 169–181 (1994).
4. Miller J.D., Hupka J., Water Deoiling in an Air-Sparged Hydrocyclone, Filtr. Sep., **20**(4), 279–282 (1983).
5. Miller J.D., Upadrashta K.R., Kinneberg D.J., and Gopalakrishnan S., Fluid-Flow Phenomena in the Air-Sparged Hydrocyclone, Proceedings of the XV International Mineral Processing Congress, Cannes, France, Volume II, June 1985, 1575.
6. Miller J.D., and Ye Y., Froth Characteristics in Air-Sparged Hydrocyclone Flotation, Mineral Processing and Technology Review, **5**, 307 (1989).
7. Bokotko R., Waste Printing Ink Determination in Process Water, Abstract Proceedings of the 5th Polish Conference on Analytical Chemistry, Sept. 3–9, 1995, Gda_sk, Poland, vol. III, 515.
8. Miller J.D., Misra M., Hot Water Process Development for Utah Tar Sands, **6**, 27–59 (1982).
9. Hupka, J., Miller, Cortez. A., Importance of Bitumen Viscosity in the Hot Water Processing Domestic tar Sands, Mining. Eng. **35**(12), 1635–1641 (1983).
10. Hall E.S., Tollefson E.L., Fines-Bitumen-Water Dispersions in Tailings from Oil Sand Extraction Plants that use the Hot Water Process, Can. J. Chem. Eng., **60**(12), 812–821 (1982).
11. Hall E.S., Tollefson E.L., Recovery of Residual Bitumen from the Aqueous Tailings from Hot Water Extraction of Oil Sands by Air Sparging, Energy Processing, **72**, 39–46 (1980).
12. Smith R.J., Miller J.D., The Flotation Behavior of Digested Asphalt Ridge Tar Sands, Mining Eng., **33**(12), 1719–1724 (1981).
13. Bowman C.W., Molecular and Interfacial Properties of Athabasca Tar Sands, Proc. 7th World Petroleum Congress, Mexico, 1967, Vol. 3, Elsevier, Amsterdam, 583–603 (1968).
14. Hupka J., Leliński D., and Miller J.D., Air Bubble Filming during Bitumen Flotation from Digested Tar Sand Slurry, Abstract Proc. 22nd Annual Meeting of the Fine Particle Society, July 29–August 2, 1991, 83–84.
15. J. Hupka, and J.D. Miller, Moderate Temperature Wâter-Based Bitumen Recovery from Tar Sand, Fuel, **70**, 1308–1312 (1991).
16. Sanford E.C., Processibility of Athabasca Oil Sand: Interrelationship Between Oil Sand Fine Solids, Process Aids, Mechanical Energy and Oil Sand Age after Mining, Can J. Chem. Eng., **61**, 554–567 (1983).
17. J.Hupka, J.F. Biernat, M. Hupka, J.P. Dworzanski, J.D. Miller, Isolation and Characterization of Surfactants From Tar Sand Process Water, Proceedings of the 1993 Eastern Oil Shale Symposium • Oil Shale • Oil Sands/Heavy Oil, Nov. 16–19, 1993, Lexington, KY, USA, 110–116.

18

POST-DYEING WASTEWATER TREATMENT IN CO-PRECIPITATION AND SORPTION PROCESS

Anna Maria Anielak

Technical University of Koszalin
Poland

ABSTRACT

In this work adsorption of selected 14 dyes on an adsorbent surface with surplus of negative charges has been investigated. Research work being carried out showed that the process of dyes adsorption is strictly connected with the dye type chemical composition), adsorbent surface charge and its electrokinetic potential; it is possible to eliminate all dyes from a solution by joint application of co-precipitation and sorption processes.

1. INTRODUCTION

In production processes of textile factories considerable volumes of wastewater are generated particularly when wet treatment of fibre takes place. Composition of textile industry wastewaters is diverse and depends on: technological operations, type of raw material and dyes being applied[12,13,17]. The most difficult to treat are post-dyeing wastewaters generated in dyeing of fibre and fabric processes[5,6,8,19]. They comprise some dyes which are very difficult to remove because of complexity of their chemical composition. In most cases the post-dyeing wastewater goes to a biological effluent treatment plant and often after biological treatment it does not change colour or its discolouration is insignificant[1,11,15]. Therefore, it is necessary to get rid of dyes before effluents are supplied to the biological treatment stage [7,9,10,14,18,21].

In the process of co-precipitation the ability of dyes to precipitate is strictly connected with their chemical composition, reagents type and order of dosing, adsorption capacity of co-precipitated sediment and its surface potential [2,3,4]. To obtain good results in dye precipitation surplus of chemicals or sorption process are necessary[16,20]. If negative surface potential adsorbent is used only very few dyes showing surplus of positive charge (those having ionised positively charged functional groups) will adsorb to its surface. Other dyes will show surplus of positive charges and affinity towards the negatively charged surface of adsorbate only after formation of hydro-iron-dye complexes which takes place during co-precipitation.

Therefore, the objective of this work was to investigate adsorption of selected dyes on a negatively charged surface of adsorbate before and after their co-precipitation.

2. INVESTGATION OF CO-PRECIPITATION AND SORPTION OF SELECTED DYES

2.1. Method Applied

Investigation of sorption was carried out for two sorbate solutions:

- aqueous solution of the dye,
- solution obtained after co-precipitation of the dye.

As sorbent active carbon 0–3 was used. This was a high quality and selected vegetable origin raw material. It was activated in a one-step physico-chemical way and neutralized in a gas method. The sorbent had the following properties:

- discolouration force MG (milligrame value) 270–370 mg
- degree of adsorption MV (methylene value), at least 14 ml
- moisture content max 8%,
- pH = 7,
- grain size - dusty,
- filtration ability - very good.

During tests carried out the variable factors were:

- dose of the active carbon expressed as the active carbon to dye concentration ratio, (C/B),
- dye and active carbon contact time, t_c,
- dose of calcium hydroxide added during process of dye co-precipitation.

As the process describing factor a percent transsmision of light and relative light transmission (having wavelength for which highest differences in percent transmission of light was obtained) has been adopted. The relative light transmission was calculated from the following formula:

$$rT = \left[\frac{T - T_o}{100 - T_o}\right]100 \tag{1}$$

where: rT - relative light transmission, %
T - transmission of light for tested sample, %
T_o - transmission of light for the initial dye solution, %

The sorbent and sorbate were being stirred mechanically with 60 rpm velocity. Other factors like temperature and pressure were constant. The sorbent electrokinetic potential of the electrical double layer was found.

2.2. Sorption of Dyes on Active Carbon

To check the selection of sorbent the electrokinetic potential for active carbon 0–3 suspension in distilled water was found and it was - 17.58 mV in electric field intensity E

= 13.86 V/cm. It has to be concluded, therefore, that the active carbon surface has a surplus of charges.

The test results showing influence of sorbent dose on the process of sorption for dyes investigated have been shown in Table 1. During testing, the contact time of the sorbent and sorbate was constant and was 60 seconds. The process was being evaluated on

Table 1. Influence of active carbon 0–3 dose on sorption of dyes tested

Dye	λ nm	Light transmission, T %			
		C/B=0	C/B=5	C/B=10	C/B=20
Direct Pure Blue	410	0.7	0.9	9.4	18.2
	550	0.0	0.0	0.0	0.5
	610	0.0	0.0	0.0	0.1
Direct Green GB	410	0.0	0.0	0.0	0.0
	550	0.0	0.0	0.0	0.0
	610	0.0	0.0	0.0	0.4
Direct Black RT	410	0.0	0.0	0.0	0.1
	550	0.0	0.0	0.0	0.0
	610	0.0	0.0	0.0	0.0
Chryzophenine G	410	0.0	0.0	0.8	1.2
	550	1.3	1.5	6.1	8.1
	610	0.6	0.9	3.5	4.8
Helion Red 3RL	410	0.0	0.0	0.7	1.2
	550	1.5	1.6	6.5	7.9
	610	1.0	1.1	3.6	4.9
Helion Red BB	410	0.0	2.3	2.5	3.2
	550	0.0	0.2	0.2	0.4
	610	0.0	1.7	1.7	20.3
Helion Black GF	410	0.0	0.0	0.0	0.0
	550	0.0	0.0	0.0	0.0
	610	0.0	0.0	0.0	0.0
Pofalen Green 3GL	410	0.0	0.3	1.1	1.2
	550	1.3	1.5	5.3	8.1
	610	0.3	0.8	4.6	4.8
Pofalen Yellow GR4	410	0.0	0.0	0.1	0.4
	550	9.3	9.4	10.2	31.0
	610	27.0	28.0	30.3	37.0
Acid Brown EG	410	0.0	0.8	38.9	67.2
	550	0.5	5.0	56.7	70.3
	610	2.4	16.8	57.0	74.4
Acid Red EGB	410	0.0	0.9	50.3	69.3
	550	0.0	0.0	42.0	42.6
	610	23.6	2.7	56.1	79.8
Acid Yellow G	410	0.0	35.0	83.1	89.5
	550	83.5	83.0	92.0	92.0
	610	89.1	90.5	92.3	92.3
Acid-Chrome Black ETN	410	0.0	0.0	0.1	3.1
	550	0.0	0.0	0.7	2.5
	610	0.0	0.0	2.8	4.5
Helactine Orange D2GE	410	0.0	0.0	0.3	0.6
	550	27.7	28.1	30.7	42.6
	610	91.6	90.9	91.0	91.6
Helactine Scarlet D2GE	410	0.0	0.0	0.0	0.0
	550	0.0	0.0	2.4	2.6
	610	88.0	11.2	30.1	49.0

the basis of the percent transmission of light having wavelengths λ = 420, 550, and 610 nm. The carbon to dye concentration ratio was being changed from 0 to 20.

Results of tests show that despite increase of active carbon dose up to the amount of C/B=20 not all dyes were sorbed by the sorbent or were sorbed insignificantly. The difficult to sorb were the direct, helion, acid-chrome, helactine, polfalane and helasol dyes. They were sorbed with difficulty even with time of sorption increased up to 60 minutes. However, sorption of remaining dyes on the active carbon increased with contact time increase (see Table 2). The sorbed dyes were: Acid Brown EG, Acid Red EGB, Acid Yellow G. These are acid dyes. The structural composition of e.g. Acid Yellow G

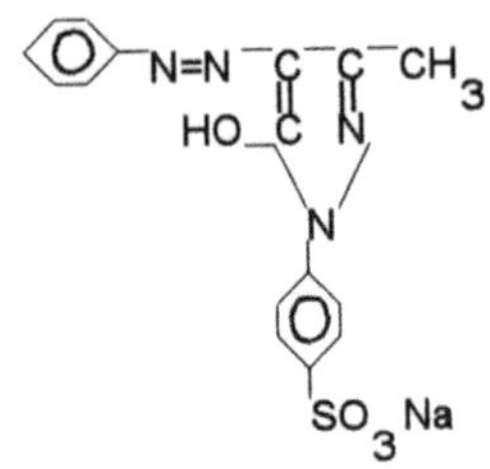

(2)

indicates that it is an organic compound with three functional groups: sulfo, hydroxyl (phenol) and methyl and complex chromophore groups with double bonds. The relative transmission of light having wavelength λ = 550 nm for acid dyes solutions varied within the limits of 42% to 70%.

Table 2. Influence of sorbent/sorbat contact time t_c on dye sorption on the active carbon, C/B= 10

Dye	λ nm	Light transmission, T% $t_c = 0$ sec	$t_c = 30$ sec	$t_c = 60$ sec	$t_c = 300$ sec	$t_c = 1200$ sec	$t_c = 3600$ sec
Helion Red 3RL	550	1.5	2.8	6.5	6.5	6.7	8.9
	610	1.0	3.1	3.6	3.5	3.7	5.6
Helion Red BB	550	0.0	0.2	0.2	0.2	6.6	18.2
	610	0.0	0.7	1.7	1.7	19.0	27.4
Helion Black GF	550	0.0	0.0	0.0	0.0	0.0	0.0
	610	0.1	0.1	0.1	0.1	0.0	0.0
Pofalen Green 3GL	550	9.3	9.1	10.2	10.4	14.3	14.5
	610	27.0	29.2	30.3	31.6	35.2	35.6
Pofalen Yellow GR4	550	9.3	9.1	10.2	10.4	14.3	14.5
	610	27.0	29.2	30.3	31.6	35.2	35.6
Acid Brown EG	550	0.5	46.0	56.7	76.0	75.0	85.0
	610	2.4	49.9	57.0	78.4	78.0	88.0
Acid Red EGB	550	0.0	23.1	42.0	44.6	56.0	56.0
	610	23.6	32.3	56.1	56.0	63.0	64.0
Acid-Chrome Black ETN	550	0.0	0.0	0.7	0.7	0.9	0.9
	610	0.0	0.0	2.8	2.8	3.3	3.3
Helactine Orange D2GE	550	27.7	27.5	30.7	30.5	30.5	32.8
	610	91.6	91.7	91.0	93.3	89.7	93.2
Helactine Scarlet D2GE	550	0.0	2.0	2.4	2.9	3.2	3.8
	610	88.0	20.9	30.1	28.3	28.5	29.1
Helasol Violet 4RN	410	0.0	0.0	0.0	0.0	0.0	0.0
	550	0.0	0.0	0.0	0.0	0.0	0.0
	610	0.0	0.0	0.0	0.0	0.0	0.0

2.3. Co-precipitation and Sorption of Dyes

Assuming contact time of t_c = 60 seconds and active carbon dose calculated in relation to the dye concentration C/B=20 (where maximum difference in light transmission results was obtained) sorption of dyes remaining in the solution after co-precipitation was tested i.e processes of co-precipitation and sorption together were employed.0

Results obtained were shown in figs 1–6. They illustrate change of transmission of light with wavelengths λ = 410 & 550 nm in dye solution after co-precipitation (continuous line) and then, change of transmission of light in post-coprecipitation solution subjected to active carbon sorption (broken line). The dyes were precipitated with constant dose of ferric chloride 500mg/dm^3 and varying doses of calcium hydroxide. Combining co-precipitation with sorption complete or almost complete elimination of dye from the solution was effected, which was in accordance with expectations.

In Figure 1 isotherms of relative adsorption of direct dyes have been showed. The first tested dye was Direct Pure Blue (Fig. 1A). As is well known, this dye precipitated

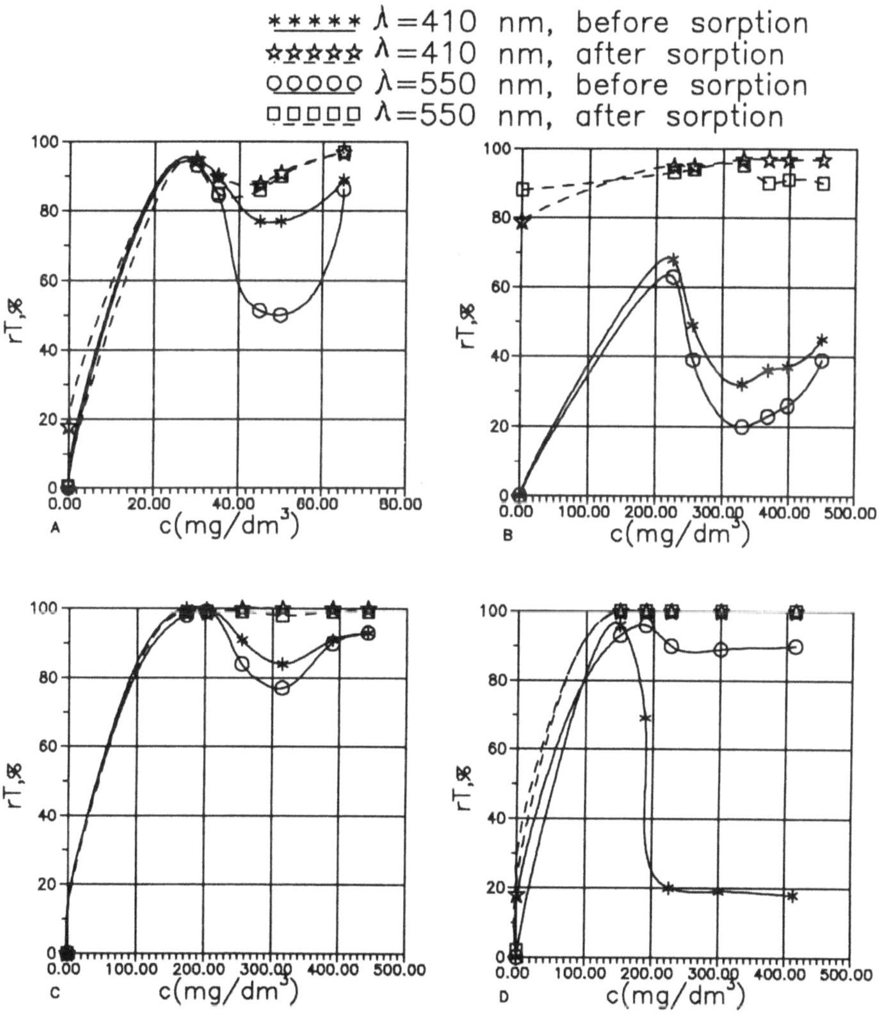

Figure 1. Isotherms of relative adsorption of direct dyes on active carbon after their co- precipitation in function of calcium hydroxide dose. A- Direct Pure Blue, B - Direct Green GB, C - Direct Black RT, D - Chryzophenine G.

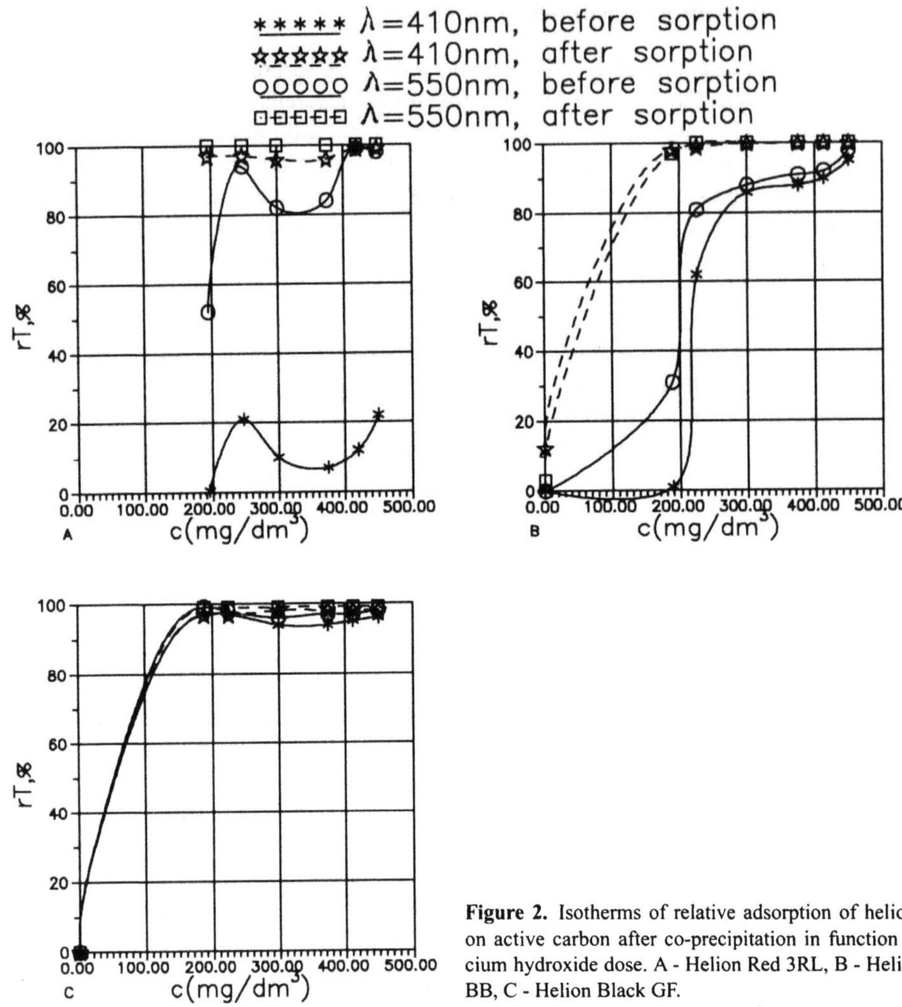

Figure 2. Isotherms of relative adsorption of helion dyes on active carbon after co-precipitation in function of calcium hydroxide dose. A - Helion Red 3RL, B - Helion Red BB, C - Helion Black GF.

easily gaining both in acid and alkaline reaction almost ninety percent relative precipitation ability. By subjecting the post-precipitation solution to active carbon sorption further increase of light transmission, approximately by 10%, was gained. Similar results have been obtained for remaining direct dyes. After co-precipitation the relative transmission of light in Direct Green B solution (Fig. 1B) was, on the average, 40% whilst after active carbon sorption it increased to almost 100%. On the other hand, increase of relative transmission of light for Direct Black RT (Fig. 1C) and Chryzophenine G (Fig.1D) solutions after execution of processes tested was 100%. It should be noted, however, that above mentioned direct dyes did not adsorb on the active carbon without the precipitation stage being carried out.

Subsequently, helion dyes were tested (Helion Red 3RL, Helion Red BB); they are considered direct dyes - see Figure 2. The analysis of results indicates that these dyes are completely eliminated from a solution when the first stage process is co-precipitation followed by sorption.

Pofalen dyes (Pofalen Green 3GL, Pofalen Yellow GRL,) co-precipitate with difficulties, particularly in neutral reaction, where dyes relative precipitation ability was al-

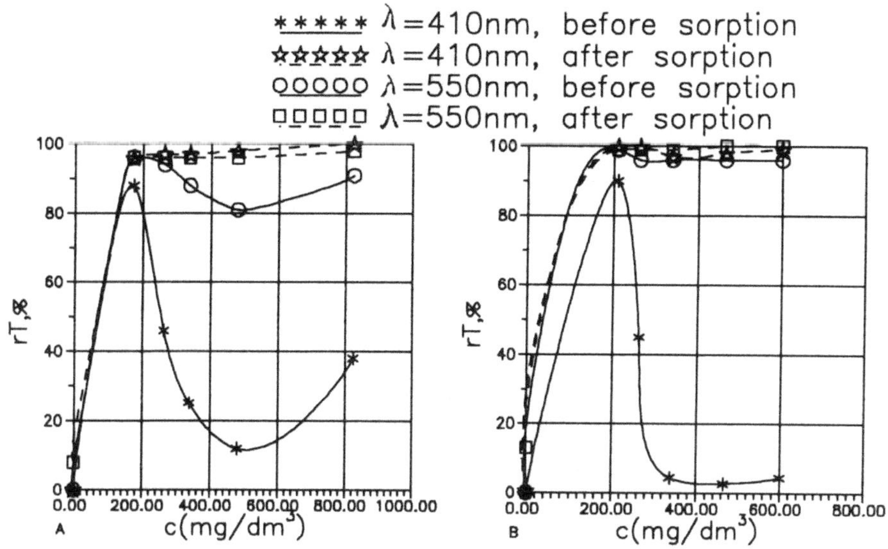

Figure 3. Isotherms od relative adsorption of polfalane dyes on active carbon after co- precipitation (c - dose of calcium hydroxide). A - Pofalen Green, B - Pofalen Yellow GRL.

most zero percent. The pofalen dyes did not adsorb on active carbon either in the process of "pure sorption". However, when sorption followed co-precipitation, increase of transmission of light up to 100% or almost 100% was gained - see Figure 3.

The acid dyes were easily sorbed by active carbon in "pure sorption" process. Increase of the relative transmission of light having wavelength λ = 550 nm in acid dyes solutions was within the limits of 42 - 70%. Applying sorption after co-precipitation (Figure 4) the relative increase of transmission of light in Acid Brown 3R (Fig. 4 A) and Acid Brown EG (Fig. 4B) solutions went up to almost 100%.

The positive influence of co-precipitation on the process of sorption can also be seen in case of acid-chrome dyes which were insignificantly sorbed in the active carbon "pure sorption" process. After co-precipitation and sorption of Acid-Chrome Black the solution transmission of light increased, within scope of change of pH values tested, up to 100%.

Worse results have been obtained for Acid-Chrome Brown 2R solution. The relative transmission of light having wavelength λ = 550 nm increased in acid solution up to 100% whilst in pH = 7 only to 50%. The last tested were helactine dyes. In the "pure sorption" process these dyes did not adsorb on active carbon at all whilst after co-precipitation, increase of relative adsorption measured as transmission of light λ = 550nm was 60 - 80% (Fig. 6).

To sum up the research results have showed that by application of co-precipitation as a first stage process followed by sorption, almost full elimination of dyes from solution is achieved. All tested dyes adsorb on active carbon including those which didn't in the "pure sorption " process.

3. INVESTIGATION OF PRECIPITATION ABILITY AND SORPTION OF DYES MIX

The research work carried out so far evaluates precipitation ability and sorption of respective dyes but does not specify the influence of a dye mix on processes investigated.

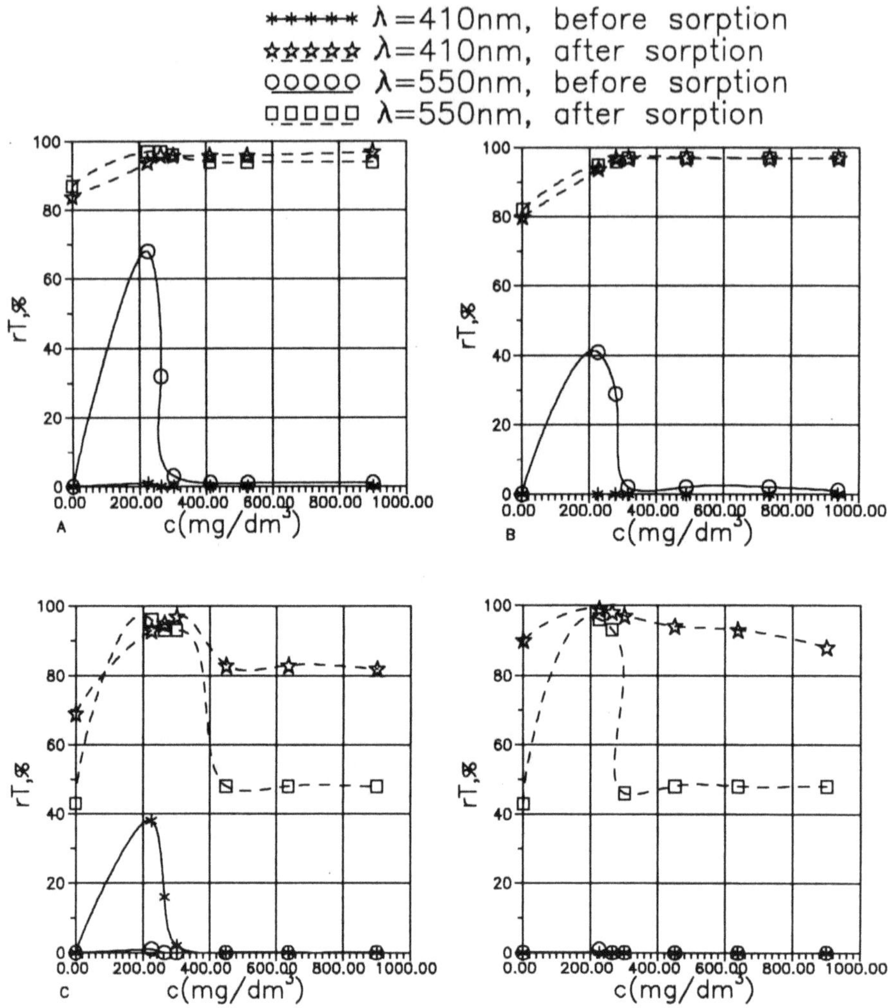

Figure 4. Isotherms of relative sorption of acid dyes on active carbon after co-precipitation (c - dose of calcium hydroxide). A - Acid Brown 3R, B - Acid Brown EG, C - Acid Red EGB, D - Acid Yellow G.

It can't be foreseen what will be the result of dyes co-precipitation and sorption if there is more than one dye in a solution. Being chemical substances dissociating in water, dyes can react chemically between themselves giving a new dye. It can be assumed that interaction of dyes impedes or facilitates the earlier described processes. Therefore, in order to determine importance of main effects and occurrence of interaction between main factors being investigated the following tests were carried out.

3.1 Method of Investigation

The simplest experimental scheme i.e. complete randomization system has been applied. The problem lies in making comparison of c different experimental objects basing on n experimental units described by y feature (resultant variable). Eight different factors and two levels of their values were adopted in this system. The value being at the lower

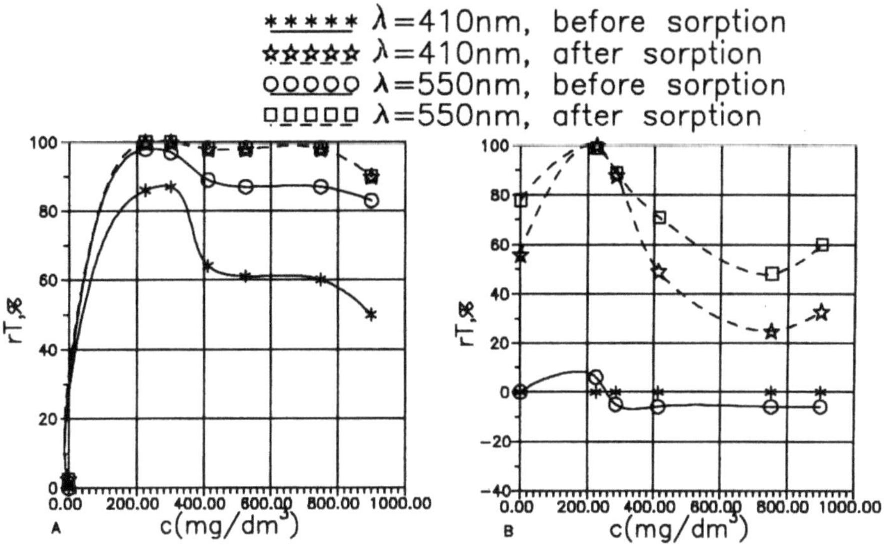

Figure 5. Isotherms of relative adsorption of acid-chrome dyes on active carbon (c - dose of calcium hydroxide). A - Acid-Chrome Black, B - Acid-Chrome Brown 2 R.

level for all factors was constant and equal to zero whilst values at the higher level together with their respective factors were as follows:

W - active carbon, 250 mg/dm^3,
R - reagents, $Ca(OH)_2$ - 263 mg/dm^3, $FeCl_3 \times 6H_2O$ - 500 mg/dm^3,
B$_1$ - Helion Red BB - C/B = 20,
B$_2$ - Direct Green GB - C/B = 20,
B$_3$ - Pofalen Yellow GR$_4$ - C/B = 20,

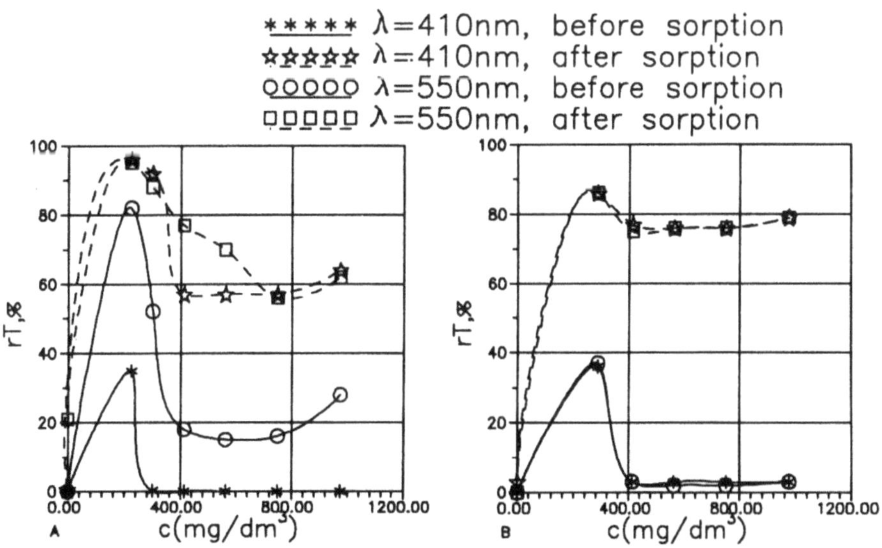

Figure 6. Isotherms of relative adsorption of helactine dyes on active carbon after co- precipitation (c - dose of calcium hydroxide). A - Helactine Orange D2GE, B - Helactine Scarlet D2GE.

B_4 - Acid Brown EG - C/B = 20,
B_5 - Helactine Scarlet D2GE - C/B = 20.

The resulting variable here was transmission of light λ = 550 nm in solution obtained after completing the process being tested. Investigations were carried out in two completely randomized sets (with full number of combinations). The number of all experimental objects in one set was 2^8. Therefore, number of experimental units was n = 2 x 2^8.

The analysis of variance for factorial experiment 2^8 assumed with a completely randomized sets method in k = 2 sets has. The following calculations have been used in the analysis:

• Correction: $$C_y = \frac{\left(\sum y\right)^2}{n} = 865401.7 \quad (3)$$

• Sum square for the whole: $$nS_y^2 = \sum y^2 - C_y = 970885 \quad (4)$$

• Sum square for the sets: $$nS_{bl}^2 = \frac{sets \cdot sums \cdot sum \cdot square}{number\, of\, elements\, in\, a\, set} - C_y = 1.785 \quad (5)$$

• Sum square for the error: $$nS_e^2 = nS_y^2 - nS_{bl}^2 - nS_{komb}^2 = 106.33 \quad (6)$$

• Deviations sum square for level combinations of factors investigated:

$$nS_{komb}^2 = \frac{W^2}{n} + \frac{R^2}{n} + \ldots = nS_W^2 + nS_R^2 + \ldots = 970{,}776.885 \quad (7)$$

• Mean square for sets: $$V_{bl} = \frac{nS_{bl}^2}{\nu} = \frac{nS_{bl}^2}{k-1} = 1.785 \quad (8)$$

• Mean square for combination: $$V_k = \frac{nS_{komb}^2}{\nu} = \frac{nS_{komb}^2}{c-1} = 3{,}806.968 \quad (9)$$

• Mean square for error: $$V_e = \frac{nS_e^2}{\nu} = \frac{nS_e^2}{(c-1)(k-1)} = 0.42 \quad (10)$$

where: n = 512 - denotes number of all observations in the experiment,
c = 256 - denotes number of objects,
k = 2 - denotes number of sets,
v - denotes the degree of freedom.

Then the test function values have been calculated:

$$F^o = \frac{V_k}{V_e} \quad (11)$$

which were being compared with boundary value $F_{0.05}$ taken from Snedecore'a F tables assuming $vk = c-1$ and $ve = (c-1)(k-1)$ degrees of freedom to determine the importance of main effects and interaction. To show that there were no differences between sets a test function for sets has been determined. The above statistical characteristics were being calculated by Yates method with application of computer technique.

3.2. Description and Analysis of Results Obtained

The results of investigations for all combinations according to type 2^8 matrix of experiments planning and results of computer calculations of main effects and interactions have been omitted in this paper because of their huge volume. Calculations result obtained indicates occurrence of two particularly important main effects - Reagents (i.e. co-precipitation) having value of test function $F^o = 1709859$ and Active Carbon (i.e. sorption) having value of test function $F^o = 132488$. Both main effects are positive which means that their presence influences increase of transmission of light in dye solution. More important effect is obtained in presence of Reagents (co-precipitation). Next value of test function in the table $F^o = 88076$ is connected with interaction WR and is positive.This means that action of Active Carbon in presence of Reagents is stronger that in their absence. The main effect B_1 has a negative value proving that presence of this dye, namely Helion Red BB, causes the transmission of light in the dye solution tested to decrease. Also participation of remaining dyes causes negative main effects to happen. The highest negative result has been obtained for Acid Brown EG where test function value was 41485. Important for the process, appeared to be positive interactions WB_4 $(F^o = 45338)$ and WRB_4 $(F^o = 30983)$ and also a negative interaction $RB4$ $(F^o = 33043)$. The positive interactions should be interpreted as follows: dye B_4 is sorbed by active carbon both in presence and absence of effect R. Negative interaction confirms importance of the positive ones because it proves that dye B_4 in presence of R effect and absence of W effect is more difficult to precipitate. The analysis of remaining interactions indicates that despite the fact they are numerous, they do not reach so high numeral values of the test function as those described above. They are generaly positive interactions between R effect and dyes and between R effect, W effect and dyes or negative between W effect and dyes. Besides, there are interactions between dyes like: B_1B_2 $(F_o = 2321)$, $B_1B_2B_5$ $(F_o = 2321)$, $B_2B_3B_5$ $(F_o = 4395)$, $B_2B_4B_5$ $(F_o = 2731)$, B_1B_6 $(F_o = 2926)$, B_2B_6 $(F_o = 2429)$, $B_1B_2B_6$ $(F_o = 3890)$, $B_2B_5B_6$ $(F_o = 8688)$, $B_1B_5B_6$ $(F_o = 4486)$, $B_2B_3B_5B_6$ $(F_o = 4836)$, $B_1B_2B_3B_5B_6$ $(F_o = 1510)$, $B_2B_3B_4B_5B_6$ $(F_o = 3171)$ These interactions are negative and positive which should be interpreted as variable influence of dyes on one another giving solutions of varied transmission of light. The variance analysis did not show a significant interaction existing between R effect and dyes and between R, W effects and the dyes. It proves that adopted system of processes eliminating dyes from solutions (co-precipitation and sorption) works also for a solution containing a mix of dyes.

4. SUMMARY

Research work carried out by the author showed that active carbon 0–3 has a negative surface potential. According to theoretical assumptions made, ions, complexes and positively charged particles will be attracted to such surface affected by electrostatic forces. Analysis of structural composition of dyes indicates that most of them has got sulfo functional groups which, being strong acids groups, are dissociated in the whole range of pH change. Dissociated sulfo functional groups have a negative charge, therefore, they will be repelled by active carbon due to action of electrostatic forces. Apart from sulfo groups the dyes also have auxochromes showing positive charge in acid reaction. The dyes through these ionised functional groups will be attracted by active carbon. Finally, a resultant charge of all ionised functional groups will decide about dye adsorption on a sorbent surface. Research work has confirmed these assumptions. The following dyes adsorbed on active carbon surface: Acid Brown EG, Acid Red EGB, Acid Yellow G, Acid-Chrome Brown 2R. Remaining dyes practically were not adsorbed. Complete elimination of all tested dyes from a solution can be achieved by application of sorption which follows co-precipitation. In accordance with theoretical assumptions of this work it should be a sorbent with a negative surface potential because to such surface complexes of hydro-iron-dye adsorb. After addition of an iron salt to a dye solution some hydro-iron-dye electropositive complexes are formed. Having added a sorbent with negative surface potential (active carbon) to a solution, one observes displacement of water dipoles from hydro-iron complexes due to stronger electrostatic action of sorbent and adsorption of complexes to the negatively charged surface of adsorbent. Thence, it has to be concluded that adsorption proceeds in accordance with certain series of affinities of ions to the surface. The research work showed that existence of different dyes in a solution does not impair the processes of co-precipitation and adsorption.

REFERENCES

1. Alekceev E.V.A. Pavlinova I.I.:*Issledovanija komplieksnoj shemy očistki stočnyh vod predprijatij trikotažnoj promyšlennosti. Meževuzovskij sbornik naučnyh trudov. Moskva 1984.*
2. Anielak A., Piecuch T.: *Varsuche zur Reinignung der Farbereiablufe von Textiliefabriken. 20 Diskussionstagnung Mechanische Flüssigkeitsabtrennung. Magdeburg 1983r*
3. Anielak A.: *Archiwum Ochrony Środowiska PAN 3–4/1984r.*
4. Anielak A.: *Monograficzny zeszyt 7. Badanie i opracowanie technologii oczyszczania œcieków pofarbiarskich i komunalnych Sp-ni CEPELIA - "Koronka" w Bobowej, II Ogólnopolska Konf. Naukowa nt "Kompleksowe i szczegółowe problemy inżynierii środowiska".Ustronie Morskie .Politechnika Częstochowska.1987r.*
5. Amadeu F. da Rocha, Luca Sercio Joao : *Utilizacao de carvoes activados nacionals no tratamento de aguas residuais. (Porto Alegre, Brasil) Revista DAE; 44(136): 57–62, 1984 .*
6. Buhl F. i Mikuła B.:*Chemia analityczna,tom 32, 307 /1987 /.*
7. Bergk K.H., Jurkschat J., Schwieger W.: *Acta hydrochim.hydrobiol., Berlin 14(1986) 5, 449–570.*
8. Bredereck K. Schumacher Ch.. *Dyes and Pigments 21 (1993) 45–66.*
9. Davis R.J., Gainer G. O'Neal, Wu I.W.: *Photocatalyticdecolorization of wastewater dyes. Water Environment Research. January/Febraury 1994*
10. Grigorjewitsch Spirkin W.: *Technologie zur Abfallverarbeitung von Textilfarbstoffen - ine neue Richtung bei der Einsparung chemischer Rohstoffe. Textiltechnik 34 (1984) 3, 149–151*
11. Holdsworth J. Thomas, Shaul M. Glen, Depsey R. Clyde. *Environmental Protection Agency (Cincinnat) 1991.*
12. Laskov Ů.M.: *Očistka stočnyh vod na predprijatijah tekstilnoj promyšlennosti. legkaja industrja .Moskva 1977.*
13. Laskov Ů.M.: *Glubokja očistka i povtornoe ispolzovanie stočnyh vod predprijatij tekstilnoj promyšlennosti. Mehanika i Energetika.Vypusk 2/1980.Moskva.*
14. Kamel M.M.,Youssef B.M., Kamel M.M.: *Dyes and Pigments 15(1991) 175–182.*
15. Krasnoborodko I.G.: *Racionalnaja shema kanalizovanija i lokalnoj očistki stočnyh vod krasilno-otdeločnyh fabrik. Tekstilnaja promyšlennost. Nr 11.Leningrad 1974.*
16. Kurbiel J., Korczak M.:*Wpływ specyficznych własności adsorbentu mineralno-węglowego na adsorpcję wybranych barwników i związków powierzchniowo-czynnych. Węgiel aktywny. III Sympozjum Sekcji Inżynierii Sanitarnej PAN. Kraków 1987*
17. Patterson W.J.: *Waste water treatment technology (Chicago). Ann Arbor Science, 1975, 265*

18. Postiglioni O.J. *Tratamiento conjunto de efluentes industriales y domesticos. Vivienda; 20(225):43–48, 1981 r.*
19. Orozco Jaramillo Alvaro. *Tratamiento de residuos liquidos industriales con lagunas facultativas aeradas. Contaminacion ambiental;4(7):23–39,1980.*
20. Qiniji Peng , Mujie Li , Kunyu Gao , Lubai Cheng : *Dyes and Pigments 15 (1991) 263–274*
21. Winnicki T., Majewska-Nowak K., Wiśniewski J.: *Przydatność różnych typów polisulfonowych modułów membranowych do ultrafiltracyjnego odbarwiania cieczy. Wysokoefektywne metody oczyszczania ścieków i odnowa wody. Politechnika Krakowska. Monografia 34, Kraków 1985 r*

PHENOMENA OCCURRING ON PHASE BOUNDARY IN A PROCESS OF COAGULATION AND CO-PRECIPITATION

Anna Maria Anielak

Technical University in Koszalin
Poland

1. INTRODUCTION

Coagulation is defined as a process of colloidal, suspended particles combination into aggregate particles (flocks) due to action of molecular forces, which can be effected by:

- addition of an electrolyte decreasing the electrokinetic potential of the coagulating particles ζ;
- formation of metal hydroxides on which the colloids adsorb;
- mutual attraction and agglomeration of the negative and positive colloids.

Colloidal and suspended particles existing in water and wastes in most cases have hydrophilic surface. Such particle surface has a specific surface charge and, therefore, the electrodynamic potential ψ_o. Some ions adsorb directly at the surface in the adsorption layer. On the adsorption layer boundary develops ψ_δ potential, called the Stern potential. The remaining counter-ions form a diffused cloud of charge, the diffusion layer. The potential in the diffused electric double layer changes with increase of distance from the phase boundary. To describe electrical features of the disperse system particles, the surface or Stern potentials, generally, are not used: instead electrokinetic potential ζ is applied - it develops, for example, during movement of particles in relation to the medium. In most cases it is assumed that the shear plane developing during movement of particles is located directly behind the Stern layer, ie potentials ψ_δ and ζ are identical.[5,6]

Particles having homopolar surface charge will repel each other, whilst heteropolar ones will combine forming aggregates. There are also molecular forces of mutual attraction, called van der Waals-London forces (dispersion forces), acting between the colloidal particles. Significance of these forces decreases with increase of the distance between the particles. It is considered, in general, that resultant interaction of intermolecular and electrostatic forces determines combination of particles into larger agglomerations. It's known

Chemistry for the Protection of the Environment 2, edited by Pawłowski et al.
Plenum Press, New York, 1996

that coagulation proceeds at the highest rate when the electrokinetic potential equals zero.[8,14]

The most frequently used coagulants are salts of iron, aluminium and salts of polyvalent metals which hydrolyze easily.

Both colloidal, suspension particles and flocks of hydroxides and other chemically co-precipitated sediments have the electric double layer. Thus, describing the electric double layer with equations, description of coagulation is achieved.

Many formulae describing the electric double layer are known. In most cases, they are complex, analytic-empirical equations. Thickness of a diffused electric double layer is designated as $1/\kappa$, called the Debye-Hückel parameter[10].

$$\kappa = \left[\frac{4e^2 \sum n_i(\infty) z_i \pi}{kT\varepsilon_o \varepsilon} \right]^{1/2} \tag{1}$$

In this equation e denotes the elementary charge, $n_i(\infty)$ - number of ions in 1 ccm outside the double layer, z_i - number of charges, ε_o - permittivity of free space (8.859 10^{-12} $AsV^{-1}m^{-1}$), ε - relative permittivity, k - Boltzmann constant (1.3805 10^{-23} JK^{-1}), T - temperature degrees Kelvin.

In 1966, using potential ζ, Colin[2] formulated the energy of repulsion of solid particles with the following equation:

$$W_r = k\varepsilon\zeta^2 e^{-2d\kappa^2} \tag{2}$$

where: $2d$ - distance between two particles.

From equation "2" it comes out that if potential ζ approaches zero, the energy of repulsion decreases and agglomeration of particles occurrs; the rate of agglomeration increases with increase of the double layer thickness.

Applying the Poisson and Boltzmann equations, Sonntag formulated the relationship between surface density σ and surface potential ψ_o in the following equation:

$$\sigma = -\frac{\varepsilon_o \varepsilon}{4\pi} \left(\frac{d\psi}{dx} \right)_{x=0} = \left[\frac{nkT\varepsilon_o \varepsilon}{2\pi} \right]^{1/2} \left[\exp\left(\frac{ze\psi_o}{2kT} \right) - \exp\left(-\frac{ze\psi_o}{2kT} \right) \right] \tag{3}$$

For low values of potential, equation "3" can be reduced to the following form:

$$\sigma \approx \frac{\varepsilon_o \varepsilon \kappa}{4\pi} \psi_\delta \tag{4}$$

From equations "3" and "4" it appears that the surface and Stern potentials depend both on the charge surface density and on the ionic composition of the solution. Sonntag claims that once the double layer gets more concentrated by addition of a neutral electrolyte, σ must increase or ψ_o (or ψ_δ) must decrease or both have to change respectively[10].

To calculate the Stern layer charge density, Gaudin[9] developed the following formula:

$$\sigma = 1.42 \cdot 10^{15} \zeta \sqrt{c} \tag{5}$$

where ζ is expressed in Volts, c - moles/litre and σ - charge density expressed in electron charges per square centimetre.

From this equation appears, Gaudin claims, that the contact surface contains few ions or number of ions of different polarity is almost equal, and the results indicate only the surplus of one type ions over the other, or eventually, the mathematical calculations have been effected with application of wrong assumptions[9].

Other authors[7], assuming flat phase boundary surface, described the relationship between a charge and the surface potential for colloidal particles where $\kappa a >> 1$ (ie radius of spherical particles a is much bigger that thickness of the electric double layer) with the following equation:

$$\sigma = \sqrt{\frac{2\varepsilon\varepsilon_o \cdot kTC}{\pi}} sh(ze\psi_o/2kT) \tag{6}$$

and for smaller particles where $\kappa a > 1$, the surface charge is as follows:

$$\sigma = \sqrt{\frac{2\varepsilon\varepsilon_o \cdot kTC}{\pi}} \left[sh\frac{ze\psi_o}{2kT} + \frac{2}{\kappa a} th\frac{ze\psi_o}{4kT} \right] \tag{7}$$

In 1980 Cuisset showed, whilst examining particles of various clays, that their agglomeration proceeded at a faster rate with increase of ions concentration and their valency, with decrease of the relative permittivity and solution temperature and with decrease of pH (ie in case of clays, Cuisset[3] claims, with increase of the electric double layer thickness).

According to the author, applying the following assumptions:

- coagulation occurrs due to electrostatic interaction between water dipoles adsorbed in the Stern layer, counter-ions and surface charge of coagulating particles and not because of the van der Waals-London forces;
- counter-ions in the Stern layer adsorb on the ionic surface of a particle with one valence, whilst the other valencies (if polyvalent ions adsorb) produce a resultant charge of the adsorption layer;
- during observations being carried out agglomeration of solid particles, existing in the solution and having the same homopolar surface charge, occurs (no hydrolysis of coagulant takes place and no hydroxides precipitate);

The electric double layer may be described with simple physical equations.

2. CHARGE AND POTENTIAL OF THE ELECTRIC DOUBLE LAYER

For the above mentioned assumptions, the author has accepted the following model of coagulation:

1. It is assumed, for example, that the surface of a solid particle with surplus of negative charges has a polar character. Such surface in water will be subject to hydration. Even chemically pure water, containing no acids nor alkalis dissociates in a very low degree (autodissociation). Therefore, solid particles in distilled water will have in the

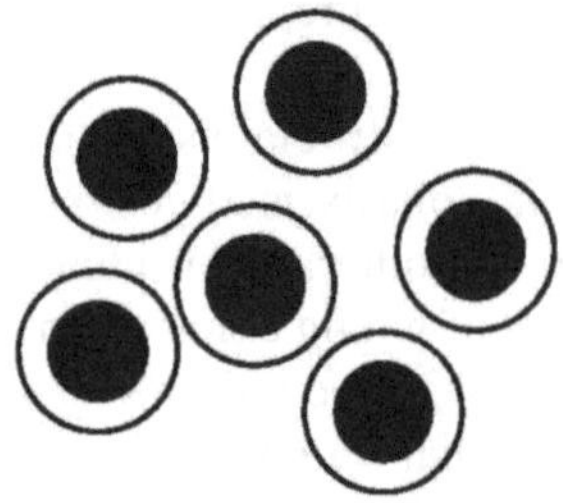

Figure 1. Aggregation of solid particles after addition of a coagulant solution.

Stern layer adsorbed water dipoles directed towards the negatively charged solid particle surface with their positive ends. Theoretically, the adsorption layer charge (not the surface charge density) should equal the solid particle surface charge and potential $\psi_\delta = \psi_o(R/R+\delta)$. Such solid particles will not coagulate having in mind a substantial surface potential (eg above - 30 mV).

2. A coagulant, eg a bivalent cation salt, is added to the above described solution. The salt dissociates in water. Cations, which have higher charge than water dipoles, repel the latter from the solid particle surface and bind themselves with the solid surface using one bond. The electric potential in the Stern layer decreases in proportion to the number of cations adsorbed there. Electrokinetic potential ζ decreases but still retains the negative value and coagulation does not occur yet.
3. The dose of coagulant is being increased. Concentration of positive ions in the adsorption layer increases. Water dipoles are still present in the layer with their negative ends compensating the positive charge of the cation's last orbital. The Stern layer potential drops to zero. Theoretically there is no diffusion layer there. The solid particles do not repel but approach one another for the distance equal to the Stern layer and aggregation of particles occurrs (Fig. 1).
4. The solid particles Stern layers arrange themselves due to electrostatic attraction of water dipoles negative ends towards the cations, which have uncompensated positive charge and coagulation occurrs.
5. Cations existing in the Stern layer show greater affinity to the solid surface than water dipoles. Therefore, water dipoles existing on the phase boundary are displaced by cations, which eventually combine the solid particles thus forming stable flocks and permanent coagulation occurs.
6. Further increase of the salt solution concentration will increase the concentration of cations in the Stern layer. The electrokinetic potential will change its polarity from negative to positive. It's value will be increasing together with the amount of cations being adsorbed in the Stern layer. The maximum positive potential at a

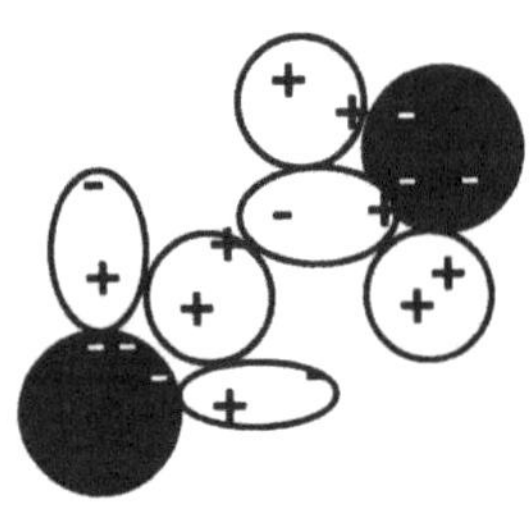

Figure 2. Arrangement of the Stern layers caused by attraction of heteropolar charges.

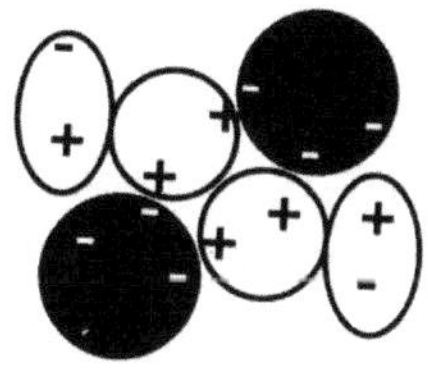

Figure 3. Displacement of water dipoles from the Stern layer and aggregation of solid particles by cations.

bivalent cation adsorption should equal the absolute value of surface potential ψ_o. If a trivalent cation adsorbs, the Stern potential will equal $2\psi_o$ (Fig.4). The concentration of negative counter-ions in thediffusion layer increases. Coagulation does not occur. The coagulant has been overdosed.

7. Due to further increase of coagulant dose, concentration of the negative ions on the inner surface of the diffusion layer increases.

Such model of coagulation can be described easily with simple physical equations.

There are surface charges, the Stern layer charges and the diffusion layer charges in the electric double layer. Each layer charge is a sum of single point charges:

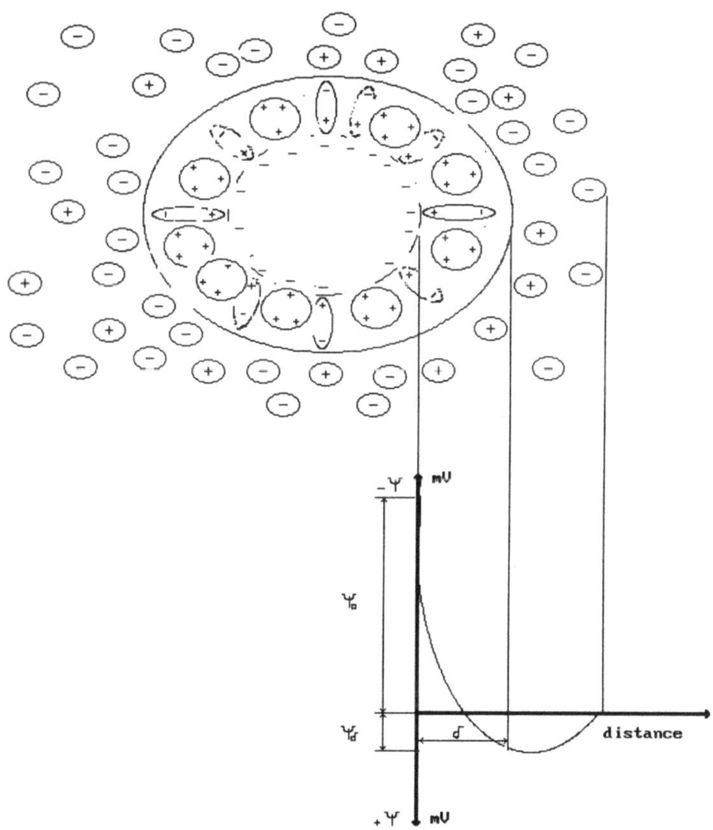

Figure 4. The electric double layer the presence of trivalent iron ions. ψ_o - surface potential, ψ_δ - potential on the boundary between the Stern and diffusion layers.

$$\sigma = \sum_{i=1}^{n} q_i \tag{8}$$

When the solution charge is zero, the sum of surface σ_o, the Stern layer σ_δ and the diffusion layer charges also equals zero. Thus, the solid particle surface charge is:

$$-\sigma_o = \sigma_\delta + \sigma_s \tag{9}$$

It comes out from the model applied that the surface charge is being compensated by single charges of the Stern layer, therefore, its total charge will depend on the surface charge of the solid particle and valency of ions W in this layer. In case of full dehydration of the Stern layer, its charge will be as follows:

$$\sigma_\delta = -\sigma_o \cdot W \tag{10}$$

The diffusion layer charge can be calculated from the following relationship:

$$\sigma_s = -\sigma_\delta - \sigma_o = \sigma_o W - \sigma_o = \sigma_o (W - 1) \tag{11}$$

In case of incomplete dehydration of the Stern layer, dehydration coefficient α has to be taken into account in formulae for the Stern layer and diffusion layer charges respectively; a takes values from the range of $0 \subseteq \alpha \subseteq 1$. Finally, the Stern layer charge will be as follows:

$$\sigma_\delta = -\sigma_o W\alpha \tag{12}$$

and the diffusion layer one:

$$\sigma_s = \sigma_o (W\alpha - 1) \tag{13}$$

Therefore, for the surface charge an identity equation will be obtained:

$$\sigma_o = \sigma_o W\alpha - \sigma_o (W\alpha - 1) \tag{14}$$

which confirms correctness of deduction performed. The potential of a field created by a solid particle surface charge in R distance, being radius of given solid particle is:

$$\psi_o = \frac{\sigma_o}{4R\pi\varepsilon_o\varepsilon} \tag{15}$$

If the field has been created by several point charges, the resultant potential in each point of the field is an algebraic sum of potentials created by particular charges. Thus, if solution potential is $\psi_{r\text{-}ru} = 0$, drop of the potential at the diffusion layer and solution boundary equals:

$$\psi_o + \psi_\delta + \psi_s = 0 \tag{16}$$

Assuming that during movement of a solid particle the shear plane is being formed on the Stern and diffusion layers boundary, ie in $R + \delta$ distance, it can be assumed that electrokinetic potential ζ equals:

$$-\zeta = -\psi_o - \psi_\delta = \psi_s \tag{17}$$

Thence, taking into account the distance of interaction of the surface charge and the Stern layer equal $R + \delta$, the following relationship for the electrokinetic potential can be developed:

$$-\zeta = \frac{\sigma_o(W\alpha - 1)}{4(R+\delta)\pi\varepsilon_o\varepsilon} = \psi_s \tag{18}$$

It can be noted for α that:

$$\alpha = \frac{a}{a_m} \tag{19}$$

where:,

a - amount of ions adsorbed,

b - capacity of the spherical monolayer,

and the following equation is obtained therefrom:

$$\zeta = \frac{-\sigma_o\left(W\dfrac{a}{a_m} - 1\right)}{4(R+\delta)\pi\varepsilon_o\varepsilon} \tag{20}$$

where electrokinetic potential ζ has been expressed in function of amount of counter-ions adsorbed in the adsorption layer.

Assuming, for low concentration solutions, that $\varepsilon\varepsilon_o \approx$const and designating constant values with B

$$B = -\frac{\sigma_o}{4(R+\delta)Gp\varepsilon\varepsilon_o} \tag{21}$$

a simplified form of equation "21" is produced:

$$\zeta = B\left(W\frac{a}{a_m} - 1\right) = BW\frac{a}{a_m} - B = Aa - B \tag{22}$$

This is an equation of a straight line the positioning of which in relation to the coordinates depends on the type of surface charge (positive or negative) and size of the spherical monolayer - Fig.5.

It appears from equation "20" that electrokinetic potential ζ is directly proportional to the amount of the surface charge and absolute valency of ions adsorbed in the Stern layer. The absolute value of the electrokinetic potential decreases with thickness of the

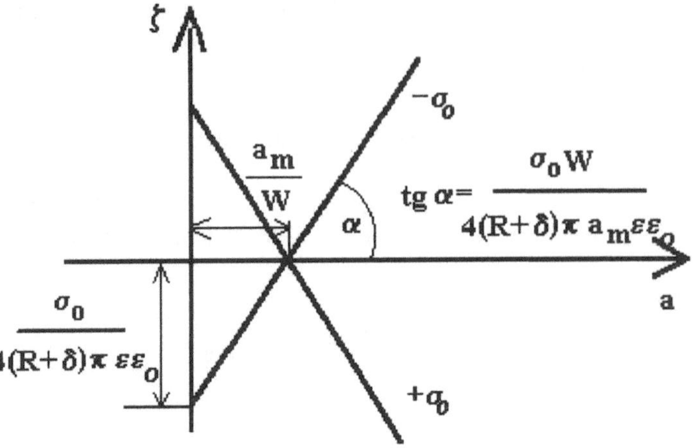

Figure 5. Graphical solution of the adsorption equation "20".

Stern layer (which is consistent with the relationship obtained by Colin) and size of the solid particle as well as increase of solution relative permittivity ε (which corresponds to results obtained by Cuisset).

Conditions, in which the electrokinetic potential assumes zero value, can be calculated from equation "18":

- if solid particle surface charge equals zero, the surface is hydrophobic;
- if solid particle surface charge differs from zero, the absolute valency of ions adsorbed equals 1 and surface dehydration is complete ($\alpha = 1$);
- if solid particle surface charge differs from zero, the absolute valency of ions adsorbed equals 2 and dehydration coefficient $\alpha = 0.5$;
- if solid particle surface charge differs from zero, the absolute valency of ions adsorbed equals 3 and dehydration coefficient $\alpha = 0.333...$

3. AN ATTEMPT TO VERIFY THE APPLIED ASSUMPTIONS AND EQUATIONS DESCRIBING THE ELECTRIC DOUBLE LAYER

3.1. Methodology of Research

The examined dispersed phase was quartz SiO_2 grain size d<45 μm and hydroxides formed in a chemical reaction of an iron salt with calcium or sodium hydroxide. The dispersion medium applied was distilled water. The salts applied in survey were: ferrous chloride, ferric chloride and ferrous sulphate.

Changes occurring on phase boundary were evaluated by electrokinetic potential ζ. The potential was being measured on a zetameter the block diagram of which has been shown in Fig. 7.

The tested suspension medium was being located in a quartz measurement cell provided with electrodes - platinum cathode and molybdenum anode. The measurement of electrophoretic velocity of solid particles was being effected by the so called parabola method applying measurement of particles movement velocity on several levels of the cell

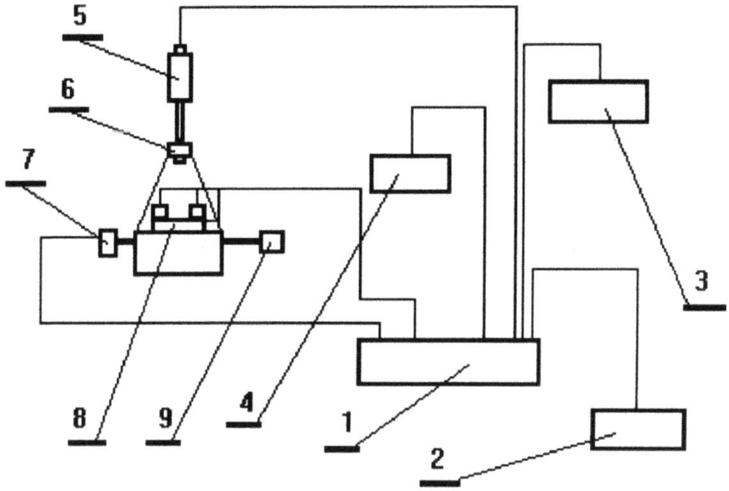

Figure 6. Zetameter block diagram : 1 - microcomputer, 2 - keyboard, 3 - monitor, 4 - printer, 5 - camera, 6 - microscope, 7 - stepper motor, 8 - casing for the cell containing electrodes and temperature sensor, 9 - microscope power supply.

depth. This method takes into account variations in the solid particles velocity at different levels of the cell arising from concurrent occurrence of the electroosmotic and electrophoretic velocities. Due to variable directions of solvent movement occurring in the cell transversal axis, distribution of observed velocity takes shape of a parabola. The actual electrophoretic movement can be watched at the so-called stationary levels, where the electroosmotic velocity of the liquid equals zero (the plane where reversion of solvent direction occurs).

The parabola method is the most accurate one from all other type of microelectrophoretic measurements. Because of measurements being carried out on several levels of the liquid in cell and complex calculations, this method can be applied only if computer assisted.

3.2. Description and Results Analysis

During investigation of quartz coagulation, the amount of iron cations adsorbed on the quartz surface was determined. The results are showed in Table 1.

The results indicated that a value of electrokinetic potential depends on a dose of salt, valency and amount of iron ions adsorbed. The higher the dose of salt and valency of iron ions, the greater the change of ζ potential. Also, change of charge, from negative to positive, can be observed. The maximum positive value of the electrokinetic potential depends not only on concentration, amount of counter-ions adsorbed, but also on their valency. Trivalent iron will adsorb on the quartz surface with one bond. The remaining two bonds will render the surface a positive ionic character.

In case of bivalent iron, only one bond will remain free. Therefore, with the same amounts of Fe^{3+} and Fe^{2+} ions adsorbed, higher values of the electrokinetic potential were obtained for trivalent iron ions.

Approximating the graph given in Fig.7 with function: $y_{Fe}3+ = 10{,}87x - 38{,}04$ and the graph given in Fig.8 with function: $y_{Fe}2+ = 6.92x - 38.04$ constants for equation “22”

Table 1. The value of electrokinetic potential ζ in function of amount of iron cations adsorbed

	pH	ζ mV	E V/cm	t K
Fe^{3+}				
0 (mg/dm^3)	6.2	−38.04	13.87	298.8
0.21	6.1	−32.51	13.87	298.7
2.50	6.0	−5.09	13.87	298.6
4.40	5.8	25.40	13.87	298.7
5.64	5.4	45.53	13.87	298.6
8.50	4.4	60.02	13.87	296.6
21.40	3.8	73.23	13.86	298.4
Fe^{2+}				
0 (mg/dm^3)	6.2	−38.04		
0.8	6.1	−33.99	13.87	298.8
4.9	5.7	−1.00	13.87	298.7
6.0	5.4	0.00	13.87	298.6
7.1	5.2	9.80	13.86	298.3
10.2	4.9	16.47	13.86	298.4
11.7	4.8	31.92	13.86	298.5
12.8	4.5	38.96	13.86	298.4

were determined and then, the spherical monolayer capacity was calculated; for Fe^{2+} adsorption isotherm it amounts to a_m = 11 mgFe/g and for Fe^{3+} adsorption isotherm equals a_m = 10.5 mgFe/g. The amounts obtained differ only slightly, therefore, it can be assumed that the method applied to determine the quartz particles spherical capacity is accurate and an average capacity of the monolayer amounts to 10.75 mgFe/g. The amount of iron ions adsorbed in surplus to the spherical capacity of particles can be attributed to the adsorption capacity of pores with their charge having no effect on the spheric surface charge of particles.

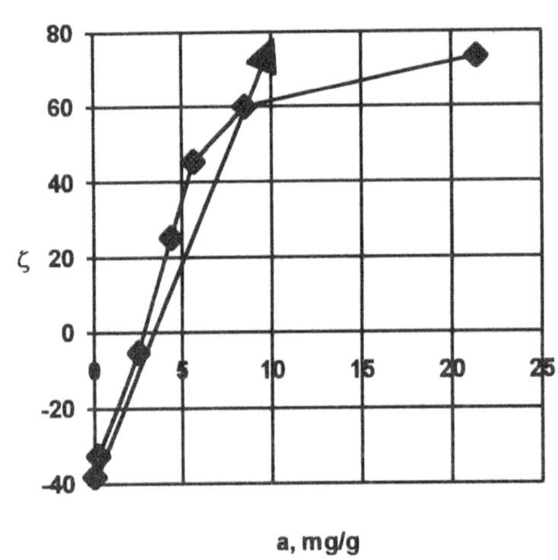

Figure 7. Potential ζ in function of the adsorbed Fe^{3+} ions.

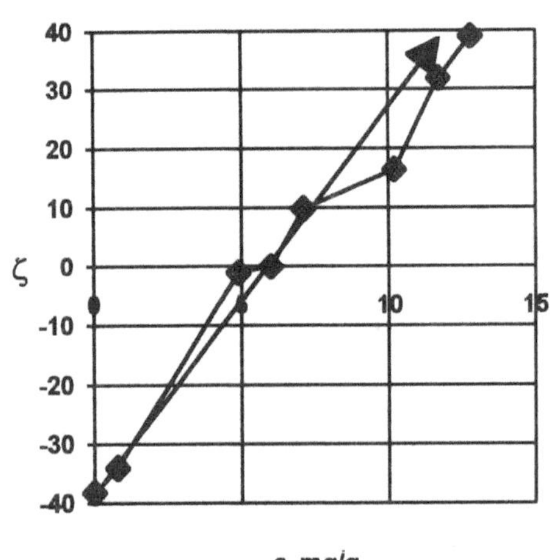

Figure 8. Potential ζ in function of the adsorbed Fe^{2+} ions.

Table 2. The influence of ferric hydroxide formation conditions on the value of electrokinetic potential ζ

Reagents added	pH	Ref µm cm/V s	E V/cm	ζ mV
$FeCl_3 + NaOH$				
	2.9	0.86	13.85	+ 10.3
	4.6	−1.72	13.85	−19.33
	8.0	−5.50	13.85	−59.15
	10.1	−20.02	13.85	−206.0
$FeCl_3 + Ca(OH)_2$				
	2.6	1.82	13.85	+ 21.60
	3.3	0.96	13.85	+ 12.58
	6.8	0	13.85	0
	9.3	0.4	13.85	0.53
	5.0	1.19	13.85	−14.04
$FeSO_4 + NaOH$				
	6.3	−0.78	13.85	−9.17
	10.3	−0.48	13.85	−5.15
	4.2	−0.77	13.85	−9.17
$FeSO_4 + Ca(OH)_2$				
	6.4	−0.88	13.85	−10.77
	9.7	−0.51	13.85	−6.25

The analysis of results confirms correctness of assumptions applied and experimental ability to ascertain equation "20".

The adopted model of coagulation is compatible with coagulation structures discovered in photographs taken by an electron microscope (Thiessen)[6]. They are oriented structures of "house of cards" or chain type and shaking can disperse them again.

In practice, to coagulate wastes and water some easily hydrolizing coagulants are applied whilst maintining pH providing for rapid precipitation of hydroxides. Likewise, in a co-precipitation process a sediment being a product of a chemical reaction between reagents added to the solution is obtained. The precipitated sediment has a specific surface charge and the Stern layer composed of counter-ions. The sediment shows, during precipitation, a substantial surface charge, it adsorbs counter-ions and coagulates. The type of counter-ions adsorbed and the electrokinetic potential depend on the type of products formed as well as on the type of reagents added.

By precipitation of ferric hydroxide with ferric chloride and sodium hydroxide, flocks with adsorption layer are obtained where, in low pH conditions, water dipoles and chloride ions can adsorb. In alkali reaction the electrokinetic potential of flocks shoul take on negative values - see Table 2.

Ferric hydroxide precipitated with ferric chloride and calcium hydroxide, in strongly acid reaction, will have positive surface and electrokinetic potentials. In reaction providing a negative surface potential, bivalent calcium ions displacing water dipoles will adsorb in the Stern layer. Due to their bivalency, calcium ions will cause a substantial drop of potential. Investigations showed that in neutral and alkali reactions, the flocks have a potential which equals, or almost equals, zero mV. Precipitating ferrous hydroxide with ferrous sulphate and sodium hydroxide in acid reaction, flocks with binegative sulphate ions adsorbed in the Stern layer are obtained; in acid reaction, they can increase electronegative character of the electrokinetic potential. Ferrous hydroxide in relation to ferric hydroxide, will have lower surface potential in the full scope of pH change. It is consistent with the results obtained. Ferrous hydroxide precipitated with ferrous sulphate and calcium hydroxide will have lower surface potential than that of ferric hydroxide. In strongly acid medium, where the surface potential has positive value, bivalent sulphate ions will be adsorbed in the Stern layer; they can change the Stern potential value from positive to a negative one. In reaction, where the surface potential takes on negative values, in the Stern layer bivalent calcium ions will be adsorbed; they change the electric potential substantially even up to positive values. This was confirmed by the research work carried out. The above analysis indicates that in coagulation and co-precipitation processes the first seconds are the most important, when the adsorption and diffusion layers are formed, until equilibrium of the system is reached. At that time adsorption of organic high-molecular substances and colloids in the Stern layer and in the ions diffusion layer proceeds and solid particle aggregates are formed. After reaching equilibrium the system becomes appropriately stable. It has to be supposed, therefore, that agglomerates of flocks develop due to the electrostatic and not van der Waals-London forces attraction (forces of dispersion acting between induced transitory dipoles).

The results of investigations showed in Table 2, and the analysis of the process presented hereby, indicate that adsorption on hydroxides must not be considered independenly of the conditions of hydroxides formation and of the electrokinetic potential ζ value.

On the grounds of investigations performed, it should be considered that adsorption in the Stern layer proceeds as per affinity of ions specified for given surface. The multivalent cations displace water dipoles and monovalent cations. The series of affinities can be developed in further investigations. Depending on the type of ions adsorbed, the adsorbent particles electrokinetic potential ζ changes. This relationship explains why the coagulation ability of Al^{3+} is 1000 times higher than that of K^{+}.

4. SUMMARY

Making the assumption that electrostatic forces, and not van der Waals-London forces, take part in coagulation, the electric double layer can be described with a simple physical equation. The equation allows for determination of the spheric monolayer and pores adsorption capacities. The equation can also be used for calculation of an optimum dose of coagulant to be applied in given process of coagulation.

5. REFERENCES

1. Anielak A.M.: Post-dyeing wastewater treatment in co-preciption and sorption process. Monografia Nr 50, WSInż. Koszalin. 1995
2. Colin F. : Elimination de la pollution colloidale. Traitements de coagulation et loculation Thé se Institut de la Recherche Hydrologique Nancy 1966
3. Cuisset O.: Raport de recherche LPC Nr 96 Proprietes ectrocinetigues des particules argileuses Aplication de la methode electrophoretique aux problemes d'envionnement et d'identification des sols.Septembre 1980 r. France Ministere de l'environnement et du cadre de vie. Ministere des transports.
4. Jung-il Yang: Recovery and selectivity improvements in fine Coal Cleaning. XII International Coal Preparation Congress 1994 Cracow.
5. Laskowski J.: Chemia fizyczna w procesach mechanicznej przeróbki kopalin. Śląsk. Katowice 1969
6. Sonntag H.: Lehrbuch der Kolloidwissenschaft. VEB Deutscher Verlag der Wissnschaften, Berlin 1977
7. Zapolskij A.K., Baran A.A.: Koaguljanty i flokulanty v processah očistki vody Himija. Leningrad 1987

8. Nikoladze.G.I., Minc D.M., Kastalskij A.A.: Podgotovka vody dlja pitego i promyšlennogo vodosnabženja. Vysšaja Škola. Moskva 1984
9. Gaudin A.M.: Flotation. McGraw-Hill Book Company, Inc. New York, Toronto, London, 1957
10. Debye P., Hückel E.: Zur Theorie der Elektrolyte, Physik. Z. 24, 185–206, 305–325, 1923
11. Winnicki T.: Polimery czynne w inżynierii ochrony środowiska. Arkady. Warszawa 1979
12. Choi, W.Z.,Cohung H.S., Yang J.I., Kim S.B. : Shcar Coagulation Process for Selective Uprgading of Ultrafine Coals. J. of Korean Inst. of Mineral anal Energy Resources Engineers, Vol. 29, No. 3, 1992
13. Roza Kintiszewa-Dimitrowa, Stojczo M. Stoew: Flokulacja selektywna. Śląsk 1987
14. Hag Seong Kim, Duk Jong Joo, Young Do Jung: Flocculant properties of the polycondensate of propanoamine and epichlorohydrin. Environmental Technology, Nr 5, 1995

20

REMOVAL OF SULPHATES FROM TAP WATER BY PRECIPITATION OR CRYSTALLIZATION OF ETTRYNGITE

A. Woroszyńska,* K. Van Leuven, and D. Wilms

Department of Chemical Engineering
K. U. Leuven
De Croylaan 46, B 3001 Heverlee (Leuven), Belgium

ABSTRACT

In this paper a new method of sulphate removal from drinking water is described. The process is based on the crystallization or precipitation of ettryngite in a fluidized bed reactor. The major advantage of the process is the fact that no sludge is produced, but small water-free pellets.

1. INTRODUCTION

The WHO European and WHO International limit standard parameters of sulphates for drinking water to 250 mg/l and 400 mg/l [1] The public water supplies of the 100 largest cities in the United States were found to contain a median sulphate concentration of 26 mg/l, and maximum of 572 mg/l. Greater concentrations were found in many ground water supplies for smaller communities in the Midwest [2]. Also in Europe, i.e. in the United Kingdom, water supplies contain sulphate levels from 4 to 303 mg/l. Water containing high concentrations of sulphate can have a laxative effect and even cause damage to organs. The corrosion of distribution conduits is increased by high sulphate levels.

Sulphates are not removed from tap water by any traditional method. This research tried to present and to develop sulphate removal by ettryngite crystallization in a pellet reactor. First, the method was examined by jar tests to get some information about ettryngite precipitation and the parameters of the process. Then the experiments were performed on a lab-scale reactor.

* Visiting scholar at Department of Chemical Engineering, K.U.Leuven, De Croylaan 46, B 3001 Heverlee (Leuven), Belgium

Chemistry for the Protection of the Environment 2, edited by Pawłowski et al.
Plenum Press, New York, 1996

2. ANALYTICAL METHODS FOR DETERMINATION OF IONS AND C_t

Water samples were analysed for cations: $[Ca^{2+}]$ by titration with EDTA, and Al by ICP.MS; and for anions: $[SO_4^{2-}]$ by gravimetry. The total carbonic acid C_t was determined by measuring pH and m-alkalinity with Titroprocesor 686—Dosimat 665. For the majority of these tests, the NTU—nephelometric number of turbidity units—was determined and the solubility product K_{sp} and saturation index SI were calculated for ettryngite and calcium carbonate. Table 1 presents the parameters of the investigated water (pH = 7.0).

3. REMOVAL OF SULPHATES FROM TAP WATER BY PRECIPITATION OR CRYSTALLIZATION OF ETTRYNGITE

It was found that sulphate removal is very effective by precipitation of ettryngite. The formation of ettryngite takes place when the following ions are present: Ca^{2+}, SO_4^{2-}, Al^{3+}.

$$6Ca^{2+} + 2Al^{3+} + 3SO_4^{2-} + 12OH^-$$

$$\downarrow\uparrow$$

$$3Ca(OH)_2\,2Al(OH)_3\,3CaSO_4\,32H_2O$$

According to Xie Ping [6] the solubility product of ettryngite is defined as:

$$\text{for pH} <8\ K_{sp}=[Ca^{2+}]^6[Al^{3+}]^2[SO_4^{2-}]^3[OH^-]^{12}=10^{-111.3}\ (M/l)^{23}$$

Since $Al(OH)_4^-$ is the principal form of Al at pH >8.0

$$\text{for pH} >8\ K_{sp}=[Ca^{2+}]^6[Al(OH)_4^-]^2[SO_4^{2-}]^3[OH^-]^4=10^{-43.13}\ (M/l)^{15}$$

Ettryngite crystallises when:

$$[Ca^{2+}]^6[Al(OH)_4^-]^2[SO_4^{2-}]^3[OH^-]^4 > K_{sp}$$

Table 1. Concentration of parameters in tap water

Parameter	mg/l	mmol/l
$[SO_4^{2-}]$	70.0	0.729
Al	< detection limit	< detection limit
$[Ca^{2+}]$	134.0	3.342
Na	14.0	0.609
C_t	—	7.450

4. EXPERIMENTS

4.1 Jar Test Experiments in Beakers

In the beginning the experiments on sulphate removal by ettryngite formation and precipitation were performed on a synthetic water in beakers to get more information and to design the research with the pellet reactor.

Ettryngite has calcium, sulphate and aluminium molecules in the proportion of: 6:3:2

$$6[Ca^{2+}]\ 3[SO_4^{2-}]\ 2[Al^{3+}]$$

The folowing parameters of the process have been varied:

- pH
- amount of chemical compounds
- chemical compounds

The chemical compounds used for sulphate removal by ettryngite precipitation:

- for test 1–6 :

$Na_2SO_4 \rightarrow [SO_4^{2-}]$
$NaAlO_2 \rightarrow [Al^{3+}]$
$CaCl_2 \rightarrow [Ca^{2+}]$
$NaOH \rightarrow pH$

In the tests 1–6 a higher concentration of sulphates in the synthetic water was obtained by adding Na_2SO_4. The stoichiometric doses of aluminium and calcium were added for tests 1 and 3.

- for test 7–8 :

$Al_2(SO_4)_3 \rightarrow [SO_4^{2-}], [Al^{3+}]$
$AlCl_3 \rightarrow [Al^{3+}]$ (only for test 7)
$CaCl_2 \rightarrow [Ca^{2+}]$
$NaOH \rightarrow pH$

In tests 7 and 8 synthetic water with a higher concentration of sulphates was obtained by adding $Al_2(SO_4)_3$. The stoichiometric dose of aluminium in the test 7 was obtained by dosing $[Al^{3+}]$ from $AlCl_3$.

4.1.1 Results and Discusion. Figures 1–3 shows the results of the tests. The experiments were carried out in the pH range 11–12.5 by dosage of sodium hydroxide. The best effect of removal of sulphates from tap water by precipitation of ettryngite were obtained for pH values above 12 (Figure 1).

Figure 1 shows that using a relative overdose of Ca^{2+}, the sulphate concentration is decreased. The reduction of the sulphate concentration was measured from 5.20 mmol/l till about 3.70 mmol/l. Similary results were received by simultaneous increase of the dose of Ca^{2+} and Al^{3+} (Figure 1, test 6). It was shown from tests 1–6 that using $NaAlO_2$ as an Al source, does not give the some result as using $Al_2(SO_4)_3$ with the respect to a low sulphate concentration. The reduction of the sulphate concentration was observed from 5.20 mmol/l till 2.90 mmol/l for tap water and till 1.40 mmol/l for distilled water (Figure 2).

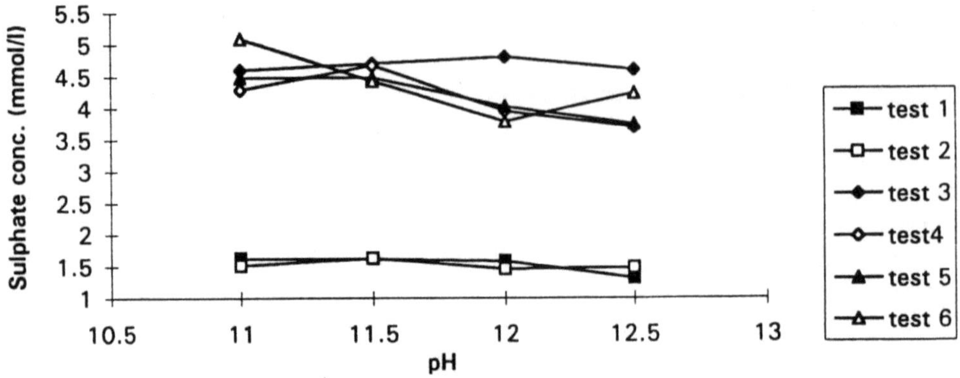

Initial sulphate conc. (mmol/l) for tests: 1, 2 - 1.67, for tests:3, 4, 5, 6 - 5.20

Test 1 $(Ca^{2+}/SO_4^{2-}/Al^{3+})_{init.}$=6:3:2 (mol/mol)

Test 2 100% more Ca^{2+} in relation to test 1

Test 3 $(Ca^{2+}/SO_4^{2-}/Al^{3+})_{init.}$=6:3:2 (mol/mol)

Test 4 32% more Ca^{2+} in relation to test 3

Test 5 92% more Ca^{2+} in relation to test 3

Test 6 32% more Ca^{2+} in relation to test 3,
50% more Al^{3+} in relation to test 3

Figure 1

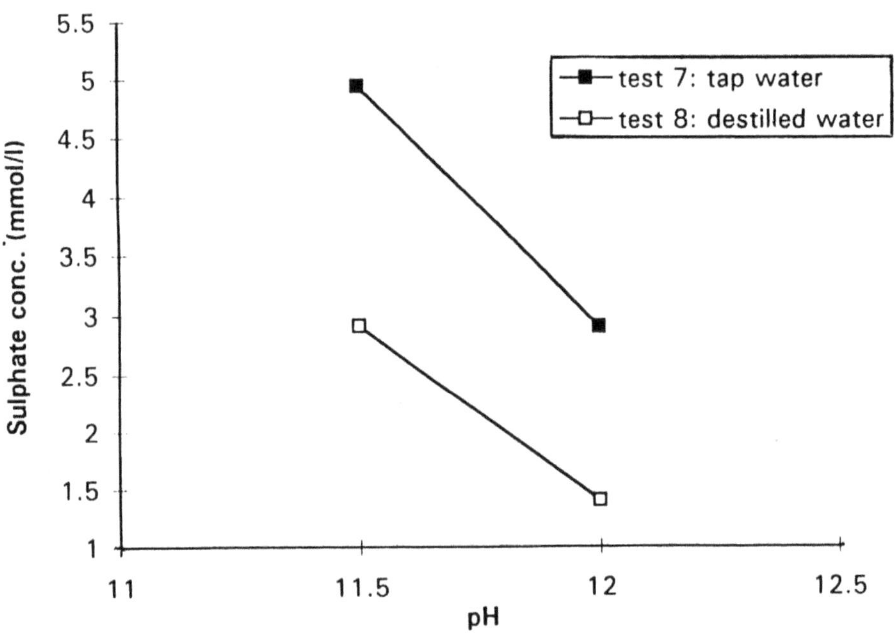

Initial sulphate conc. 5.2 mmol/l

Figure 2

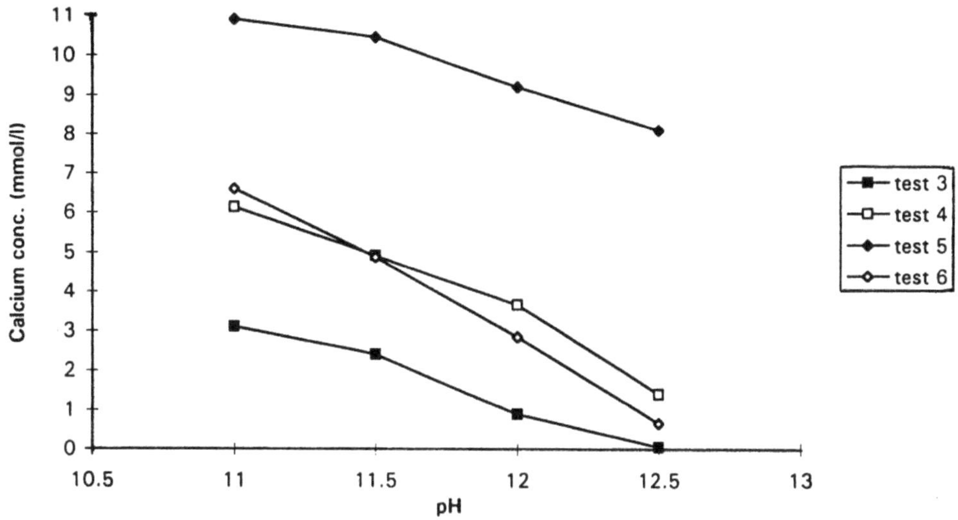

Initial calcium conc. (mmol/l) for: test 3 - 10.41, test 4 - 13.74, test 5 - 20.00, test 6 - 13.74

Figure 3

The higher sulphate concentration for tap water can be due to the reaction of Ca^{2+} ions with CO_3^{2-} ions and precipitation of calcium carbonate $CaCO_3$.

Also the second effect of the process of reduction of calcium was noticed (Figure 3) the water is also softened.

4.2 Pellet Reactor Experiments

The experiments for the removal of sulphates from drinking water were carried out in a pellet reactor.

The scheme of the reactor is shown in Figure 4.

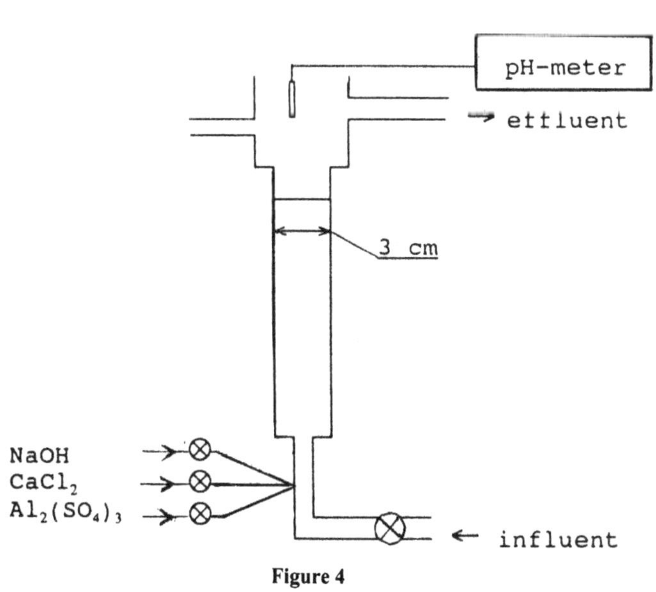

Figure 4

Table 2. The parameters of the process for tests: 9, 10, 11

Number of test	Total concentration of ions - c_{tot}						Flow rate for pumps q	
	$[SO_4^{2-}]$		$[Al^{3+}]$		$[Ca^{2+}]$		Pump 2	Pump 3
	mg/l	mmol/l	mg/l	mmol/l	mg/l	mmol/l	ml/s	ml/s
9	500	5.20	80.73	2.99	550.83	13.74	0.00815	0.05728
10	500	5.20	80.73	2.99	688.54	17.18	0.00815	0.09526
11	750	7.81	127.51	4.72	626.25	15.625	0.01290	0.08651

The main part of the reactor is a glass cylinder with an inner diameter of 3 cm and a total length of 170 cm. The reactor was filled with sand grains (0.2–0.3 mm). The water was pumped through the reactor in an upward direction by pump 1 and the chemicals ($Al_2(SO_4)_3$, $CaCl_2$, NaOH) by the next three pumps in the same direction.

The pH of the process was observed by means of a pH meter.At the top of the glass reactor the effluent was analyzed for the following ions SO_4^{2-}, Ca^{2+}, Al^{3+}.

The higher concentration of sulphate in the synthetic water was obtained by adding $Al_2(SO_4)_3$. Because the aluminium was applied in the same compound, aluminium was always dosed 14% less than stoichiometric dose.

The water was always injected with a flow rate of Q=5.45 ml/s

4.2.1 Results and Discussion. Figures 6 and 7 show the results of the tests. The experiments were performed in the pH range 12.0–12.40. Ettryngite crystallized as pellets in the pellet reactor. It was found that the pellet diameter was increased along the reactor from bottom to top, as the density of ettryngite is much lower than the density of quartz sand.

Figure 6 shows that a higher dose of Ca^{2+} decreases the sulphate concentration in the effluent.

Starting with a sulphate concentration in the tap water of 7.81 mmol/l, the lowest effluent sulphate concentration was 1.45mmol/l. Figure 7 shows the reduction of calcium for all tests was always much below limit standard parameters of calcium [1].

A mix of ettryngite and calcium carbonate was obtained in all tests (tab. 3, 4).

The negative effect of this process is high from a pH-value above 12. To decrease it is necessary to use a large amount of acid; also we added some new ions to the water.

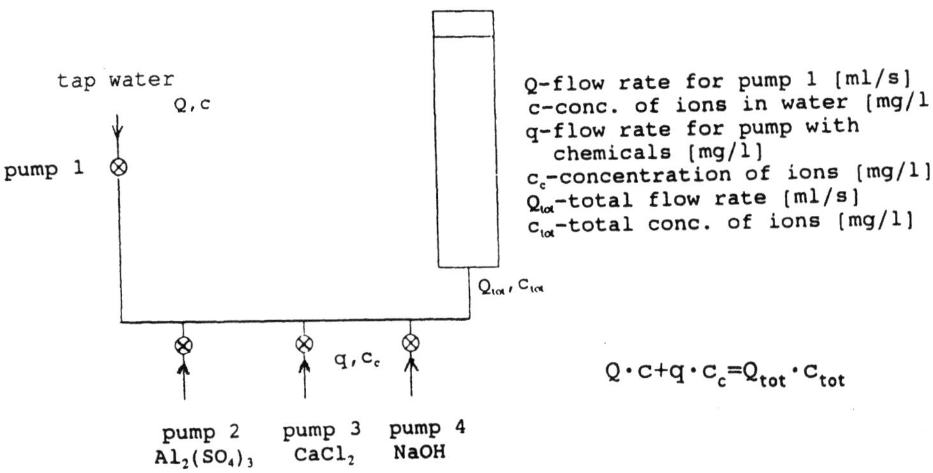

Figure 5

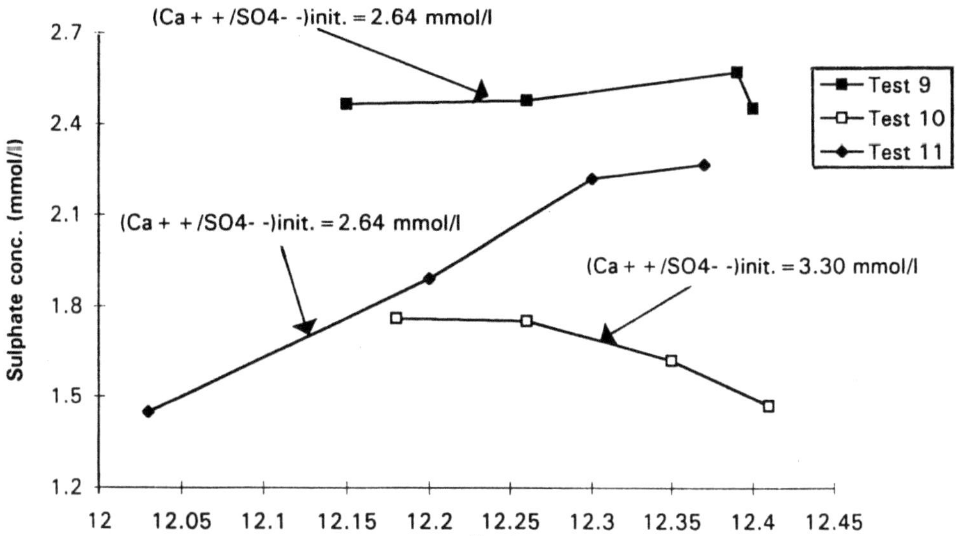

Initial sulphate conc. (mmol/l) : for tests 9, 10 - 5.20 , for test 11 - 7.81

Figure 6

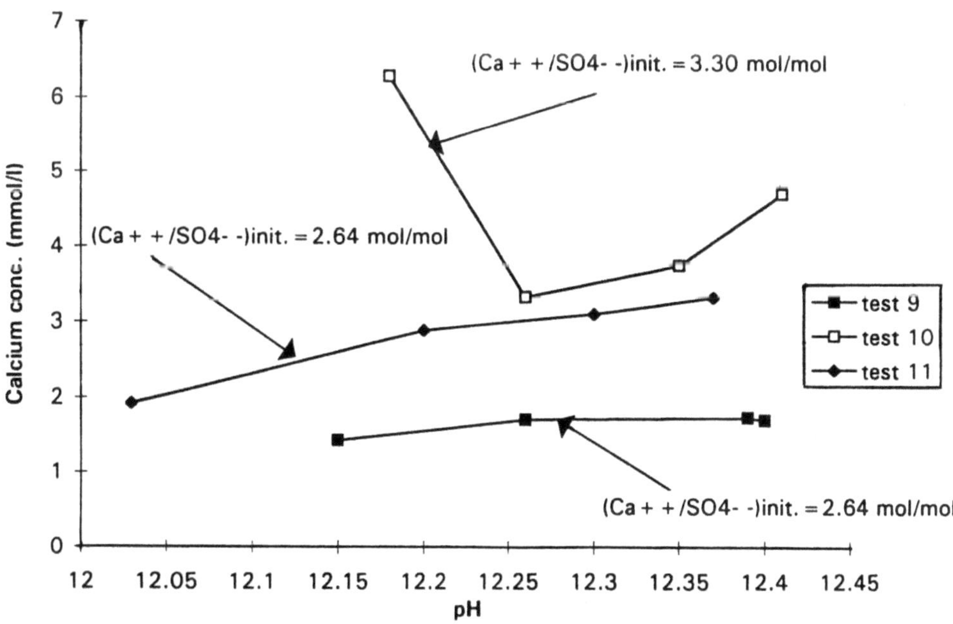

Initial calcium conc. (mmol/l) for test: test 9 - 13.75, test 10 - 17.18, test 11 - 15.63

Figure 7

Table 3. The conc. of ions in pellet for tests: 9, 10, 11

Number of test	Determinant		Concentration in pellet			
9						
	$[SO_4^{2-}]$	mmol/l	2.744	2.731	2.637	2.757
	$[Ca^{2+}]$	mmol/l	12.315	12.040	12.010	12.045
	$[Al^{3+}]$	mmol/l	1.224	2.157	2.367	2.344
	$Ca^{2+}:SO_4^{2-}:Al^{3+}$		12:3:1	12:3:2	12:3:2	12:3:2
10						
	$[SO_4^{2-}]$	mmol/l	3.440	3.450	3.580	3.730
	$[Ca^{2+}]$	mmol/l	10.905	13.855	13.430	12.480
	$[Al^{3+}]$	mmol/l	2.906	2.972	2.968	2.978
	$Ca^{2+}:SO_4^{2-}:Al^{3+}$		11:3:3	14:4:3	13:4:3	13:4:3
11						
	$[SO_4^{2-}]$	mmol/l	6.361	5.920	5.590	5.543
	$[Ca^{2+}]$	mmol/l	13.705	12.745	12.525	12.305
	$[Al^{3+}]$	mmol/l	4.163	4.564	4.650	4.639
	$Ca^{2+}:SO_4^{2-}:Al^{3+}$		14:6:3	13:6:5	13:6:5	12:6:5

Table 4. The solubility product K_{sp} and saturation index SI for ettryngite and calcium carbonate

Number of test	pH	Ettringite		Calcium carbonate	
		K_{sp} $(M/l)^{15}$	SI	K_{sp} $(M/l)^2$	SI
5					
	11.0	$1.30 \cdot 10^{-41}$	2.12	$3.91 \cdot 10^{-5}$	3.91
	11.5	$8.54 \cdot 10^{-47}$	−3.07	$4.41 \cdot 10^{-5}$	3.96
	12.0	$1.11 \cdot 10^{-38}$	5.04	$4.95 \cdot 10^{-6}$	3.01
	12.5	$6.57 \cdot 10^{-38}$	5.82	$5.52 \cdot 10^{-5}$	4.06
6					
	11.0	$1.79 \cdot 10^{-38}$	5.25	$1.80 \cdot 10^{-5}$	3.57
	11.5	$4.34 \cdot 10^{-37}$	6.64	$1.46 \cdot 10^{-5}$	3.48
	12.0	$1.85 \cdot 10^{-36}$	7.27	$7.09 \cdot 10^{-6}$	3.17
	12.5	$7.05 \cdot 10^{-39}$	4.85	$1.77 \cdot 10^{-6}$	2.56
7					
	11.5	$7.37 \cdot 10^{-37}$	6.87	$1.76 \cdot 10^{-5}$	3.58
	12.0	$5.41 \cdot 10^{-37}$	6.73	$9.81 \cdot 10^{-6}$	3.31
8					
	11.5	$2.36 \cdot 10^{-37}$	6.37	$1.27 \cdot 10^{-5}$	3.42
	12.0	$5.65 \cdot 10^{-41}$	2.75	$1.84 \cdot 10^{-6}$	2.58
9					
	12.15	$1.55 \cdot 10^{-38}$	5.19	—	—
	12.26	$2.77 \cdot 10^{-38}$	5.44	—	—
	12.39	$6.36 \cdot 10^{-38}$	5.80	—	—
	12.40	$5.75 \cdot 10^{-38}$	5.76	—	—
10					
	12.18	$1.15 \cdot 10^{-37}$	6.06	$7.43 \cdot 10^{-6}$	3.19
	12.26	$1.79 \cdot 10^{-40}$	3.25	$2.62 \cdot 10^{-6}$	2.74
	12.35	$1.07 \cdot 10^{-39}$	4.03	$4.72 \cdot 10^{-6}$	2.99
	12.41	$1.21 \cdot 10^{-39}$	4.08	$5.36 \cdot 10^{-6}$	3.05
11					
	12.03	$6.30 \cdot 10^{-40}$	3.80	$3.16 \cdot 10^{-6}$	2.82
	12.20	$6.15 \cdot 10^{-39}$	4.79	$2.15 \cdot 10^{-6}$	2.65
	12.30	$8.20 \cdot 10^{-39}$	4.91	$3.22 \cdot 10^{-6}$	2.82
	12.37	$5.43 \cdot 10^{-38}$	5.74	$3.55 \cdot 10^{-6}$	2.87

5. CONCLUSIONS

1. Sulphates are effectively and quickly removed from tap water by dosing aluminium and calcium salts at room temperature.
2. The best effects of sulphate removal were obtained for a pH value greater than 12.
3. During the process calcium hardness is simultaneously removed from tap water.
4. As a final product water-free pellets are obtained.

6. REFERENCES

1. W.J.Masschelein, Unit processes in drinking water treatment, Marcel Dekker, Inc., 1992
2. Water quality criteria 1972—Ecological Research Series, EPA, 1972
3. Guidelines for drinking-water quality, WHO, 1984
4. J.C.Van Dijk, D.A.Wilms, Water treatment without waste material: fundamentals and state-of-the-art of pellet reactor, KU Leuven, 1991
5. D.A.Wilms, K.Van Leuven, F.Verslype, Removal of sulphates from waste water by precipitation or crystallization of ettryngite, Environmental Platform May 1993—proceedings, 1993
6. Xie Ping, J.J.Beaudoin, Mechanism of sulphate expansion, Cement and concrete research, vol.22 p. 631–640, 1992

21

THE ELIMINATION OF PESTICIDES FROM DRINKING- AND WASTEWATER

I. Raisz,[1] J. Emmer,[1] T. Montovay,[2] M. Assenmacher,[3] and F. H. Frimmel[3]

[1] University of Miskolc
Department of Physical Chemistry
[2] University of Miskolc
Department of Analytical Chemistry
[3] University of Karlsruhe
Department of Water Chemistry

1. INTRODUCTION

Several pesticides and their derivatives mean large potential risk for the living nature and for the whole mankind. These chemicals can get in to natural systems basically in two ways: by the wastewater from production, and during the agricultural application.

This project is focused on the removal of thiocarbamate derivatives from aqueous emissions of thiocarbamate based pesticide manufacturing. The transformation of these odorous and harmful compounds was attempted by oxichlorination, photochlorination, photooxidation with air, photocatalic oxidation with air over TiO_2 and electrooxidation. Analytical methods using gas-chromatography (GC), high-performance-liquid-chromatography (HPLC), mass-spectrometry (MS) coupled with GC were developed to detect intermediates of degradation. The investigations with appropriate theoretical background comprise a wide range of solutions for this crucial problem of environmental protection. The electrooxidation procedure, can be applied for the decontamination of wastewaters from production, where these compounds can be found in higher concentrations. For the elimination of pesticides of lower concentrations (< 10 μg/l) the nanofiltration method can be applied.

In our experiments we have used laboratory scale nanofiltration to investigate the elimination of atrazin and metazachlor from dilute aqueous solutions. We have chosen these two compounds from the huge number of different pesticides for our work because of their frequent occurrence in several drinking-water sources.

Our experimental work has consisted of two parts:

- In the first step we have determined the parameters of the HPLC analysis for the two examined pesticides.
- We studied the potential of the elimination by means of nanofiltration under several experimental conditions.

Chemistry for the Protection of the Environment 2, edited by Pawłowski et al.
Plenum Press, New York, 1996

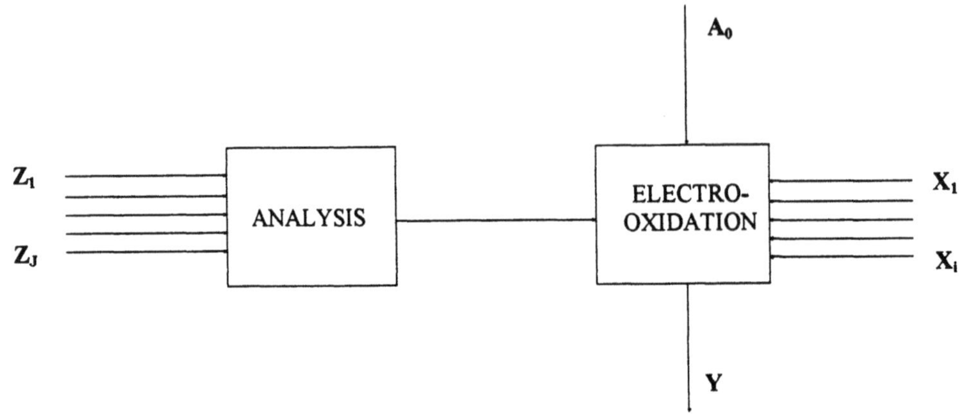

Figure 1. System and its parameters.

2. ELECTROOXIDATION

One of the most important aspects in the treatment of highly toxic compounds is the degradation of pesticides. Wastewater decontamination by using halogens reduces the smell level of the treated wastewater, however this procedure often leads to the formation of new toxic substances. Numerous modern methods have been developed using electrochemical processes [1]. Some advantages of these methods include:

- the method is exempt from secondary materials
- separation is effective
- the possibility for direct recovery of compounds (for example metals)
- easy process control.

We have developed an electrooxidation procedure for wastewater decontamination. First we calculated the change of free enthalpy for each important reaction by the determination of electrode potential and formation of degradation products, as written by Robert A. Alberty [2]. No thermodynamic data for calculations were available. Therefore the basic information was calculated by approximation methods. We have determined the goodness of the methods on the basis of electrooxidation experiments [3]. Some possible products of degradation of EPTC from theoretical considerations and experimental results include:

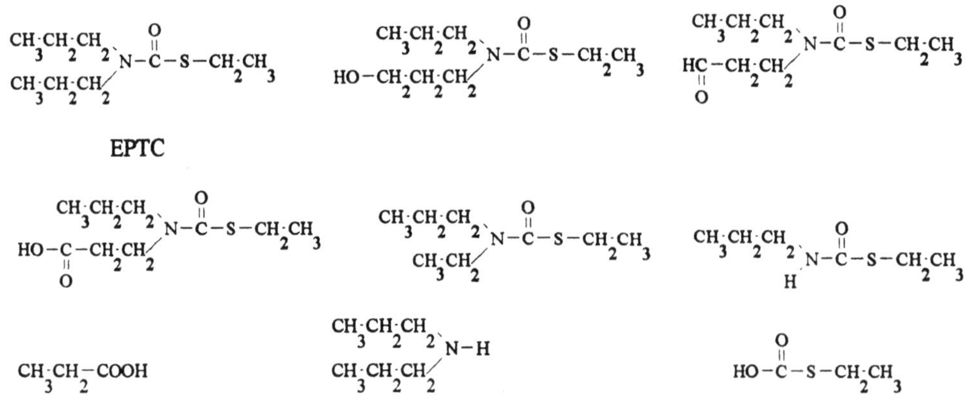

Experimental planing was carefully performed to ensure effective experiments.

2.1. The Planning of Experiments

For the planning of experiments and assessment of experimental results we used computer methods. It was absolutely necessary for the planning of experiments because of numerous degradation reactions and the complexity of parameters of oxidation and analysis.

The block schemes of subsystem analyses illustrates the complexity of the system (Figures 2 and 3).

The following formulas show the method of calculation for parameter centres. $\left(z_1^0\right)$

$$z_i^o = \frac{z_i^{max} + z_i^{min}}{2}$$

where z_i^{max}: maximum value of parameter; z_i^{min} : minimum value of parameter.

Range of parameters:

$$\Delta z_i^o = \frac{z_i^{max} - z_i^{min}}{2}$$

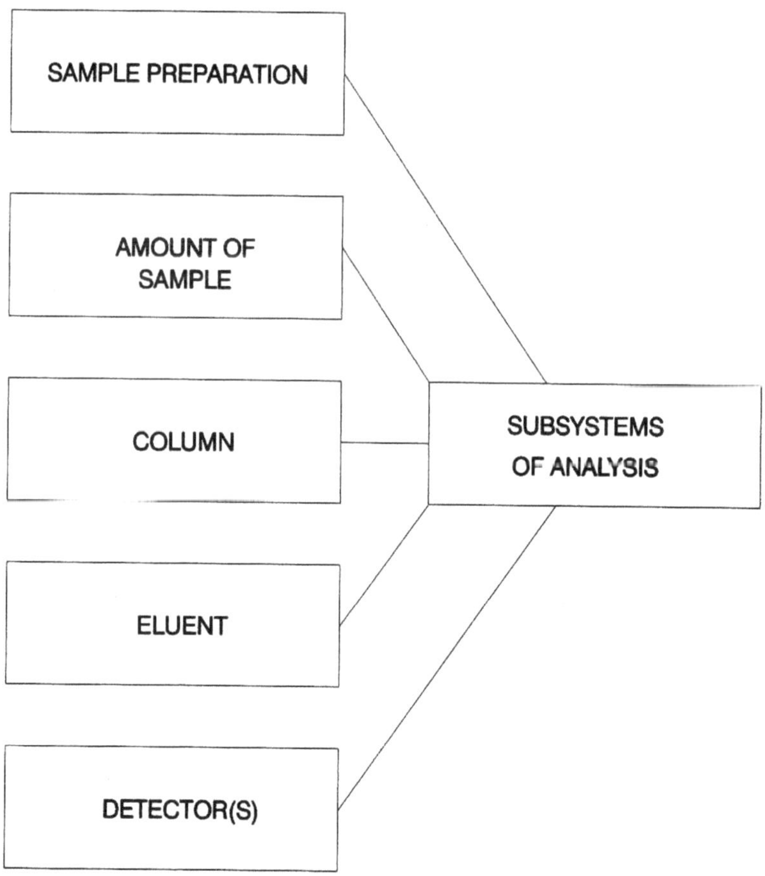

Figure 2. Subsystem of analysis.

Factors of analysis

Extraction	0, 1
Polarity of solvent	4 levels
Flow rate of eluent	4 levels
Detector	0, 1
Wavelength	4 levels
Electrode potential	4 levels
Amount of sample	0, 1
Column	0, 1

Factors of electrooxidation

Electrode potential	4 levels

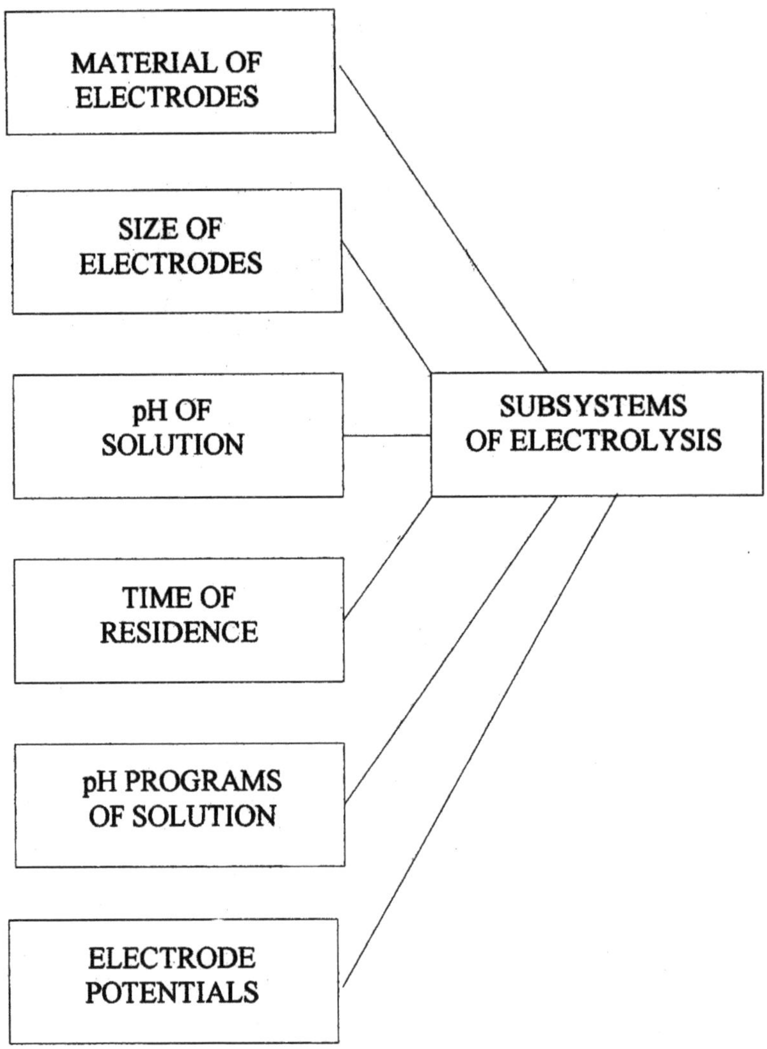

Figure 3. Subsystem of electrolysis.

Material of electrode	4 levels
pH	4 levels
Residual time	0, 1
Electrode	0, 1
Auxiliary material	4 levels

2.2. Experiments

The scheme of the experiment can be seen in the Figure 4.

The level of impurities in the artificial wastewater (starting samples) was in the range of about 75–125 ppm. We have used different cells and electrodes for the electrolysis. One had a diaphragm, the others did not. The electrodes were Pt, Ti, Sn and Cu in different pairs. An HPLC system was been used for investigation of EPTC degradation. Sampling methods were both periodical and continuous. For the preparation of samples we used SPE methods with Baker cartridges and filtration.

The parameters of analysis are the following:

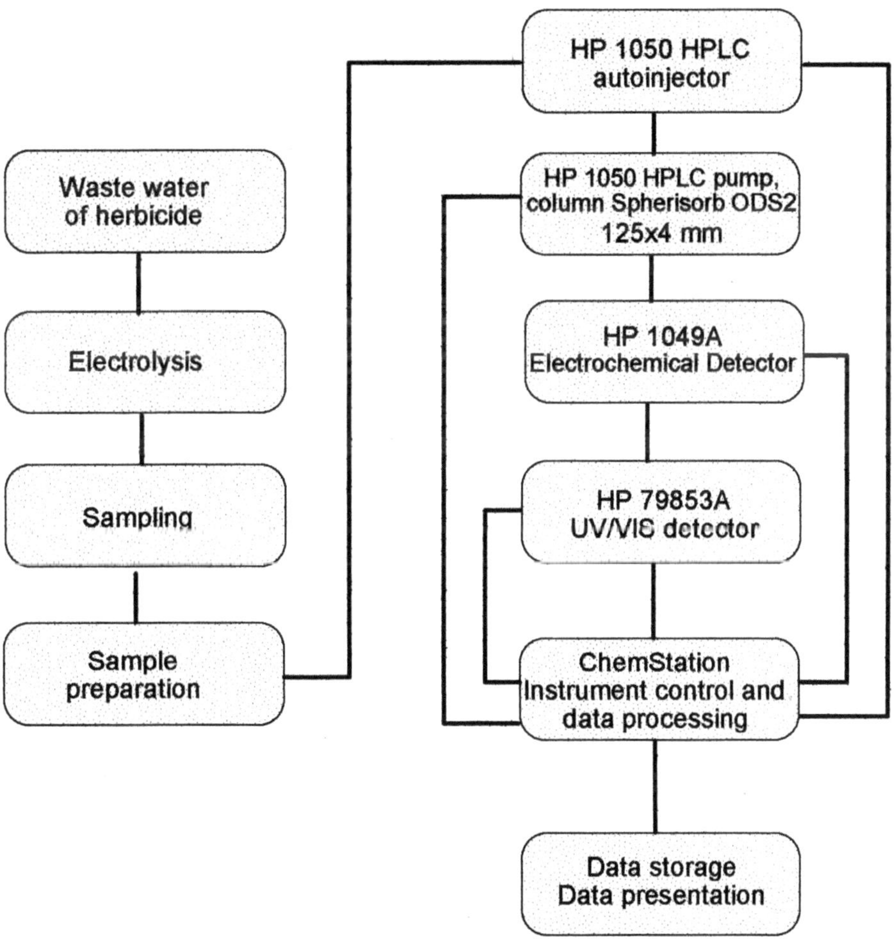

Figure 4. Method.

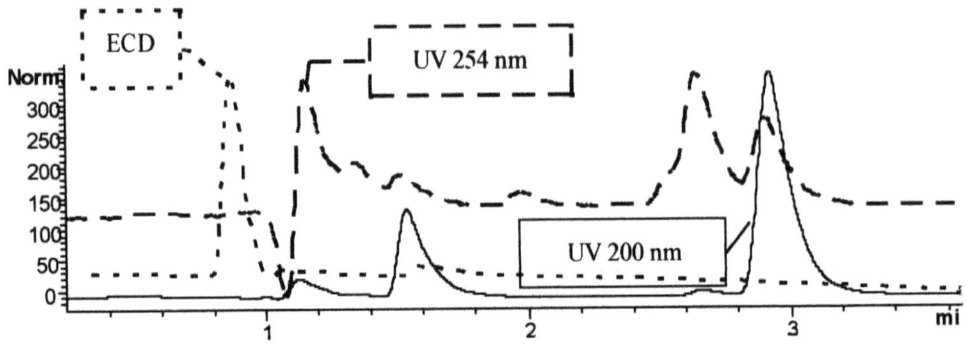

Figure 5. Comparison of detectors.

• HPLC	Hewlett Packard (HP)1050
• Detectors	HP 79853 UV/VIS
	HP 1049A ECD
• Columns	HP Spherisorb ODS2 5μm, 250 x 4 mm
	HP Spherisorb ODS2 5 μm, 250 x 2 mm
• Temperature of columns	40 °C.
• Eluents	A: Acetonitril
	B: sodium-acetate 0.02 mol/l
	at optimum A/B = 70/30
• Flow rate	1 cm^3/min
• Sample	20 μl

For the determination of optimum of analytical methods, we measured the UV spectrum of a compound by a UV detector and voltammogram by ECD. The same sample signal was registrated in Figure 5 with different methods.

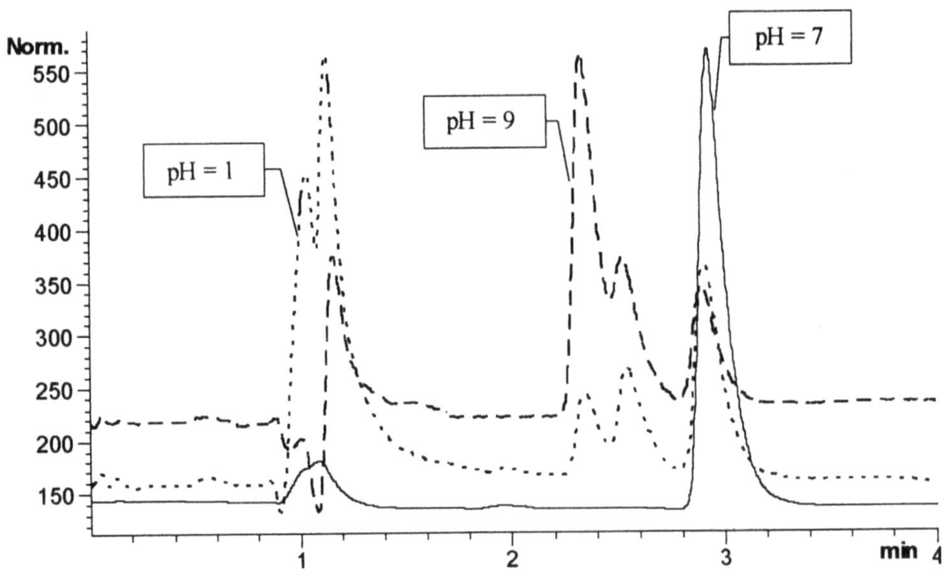

Figure 6. Effect of pH.

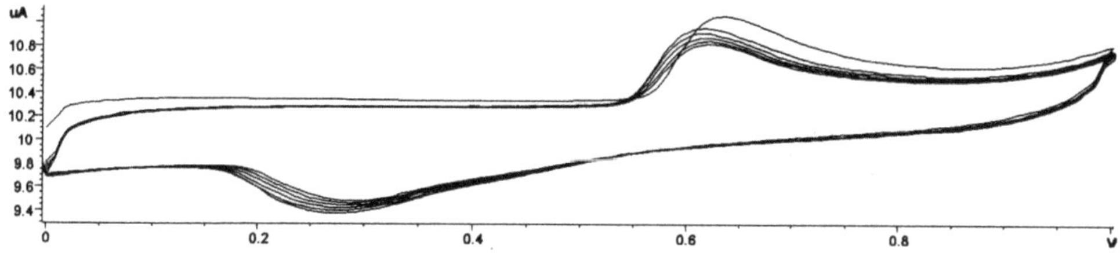

Figure 7. Voltammogram.

As shown in the Figure 5, exact results are available only with programmed UV and ECD detectors simultaneous but the first ECD detector has a small dead volume, and in this case the resolution of the UV detector is worse.

The effect of pH is important. It can be seen in Figure 6, that the EPTC peak (retention time: 2.9 min) has decreased significantly in both acidic and alkali solutions. Also the first visible products of degradation appeared at retention times around 1 min. and 2.4 min.

We determined the optimum potential with the help of the electrochemical detector. Without a column, the starting sample was filled using the HPLC pump, after that in the stationary state we registered the voltammograms (Figure 7). So we determined the optimum potential and we had the maximum signal from degradation products.

We performed a large number of kinetic investigations with electrodes made of different materials. Two of them are presented in Figure 8.

The differences between the two electrode materials a induced half-life time of more than one order of magnitude. The cause of differences is the oxide, developed from electrode material.

With a full knowledge of the experimental data, we set up the complex system for large scale laboratory experiments shown in Figure 9.

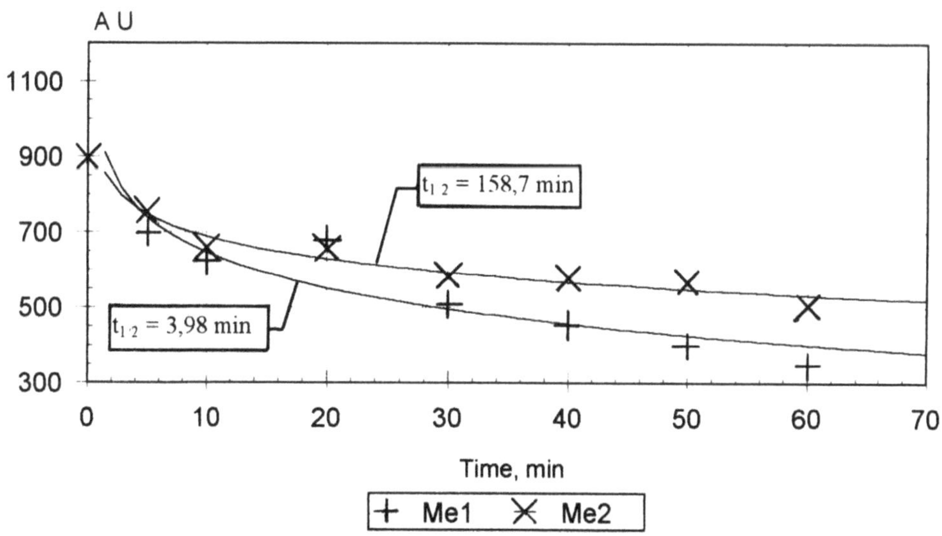

Figure 8. Effect of electrode material.

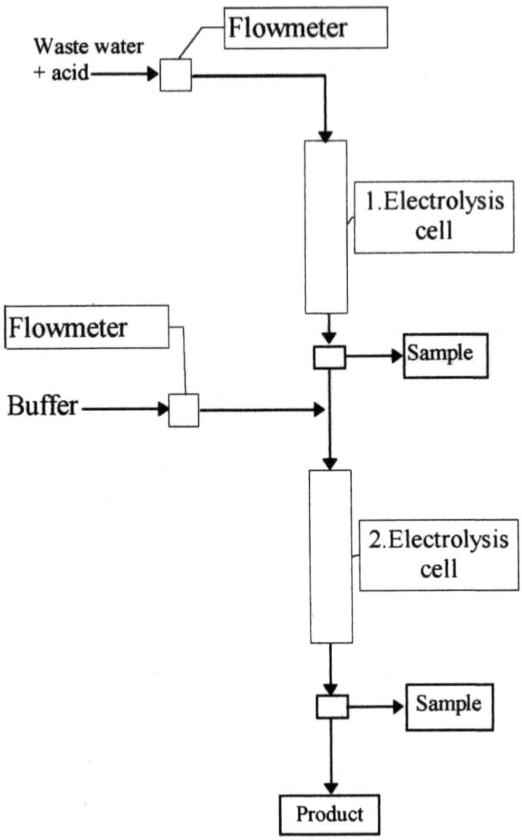

Figure 9. Pilot system.

The acidic wastewater is conditioned and electrolysed in the first cell. After this, the pH is adjusted to 10 with an alkali buffer and electrolysis performed again. The degradation products can be seen in Figure 10 as recorded by ECD. The different curves belong to the samples following 10 minutes after each other in the secondary cell.

Transformations of products are shown in Figure 10 and 11. There can see the transformation of alcoholic - aldehyde - carboxylate groups. With small electrode distances the process was successful.

The key to the process electrooxydation was the use of proton-exchange membranes by A. K. P.Chu[4] and O. J. Murphy [5]. If the degradation process is a series of oxidation and reduction reactions, the small electrode distances function better than membranes.

We have an electrochemical degradation method for pesticides plant, with more then 50% current efficiency. With 2 hours residence time the mineralisation of EPTC is more then 80%.

3. THE ELIMINATION OF PESTICIDES BY THE MEANS OF NANOFILTRATION

Because of the very low pesticide concentrations, direct liquid chromatographic analysis of the compounds has not been possible. For this reason we have enriched our

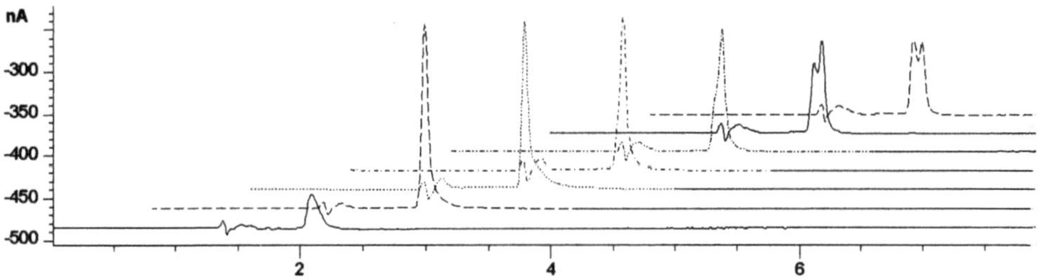

Figure 10. Effect of time for degradation products.

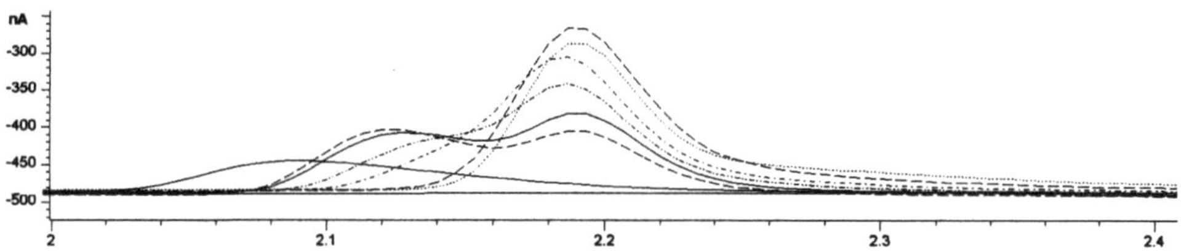

Figure 11. Change of degradation products.

samples by a factor 500 by the means of solid phase extraction on a RP C-18 chromatographic adsorbent. The retention time for the atrazin was about 8.6 min, for the metazachlor about 10.5 min.. At a concentration factor of 5000 the detection limit for the both substances was lower than 0.05 µg/l.

The equipment schematic we used for the filtration experiments can be seen in Figure 12. The water stream can be regulated between 0 and 4 m^3/h, and the effective pressure between 0 and 15 bar. The effective surface of the applied filtration module is 2.5 m^2, and it can be used between pH 4–12. During measurement the concentrate and the filtrate

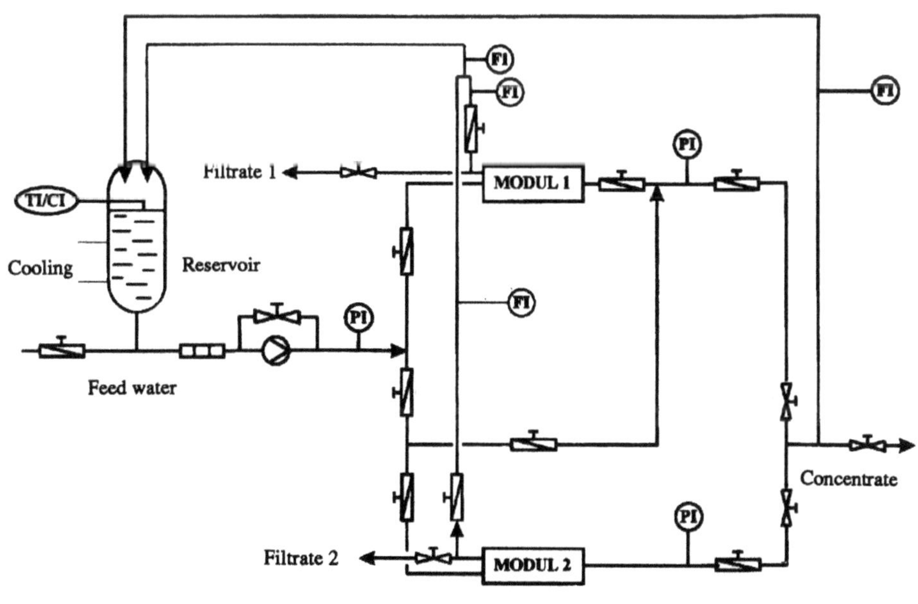

Figure 12. The scheme of the membrane filtration equipment.

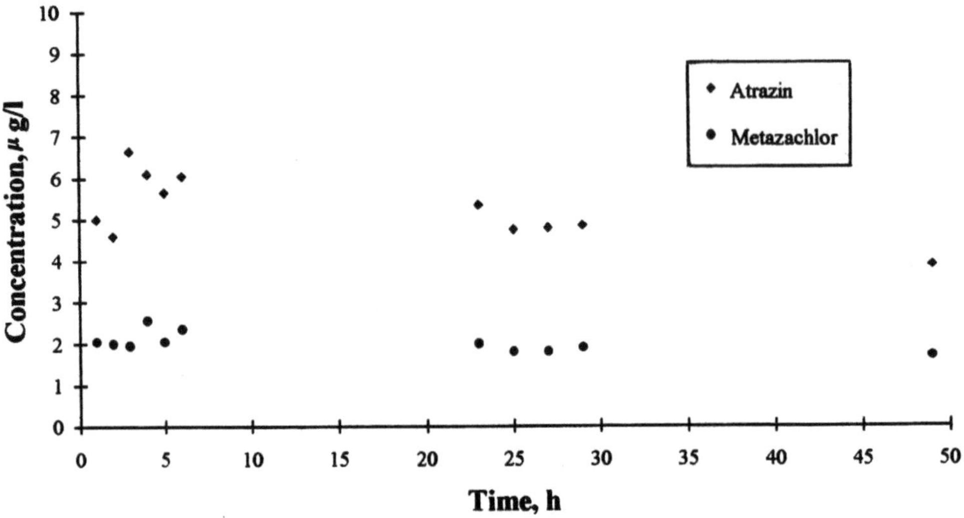

Figure 13. The adsorption of the atrazin and the metazachlor in the membrane filtration equipment.

are recirculated, and has made possible an optional period to measure with 50 l sample solution. The composition of the sample solution was about 1 mmol/l $CaCl_2$, 1 mmol/l NaCl 10 µg/l atrazin and 4 µg/l metazachlor.

Before the membrane filtration experiments we measured the adsorption of the both substances in the equipment at 0 bar effective pressure (Figure 13.). The beginning concentrations of the two pesticides decreased rapidly to half, and then it changes only relative slowly. Since the adsorption was very significant, and the adsorption-desorption processes depend on the pressure and temperature, we have determined before each filtration experiment the actual concentrations of the two pesticides in the feed water.

Figure 14 shows the rate of the atrazin and metazachlor elimination as a function of the effective pressure. As it can be seen from the results of two parallel experiments, the

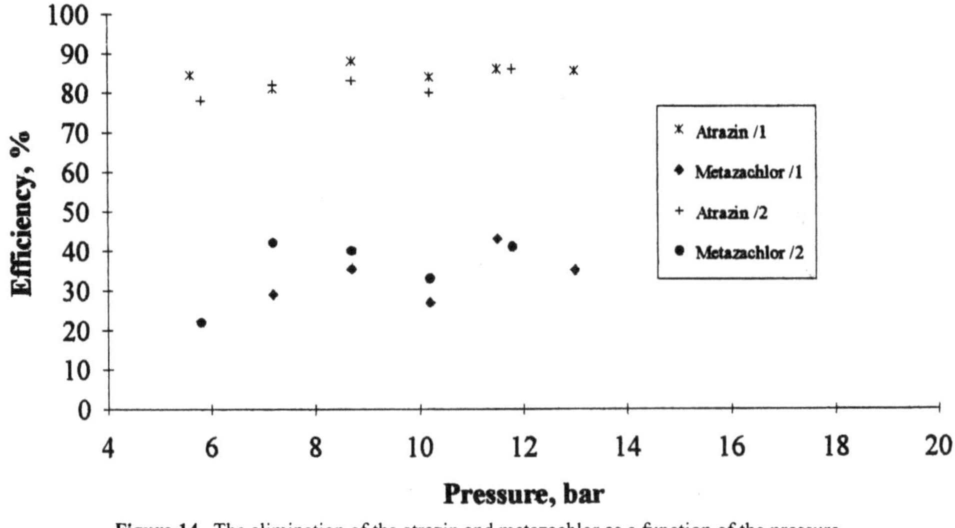

Figure 14. The elimination of the atrazin and metazachlor as a function of the pressure.

atrazin can be eliminated in a significantly higher rate (about 80%) than the metazachlor (30–40%). The efficiency of the elimination is dependent on the pressure in the case of the both substances at only at low rate.

4. ACKNOWLEDGMENT

The authors are indebted to Ms. G. Lengyel for her helpful experimental work and to Mr. S. Gál-Sémer for the successful data-processing. The cost of this research was supported by PHARE ACCORD H-9112 0090.

5. REFERENCES

1. Szebényiné, Győri Enikő, Magyar Kémikusok Lapja, **92**, 53, (1992)
2. R. A. Alberty, J. of Electrochem. Soc., **142**, 120 (1995).
3. I. Raisz, J. Emmer, Study of the Electrochemical Oxidation of Thiocarbamat Herbicides. 13th IUPAC Conference on Chemical Thermodinamics, Clermont-Ferrand, France, July 17–22, 1994.
4. Chu A. K. P. at al: J. Appl. Electrochem., **4**, 323, (1974)
5. Murphy O. J. at al: 20th Intersociety Conference on Environmental Systems, Williamsburg, July 9, 1991.

DIFFERENCES OF OPINION IN THE APPLICATION, USAGE, AND RESEARCH OF LCA

Nebiyeleul Gessese, Andreas Windsperger, and Mashid Sotoudeh

Research Institute of Chemistry and Environment*
University of Technology, Vienna
A-1060 Vienna, Getreidemarkt 9/191

This paper briefly presents the current practicable steps for carrying out LCA, and points out the reasons which lead different LCA studies to different conclusion, although they describe the same type of studies referring to the assessment of the same products. As an example results of two different studies for different packing materials, the BUWAL packaging study and the study conducted by European Centre for Plastic in the Environment, has been considered

1. GENERAL DESCRIPTION

Differences of different studies start with on the term used to describe LCA such as Cradle-to-Grave, Life Cycle Analysis, Life Cycle Assessment, Eco-balance... etc., although it is used to describe the same type of study referring to the assessment of the complete life of a product. In this article the term LCA is used.

An LCA of a product begins with the extraction of raw material, processing and then looks at the conversion to a product, distribution and use of a product, and its ultimate disposal as waste or by recycling the product.

The five fundamental steps which should be carried out in any LCA studies are:

1. Goal definition
2. Preparation of an inventory of data
3. Analysis of impact on the environment
4. The evaluation of data from inventory in terms of their impact on the environment
5. Implement analysis which defines method for reducing the environmental load (Life Cycle Improvement Assessment).

* Tel. (0043- 1) 58801 / 5189; Fax. (0043 1) 581 29 52

Chemistry for the Protection of the Environment 2, edited by Pawłowski et al.
Plenum Press, New York, 1996

We will examine each step in detail and which points should be considered in the research, usage and application of LCA:

2. GOAL DEFINITION

The exact system to be considered in the studies should be defined at this stage. The major factors which are going to be determined are:

- the functional unit
- the time horizon
- the depth of the investigation of the inventory data
- the boundaries of the investigation
- the way of interpreting the inventory data

3. PREPARATION OF AN INVENTORY OF DATA

An inventory of LCA can be used to quantify the flow of materials and energy through the life cycle of a product. It provides the basic information needed to determine environmental impact and to assess ways to improve different product and process.

Major activities which should be considered at this stage are:

- definition of the process of the product system
- definition of the boundary of production system by following the life cycle
- specification of all processes
- identification of input and output of material, energy, emission,..etc.

4. ANALYSIS OF IMPACT ON THE ENVIRONMENT

At this stage the input and output data which are collected during the inventory will be characterised and classified according to their impact in the environment.

5. THE EVALUATION OF DATA IN TERMS OF IMPACT ON THE ENVIRONMENT

This step of LCA deals with the ranking of environmental impacts. We follow with a detailed examination of points defined under each step:

5.1. Goal Definition

5.1.1 The Functional Unit. This type of applied unit influences the whole procedure. The application always involves some kind of comparison for which a unit of use should be specified forming the basis for the comparison. The unit based on the function of the products to be compared is called the functional unit of a product.

5.1.2 The Time Horizon. This determines the type of technology to be specified. A well defined time horizon makes the case studies have a uniform technological level.

5.2 Preparation of an Inventory of Data

5.2.1 Definition of the Process of the Product System. The first activity of the inventory is the definition of the product system of all processes of a product. In determining the product system, it is important to define what belongs to the system!

Data collected at the inventory stage for LCAs fall into four main groups:

- quantitative primary data for unit operations
- data describing the linkages between unit operations
- qualitative or description data for unit operation and
- calculated data

Quantitative data for unit operations relate to:

- the quantities of the input material

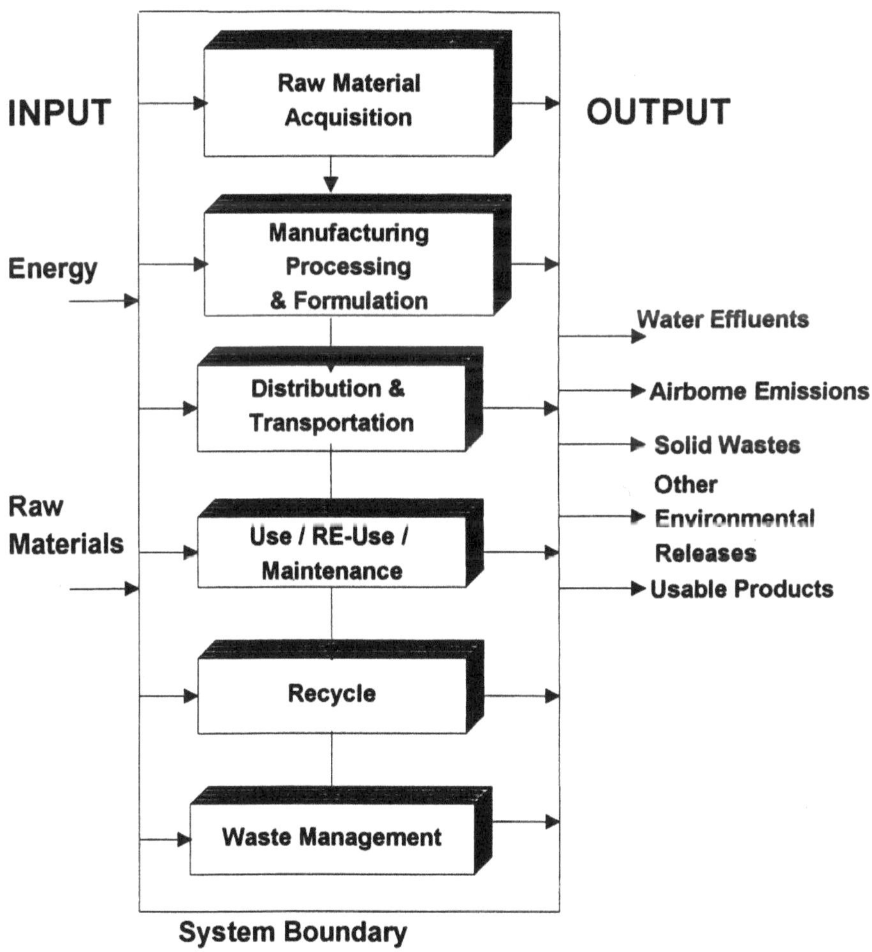

Figure 1. Life-cycle inventory.

- the quantities of the various energy inputs
- the quantities of the different solid wastes
- the quantities of the different air emission
- the quantities of the different water emission

5.2.2 Definition of the Boundary of Product System. The boundaries between the product system and the environment i.e. the input and the output to the environment, and the boundaries between the product system under study and the unit operation should be defined at equal level by considering a relevant contribution to the environmental input or output.

5.2.3 Specification of All Process.

5.2.4 Dentification of Input and Output of Material, Energy, Emission,..Etc.

5.2.5 What Are the Problems Raised by Inventory Data?

Data quality: Some ***data*** *are* in aggregated form. It becomes difficult to identify and which information belongs to which stage of the process.

In some cases, the data are difficult to determine, i.e., where they came from, therefore it is necessary in doing a survey, making assumptions, estimating, and experiencing data. All of these factors, of course, have an influence on the quality of the data.

- It is sometimes difficult to analyse all ***Recycling*** conditions. Recycling includes the collection and handling of the material, plus any processing like washing, crushing, baling etc. which is needed to prepare and deliver the materials to re-enter a manufacturing system as a raw material.
- In some cases, addition to the main necessary product, the process under study is forced to include the emissions, energy consumption...etc. of the ***couple product.*** These may be a mislead to the conclusion of the study.
- As per the definition of SETAC ***transportation*** includes moving raw materials from a production facility to a processing facility, moving finished goods from a production facility to a warehouse, moving products from point of sale to point of use, and moving electricity from a generation site to a use site. Transportation is not an independent portion of LCA because it is so intimately tied to various other activities. The boundaries of the numerous transportation activities, for instance, are defined by those activities with which they connect and depend on analysis of these activities for quantification of the amount and type of material or energy being transported. Therefore, taking the boundaries into consideration should be considered as one factor in LCA.

5.3 Analysis of Impact on the Environment

At this stage, extraction, aggregation of emission will be conducted on the basis of their potential effect on a number of assessment endpoints. The basic question raised by the impact assessment is how to proceed in the assessment after the inventory analysis. Some effects can not be easily characterised or classified as one definite environmental problem. Therefore, the characterisation or classification "scientific knowledge of environmental processes" is used to estimate the contribution of all effects to the selected environment problems.

5.4 The Evaluation of the Data in Terms of Impact on the Environment

The fourth component of environmental LCA deals with the final environmental problem appraisal based on the environment of the product under study.

5.5 Life Cycle Improvement Assessment

A systematic evaluation of the needs and opportunities to reduce the environmental burden associated with energy and raw material use and environmental releases throughout the whole life cycle of the product, process or activity by improving and changing the product, process and activity design, raw material use, industrial processing, consumer and waste management...etc. are defined and analysed under this stage of LCA.

6. WHAT ARE THE REASONS WHICH LEAD DIFFERENT LCA STUDIES TO DIFFERENT CONCLUSION ALTHOUGH BASIC ASSUMPTION OF CASE STUDIES ARE THE SAME?

Different interpretations and opinions related to LCA can have a decisive effect on the final results of the study. Explanations for the discrepancies between the results of different studies is the fact that the specification of the basis of comparison, the system boundaries,the classification. In general the assessment methods and the data used vary considerably from study to study. To give some idea of this effect, an example is illustrated below.

7. EXAMPLE OF CASE STUDIES

One of the important application field of LCA today is the packing materials, even through other fields uses in evaluating their products.

Results of two different studies for different packaging materials are compared in this paper. The BUWAL packaging study, Oekobilanz von Packstoffen Stand 1990 and the study conducted by the European Centre for Plastic in the Environment, PWMI.

In most case the basis of comparison, *functional unit*, of many studies is the same, however the *specification* of the product difference one to the other, (e.g. Crude oil input for the production of 1 kg PVC 0.90635 kg and 0.471 kg in the study of BUWAL and PWMI has been used respectively.)

The crude oil input in the case of BUWAL is almost double compared to that of PWMI, this could be by considering the other couple products. This consideration causes a substantial difference in the Energy consumption and the emissions during processing.

The other differences between these two studies is that they use different energy data as a basis for their assessment, therefore the comparison of the differences between these two studies is based principally on the energy figure.

- The PWMI study uses data from North Sea Natural Gas production, Oil input are calculated as a mix of 15% derived from North Sea and 85% from Non-North Sea production. From Non-North Sea production, though, little information was available and data had been combined with the information from the North Sea.

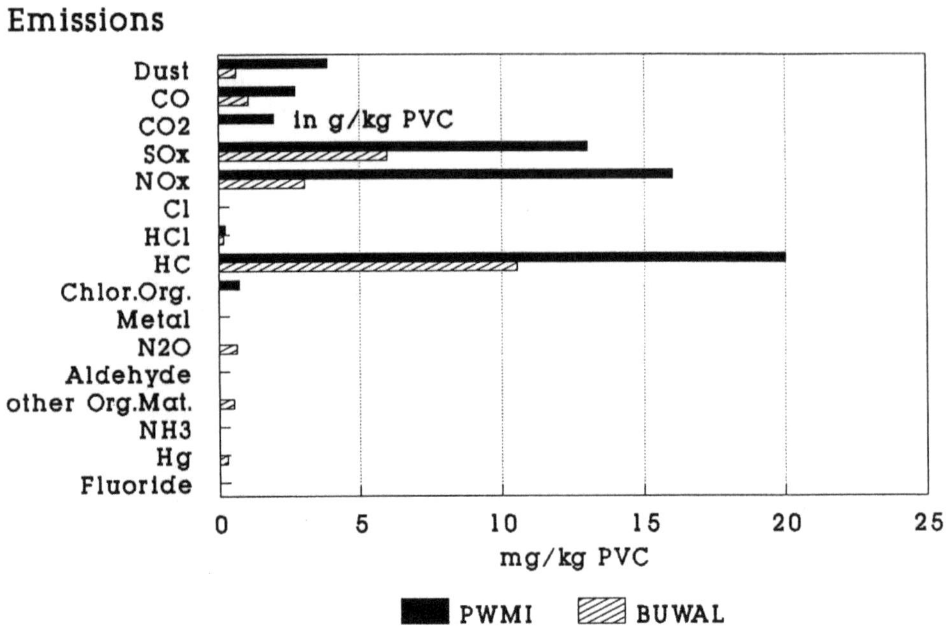

Figure 2. Air emission by PVC-production.

- The BUWAL study based on data from the study conducted by Franklin Associates and the Öko-Institute to describe environmental impact from oil and gas production. Natural gas is derived a mix of Dutch and Russian production with an average transportation distance of 4000 km. Oil products are derived as a mix of 50% Offshore North Sea production and 50% Onshore OPEC production. For electricity production data from "Union Paur Coordination de la Production et du Transport de l'Electricite (UCPTE)" of the year 1988 is taken.

Now the question is that "*how do the difference in the energy data sources affect the final result of the product under study?*" In this case the packing material.

Taking the example of Poly Vinyl Chloride (PVC), it can be seen that the differences in the ultimate conclusion of these two studies are remarkable. Most air emissions, reported in the PWMI, are considerably greater than those reported in BUWAL, Fig 2. In the BUWAL study the value for CO_2, Chlorinated Organic, and Metal emissions are not reported. On the other hand, N_2O, Aldehyde, other Organic, Ammoniac, Hg and Fluoride are not recorded in the study of PWMI. Dust, CO, and HC values are higher in the PWMI study than those of in the BUWAL study. The reason is unknown. It can be because of different *energy data used or different definitions of the boundaries,..etc.*

Let's now change to the Water emission Fig 3, Just like the air emissions, some emissions are not indicated, such as, Acid as H^+, Metals, Chloride Ions, Sulphate Ions, Sodium Ions in the BUWAL study and Phenol, TOC, Fluoride and Hg in the PWMI study.

The crude oil input of BUWAL is greater than that of PWMI, It makes therefore all burdens to the environment caused by this factor greater than those caused by PWMI. On the other, hand the transport assumptions taken by PWMI is greater than that of BUWAL, therefore the burden caused by this factor is greater by PWMI. Although the type of emis-

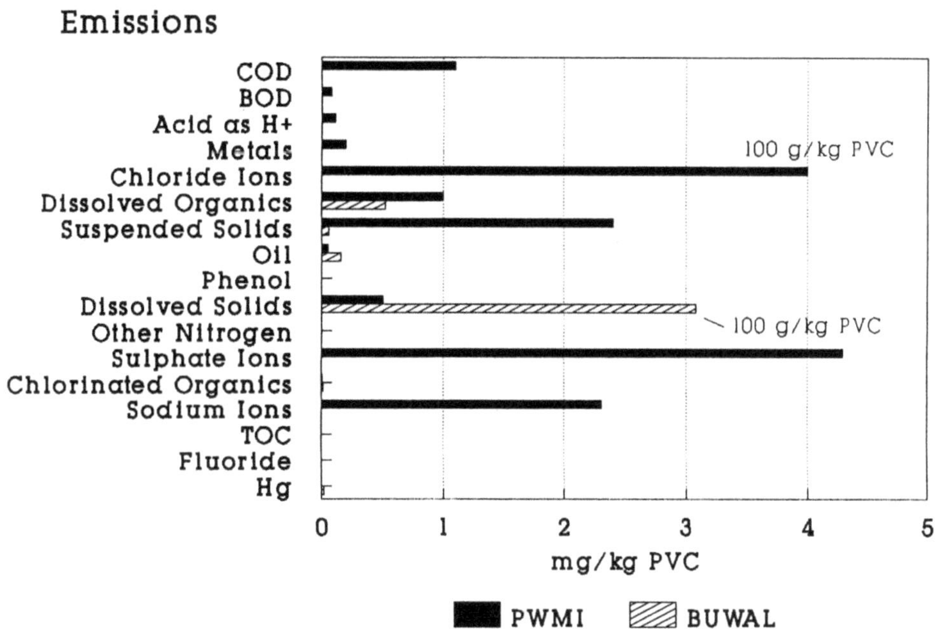

Figure 3. Water emission by PVC-production.

sions caused by these two facts are different, the ultimate result normally should not show such a big differences as shown in the above two Figures.

Basically LCA has the objective of identifying the most environmental sounded products. If so, "*Is it possible that differences in the energy data consideration change the ranking of different products?*" The answer is *yes.*

Regarding one of the important emissions, CO_2, is recorded only in one study, therefore comparison of this important factor is impossible. NO_x and Oil in water emissions normally should not allowed be to have such big differences

8. CONCLUSION

It was shown above that LCA studies by using different data and assessment methods for their analysis to their conclusions lead to considerable diverging results. This has not only an impact on the end value of the emission, but can also lead to different evaluation rankings of materials with regards to their environment burden and to misleading signal for choices that have been made in industry, Therefore, in order to compare products, systems or process by using LCA as analysis tools all the above mentioned fundamental steps of LCA should be carried out at equal levels for those compared products systems or processes and the data information should be as deep as possible until relevant data is available.

9. REFERENCES

1. Ökobilanzen für Unternehmungen, eine Wegleitung für die Praxis, 1993.

2. A Technical Framework for Life Cycle Assessment Published by SETAC, 1991.
3. General principles for life cycle assessment databases, Ian Boustead, J. Cleaner Prod. 1993 Volume 1 Number 3–4.
4. Application of life cycle assessment. expectation, drawbacks and perspectives, Helias A. Udo de Haes,J. Cleaner Prod. 1993 Volume 1 Number 3–4.
5. Choice of system boundaries in life cycle assessment, Anne-Marie Illlman, T.Kvall, Henrikke Baumann, J. Cleaner Prod. 1994 Volume 2 Number 1.
6. Life cycle methods and applications: issues and perspectives, Carlenrico Pesso J. Cleaner Prod. 1993 Volume 1 Number 3 - 4.
7. Kunststoffe genauer betrachtet, APME, PWMI.
8. Eco- balance methodology for Commodity Thermoplastics, PWMI, 1992.
9. Eco-Profiles of the European polymer industry, Report 5: Co- Product allocation in Chlorine Plants, APME, 1994.
10. Eco-profiles of the European Plastics Industry, Report 2 : Olefin Feed stock Sources, PWMI, 1993.
11. Eco-profiles of the European Plastics Industry, Report 6 : PVC, PWMI, 1993.
12. Quantitative life cycle assessment of products, 1. goal definition and inventory J.B. Guinee, H.A. Udo de Haes and G. Huppes. J. Cleaner Prod. 1993 Volume 1 Number 1.
13. Quantitative life cycle assessment of products 2. Classification, valuation and improvement analysis, J.B. Guinee, R. Heijungs, Helias A. Udo de Haes and G. Huppes J. Cleaner Prod. 1993 Volume 1 Number 2.
14. The importance of Energy in Environmental Life Cycle Assessment of Packing Materials, I. H. Knoepfel, Packing Technology and Science Vol. 7 261 271 (1994).
15. Industrielle Abfall vermeidung und deren Bewertung am Beispiel der Leiterplattenherstellung, Verlag für die Technische Universität Graz, 1993.
16. PVC and the Environment, Norsk Hydro a.s, Petrochemical division, Oslo, Norway 17. Life cycle assessment, A comparison of three methods for impact analysis and evaluation, H. Baumann and T. Rydberg, J. Cleaner Prod. 1994 Volume 2 Number 1.
18. Oekobilanz von Packstoffen Stand 1990, BUWAL Schriftereihe Umwelt Nr.132, 1991.

23

CHEMICAL CONDITIONING FOR REMOVAL OF WATER AND PHOSPHATE FROM ENHANCED SLUDGE

Jih-Gaw Lin,[1] Cheng-Nan Chang,[2] and Hsin-Chang Hung[1]

[1] Institute of Environmental Engineering
Chiao Tung University
75, Po-Ai Street, Hsinchu, Taiwan, ROC
[2] Graduate Institute of Environmental Science
Tunghai University
Box 5–819, Taichung, Taiwan, ROC.

ABSTRACT

The chemical conditioning required to improve water removal from waste activated sludge and phosphorus from the filtrate was examined in this study. The sludge, collected from a local municipal wastewater treatment plant, was adjusted to 10 g/L total solids (TS) before testing. The conditioning chemicals were ferric chloride ($FeCl_3$) and cation polymer (T-3086). The free water content in the sludge was determined by measuring the capillary suction time (CST) and specific resistance filtration (SRF). Experimental results indicated that CST had a large correlation coefficient of 0.92 with SRF. Also, adding either $FeCl_3$ or T-3086 alone did not simultaneously improve both sludge dewatering and phosphorus removal. A combined addition of $FeCl_3$ (750 mg/L) and T-3086 (20 mg/L) resulted in significant improvement in sludge dewatering as well as phosphorus removal from the filtrate. The optimum process to condition sludge was proposed as follows: the waste activated sludge was mixed with $FeCl_3$ (750 mg/L) at 200 rpm for 1 min, then immediately mixed with T-3086 (20 mg/L) for 3 min, and then slowly mixed at 30 rpm for 10 min. Under such conditions, the minimum values of CST (6 s) and supernatant turbidity (1 NTU) were achieved. Moreover, combined dosages of $FeCl_3$ and T-3086 also produced phosphorus removal in either supernatant or filtrate up to 90%.

INTRODUCTION

Efficient treatment and disposal of sludge generated on biological treatment of wastewater have received a significant amount of attention. Cost estimates of a waste treatment plant indicate that sludge handling represents as much as 30 to 40 per cent of the

Chemistry for the Protection of the Environment 2, edited by Pawłowski et al.
Plenum Press, New York, 1996

capital cost and approximately 50 per cent of the operating costs. Stringent pollution control regulations have promulgated necessary secondary treatment, having subsequently increased the quantity of biological sludge produced. The large fraction of bound water in waste activated sludge (WAS) makes it increasingly difficult to dewater, thereby increasing the volume to be handled.

Commercial cationic[1] or synthetic organic[2] polyelectrolytes were used to condition sludge to improve their dewaterability. Ferric chloride conditioned biological sludge is generally more effective than alum conditioned sludge to dewater[3]. Conditioning and dewatering reduces the sludge volume and subsequently lowers transportation and handling costs. The dewatering processes also produce filtrates that must be disposed of[4]. When conditioning and dewatering anaerobically digested sludge from a full-scale biological phosphorus removal treatment facility, an inorganic chemical addition using both lime and ferric chloride would promote effective sludge conditioning while concurrently reducing the filtrate phosphorus concentration[5].

In this study, the optimum dosages of chemicals for conditioning waste activated sludge are determined to improve water removal from the sludge and phosphorus removal from the filtrate. An optimum process to condition sludge is also proposed.

EXPERIMENTAL

Waste activated sludge, collected from a local municipal wastewater treatment plant, was adjusted to 10 g/L total solids (1% TS). The conditioning chemicals were ferric chloride ($FeCl_3$) and cationic polymer (T-3086). The chemical dose was determined by a jar test. A conventional multiple jar-stirring device was used to mix the sludge. The operating conditions, optimized through a series of tests, are listed in Table 1. The sludge was mixed for 10 s at a given rapid stirring rate to ensure sample homogeneity. Next, the chemical was added, and rapid stirring continued for a given duration. After the rapid mixing, the samples were stirred gently at 30 rpm for 10 or 20 min. Finally, the mixing was ceased for a given time and the supernatant turbidity was analyzed.

The phosphate and free water content in the conditioned sludge were determined directly after the slow stirring. Both supernatant and filtrate were used during total phosphate analysis. The supernatant was filtered through a glass fiber filter with a pore size of

Table 1. Operating conditions of jar test

Chemical	Stirring rate, rpm	Duration
$FeCl_3$		
	Rapid, 300	30 s
	Slow, 30	20 min
	Quiescent	120 min
T-3086		
	Rapid, 200	180 s
	Slow, 30	10 min
	Quiescent	30 min
$FeCl_3$/T-3086		
	$FeCl_3$, rapid, 200	60 s
	T-3086, rapid, 200	3 min
	Slow, 30	10 min

0.45 μm, as the soluble portion, and the filtrate with a pore size 11 μm, as dewatering filtrate, respectively. Turbidity and phosphorous analyses followed the procedure recommended in the 18th edition of Standard Methods[6]. Free water content in the sludge was determined by measuring capillary suction time (CST) and specific resistance filtration (SRF). The CST testing instrument used was an apparatus manufactured by Triton Electronics Ltd., England. This device consists of a timer, a sludge holder (18-mm cylinder), and blotter paper (Whatman No. 17, size 6.0 cm x 9.8 cm). SRF tests were performed by applying the Buchner funnel assembly apparatus. Sludge pH value was also monitored before and after chemical conditioning using a Schott Model CG 841 pH meter.

RESULTS AND DISCUSSION

Sludge samples for the various runs were collected at different time intervals from a municipal treatment plant with fluctuating operating parameters and inflows. The sludge sample, adjusted to 1% TS, was stored at 4°C and all tests were performed within 48 h. Although volatile solids (VS) varied (0.75–0.85%) and capillary suction time (CST) increased during the sludge storage (12 d), analytical parameters of the sludge, e.g., VS and CST, remained almost constant up to 2 d (Fig. 1).

Correlation of CST with SRF and TS

SRF test is a relatively difficult test to conduct although it is based on sound theory and generally accepted as an useful measure of sludge dewaterability. The CST test is not based on a theoretical analysis of sludge dewaterability, but is simple to run. Therefore, a correlation can be made between CST and SRF results for a range of total solids concentration (% TS) of waste activated sludge. Both TS vs. CST (Fig. 2a) and CST vs. SRF (Fig. 2b) displayed a linear relationship. Moreover, TS had a large correlation coefficient of 0.95 with CST; CST had that of 0.92 with SRF. Vesilind[7] also indicated that CST linearly increased with solids concentration for an inert Minusil slurry and an anaerobically digested sludge.

Optimum Chemical Concentration

Two parameters used to evaluate the chemical conditioning were sludge dewaterability and supernatant quality. Employing capillary suction time (CST) measurement, Cole and Singer[2] investigated the most effective type of polymer product to condition anaerobically digested sludge prior to dewatering. The effects of T-3086 and $FeCl_3$ dosages on the sludge dewaterability (CST) and supernatant quality in terms of turbidity (NTU) are presented in Fig. 3. This figure indicates that as T-3086 dose increased, both CST and NTU decreased rapidly, then gradually. CST and NTU remained almost constant at a T-3086 dose in the range of 40–80 mg/L (Fig. 3a). The minimum values of CST (7 s) and supernatant turbidity (15 NTU) were achieved at an optimum dose of 50 mg/L T-3086. For $FeCl_3$ (Fig. 3b), at a dose less than 700 mg/L, NTU decreased as $FeCl_3$ dose increased. However, $FeCl_3$ produced an inverse effect in supernatant turbidity at larger dosages. The effect of $FeCl_3$ dose on CST was similar to that of T-3086. Knocke et al.[5] reported that addition of $FeCl_3$ alone was insufficient to achieve the desired degree of sludge improvement. According to Fig. 3b, the optimum dosage of $FeCl_3$ was 700 mg/L which made the sludge dewaterability (CST) 5 s and supernatant turbidity 10 NTU. A comparison of T-

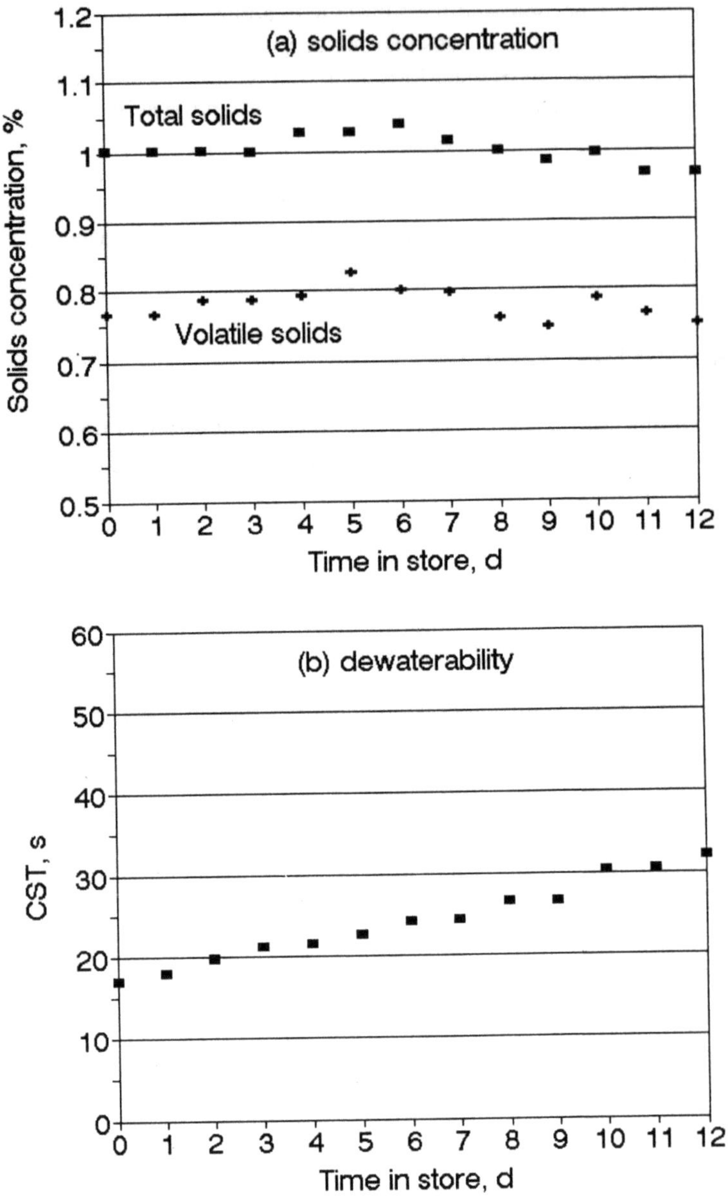

Figure 1. Effects of time in storage on solids concentration and sludge dewaterability.

3086 with $FeCl_3$ revealed that the amount of T-3086 dose (0–100 mg/L) is obviously less than the amount of $FeCl_3$ dose (0–1700 mg/L) required to sludge dewaterability and supernatant quality.

Combination of $FeCl_3$ and T-3086

Results obtained from chemical conditioning of sludge on phosphate removal are shown in Fig. 4. Phosphorus concentrations in supernatant and filtrate (without dosage of

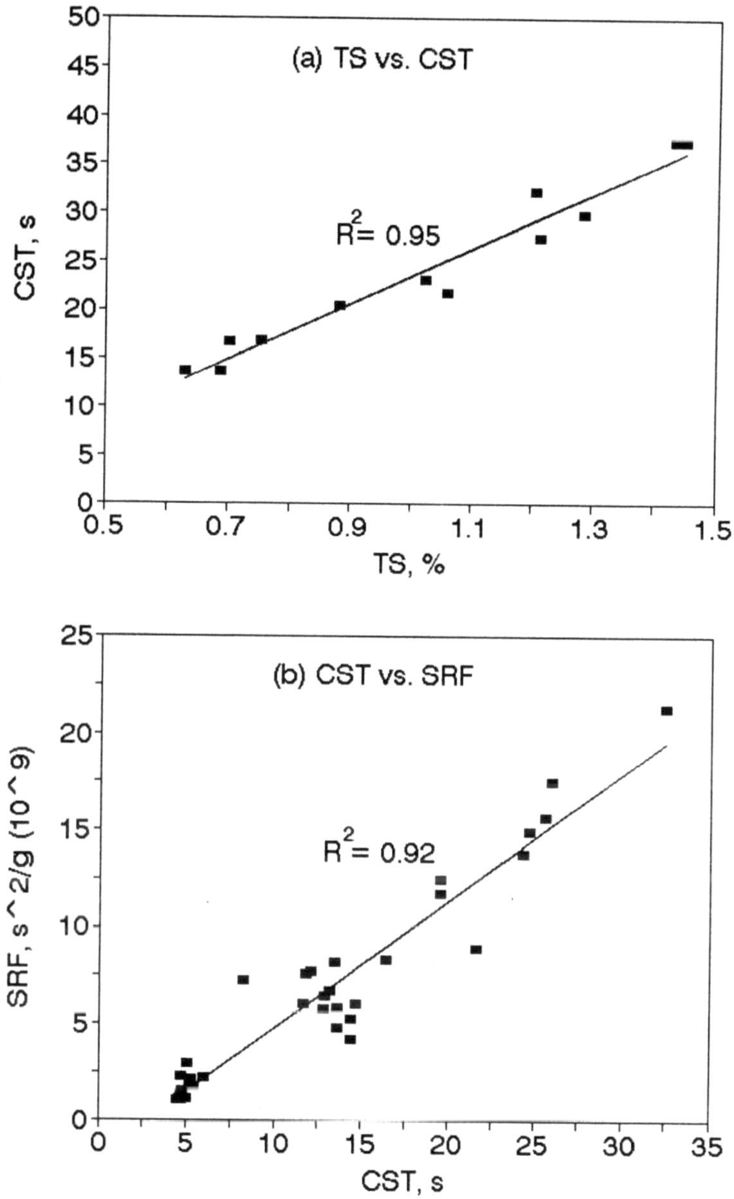

Figure 2. CST versus solids concentration (TS) and SRF plots as a linear correlation.

chemical) were 53 and 73 mg/L, respectively. The cationic polymer (T-3086) did not markedly increase the phosphate removal rate, which amounted to 15 per cent of the supernatant and 21 per cent of the filtrate (Fig. 4a). Adding $FeCl_3$ (750–1250 mg/L) to the sludge improved the phosphate removal rate up to 85 per cent for both supernatant and filtrate (Fig. 4b). Knocke et al.[5] found that inorganic chemicals such as $FeCl_3$ and $Ca(OH)_2$ are necessary to remove soluble phosphorus from the filtrate; however, polymer conditioning alone did not achieve any significant phosphate removal. They indicated that a coupled addition of both lime and ferric chloride resulted in significant improvements in

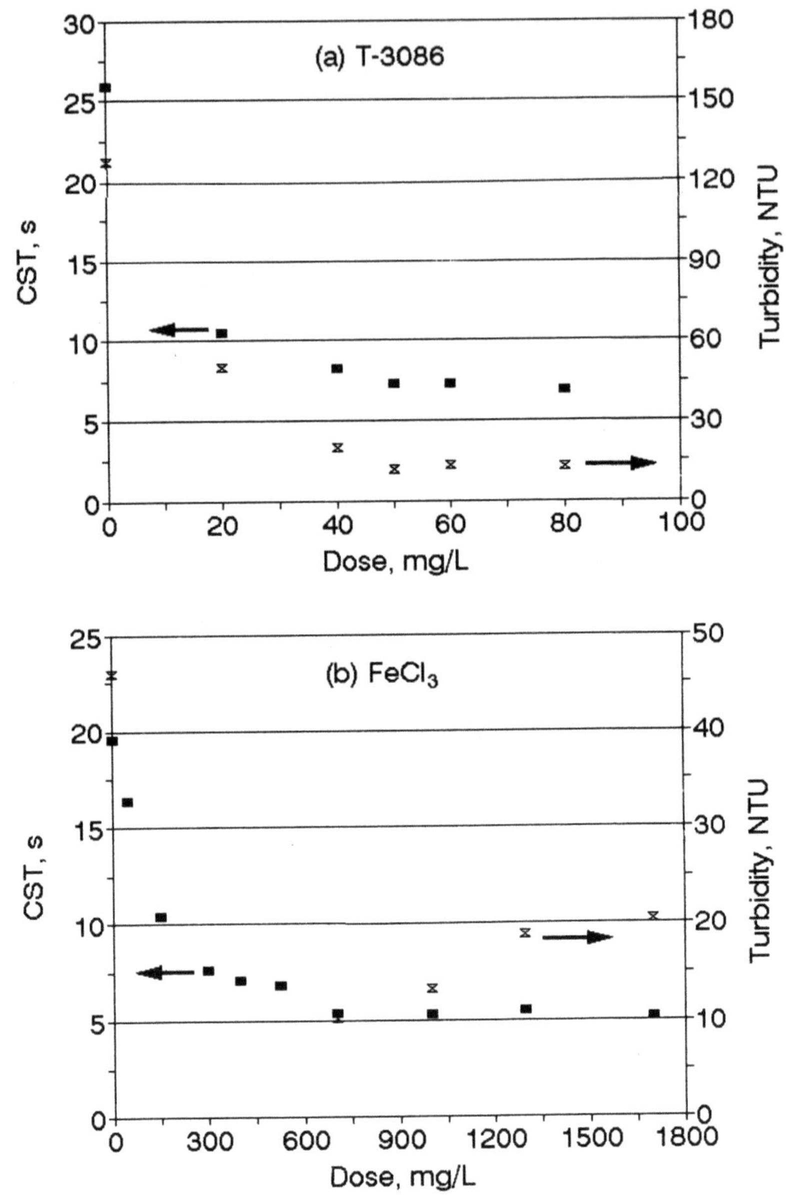

Figure 3. Effects of T-3086 and $FeCl_3$ conditioning on dewaterability and supernatant quality of sludge.

sludge dewatering rate as well phosphorus removal from the filtrate. Thus, to achieve a low as possible CST and phosphorus content, $FeCl_3$ with T-3086 would seem to be an optimal condition.

Combining $FeCl_3$ with T-3086, several experiments were undertaken to assess sludge conditioning for both sludge dewaterability and phosphorus removal. The impact of addition order of $FeCl_3$ (0–1000 mg/L) and T-3086 (20 mg/L) on sludge dewaterability is shown in Fig. 5. The sludge conditioning with addition of $FeCl_3$ prior to T-3086 resulted in better sludge dewaterability than with an addition of T-3086 prior to $FeCl_3$.

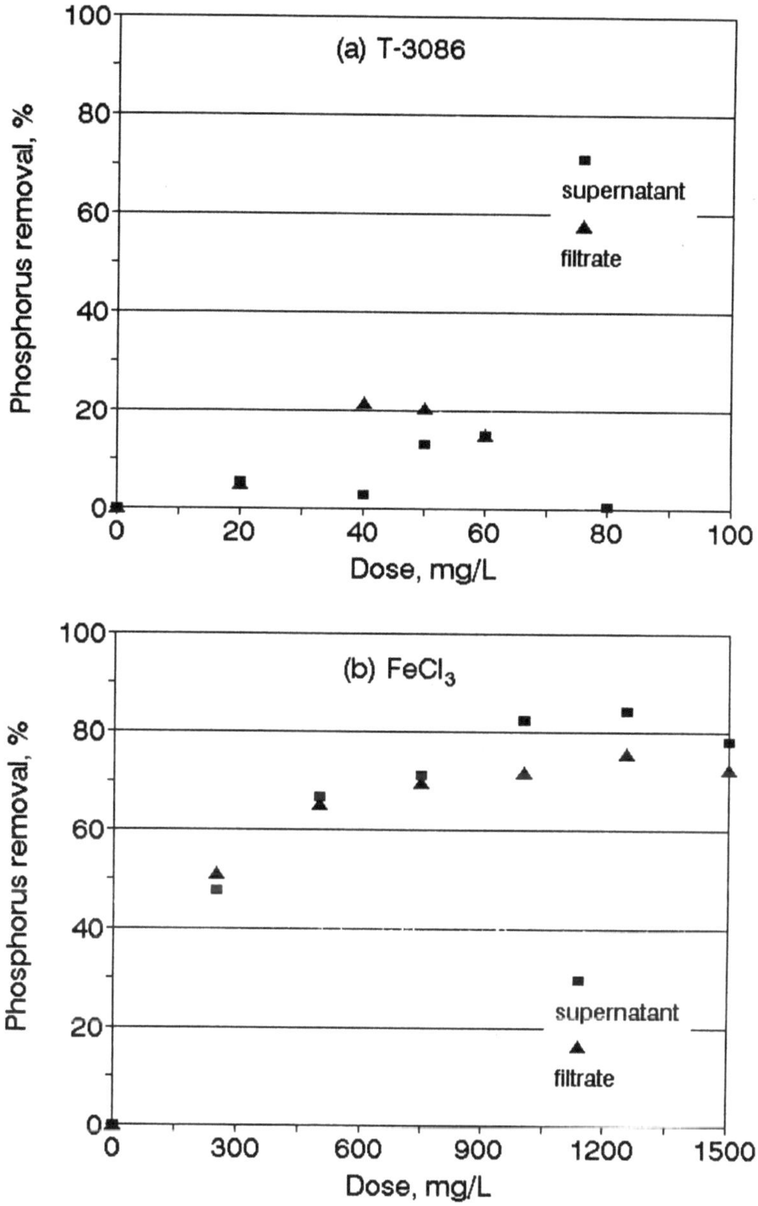

Figure 4. Effects of T-3086 and $FeCl_3$ conditioning on supernatant and filtrate phosphorus removal of sludge.

$FeCl_3$ was dosed during an initial rapid mixing (200 rpm) period of 1 min and then immediately with T-3086 for 3 min. This was followed by 10 min of slow mixing (30 rpm). Under such conditions, the minimum values of CST (6 s) and supernatant turbidity (1 NTU) were achieved at an optimum dose of 750 mg/L $FeCl_3$ with 20 mg/L T-3086 (Fig. 6). Combined dosages of $FeCl_3$ with T-3086 also produced phosphorus removal in either supernatant or filtrate up to 90% (Fig. 7). A combined addition of $FeCl_3$ (750 mg/L) and T-3086 (20 mg/L) obviously resulted in significant improvements in sludge dewatering as well as removal of phosphorus from the filtrate. The addition of either $FeCl_3$ or T-

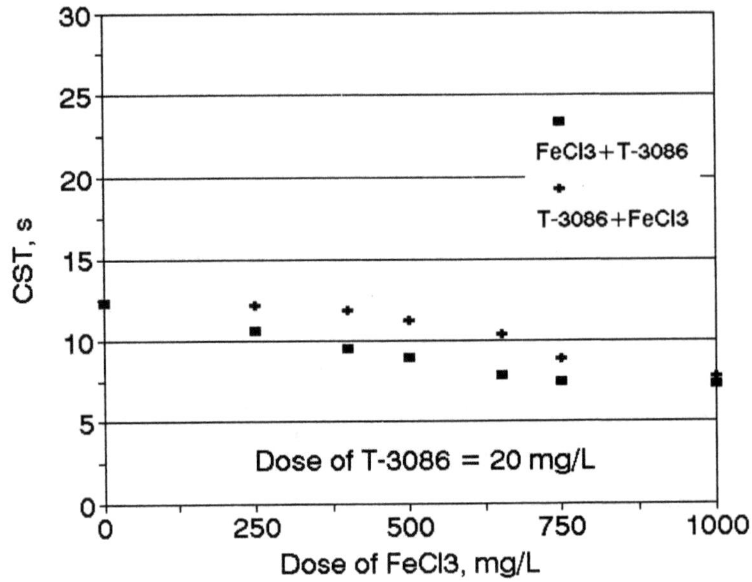

Figure 5. Effect of addition order of $FeCl_3$ and T-3086 on sludge dewaterability.

3086 alone did not simultaneously improve both sludge dewatering and phosphorus removal (Figs. 3 and 4). Results obtained from a combination of ferric chloride and Percol 757 (a cationic polymer) did not suggest a significant improvement in the sludge dewatering, even though phosphorus in the filtrate was effectively removed[(5)].

The pH results of chemical conditioned sludge are listed in Table 2. $FeCl_3$ dosages in the range of 250–1000 mg/L produced a sludge pH value between 6.5 and 4.5. Knocke

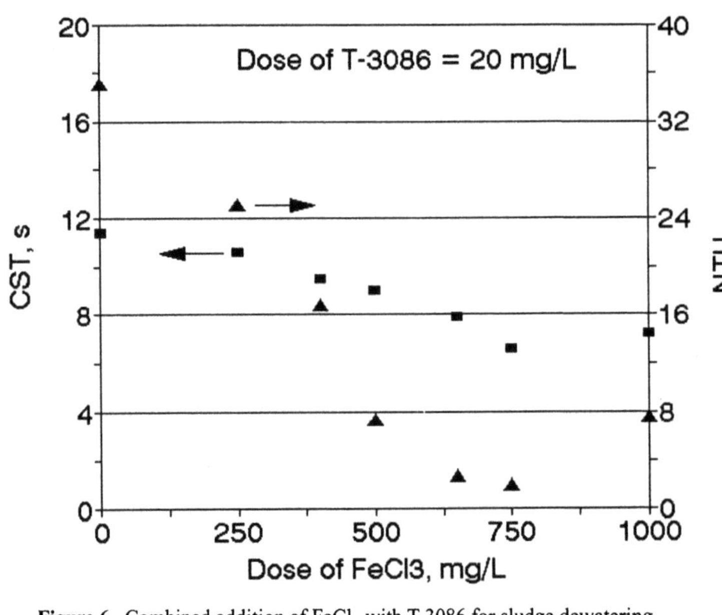

Figure 6. Combined addition of $FeCl_3$ with T-3086 for sludge dewatering.

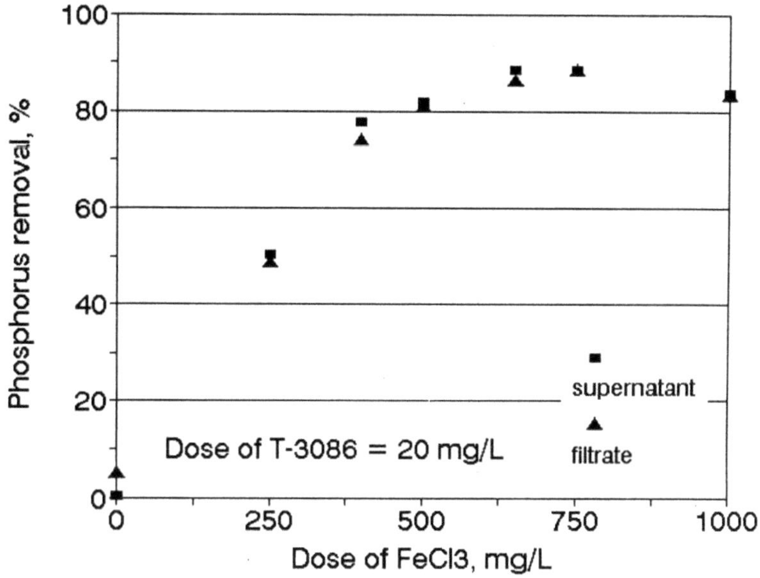

Figure 7. Combined addition of $FeCl_3$ with T-3086 for phosphorus removal.

et al.[5] indicated that $FeCl_3$ dosages in excess of 10% by weight would depress sludge pH to values below 6.5.

CONCLUSIONS

The following conclusions can be made on the basis of results obtained in this study.

Analytical parameters of the sludge (VS and CST) remained almost constant up to 2 d, even though VS varied (0.75–0.85%) and CST increased during the sludge storage (12 d). Both TS vs. CST and CST vs. SRF for sludge samples showed a linear relationship having excellent correlation coefficients of 0.95 and 0.92, respectively.

Adding either $FeCl_3$ or T-3086 alone did not simultaneously improve sludge dewatering and phosphorus removal. A combined addition of $FeCl_3$ (750 mg/L) and T-3086 (20 mg/L) resulted in significant improvements in sludge dewatering as well as removal of phosphorus from the filtrate. The optimum process to condition sludge was proposed as follows: the waste activated sludge was mixed with $FeCl_3$ (750 mg/L) at 200 rpm for 1 min, then immediately mixed with T-3086 (20 mg/L) for 3 min, and then slowly mixed at 30 rpm for 10 min. Under such conditions, the minimum values of CST (6 s) and super-

Table 2. Sludge pH before and after chemical conditioning

$FeCl_3$, mg/L	0	250	400	500	650	750	1000
pH[a]	6.91	6.40	6.06	5.85	5.53	5.29	4.70
pH[b]	6.91	6.40	6.08	5.82	5.44	5.16	4.61

[a] $FeCl_3$ only.
[b] Combined with T-3086 (20 mg/L).

natant turbidity (1 NTU) were achieved. Combined dosages of $FeCl_3$ and T-3086 also produced phosphorus removal in either supernatant or filtrate up to 90%.

ACKNOWLEDGMENTS

The authors would like to thank the National Science Council of the Republic of China for partial financial support of this manuscript under Contract No. 83–0410-E-009–070.

REFERENCES

1. Tiravanti, G., Lore, F. and Sonnante, G. (1985) Influence of the charge density of cationic polyelectrolytes on sludge conditioning. *Water Res.*, Vol. 19, No. 1, pp. 93–97.
2. Cole, A. I. and Singer, P. C. (1985) Conditioning of anaerobically digested sludge. *J. Environ. Eng*, Vol. 111, No. 4, pp. 501–510.
3. Sedlak, R. I., eider (1991) Phosphorus and nitrogen removal from municipal wastewater: principles and practice. Lewis publishers, 2nd ed., pp. 111–140.
4. Zhou, H., Smith, D. W. and Stanley, S. J. (1992) Characterization and treatment of lime sludge dewatering effluents. *Can. J. Civ. Eng.*, Vol. 19, pp. 794–805.
5. Knocke, W. R., Nash, J. W. and Randall, C. W. (1992) Conditioning and dewatering of anaerobically designed BPR sludge. *J. Environ. Eng.*, Vol. 118, No. 5, pp. 642–656.
6. Standard Methods for the Examination of Water and Wastewater (1992) 18th ed., Am. Public Health Assoc., Washington, D. C.
7. Vesilind, P. A. (1988) Capillary suction time as a fundamental measure of sludge dewaterability. *J. Water. Pollut. Control Fed.*, Vol. 60, No. 2, pp. 215–220.

24

ENVIRONMENTALLY SAFE PHOSPHATE FEED PRODUCTION PROCESS

Technology Progress

Zbigniew Wzorek and Zygmunt Kowalski

Institute of Inorganic Chemistry
11 Sowinski Str., 44–101 Gliwice

1. INTRODUCTION

Until 1993, only one phosphate feed was produced in Poland: DFP(defluorinated feed phosphate) as produced in Cracow Inorganic Works "Bonarka" with a conventional method based on the thermal defluorination of apatite, soda ash and phosphoric acid mixture in rotary kilns at a temperature 1723–1773K. This process come be summaryd by the following reactions:

$$Na_2CO_3 + H_3PO_4 \longrightarrow Na_2HPO_4 + H_2O + CO_2 \quad (1)$$

$$2Na_2HPO_4 \longrightarrow Na_4P_2O_7 + H_2O \quad (2)$$

$$Na_4P_2O_7 + 2Ca_5F(PO_4)_3 + H_2O \longrightarrow 2Ca_3(PO_4)_2 + 4CaNaPO_4 + 2HF \quad (3)$$

The emitted HF is absorbed in a two staged wet absorption unit with white wash. The neutral calcium fluoride slurry is transferred for disposal.

A basic flow-sheet of thermic method is presented in fig.1.

The obtained DFP(CaNaP) is composed of 18 % P, 46 % CaO, 5% Na to 0.3%F. The product is used for manufacturing animals feeds (0.5–1.5% in different feeds) [1].

Analysis of the process presented in fig. 1 indicates that the emission of dust from apatite drying, product milling and acidic fumes from the phosphoric acid reactor and the soda-apatite mixture reactor were the basic ecological problems. Another problem is fluor emission (to 8g/1t of product) and calcium fluoride slurry disposal. Only a small quantity of the calcium fluoride slurry is utilized in calcium fluoride production (2000 t/y). A longe disposal only 1 km from the centre of Cracow contains over 200,000 t of waste (disposed over a 20 years DFP production) is a period very serious problem for the city [2,3].

Chemistry for the Protection of the Environment 2, edited by Pawłowski et al.
Plenum Press, New York, 1996

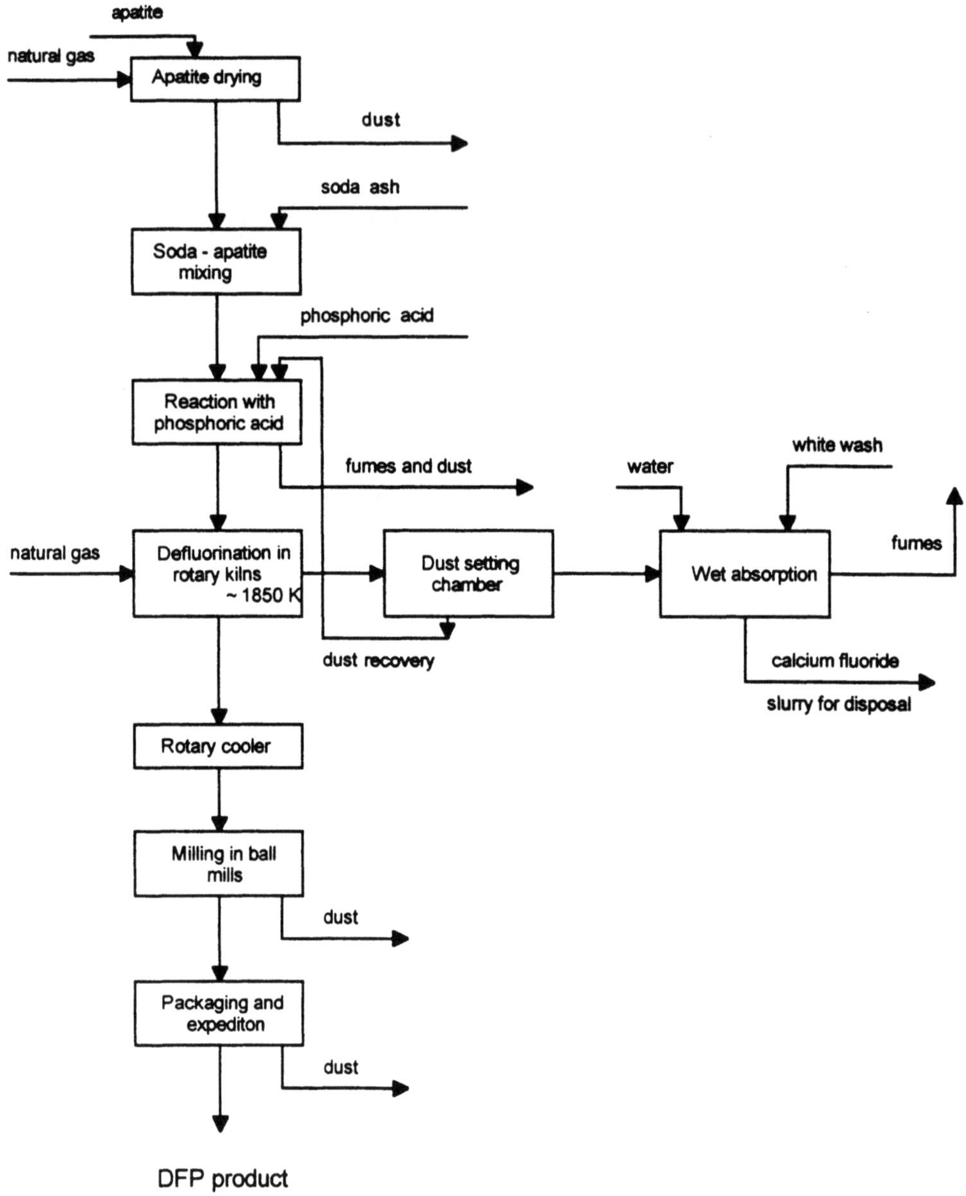

Figure 1. Conventional thermic method of DFP production process.

2. MODERNIZATION OF DFP PRODUCTION PROCESS

Our first efforts were connected with technology and equipment changes in the DFP process. At first apatite wetting was eliminated during transportation which resulted in the elimination of the apatite drying operation before its use as a raw material. A basic technology change involved using sodium phosphate instead phosphoric acid and soda ash. This eliminated the first stage of mixing (soda ash with apatite) and especially the emission of acidic fumes which occured before in the second stage reactor (phosphoric acid and soda-apatite mixture reaction). Another results from using sodium phosphates was temperature decrease in the rotary kiln of about 100K and an increase in the rotary kiln capacity of 8%. These allowed a decrease of over 15% natural gas consumption per 1 t of produced DFP. Raw materials used are sodium tripolyphosphate and other phoshphates.

The defluorination reaction is:

$$Na_5P_3O_{10} + 3Ca_5F(PO_4)_3 + 2H_2O \longrightarrow 3Ca_3(PO_4)_2 + 5CaNaPO_4 + CaHPO_4 + 3HF \quad (4)$$

Implementation of trisectional vibrating screens decreased the quantity of product which was milled in ball mills to 30%. A recirculation system of DFP oversize particles was implemented (with too large fluor content) allowed better product quality (with fluor level below 0.25%). Also, changing the very old pneumatic system for raw material transport with mechanical (screw conveyors) decreased dust emission to the environment [1–4]. The modernized DFP production process is presented in fig.2.

Much research has been done to solve the problem of the utilization of the calcium fluoride slurry. Production of dried calcium fluoride (2000–3000 t/y) was modernized and its cost decreased by about 30%. The roduction of building materials on a dried calcium fluoride base was implemented [2] and a technology for synthetic fluorspare production was worked out, too [4,5,].

But these efforts were not enough. DFP produced by the thermic method even after modernization led production costs too hight to be competitive with other phosphates offered on the polish market. A new production method was needed..

3. LOW-TEMPERATURE FEED PHOSPHATES PRODUCTION PROCESS

A general process reconstruction resulted in the implementation of a new dicalcium phosphate low-temperature technology in 1993 [4,5]. This wasteless process was realized at the DFP unit with low investment costs. The new technology passed practical examination, but some old equipment, particulary the dedusting and raw materials preparation units were not adaptable to the DCP process. These resulted in inferior product quality and in rather low production yield (~90%) [6]. A flow-sheet of this method is presented in fig.3.

A new method based on the reaction of calcium hydroxide with phosphoric acid follows the formula:

$$Ca(OH)_2 + H_3PO_4 \longrightarrow CaHPO_4 + H_2O \quad (5)$$

The process was solid wasteless and fluor emissions were also eliminated. Defluorination at a temperature ~1773K changed with a drying temperature of ~400K. This resulted in very low energy consumption (natural gas consumption per 1t of product was 95% lower than in the thermic method). Elimination of product milling decreased product dust emission and electricity consumption. Total production costs for the DCP production process are comparatively about 33% lower than the thermic method.

The produced DCP was competitive on the polish market but dust emissions were still rather high and maintenance, repairs and labor demand costs were competitivly high. To resolve this situation two possibilites were considiered:

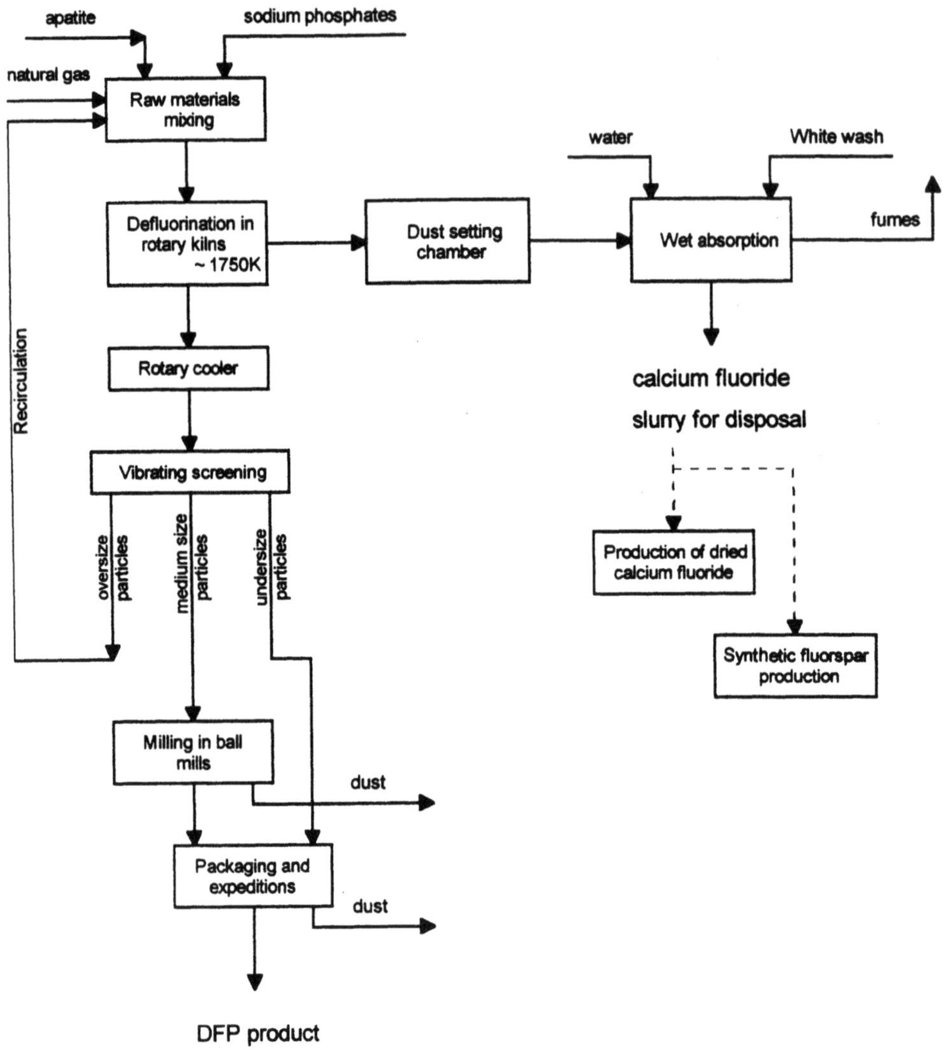

Figure 2. Modernized thermic method of DFP production process.

- the modernization of the "Bonarka" unit in Cracow,
- the building of a new unit in another factory.

Technical and economic analysis demonstrated building of the new unit to be more profitable [10].

4. NEW TYPE OF LOWTEMPERATURE DCP PRODUCTION PROCESS

Our research [6–9] has allowed us to propose new technological and engineering solutions. The changing of the drying unit, the two staged raw materials mixing rector and the four staged wet absorption units are examples of these. A new DCP unit was built in

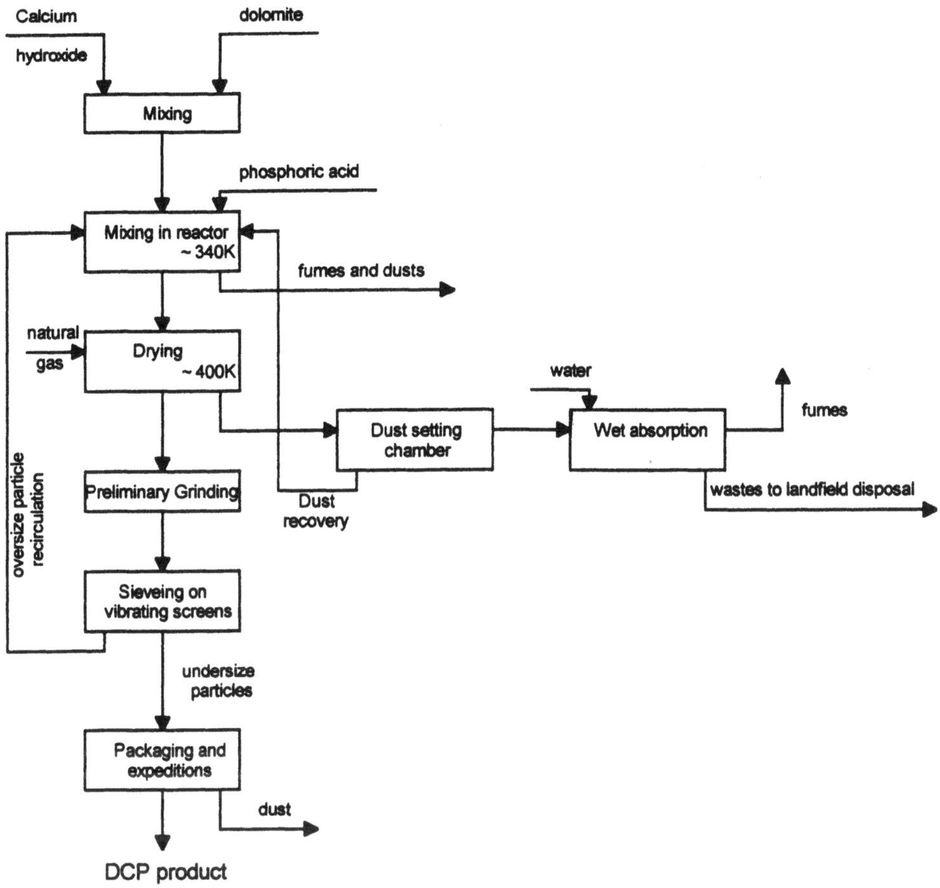

Figure 3. Low temperature method of feed phosphate production process realized on DFP unit.

the "Silikaty" Klucze factory. This project was preced detailed analysis made with the "reduction of source method" [10] and research suggested a new type of technology based on the reaction

$$CaO + H_3PO_4 \longrightarrow CaHPO_4 + H_2O \tag{6}$$

A flowsheet of this process is presented in fig.4.

New technological solutions, modern project apparatus and equipment allows high production process yield (to 99.5 %) to be achieved with very low energy consumption. Using a project time "reduction at source" analysis methodology allowed us to eliminate practically all dust emissions and fully protect the natural environment. Computerized process control results in higher product quality and low labor demand.

A detailed technical and economic analysis which were performed to investment possibilities for modernization of the "Bonarka" unit and for building a new in factory "Silikaty" Klucze demonsrated, that this second possibility had a much better NPV and IRR [10]. Production costs for the new "Silikaty" unit were estimated to be 30% lower than in the process presented in point 3.

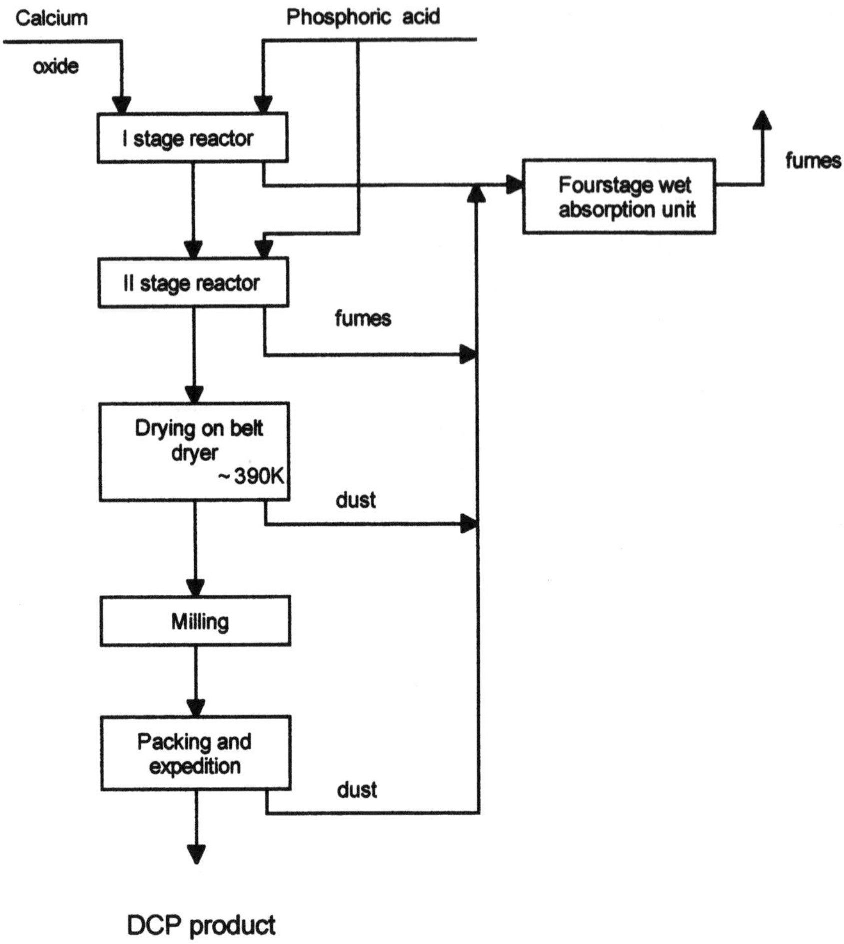

Figure 4. Feed phosphate low temperature production process on new unit.

At the new unit we could produce all dicalcium phosphate grades (with P content from 16 to 20%) and possibly monocalcium phosphate (MCP). MCP technology is now the subject of our research.

5. REFERENCES

1. Z. Kowalski; Pr. Nauk. AE Wrocław nr 610 /1992/ s. 165–172
2. Polish pat. 164 671 /1994/
3. Z. Kowalski, M. Gollinger; Ochr. Pow. nr 5(145) /1991/ s. 116–120
4. H. Górecki, Z. Kowalski, J. Hoffmann; Pr. Inst. Techn. Nieorg. Polit. Wr. nr 40 seria 22 /1994/
5. Z. Kowalski, H. Górecki; Mater. Konf. "Ekologia rolnicza", WSP Opole, listopad /1993/
6. Z. Wzorek, H. Górecki, Z. Kowalski; XXII Seminarium Polit. Wr. "Chemia dla Rolnictwa", Jarnołtówek, grudzień /1994/
7. Z. Kowalski; Prace PPUH "Alchem" Alwernia /1994/, (not published)
8. Polish patent application P.305 491 /1994/
9. Polish patent application P.307 941 /1995/
10. Z. Kowalski, M. Musioł, P. Furlepa; Arch. Ochr. Srod. /1995/ (in press)

25

INORGANIC PHOSPHORUS PROBLEM IN DOMESTIC SEWAGE

A. Montusiewicz and M. Dudzińska

Department of Environment Engineering
Technical University of Lublin

ABSTRACT

Concentration of different forms of inorganic phosphorus in domestic sewage was measured by ion chromatography. Very low content of poly- and pyrophosphates was found. Potential sources of different P-forms, like popular washing powders, as well as the degradation process kinetics in raw sewage, were investigated.

The results obtained suggest that the crucial problem seems to be the level of total inorganic P represented by orthophosphate concentration rather than polyphosphate content.

1. INTRODUCTION

Phosphorus is recognized as the main biogenic parameter, stimulating eutrophication and saprobion processes in the receivers (lakes and rivers). As a natural process, eu trophication continuously occurs in the environment, however, its rate increases due to anthropogenic influence. The equilibrium between nutrient intake and organic matter is disturbed by an inflow into the receivers insufficiently treated wastewaters containing assimilable soluble phosphorus compounds. According to the literature[1] the main source of phosphorus in domestic sewage are detergents based on sodium mono- and polyphosphates, which stabilize the dirt particles and complex Ca^{2+} i Mg^{2+} ions present in the water. The contribution of detergent phosphorus to the total phosphorus load in untreated wastewaters varies from 25 to 50 %. Estimation of the phosphorus load to a sewage treatment plant shows[2] that 2–8 g P per person per day enters the wastewater and the content of phosphorus forms is[3]: orthophosphates - 5 mg P/L; tripolyphosphates - 3 mg P/L; pyrophosphates - 1 mg P/L; organic phosphates -1 mg P/L. Because of the great problem of phosphorus content, in many countries its concentration in treated sewage, as well as in typical phosphorus sources, e.g.detergents, is regulated by law. Under Polish regulations, only the content of phosphorus in treated sewage is limited.

2. EXPERIMENTAL DETAILS

Levels of inorganic P-form in raw sewage from the city of Lublin and at several steps of the typical municipal wastewater treatment plant were measured using ion chromatograph from Waters Millipore. Ion chromatography method with the Anion HP column, three-step gradient HNO_3 elution, post-column reaction (1 % $Fe(NO_3)_3$ $9H_2O$ + 2 % $HClO_4$) in Reagent Delivery Module (Waters) and UV detection at 340 nm, was developed to analyse inorganic P-forms content in the solutions. Detection limits specific for this method are[4]: orthophosphates - 1 ppm; pyrophosphates - 1 ppm; tripolyphosphates - 2 ppm. Peak identification was performed by internal standards. Investigations were made in the spring time, samples of sewage were filtered (0.45 mm) and kept at low temperatures (277 K) until measured. Very low content of poly- and pyrophosphates was found; therefore potential sources of different P-forms, like available popular washing powders, as well as the degradation process kinetics in raw municipal sewage, were examined. For washing powder examination, a selection was made from "phosphate free" and "regular" detergents commercially available. Kinetic experiments were performed in laboratory conditions. Known amounts of pyro- or polyphosphates were added to the raw wastewater sampled from the sewage system. The content of ortho-, pyro- and polyphosphates in those mixtures was measured by IC after 1,2,4,6,8,10 and 20 hours (retention time in the method - 30 minutes). The temperature of the experiments was 293 K ±1 - characteristic for the spring to autumn period. Degradation in function of the temperature was not studied.

3. RESULTS AND DISCUSSION

A number of analytical measurements were performed and the average levels of P-forms in raw and treated municipal wastewater are shown in Table 1.

An increase of orthophosphate level in the sewage passing the several steps of the municipal wastewater treatment plant was observed. This might be caused by polyphosphate degradation and by circulation of waste streams within the plant. At certain stages of wastewater treatment primary input meets additional eluates rich in orthophosphates recirculated from thickeners, digestion tanks and sludge drying beds.

Detergents are considered to be the main source of phosphates in domestic sewage. The results for washing powders are presented in Table 2 and expressed in g P/kg of powder.

Powders 1, 2 and 3 are produced in Poland by a Western company and the first one is the most popular (about 20% of the market sector). They are not advertised as "environment friendly". Powders 6 and 7 are produced, manufactured and packaged in Western

Table 1. The average levels of P-forms in raw and treated municipal wastewater

	Concentration of phosphorus in wastewater [mg P/L]			
Wastewater	$P\text{-}PO_4$	$P\text{-}P_2O_7$	$P\text{-}P_3O_{10}$	$P_{total\ inorg.}$
Raw wastewater	3.60	undetected	0.2	3.80
Sewage after the primary sedimentation tank	4.40	undetected	undetected	4.40
Purified output effluent	6.10	undetected	undetected	6.10

Table 2. Concentration of inorganic phosphorus forms in popular washing powders commercially available in Poland

Washing powder	Concentration of phosphorus in washing powders [g/kg]			
	$P\text{-}PO_4$	$P\text{-}P_2O_7$	$P\text{-}P_3O_{10}$	$P_{total\ inorg}$
Powder 1	2.44	5.17	11.28	18.89
Powder 2	2.24	7.29	10.61	20.14
Powder 3	5.31	4.14	20.46	29.91
Powder 4	1.47	4.11	6.88	12.46
Powder 5	1.77	0.76	—	2.53
Powder 6	21.21	—	—	21.21
Powder 7*	27.06	—	—	27.06

*concentrate

countries and commercially available; however their share is not large due to high prices. Powders 4 and 5 are manufactured in other Central European countries, and only for Eastern and Central European Markets, and packaged in Poland. They are the same brand, but the latter is advertised as "phosphate free".

Each of the powders examined contains phosphorus, although the last two only in orthophosphate form. A total load of inorganic P is in a similar range for all the powders checked, in spite of the fact that Powder 5 is advertised as "phosphate free" and contains very low levels of ortho- and pyrophosphates. A total load of inorganic phosphorus from Powder 7 should be recalculated for comparison due to its concentrate status.

Polyphosphates are considered to be quickly degradable in the sewage system, which was confirmed by the authors' measurements. The results of the laboratory experiment of polyphosphates degradation in raw municipal wastewater are shown in Figure 1.

Degradation of polyphosphates in raw sewage is quick and after 10 hours only 1/10 of the initial concentration was found. After the first 30 minutes a rapid degradation with a 2.5-time decrease of input concentration was observed. The retention time of the sewage system in the city of Lublin is 4 hours,which explains the lack of polyphosphates at the entrance point of the wastewater treatment plant.

Considering the above, the form of inorganic phosphorus is not significant. The most important is the total load of inorganic P expressed as phosphate phosphorus. The most hazardous for the environment seems to be Powder 3, because of the high level of phosphates expressed as total phosphate phosphorus (see Table 2). Even powders which are advertised as "phosphate free" contain phosphorus in orthophosphate form. They are "free" from polyphosphates only. Their impact on the environment depends on the total inorganic P content.

4. CONCLUSIONS

Polyphosphates are hardly detectable in the raw wastewater entering a sewage treatment plant. The kinetic experiment in raw sewage showed a quick decrease of polyphosphates concentration, probably due to degradation to orthophosphates in the sewage system and further sorption and biosorption processes.

Polyphosphates are not as significant a problem as orthophosphates in raw domestic wastewater, because of the primary degradation in the sewage system.

The "phosphates free" detergents are neither very easily available commercially nor particularly popular due to high prices. The most common "phosphates free" detergent is

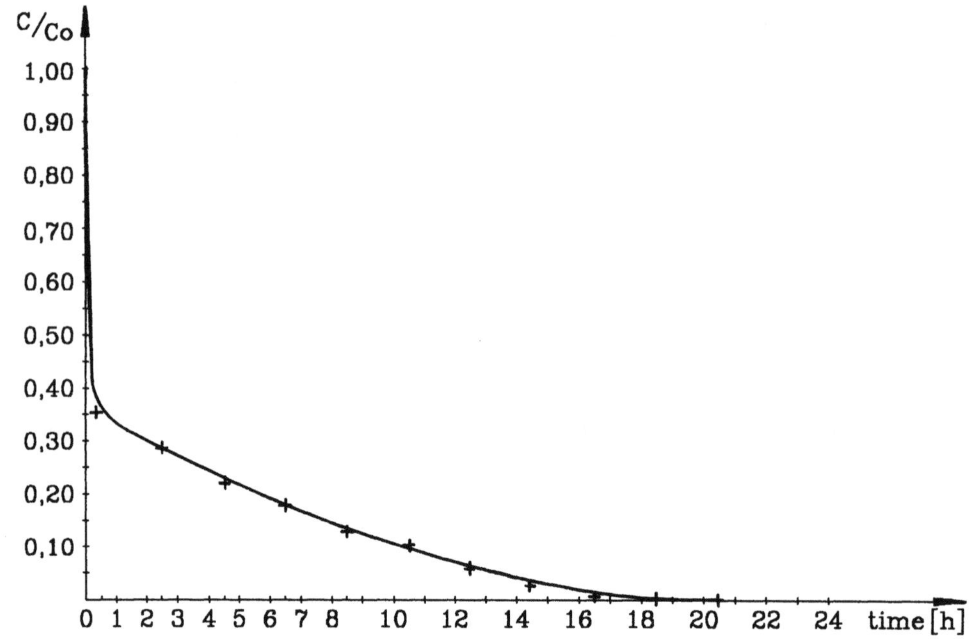

Figure 1. The polyphosphates degradation in municipal wastewater over time in laboratory conditions. Where: c- concentration of polyphosphates over time, c_0- input concentration of polyphosphates.

Powder 5 and according to the investigations it does not contain polyphosphates, but contains phosphorus in orthophosphate form. Only the limitation of the total P content in detergents and in the other orthophosphate sources can reduce P level significantly both in municipal wastewater and water receivers.

Our measurements were made during spring time. In autumn, which is the "campaign time" for different food processing factories, according to the authors' unpublished data, the deficit of inorganic assimilable P-forms occurs in raw sewage, so that the addition of nutrients is required for a satisfactory performance of biological treatment installations. This indicates, that the problem of phosphorus in municipal sewage is complicated and complex.

REFERENCES

1. Dojlido, J. and Best, G.A. (1993) Chemistry of Water and Water Pollution. *Ellis Horwood*, Chichester, 129.
2. Devey, D.C. and Harkness, N. (1973) The Significance of Man-Made Sources of Phosphorus: Detergents and Sewage. *Water Res.* 7, 35–54.
3. Snoeyink, V.L. and Jenkins, D. (1980) *Water Chemistry*. Wiley, New York.
4. Operator's Manual (1989) nr 07702 *Waters 431 Conductivity Detector. Millipore Corporation Waters Chromatography Division*, Milford.

ENVIRONMENTAL IMPLICATIONS OF SUBSTITUTION OF POLYPHOSPHATE IN DETERGENTS BY CHELATING LIGANDS

Marzenna R. Dudzińska and Agnieszka Montusiewicz

Department of Environmental Protection and Engineering
Technical University of Lublin
40, Nadbystrzycka Str., 20–618 Lublin, Poland

ABSTRACT

This paper is an attempt to describe the possible impact of substitutes for polyphosphates in popular detergents on the environment. From different substituents, chelating ligands, like EDTA and NTA, were chosen, as they are also widely used in a number of processes and technologies. Actual levels of EDTA and NTA in the raw domestic sewage and at every step of the typical municipal wastewater treatment were measured using ion chromatography. Contents of EDTA and NTA in the popular washing powders were also checked. EDTA was reported to be non-degradable and no decrease of its level in the wastes passing through the treatment plant was observed. NTA was reported to be biodegradable, which was confirmed by a very low level of NTA in the raw wastes. In Poland industrial wastes containing heavy metals are often mixed with the municipal sewage system only after primary pre-treatment, so possible implications of the presence of non-degradable chelating ligands in the wastes and water environment were discussed.

1. INTRODUCTION

Polyphosphates from washing powders were accused of enviromental degradation of surface waters, specially stimulation of eutrophication and saprobion processes. High phosphorus level causes disturbances during the normal work of the municipal wastewater treatment plant. Inorganic phosphorus, assimilable for microorganisms is needed at a certain level in the wastewater stream, but a level which is too high from antropogenic sources causes problems. Because, a contribution of phosphorus from detergents to the total phosphorus load in the raw wastewater is estimated to be 30–50%, a number of substituents for polyphosphates were proposed by different producers of commercial washing

Chemistry for the Protection of the Environment 2, edited by Pawłowski et al.
Plenum Press, New York, 1996

powders and liquid detergents[1]. Among them zeolites and organic chelating ligands like ethylenediaminetetraacetic acid (EDTA) and nitrylotriamineacetic acid (NTA) have been the most popular. Unfortunately, the latest studies have shown that none of them is as efficient as tripolyphophates and none of them appears to be neutral for the environment. The final results of zeolites as substituents for polyphosphates are sediments, very difficult for utilization, which are formed in either wastewater treatment plants or in natural receivers[2]. Some aspects of using chelating ligands will be discussed in this paper.

EDTA and NTA were also widely used for metal complexation in plating and tanning processes, boiler-scale cleaning and metal separation, stabilization of herbicides and pesticides, as well as stabilization of cosmetics, carbonated beverages and processed food[3]. There is no evidence which one of the above processes might be the main source of chelates in the environment. Although, substitution of polyphosphates in detergents by chelates has to be carefully studied, because of large amounts of household wastewaters produced every day. Both ligands have very high affinity to complex almost all metal ions. According to the literature, EDTA is non-degradable[4]. NTA is biodegradable[5], although there is evidence that some of its complexes are nondegradable[6]. The question about a possible impact of chelating ligands on the sewage treatment plant processes as well as on the water environment arises. Especially in the case of Poland, industrial wastes containing heavy metals are often mixed with municipal sewage system only after primary pre-treatment, so traces of heavy metals have to pass through the municipal wastewater treatment system.

2. EXPERIMENTAL DETAILS

Levels of EDTA and NTA in the raw domestic sewage and at every step of typical municipal wastewater treatment installation (Hajdów near Lublin) were measured. Waste samples were filtered (0.45μm) and kept at low temperatures until measured. Because, under the Polish law, the "phosphate-free" detergents as well as old detergents containing polyphosphates are commercially available, the contents of EDTA and NTA in the typical popular washing powders and liquid washing-up detergents were also checked. Different types of detergents (Polish made as well as imported) commercially common in 1994 were selected.

Ion chromathograph from Millipore-Waters (Waters Action Analyzer 625) was used and ion chromatography method with Anion HP column (Waters), gradient HNO_3 elution, post-column reaction (1% $Fe(NO_3)_3$ + 2% $HClO_4$) and UV detection at 340 nm was developed to analyse the chelating ligands content in the water solutions. This method allowed for detemination of EDTA (with accuracy up to 0.5 ppm), NTA (1 ppm), as well as different forms of phoshates: orthophosphate (1 ppm), pirophosphate (1 ppm), and polyphosphate (2 ppm).

3. RESULTS

The results of sewage examination are presented in Table 1. Our previous studies[6] showed that no polyphosphates were found (at that detection limit) in the raw sewage, so only the data for EDTA and NTA are presented. Hajdow is a mechanical-biological treatment plant, without a chemical stage, which is typical for municipal sewage treatment in Poland, and it serves the Lublin and Swidnik communities.

Table 1. Chelating ligands level in municipal seweges at every stage of sewage treatment plant

	Average concentration (mg/dm^3)	
Sewages	EDTA	NTA
Raw municipal sewages	34.13	non-detected
Sewages after preliminary sedimentation	45.85	non-detected
Sewages after biological treatment	49.98	non-detected

Lublin is a city with 400,000 inhabitants, motor plant, sugar plant, brewery and some other small enterprises. Swidnik is a small town with 20,000 inhabitants and a big motorcycle and aircraft factory. The measurements were made for raw sewage, sewage after the first stage of the treatment procedure (mechanical) and outcoming sewage (after the second step of cleaning) during the spring time.

The results of measurements for washing powders and liquid washing-up detergents are presented in Tables 2 and 3, respectively. The results for liquid detergents in Table 2 are expressed in mg/dm^3 of detergent. The results for washing powders are expressed in g/kg of detergent. *Liquids 1* and *2* were Polish products, not advertised as "environment friendly". *Liquids 3* and *4* were imported products, *Liquid 4* advertised as "environment friendly". A similar selection was made for washing powders. *Powder 1* was the most popular product (about 20% share of market), produced in Poland by a Western company. Other products are imported - *Powder 3* is manufactured in another Central European country and packaged in Poland (which means that it was produced only for the Eastern and Central European Market). *Powder 4* is the same brand name but advertised as "phosphate free". *Powder 2* is produced in Poland by a Western company, while *Powders 5* and *6* were products manufactured and packaged in Western European countries, commercially available but considerably more expensive, so their share in the market is not very big.

Together with the results for EDTA and NTA, the measurements of "phosphate' content are presented. "Phosphate" means all forms of inorganic phosphates (ortho-, piro- and poly-), expressed in mg of P/dm^3 for liquids and mg P/kg of powders. As in the sewage system polyphosphates and pirophosphates degradate quickly to ortophosphates[7,8], there was no need to distinquish these forms.

4. DISCUSSION

The results for treated and untreated wastewater are presented in Table 1. NTA was not found in the raw sewage (at method detection limit), although some of the examined

Table 2. Chelating ligand and phosphate content in popular liquid dishwashers

	Concentration (mg/dm^3)		
Liquid detergent	EDTA	NTA	P_{SUM}
Liquid 1	9.59	non-detected	8.10
Liquid 2	27.29	20.25	8.13
Liquid 3 (concentrate)	33.25	non-detected	11.41
Liquid 4	non-detected	non-detected	non-detected
Liquid 5 (concentrate)	8.89	non-detected	12.42

Table 3. Chelating ligands and phosphate content in popular washing powders

Washing powder	Concentration (g/kg)		
	EDTA	NTA	P_{SUM}
Powder 1	non-detected	23.49	18.89
Powder 2	9.17	non-detected	20.14
Powder 3	non-detected	non-detected	12.46
Powder 4	8.37	non-detected	2.53
Powder 5	non-detected	non-detected	21.21
Powder 6 (concentrate)	8.23	19.88	27.06

household detergents, including the most common "Powder 1" and expensive, not very common *Powder 6 - concentrate*, contain significant amounts of NTA. In the examined liquid washing-up detergents NTA was not found with one exception - *Liquid 2*.

The total load of NTA into domestic sewage might be too low to be observed as significant concentration of ligand in the municipal wastewater system. Wastes coming from industry into the system may dilute the stream at the entrance point of the wastewater treatment plant, so NTA was not detected.

Additionally, Thum[5] and other authors[6,9] confirmed biodegradability of NTA ligands both in natural receivers and in the wastewaters. Although degradation of NTA in the surface waters occurs in days rather than in hours. At present, in Poland the level of NTA in waters is much lower than that estimated by Swicher *et al*,[6] : to be in the range 0.1 - 0.01 mg/dm^3 . Such estimation might be resonable while all polyphospates were substituted by NTA.

Quite different results were obtained for EDTA. Almost all the examined liquid washing-up detergents contain EDTA (see Table 2). The only exception is *Liquid 4*, produced in Western Europe and advertised as "environment friendly". The results for washing powders vary. The most common Polish product, *Powder 1* does not contain EDTA, as well as *Powders 3* and 5, which are produced abroad. *Powder 4*, which is the same brand name as *Powder 3*, but advertised as "phosphate free", does contain 2.55 g/kg of phosphates, in the orthophosphate form, and also a signifficant amount of EDTA. *Powder 6 concentrate* is the only product, among those examined, which contains both EDTA and NTA, as well as "phosphates". Considerable high levels of chelates and "phosphates" might be caused by a "concentrate form" of that powder. An interesting fact was observed for Powder 2, which was advertised as "a strong detergent" and contained high levels of phosphates and the higest level of EDTA.

EDTA was also found in the raw sewage (see Table 1). The results from the wastewater treatment plant show no evidence of decreassing EDTA level while passing through the mechanical-biological cleaning system. On the contrary, the slight increase was observed which could not be explained by the treatment method used.

EDTA has been reported, unlike the NTA, to be resistant to biodegradation in the surface waters. It appears to be resistant to degradation in the sewage treatment processes (mechanical-biological), too. This makes concern about the EDTA load into the natural waters and soils reasonable. Without biodegradation, EDTA concentration in the treated wastes in the range of 40–50 mg/dm^3 could eventually lead to the 4–5 mg/dm^3 in those surface waters receiving 10% sewage, or proportionally lower for other circumstances of dilution.

Experiments were performed in the spring time. In autumn, the time of campaigns in the food industry, dilution of wastes might vary. Higher loads of wastes were put into the system, which is considered as dilution of domestic sewage, but in some food processing, EDTA is used for stabilization and conservation of products, so the level of EDTA in the autumn time required separated measurements.

Even though EDTA is itself not significantly toxic to either mammalian or aquatic life[8], nevertheless questions possible environmental consequences were raised. Some of those questions deal with the EDTA-metal interaction, because EDTA is a strong complexing agent for divalent and polyvalent metals. As was mentioned before, in Poland, industrial wastewater possibly containing heavy metals is often mixed with municipal sewage only after primary pre-treatment (or sometimes without any treatment).

In the natural environment heavy metals which form very strong complexes with EDTA ligands are found at very low concentration levels. Much higher concentrations of alkaline earth metals (Mg and Ca) might be found although Ca-EDTA and Mg-EDTA complexes have stability constants a few orders of magnitude lower. But due to their high concentrations, possible complex formation has to be taken into consideration.

5. EQUILIBRIA CONSIDERATIONS

Generally EDTA forms complexes with almost all metals in a 1:1 molar ratio. Very few authors considered the possiblility of the formation of more complicated complexes, containing more that one ligand or metal ions[10]. But, depending on pH, hydrogen ions as well as hydroxyl ions may compete with the metal ions for ligands. In the natural systems, where more than one metal cation and one ligand anion occur, equilibria considerations are complicated.

Examples of theoretical calculations using a computer program (SPE by Martell and Motekaitis[11]) are presented in Figures 1 and 2. Equilibrium in the water solution containing cadmium, calcium and EDTA ions was considered. Cadmium was chosen, due to its occurrence in the Lublin area surfaces waters at trace levels, probably from the phosphate fertilizers. The situation, where concentrations of cadmium and calcium are equal (0.001*M*) and concentration of EDTA is at the same level (0.001*M*) and is not sufficient to complex both ligands, is presented in Figure 1. Under such circumstances, only cadmium is complexed, due to the differences in stability constants. In Figure 2, the situation, where calcium concentration is 10 times higher than that of cadmium (0.01*M* and 0.001*M*, respectively) is presented. EDTA concentration is 0.0015*M*, which also is not sufficient to complex both ligands, but in the presence of higher calcium concentration, calcium ions successfully compete with cadmium for EDTA ligands (see Figure 2).

The complexation processes may lead to the release of metals from the sediments, although such release of heavy metals is expected more readily at low calcium concentration. At high levels of Ca (which is typical for the Lublin area, geologically built on lime stone) competitive formation of calcium complexes have to be considered. If such reactions may possibly influence the carbon dioxide equilibria in the natural waters based on the dynamic reaction with calcium carbonates, requires further studies.

6. CONCLUSIONS

"Phosphate free" detergents are not "neutral" for the environment, as they contain both: "phosphates" in the orthophosphate form, and chelating ligands. "Phosphate free"

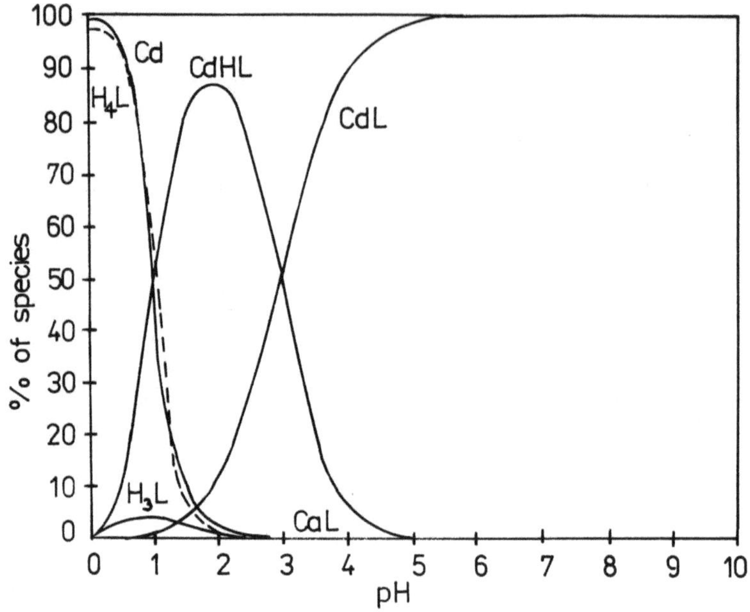

Figure 1. Species distribution diagram for 0.001 *M* Ca, 0.001 *M* Cd and 0.001 *M* EDTA at 0.1 M ionic strenght and 20^0C, calculated by the SPE program.

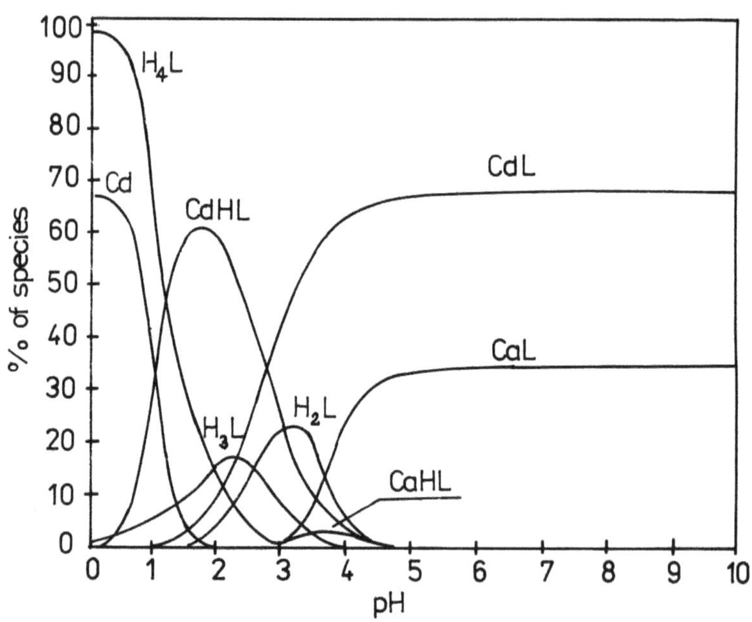

Figure 2. Species distribution diagram for 0.01 *M* Ca, 0.001 *M* Cd and 0.0015 *M* EDTA at 0.1 M ionic strenght and 20^0C, calculated by the SPE program.

means generally "free from polyphosphates" but not free from inorganic phosphorus, as we have shown in our previous research[7]. Most of the examined washing powders and washing-up liquids are a potential source of non-degradable chelates like EDTA. So, their hazardous impact has to be considered from both sides: excess of phosphorus load and amount of EDTA, which passes through the municipal wastewater treatment plant without the noticeable reduction.

Stable chelating ligands may cause mobilization of heavy metals from sediments, which means a highly negative impact on the water environment. Strong chelating capacities of EDTA may also cause reactions with magnesium and calcium, especially beacuse concentrations of Ca are much higher than the concentration of heavy metals in the natural waters and equilibria of complexation reaction may be forces in Ca-EDTA complex formation. The effect of the last process requires additional equilibria studies.

To prevent unexpected effects in the water environments limitation of non- degradable chelates is required, especially substitution of EDTA by NTA.

Possible effects of NTA and EDTA complex formation with heavy metals in soil and water environments, as well as in the sewage treatment systems should be carefully monitored.

REFERENCES

1. Dojlido, J. and G.A. Best, *Chemistry of Water and Water Pollution,* Ellis Horwood, Chechester, 1993, pp. 130.
2. Klopotek, A. and B. Klopotek, Wplyw fosforanow zawartych w w detergentach na eutrofizacje wod powierzchniowych i mozliwosci ich zastapienia kompleksonami bezfosforanowymi, *Chemia w ochronie srodowiska*, wyd. Politechniki Lubelskiej, 1993, p. 123.
3. Mark, H.F., McKetta, J.J. and D.T. Othmer, *Kirk-Other Encyclopedia of Chemical Technology*, vol.6., Interscience, New York, 1965, pp. 21–29.
4. Bunch, R.L, and M.VB. Ettinger, *Biodegrability of potential organic substitutes for phosphates,* Proc.22nd Ind.Waste Conf., Purdue Univ., Engng. Extn. Ser., No129,(1), 393–96.
5. Thom, N.S., Nitrilotriacetic acid. A literature survey, Water Res., 1971, 5, 391.
6. Swicher, R.D., Taulli, T.A. and E.J. Malec, Biodegradation of NTA metal chelates in river waters, in T*race Metals and Metal-Organic Interactions in Natural Waters,* ed. P.C. Singer, Ann Arbor Science Publishers, Ann Arbor 1973, pp. 237–263.
7. Montusiewicz, A. and M.R. Dudzińska, Inorganic phosphorus problem in domestic sewage, *this volume*.
8. Dojlido,J. and G.A.Best, *Chemistry of Water and Water Pollution,* Ellis Horwood, Chechester, 1993, pp. 129.
9. Gardiner, J., Complexation of the trace metals by EDTA in natural wates, *Water Res*., 1976, 10, 507.
10. Jawaid, M., On the complex formation between Cd(II) and EDTA, *Talanta,* 1980, 27, 95.
11. Martell A.E. and R.J. Motekaitis, *The Determination and Use of Stability Constants,* VCH Publishers, Inc., N.York 1988.

NEUTRALIZATION OF LAUNDRY WASTES BY COAGULATION

R. Tosik, S. Wiktorowski, and K. Janio

Institute of General and Ecological Chemistry*
Technical University of Łódź
36 Żwirki Street 90–924 Łódź, Poland

ABSTRACT

Laundry wastes are nuisant to individual purification if they rise anywhere the city sewage system is not. They have to be pretreated before their biological treatment because they contain high amounts of: surfacant, phosphate, etheric extract, suspension, COD and BOD. Purification of laundry wastes was performed by coagulation process with addition of flocculant. Among applied agents iron chloride was best. Optimum of coagulation condition was determined with 1000 mg/L $FeCl_3$, 0.2–0.3 mg/L of Magnafloc 336 as flocculant aid and pH in the range from 5.65 to 6.1. A high degree of reduction was obtained for all indices but residual concentration of anionic surfacants was still so high for purification by activated sludge method.

1. INTRODUCTION

The main source of laundry wastes' resulting from laundry companies are the wet washing- wastewaters. About 3 cu m of wastewaters are produced from 100 kg of the laundry. They are very burdensome with regard to the great load of organics, particularly surfacants, of which the concentration is usually about 100 mg/L , but it may also exceed 300 mg/L [1,2]. In municipal sewage the concentration does not exceed 5 mg/L. In comparision with municipal sewage, the concentration of petroleum ether, extractable substances, and also the concentration of phosphates are markedly exceeded. Thus far, the purification of laundry wastewaters has not been solved to a considerable degree. Anywhere the city sewage system isn't, there are great difficulties in the obtaininment of the required degree of purification.

* phone 48. 42. 313118, fax 48. 42.313103

Because of the changes of pH values and the big concentration of surfacants, the direct, biological purification has not been taken into consideration. Coagulation with flocculants or oxidants such as ozone and hydrogen peroxide are applied to chemical purification [3,4,5,6].

The technological difficulties induced us to apply the coagulation process to purification of laundry wastes.

2. AIM OF STUDY

The investigations were made with the aim of attaining optimization of the coagulation process prior to biological purification of laundry wastes by means of activated sludge method.

The evaluation of acting efficiency of three coagulant agents : iron sulphate $Fe_2(SO_4)_3$, iron chloride $FeCl_3$ and sodium aluminate $Na_2Al_2O_4$, was performed in our investigations. The main criteria of the evaluation was a decrease of organic contents to the required level before biological purification. We did not exceed the admissible degree of salination measured as chloride and sulphate concentration in the wastewater. After the coagulation process the content of wastewater was evaluated as meeting standards for biological treatment by activated sludge. The experiments were made under laboratory conditions.

Characteristic of laundry wastes and required degree of purification are presented in Table 1.

3. METHODS

The experiments were made on standard laboratory equipment to coagulation-flocculation studies, electrically controlled with variable speed drive. The procedure for coagulation testing was standardized by 1 liter beakers, flocculation process observation, measurements of flocculent settling velocity with volume of sediment, and determinations specific to the treatment under study. The later include indices of settled water : suspensions, color and turbidity measured by A420 - absorbance of light λ = 420 nm, COD,

Table 1. Characteristic of raw laundry wastes

Indices	Determined value	Standard value	Required degree of purification %
pH	—	6.5–9.0	—
Suspension mg/l	520	50	91
BOD mg/l O_2	570	30	95
COD mg/l O_2	1640	150	91
Phosphate mg/l	52	5.0	91
Chloride mg/l	85	1000	—
Sulphate mg/l	—	500	—
TDS mg/l	1280	2000	—
Iron mg/l	—	10	—
Surfactant mg/l	115	5.0	96

BOD, TDS, anionic surfacants, phosphates, iron, conductivity, chlorides and sulphates. The range of the determinations dependend on the aim of investigation and was differential during the experiments. The determinations were made according to Polish standards.

4. INVESTIGATIONS OF COAGULATION PROCESS

4.1. Pretreatment by Coagulation

The raw laundry wastes were pretreated by the coagulation process. Coagulant agents were applied : iron chloride (III) , iron sulphate (III) and their mixture , and also the mixture of iron chloride and sodium aluminate. The coagulation process was investigated for maximum doses of coagulant aids, at various pH.

That maximum dose was limited by standard admissible concentrations of chloride and sulphate ions in the wastewaters. The evaluation of effects was performed based on A420 and the changes COD of the settled water regarding the beginning values. The results of investigations are presented in Figure 1.

We concluded that the best finding for 1300 mg/L of iron chloride (III) was obtained because absorbance A420 and COD reduction were the lowest and highest, respectively (compartment two on Figure 1).

Therefore in the further experiments only iron chloride was used and the influence of doses was investigated. The results are presented in Table 2.

As we see from Table 2, the best coagulation effect was obtained for the dose 1300 mg/L, however, it was associated with a considerable decrease of pH to 3.8.

For the dose 1200 mg/L effects are only slight worse , while the value of pH was already 6.05. Therefore the dose 1200 mg/L was admitted as the optimal one and it was used for investigations of influence of pH on coagulation process.

4.2. Influence of pH and Optimum of Coagulation

Using the dose 1200 mg/L $FeCl_3$ the run of coagulation at various values of pH in the range from 4.95 to 9.1 was investigated. In the settled water absorbance A420 and iron concentration were measured. Table 3 presents the findings and as one may see the best effects are at pH in the range from 5.65 to 6.1 because minimum iron concentration and low absorbance A420 were obtained.

In order to attain optimization of coagulation process conditions for laundry wastes purification a series of experiments were complited. At the optimal pH 5.65, the doses of iron chloride were applied : 700, 800, 900, 1000, 1100 and 1200 mg/L. The kinetics of

Table 2. Results of coagulation by iron chloride

Dose	Absorbance λ420			Surfacant		COD		Iron	Chloride
mg/l	Value A	Reduction %	pH	mg/l	Reduction %	mg/l	Reduction %	mg/l	mg/l
0	1.467	0	8.6	115.0	0	1640	0	—	85
1000	0.206	86	6.9	33.3	61	476	71	11	680
1100	0.114	92	6.85	22.5	80	373	77	6.8	780
1200	0.090	94	6.05	16.8	85	237	86	2.8	850
1300	0.041	97	3.8	10.3	91	226	86	1.6	920

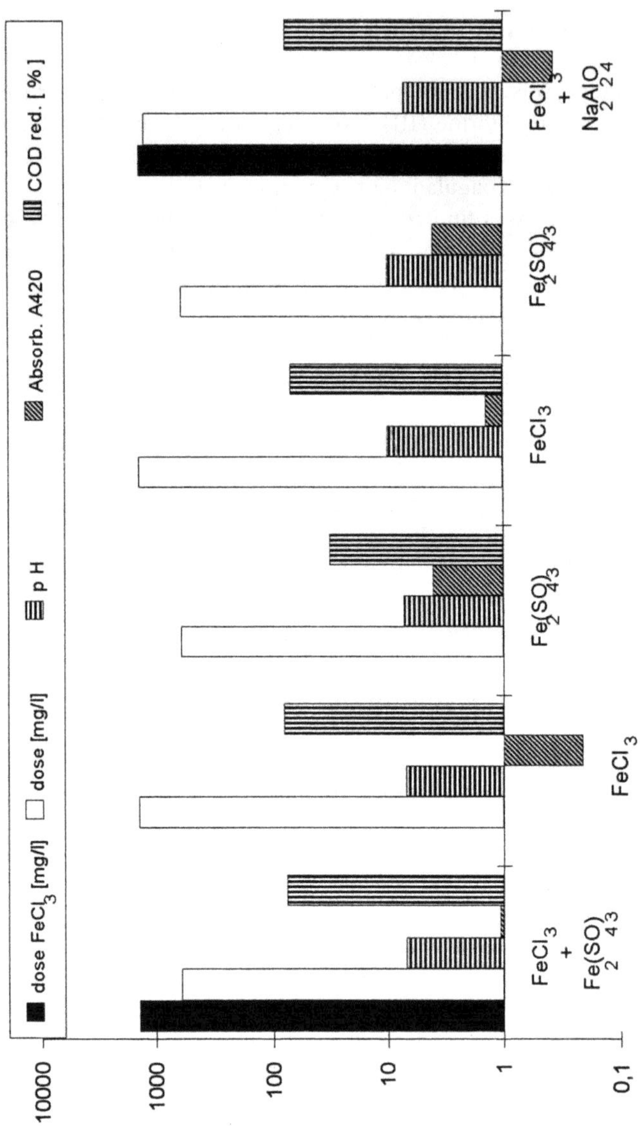

Figure 1. Coagulation findings with different agents.

Table 3. Influence of pH on coagulant's effects by 1300 mg/L $FeCl_3$

No	1	2	3	4	5	6	7	8
pH	4.95	5.55	5.65	6.1	6.95	7.95	8.75	9.1
A420	0.027	0.045	0.036	0.079	0.118	0.117	0.122	0.465
C_{Fe}	3.4	3.0	2.2	2.3	2.8	4.0	7.2	20.0

sedimentation were observed and then analysis of the settled water was made. Absorbance A420, final pH, COD and concentration of surfacant were determined. The results are presented in Figure 2 and Table 4.

After two hours of sedimentation the best result was for the dose 1100 mg/L because sludge was compacted best. Increase of the dose to 1200 mg/L caused a marked extension of the sediment's volume about 21 % without an essential improvement of rates (see Table 4).

The dose 1100 mg/L $FeCl_3$ was chosen as the optimal one.

4.3. Improvement of Coagulation by Flocculant

As we previously stated coagulation of laundry wastes requires high doses of iron chloride. In order to decrease the coagulant's agent we applied flocculant named Magnafloc 336 of Allied Colloids Firm. We investigated the influence of flocculant doses on the efficiency of coagulation process with different amounts of iron chloride.

First experiments with 1 mg/L of Magnafloc 336 allowed us to decrease the dose of $FeCl_3$ to 1000 mg/L. Next we applied doses of flocculant from the range 0 to 5.0 mg/L for coagulation performed with 1000 mg/L $FeCl_3$. The evaluation of effects was made after coagulation, floculation and sedimentation by measurement of : absorbance of settled water and sludge volume.

The results are presented in Table 5.

Flocculant gave the highest efficiency with the amount 0.2 - 0.3 mg/L because absorbances and volume of sludge were the lowest, conspicinously precipitating sedimentation and compaction of sediment. The increase of the flocculant's dose to 0.5 mg/L caused worsenig of the obtained effects.

A renewed improvement of coagulation effects occured for the dose of flocculant 1 and 5 mg/L, but instead of sedimentation, flotation of sediment occured. The flocculation process was very improved by addition of 0.2 or 0.3 mg/L of Magnafloc 336 that gave great and heavy flocs. Sedimentation of flocs was very fast and the settled water reached

Table 4. Characteristic of settled water after coagulation with different doses of $FeCl_3$

Dose $FeCl_3$	Absorbance			COD		Surfactant	
mg/l	λ420	Reduction %	pH	mg/l	Reduction %	mg/l	Reduction %
0	1.810	0	6.05	1490	0	147.2	0
700	0.398	78	5.55	525	65	45.6	69
800	0.275	85	5.70	386	74	45.5	69
900	0.151	92	5.65	294	80	35.7	76
1000	0.086	95	5.60	228	85	31.2	79
1100	0.040	98	5.50	177	88	24.9	83
1200	0.037	98	5.60	173	88	24.5	83

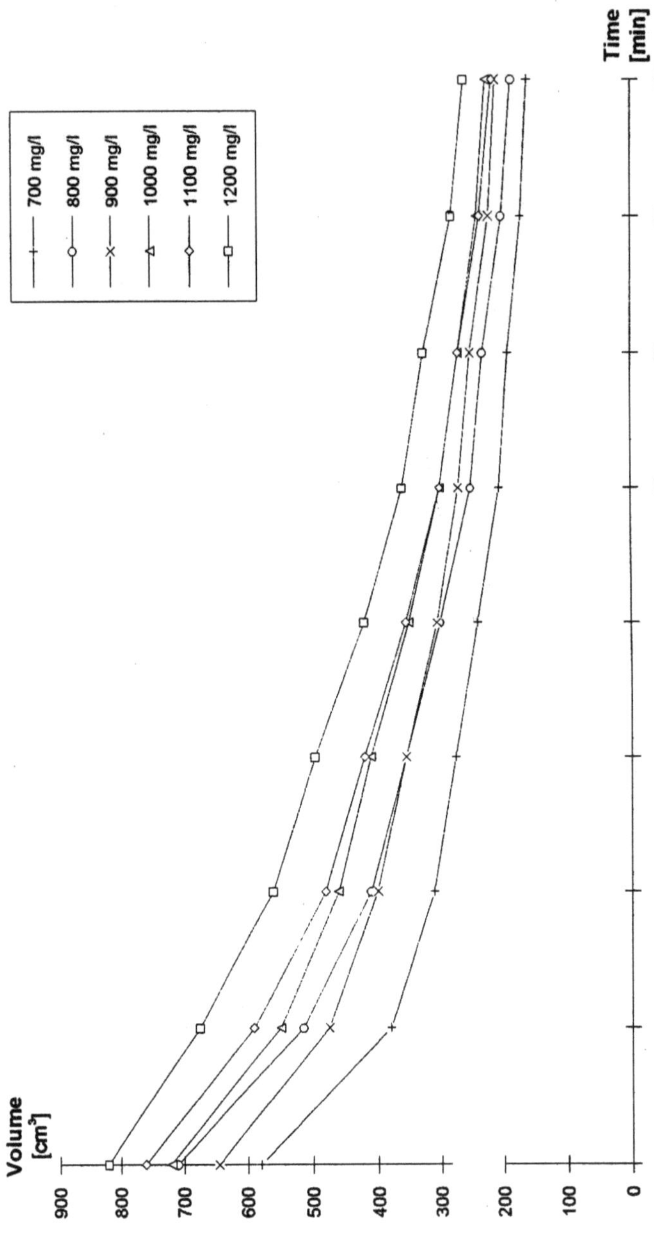

Figure 2. The kinetics of sedimentation after coagulation with different doses of $FeCl_3$

Table 5. Coagulation 1000 mg/l $FeCl_3$ with flocculant Magnafloc 336

Flocculant dose mg/l	Absorbance λ420 %	Percentage sludge volume
0	100	100
0.05	108	85
0.1	83	75
0.2	58	75
0.3	38	70
0.5	180	55
1.0	58	flotation
5.0	33	flotation

clear. Likewise clear liquid phase was obtained after flotation with 5 mg/L of Magnafloc 336 and we observed that the suspended phase was very stable.

5. DISCUSSION

Finally we stated that the best parameters of the coagulation process of laundry wastes are following :

- the dose of iron chloride 1000 g/L $FeCl_3$
- pH of reaction habitat in the range from 5.65 to 6.1
- the doses of flocculant Magnafloc 336 0.2 or 0.3 mg/L

Raw laundry wastes were purified in the aforementioned conditions by coagulation and flocculation processes and no pH adjustment. The changes of pH were caused by addition of coagulant agent, fortunately, the dose of coagulant decreased the initial pH level to required value.

The results of waste treatment are presented in Table 6.

As we see from Table 6 most indices were decreased, more than 50 per cent , particularly BOD, COD, suspension. These values are in the range of required levels for biological treatment. Phosphate, surfacant and etheric extract were more fully removed however their residual concentrations were still high. Simultaneous others indices were increased as the result of coagulant agent addition, namely conductivity, soluble salts and chloride ions. However, they were in the required range of values. After that pretreatment nitrogen compounds were not removed and it will not be a problem if the wastes should be later purified by biological methods.

6. CONCLUSIONS

We finally concluded that pretreatment of raw laundry wastes should be done before biological purification. We stated that coagulation with 1000 mg/L of $FeCl_3$ and the addition 0.3 mg/L of Magnafloc 336 gave the good results without pH adjustment.

Pretreated laundry wastes should be tested by biological methods and we hope to do so more in the future. However, because we were concerned about the influence of surfacants on biocenosis or their removal by activated sludge, we tested only the laundry wastes after coagulation on laboratory equipment with activated sludge. We stated that un-

Table 6. Final result of purification of laundry wastes Coagulant 1000 mg/l $FeCl_3$, flocculant 0.3 mg/l Magnafloc 336

No.	Indices	Unit	Raw wastes	Purified wastes	Reduction – Extension +
1.		pH	9.2	6.0	—
2.	Conductivity	μS/cm	1750	2400	+37
3.	TDS	mg/l	1280	1770	+32
4.	Suspension	mg/l	520	200	−62
5.	Chloride	mg/l	85	680	+700
6.	COD–chromate	mg/l O_2	1100	370	−66
7.	COD–permanganate	mg/l O_2	141	84	−40
8.	BOD	mg/l O_2	570	280	−51
9.	Ammonia	mg/l N	3.5	3.5	—
10.	Nitrite	mg/l N	0.007	0.006	—
11.	Nitrate	mg/l N	0.1	0.1	—
12.	Organic nitrogen	mg/l N	5.0	5.2	—
13.	Total nitrogen	mg/l N	8.6	8.8	—
14.	Surfacant	mg/l	70	23	−67
15.	Phosphate	mg/l	52	7.5	−86
16.	Etheric extract	mg/l	95.8	42	−56
17.	Iron	mg/l	—	7.1	—

der loading COD nearly 0.4 g/g dried mass of activated sludge, COD decreased to about 50 mg/L but residual concentration of anionic surfacants was still higher than 7 mg/L, foam was generating and activated slugde was escaping from the bioreactor. It is obious that this result is not fully acceptable and it is the reason why investigations should be continued.

7. REFERENCES

1. H. Meinck, H. Stoff, H. Kohlschutter , Scieki przemysłowe , ed. Arkady 1975 p 772
2. A.A. Mamontova, N.A. Klimenko, I.G. Roda, S.V. Drozdovich "removal of surfacants from wastewaters" Izobretenija , 1992 (1) 10
3. T.M. Ivkina, N.D. Barabash, T.N. Shashkova "Removal of nonionic surfacants from rayon manufacturing wastewater" Tekhnol. Fiz - Khim. Ochistki Prom. Stochnych Vod. Anal. Kontrol Processow Oczistki 1990, 35–8 ed. Vodgeo Moscaw
4. M.A. Grachok, S.A. Prokudina "Reaction of ozone with surfacants in wastewaters from laundering of protective apparel in nuclear power plants" Khim Tekhnol. Vody 1992 14 (2) 128–32
5. M.A. Grachok, S.A. Prokudina, V.G. Permakov, M.I. Shulyatev, " Possibilty of using ozone for removal of surfacants from wastewater from laudering of protective apporeal in nuclear power plants " ibid 1992 14 (2) 132–6
6. L. Stoica, C. Bejon, R. Catuneanu, M. Petrscu, D. Furmuseln "Treatment of wastewaters containing anionic detergents by coagulation and flotation" Pat. rom. RO 94263 30 05 1988 (Cl, CO 21F 1/24)

UTILIZATION OF LEATHER SCRAPS AFTER CHROME AND VEGETABLE TANNING

Zygmunt Kowalski[1] and Michał Piwowarski[2]

[1] Institute of Chemistry and Inorganic Technology,
Cracov University of Technology
Warszawska St. 24 31–155 Kraków, Poland
[2] Institute of Inorganic Chemistry
Sowińskiego St. 11 44–101 Gliwice, Poland

1. INTRODUCTION

Production of shoes, leather fancy goods and leather tanning processes produce wasted leather scraps of which quantity in Poland is estimated to be at the level 15 000–20 000 t/y.

This quantity strongly depends on the market situation in the tanning industry. Over 90% of scraps are those after chrome tanning, 5–6% after chrome-vegetable tanning and the rest is after vegetable tanning. All those types of scraps are classified in Poland among I-st category hazardous wastes.

A typical scrap has an area from 1 cm^2 to 10 cm^2, moisture averages from 30% to 60% and the bulk density of scraps, which depend on moisture content, averages 1600 kg/m^3. Large industrial tanneries cover only 60% of the total leather production and leather scraps formed in some thousand small tanneries create serious utilization problems.

Leather scraps after chrome tanning include:

- leather scraps after shoes and fancy goods production containing up to 12.5% N, up to 2% of fat, 3.5 -5% of Cr(III) and over 30% H_2O.
- tanning solid wastes (after trimming and splitting) containing up to 12.5% N, up to 2% of at, 3,5-% of Cr(III) and up to 70% H_2O.

Leather scraps after chrome-vegetable tanning contain up to 1% Cr(III), 7–9% N and up to 10% of fat. Scraps after vegetable tanning contain up to 7% N 7–20% of fat. Moisture of both mentioned types is similar and averages from 30% to 60% H_2O [1]. Until now only 30% of leather scraps after chrome tanning were utilized into fertilizing "leather powder". Research carried out during the last 15 years on new utilization technologies for

Chemistry for the Protection of the Environment 2, edited by Pawłowski et al.
Plenum Press, New York, 1996

all types of leather scraps into fertilizers and leather scraps hydrolysis were required for solving this problem [2,3].

Our paper presents three methods for leather scrap utilization

1. "semi dry" method for fertilizing "leather powder" production. This oldest technology now is still applied for utilization of 30–40% of domestic scraps after chrome tanning containing up to 4% of Cr(III).
2. new developed "dry" and "wet" methods. The "wet" method has now been implemented in industry.In applying "wet" method we could utilize all types of wasted leather scraps occurring in Poland.

2. SEMI-DRY AND DRY TECHNOLOGIES OF LEATHER SCRAPS UTILIZATION

All leather scrap utilization technologies are based on the hydrolysis process of fibrylar collagen from scleroproteins group. Collagen is insoluble in water, diluted acids, and bases, and has low hydrolysis susceptibility. Tanning processes reinforce protein structures, therefore only high pressure and high temperature hydrolysis could partially destroy collagen structure.

The oldest technology for leather scrap utilization is the "semi dry" method. The flow diagram of this method is shown in Fig.1. [1,2,3]

The batch process of hydrolysis at temperature up to 160 °C and under pressure up to 0.5 MPa for 2 hours can be carried out in autoclaves. After grinding, seasoning, secondary grinding, drying and milling in a desintegrator the leather powder is obtained containing proportionally:

N_{total}	Up to 12.2% ($N_{digestible}$ - up to 7.3%)
Fat	4.0%
H_2O	7.0%
Cr(III)	up to 3.5%
$Proteins_{total}$	up to 76% ($proteins_{digestible}$ - up to 46%)

Bulk density averages 1600 kg/m^3 and granulation: 70% between 0.5–1mm, 28% below 0,5 mm.Batch processes can result in 60% hydrolysis of total proteins. The obtained "leather powder" is used as a component in the production of different types of grass fertilizers.

A basic problem of this technology is the rather large quantity (up to $1.5 m^3/t$ of product) of liquid wastes, which contain: Cr(III)—from 0.6 to 1 g/dm^3, Fe_2O_3—0.1 g/dm^3 and CaO—from 30 to 40 g/dm^3.The BOD_5 of the wastes is between 10–20 g/dm^3. Some methods for waste treatment can be implemented [4]. Treated wastes could be used as a raw material for liquid fertilizer production, but treatment costs are rather high. Another problem is bed product quality connected with high concentrations of Cr(III) resulting in low N/Cr ratios which range from 3.5 to 6 and should be much higher.

The chromium content in the product can be decreased to 2%. Only the part of the leather scraps after chrome tanning, containing less than 4% of Cr(III) can be utilized in the "semi-dry" technology. It isn't possible to utilize in this process leather scraps after chrome vegetable and vegetable tanning. Research in this field has on not resulted in any significant progress.

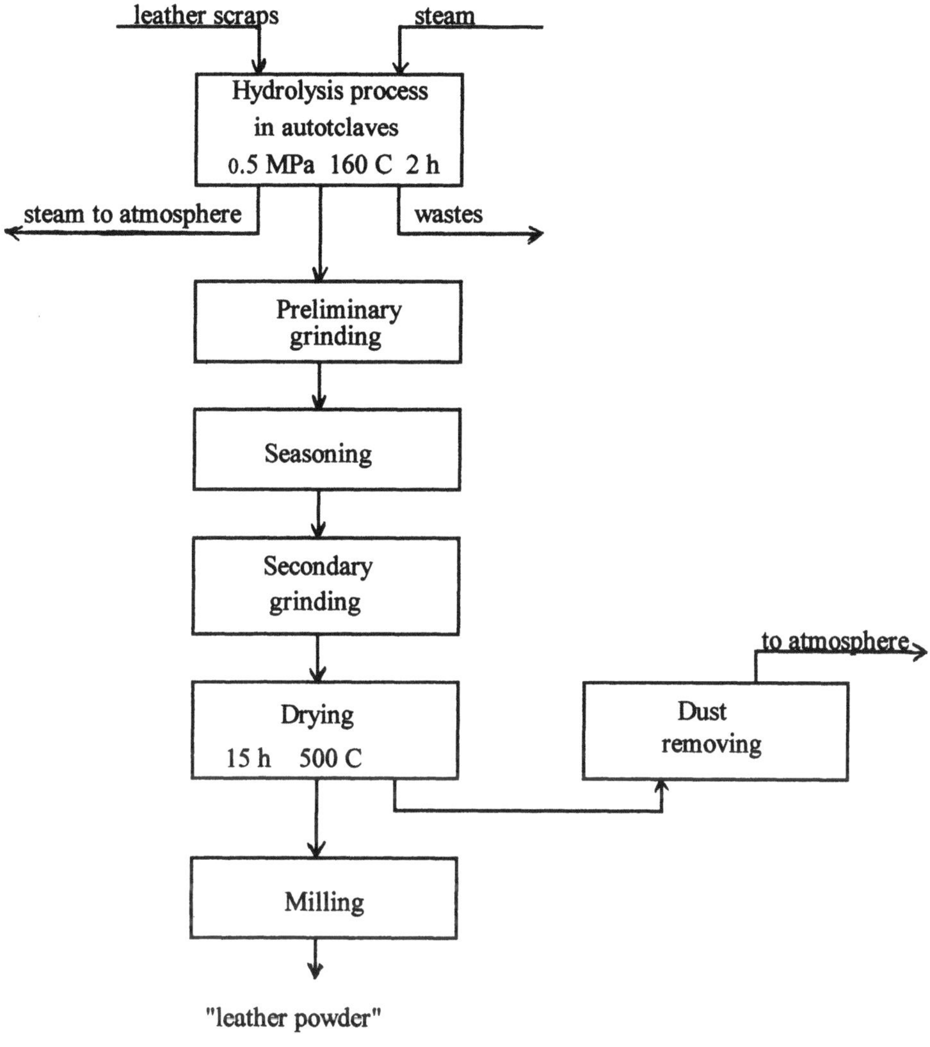

Figure 1. "Semi-dry" method of leather scraps utilization into fertilizing "leather powder".

In the years from 1987–1991 the "dry" method was developed [5,6] as presented in Fig.2.

In this method a two stage process is realized in a rotary autoclave of special construction [6].

The first stage is a hydrolysis process at the same parameters as in the "semi-dry method".

The second stage (subsequent operation) in rotary dryer under vacuum down to 0.05MPa and at a temperature of 100 °C during a 4 hours period. After preliminary grinding and milling the "leather powder" is obtained with better quality period with respect to Cr(III) contamination.

The resultant Cr(III) content is even lower than 1.8% and N/Cr ratio is 5–7. The "dry" method is wasteless.It has not been applied in industry for two reasons . One is high investment and operation costs of rotary autoclaves and some risk in implementation of

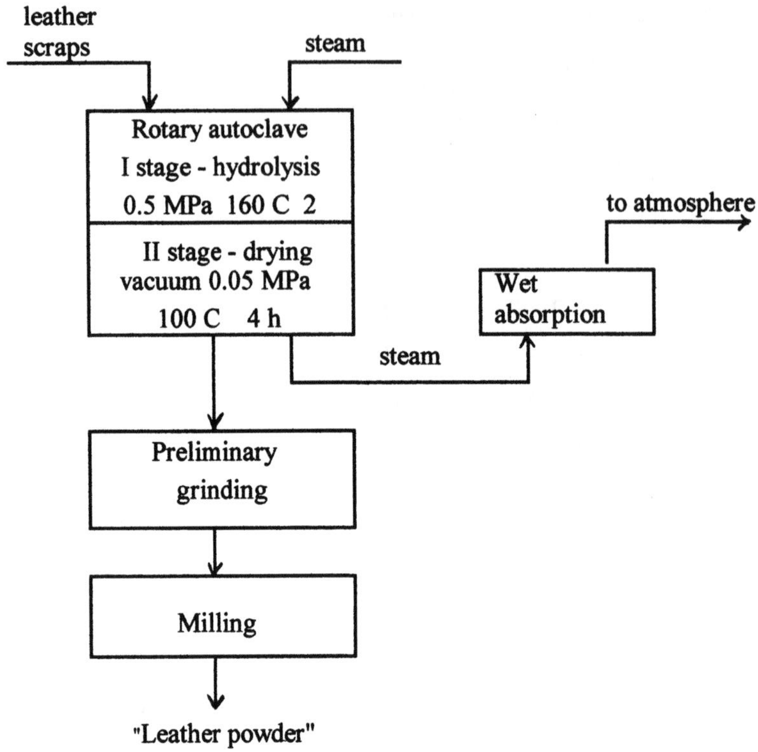

Figure 2. "Dry" wasteless method of leather scraps utilization.

new equipment, the second reason is there it is still not satisfactory, too high a Cr(III) content and too low a N/Cr ratio.

3. WET PROCESS OF LEATHER SCRAPS UTILIZATION

The solving of quality problems and utilization of all types of leather scraps has required basic research on leather scraps hydrolysis . First the possibility of hydrolysis at normal pressure with water containing acids or hydroxides was examined [2]. Results suggested that new solutions can utilize hydrolysis under high pressure [7].

Experiments on leather scraps after chrome tanning containing 2- 4% of Cr(III) and 30 -40% of moisture were carried out in laboratory autoclaves. The hydrolysis reaction time was up to 4 hours, steam was brought directly into the autoclave under pressure 1.5 MPa and at temperature c. 200 $^{\circ}$ C.

Addition of NaOH was applied in hydrolysis experiments using concentrations from 1 to 4% NaOH solutions and scraps/solution ratio ranges from 1:1 to 1:3.

Results from hydrolysis yieldid brownish liquid with disagreeable odour and visible suspension. Total suspended solids averages up to 70 g/dm^3. After filtration with a laboratory frame filter press, the filtrate contained 2–2.5% N and less than 1000 ppm Cr(III), the N/Cr ratio reached 25. Protein hydrolysis was at the level of 94% and the yield of nitrogen recovery in filtrate was as high as to 75% [3].

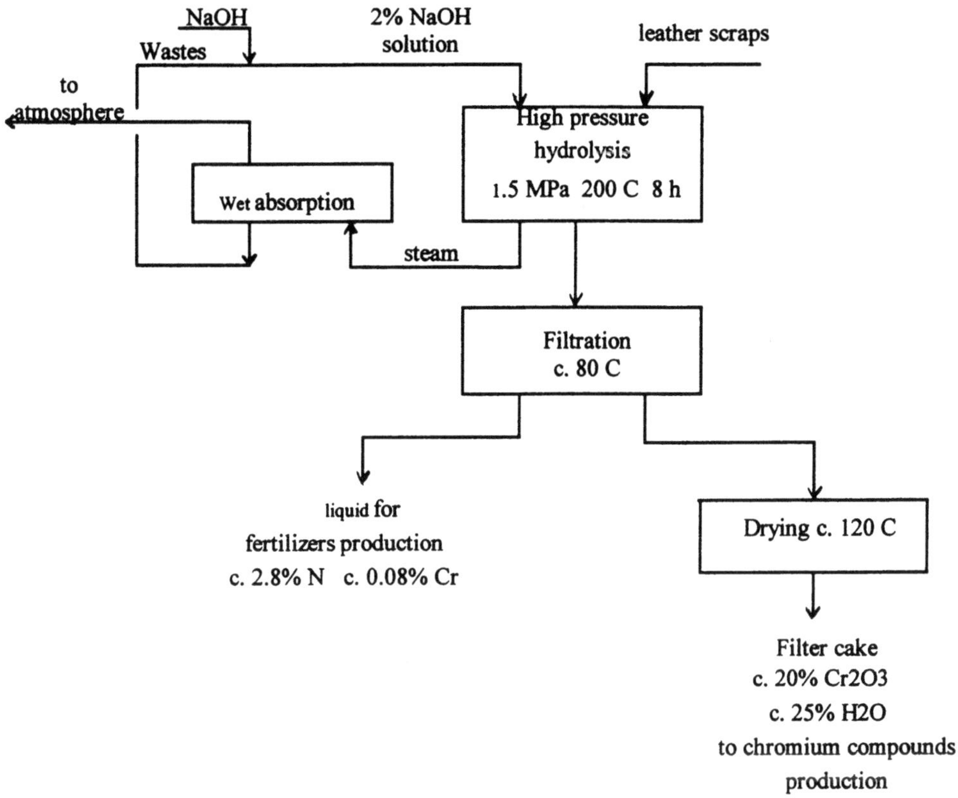

Figure 3. "Wet" method of leather scraps utilization.

Industrial trials carried out in 1994 with a of scale 3000 kg per charge gave even better results.

The product of hydrolysis during 8 hours under a pressure of 1.5 MPa contained an average of 2.8% N, 820 ppm Cr(III) and a N/Cr ratio was over 30.

These results formed the bases for a project of an industrial plant with a capacity of 1000 t/y, which is now erected near Olkusz. The basic flow diagram of this system is shown in Fig.3.

Nitrogen liquid obtained will be used as the raw material for production of mineral-organic fertilizers. This production is based on the mixing of raw materials (micro- and macroelements) in the way as in bulk blending plants.

The designed unit should produce up to 12 000 t/y of different type solid fertilizers and 1000 t/y of liquid nitrogen fertilizers.

The flow diagram of this unit is shown in Fig. 4.

The basic production program includes:

- liquid nitrogen fertilizer (16–17%N) for lawns and flowers
- mineral organic solid growth substrate (1.4–1.5 N) for flowers and greens
- ornamental plants fertilizers (1.5% N)
- mineral organic fertilizers for sugar beets (2% N, 1% P_2O_5, 1% K_2O
- mineral organic fertilizers for grassland (1.7% N, .9% P_2O_5, 0.7% K_2O

More detailed information about these are given in [1].

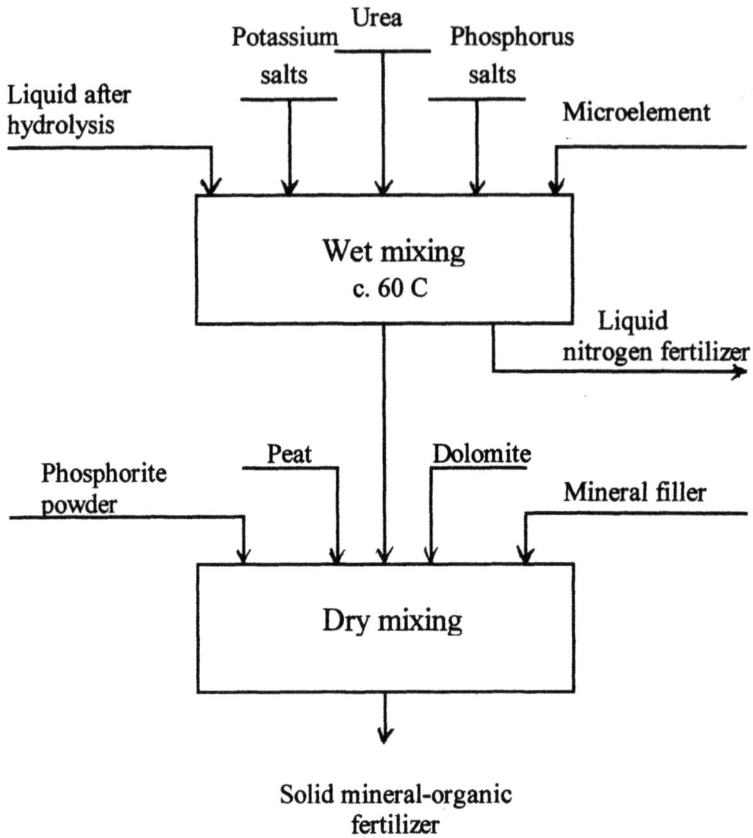

Figure 4. Flow diagram of mineral-organic fertilizer production.

Another interesting problem of wet technology is the utilization of cake from filter presses in chromium compounds production. This cake contains, after preliminary drying, up to 21% Cr_2O_3 and 25% H_2O. After calcination at temperature of 600 °C a concentration of Cr_2O_3 is increased up to 53%. Calcinated product of this type could be used as a chromic ore substitute in sodium chromate production. Research in this field is continuing.

4. SUMMARY

Research carried out during 15 years, has developed technologies for leather scraps utilization from a "semi-dry" method, which could utilize only some part of the scraps after chrome tanning, to a "wet" method for all type leather scraps utilization which occurr in Poland. So now we posseses the technology to solve the problem of over 20,000 tons of solid wastes classified as I-st category hazardous wastes. The wet method for leather scraps utilization into fertilizers with high pressure alkaline hydrolysis process is now implemented.

Obtained liquid nitrogen fertilizer will be base for production of different types mineral-organic fertilizers (for flowers, greens, sugar beets and others).

The technology is essentially wasteless because the by-product filter cake could be used together with chromic ore as a substitute in chromate production. The proposed technology is simple, cheap and should be also profitable for future producers.

5. REFERENCES

1. Z. Kowalski, Projekt utylizacji ścinek skór do produkcji mączki skórzanej i nawozów mineralno-organicznych w ZWP "Silikaty" Klucze (1994), not published
2. H. Górecki, J. Hoffman, Z. Kowalski, Prace Instytutu Technologii Nieorganicznej i Nawozów Mineralnych Politechniki Wrocławskiej (1990), not published
3. Z. Kowalski, M. Piwowarski, J. Hazard. Mater. (1995), dispatched to print
4. Z. Kowalski, J. Hazard. Mater. 37 (1994), 137–141
5. Z. Kowalski, Prace PPUH "Alchem," Alwernia (1993) not published
6. Polish utility pattern 79804 (1990)
7. Polish patent application 304 590 (1994)

29

MOLTEN SALT DESTRUCTION OF ENERGETIC MATERIAL WASTES AS AN ALTERNATIVE TO OPEN BURNING*

Ravindra S. Upadhye, César O. Pruneda, and Bruce E. Watkins

Lawrence Livermore National Laboratory
Livermore, California

ABSTRACT

The Lawrence Livermore National Laboratory in conjunction with the Energetic Materials Center (a partnership of Lawrence Livermore and Sandia National Laboratories), is developing methods for the safe and environmentally sound destruction of explosives and propellants as a part of the Laboratory's ancillary demilitarization mission. As a result of the end of the Cold War and the shift in emphasis to a smaller stockpile, many munitions, both conventional and nuclear, are scheduled for retirement and rapid dismantlement and demilitarization. Major components of these munitions are the explosives and propellants, or energetic materials. The Department of Energy has thousands of pounds of energetic materials which result from dismantlement operations at the Pantex Plant. The Department of Defense has several hundred million pounds of energetic materials in its demilitarization inventory, with millions more added each year. In addition, there are vast energetic materials demilitarization inventories world-wide, including those in the former Soviet Union and eastern Bloc countries. Although recycling and reusing is the preferred method of dealing with these surplus materials, there will always be the necessity of destroying intractable or unusable energetic materials. Traditionally, open burn/open detonation (OB/OD) has been the method of choice for the destruction of energetic materials. Public concerns and increasingly stringent environmental regulations have made open burning and open detonation of energetic materials increasingly costly and nearly unacceptable. Thus, the impetus to develop environmentally sound alternatives to dispose of energetic materials is great.

The Molten Salt Destruction (MSD) Process has been demonstrated for the destruction of HE and HE-containing wastes (1,2,3). MSD converts the organic constituents of the waste into

* This work was performed under the auspices of the U. S. Department of Energy by the Lawrence Livermore National Laboratory under contract number W-7405-ENG-48.

non-hazardous substances such as carbon dioxide, nitrogen and water. Any inorganic constituents of the waste, such as binders and metallic particles, are retained in the molten salt. The destruction of energetic material waste is accomplished by introducing it, together with oxidant gases, into a crucible containing a molten salt, such as sodium carbonate, or a suitable mixture of the carbonates, chlorides or sulfates of sodium, potassium, lithium and calcium.

We destroyed the following pure component DOE and DoD explosives in the experimental in our High Explosives Applications Facility (HEAF): Ammonium Picrate, HMX, K-6, NQ, NTO, PETN, RDX, TATB, and TNT. In addition, we destroyed the following formulations: Comp-B, LX-10, LX-16, LX-17, and PBX-9404. We continued our investigation into the destruction of the liquid gun propellant XM-46 for the US Army. In all these cases, the fractions of carbon converted to CO and of chemically bound nitrogen converted to NOx were found to be well below 1%. The details of the experiments are contained in the attached paper. In addition, we have completed construction of the next MSD unit, incorporating our advanced design. This unit, currently under installation, is nominally rated at 5 kg/hr of HMX.

INTRODUCTION

The Lawrence Livermore National Laboratory in conjunction with the Energetic Materials Center (a partnership of Lawrence Livermore and Sandia National Laboratories), is developing methods for the safe and environmentally sound destruction of explosives and propellants as a part of the Laboratory's ancillary demilitarization mission. As a result of the end of the Cold War and the shift in emphasis to a smaller stockpile, many munitions, both conventional and nuclear, are scheduled for retirement and rapid dismantlement and demilitarization. Major components of these munitions are the explosives and propellants,

or energetic materials. The Department of Energy has thousands of pounds of energetic materials which result from dismantlement operations at the Pantex Plant. The Department of Defense has several hundred million pounds of energetic materials in its demilitarization inventory, with millions more added each year. In addition, there are vast energetic materials demilitarization inventories world-wide, including those in the former Soviet Union and eastern Bloc countries. Although recycling and reusing is the preferred method of dealing with these surplus materials, there will always be the necessity of destroying intractable or unusable energetic materials. Traditionally, open burn/open detonation (OB/OD) has been the method of choice for the destruction of energetic materials. Public concerns and increasingly stringent environmental regulations have made open burning and open detonation of energetic materials increasingly costly and nearly unacceptable. Thus, the impetus to develop environmentally sound alternatives to dispose of energetic materials is great.

The Molten Salt Destruction (MSD) Process has been demonstrated for the destruction of HE and HE-containing wastes(1,2,3). MSD converts the organic constituents of the waste into non-hazardous substances such as carbon dioxide, nitrogen and water. Any inorganic constituents of the waste, such as binders and metallic particles, are retained in the molten salt. The destruction of energetic material waste is accomplished by introducing it, together with oxidant gases, into a crucible containing a molten salt, such as sodium carbonate, or a suitable mixture of the carbonates, chlorides or sulfates of sodium, potassium, lithium and calcium. (See Figure 1 for the process flow diagram, and Figure 2 for the schematic of the experimental set-up.)

The temperature of the molten salt can be between 400° to 900°C. The organic components of the waste react with oxygen to produce carbon dioxide, nitrogen and steam.

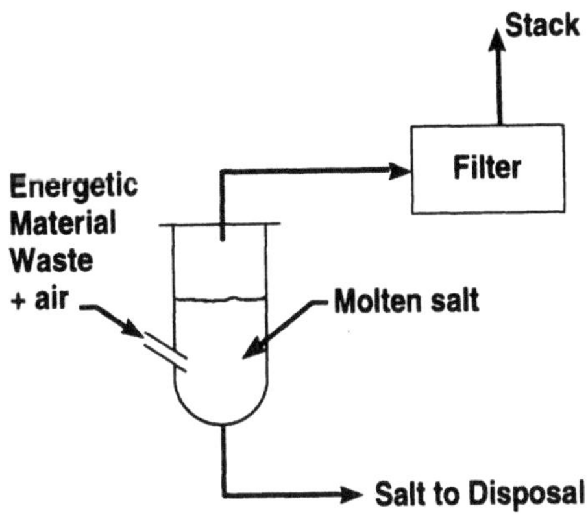

Figure 1. Process flow diagram.

The inorganic components, in the form of "ash", are captured in the molten salt bed as a result of wetting and dissolution of the ash. Halogenated hydrocarbons in the waste (which may be present as a result of halogenated solvents in the waste, or as a result of halogenated binders for the HE) generate acid gases such as hydrogen chloride during the pyrolysis and combustion processes occurring in the melt. These are scrubbed by the alkaline carbonates, producing steam and the corresponding salt, such as sodium chloride. The off-gases from the process are sent through standard off-gas clean-up processing (such as bag filters or HEPA filters) before being released to the atmosphere. At the end of the process runs, the salt is separated into carbonates, non-carbonate salts, and ash. The carbonates are recycled to the process, and the stable salts are disposed of appropriately.

LLNL has built a small-scale (about 1 kg/hr throughput) unit to test the destruction of HE using the MSD process. (See Figure 3.) In addition to the high explosive HMX (octahydro-1,3,5,7-tetranitro-1,3,5,7-tetrazocine), we have destroyed RDX (hexahydro-1,3,5-trinitro-1,3,5-triazine), PETN (2,2-bis[{nitoxy}methyl]-1,3-propanediol dinitrate), ammonium picrate, TNT (trinitrotoluene), nitroguanadine, and TATB (2,4,6- trinitro-1,3,5-benzenetriamine). We have also destroyed a liquid gun propellant comprising hydroxyammonium nitrate, triethanolammonium nitrate and water. In addition to these pure components, we have destroyed a number of commonly used formulations, such as LX-10 (HMX/Viton), LX-16 (PETN/FPC 461), LX-17 (TATB/Kel F), and PBX-9404 (HMX/CEF/nitrocellulose). Our experiments have demonstrated that energetic materials can be safely and effectively treated by MSD. We have also investigated the issue of steam explosions in molten salt units, both experimentally and theoretically, and concluded that steam explosions can be avoided under proper design and operating conditions. The relationship between residence time, temperature, concentration and throughputs, avoidance of back-burn, and determination of the products of combustion under different operating conditions are the key technical issues to be resolved in more controlled experiments.

TECHNICAL BASIS

The molten salts are typically mixtures of alkali or alkaline earth carbonates and halides. The salts provide excellent heat transfer and reaction media, catalyze oxidation of or-

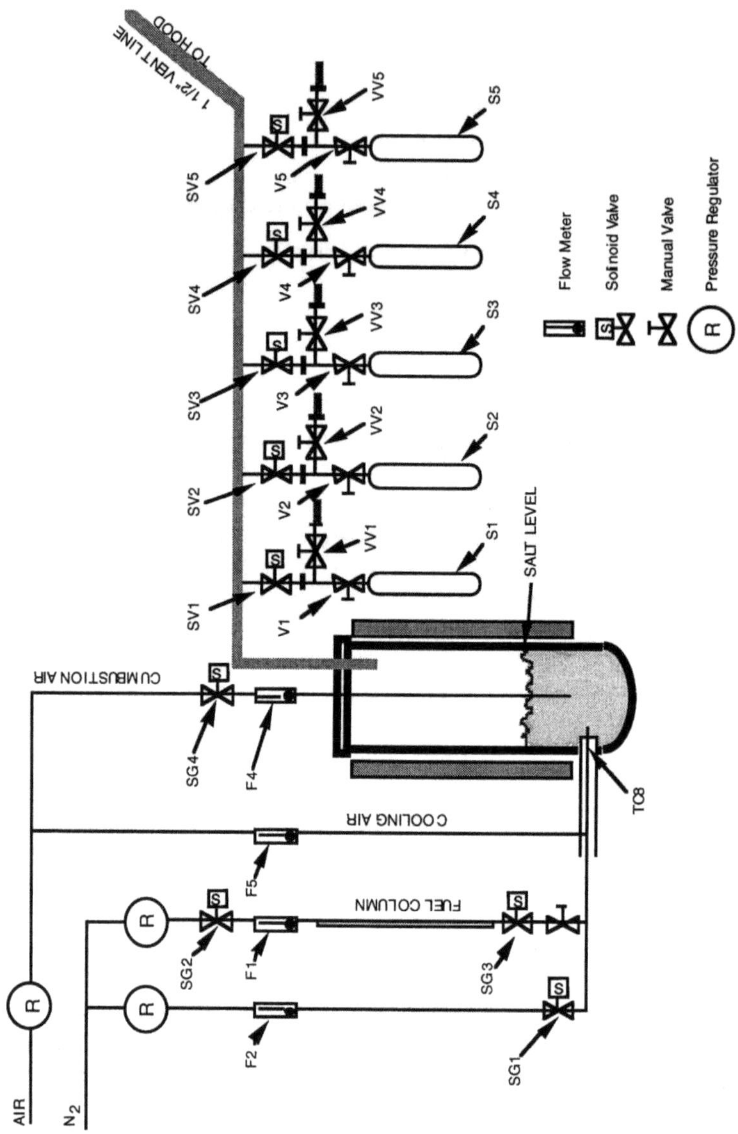

Figure 2. Experimental schematic.

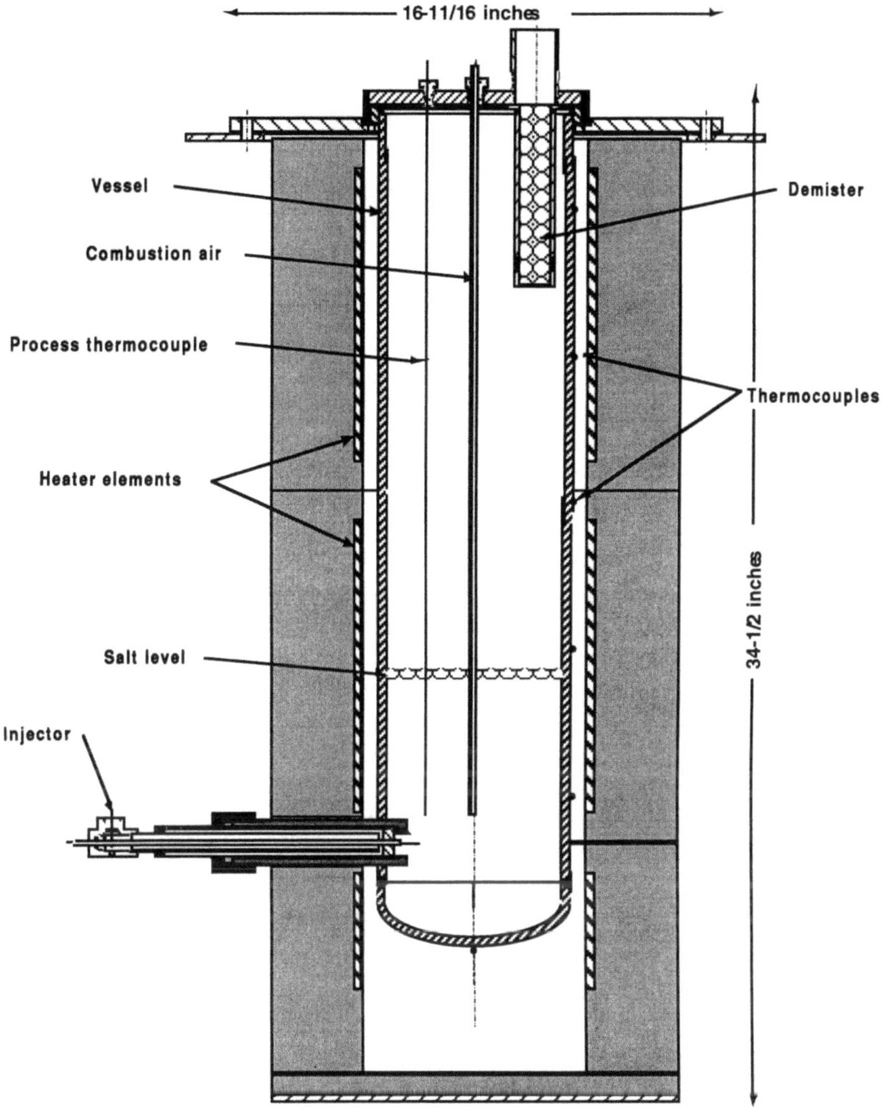

Figure 3. Details of the crucible.

ganics, and neutralize acid gases such as hydrogen chloride by forming stable salts such as sodium chloride (4). The relatively high thermal inertia of the melt resists changes in temperature resulting from sudden changes in the feed or heat transfer.

Safety is a major consideration in any process where high explosives or wastes containing high explosives are destroyed. We have successfully and safely destroyed slurries (or solution, in case of the liquid gun propellant) of 35 weight percent energetic material (high explosives and high explosive formulations) in water. The liquid, in addition to keeping the HE concentration to a safe dilution level, provides a vehicle for handling the HE.

A typical HE, such as HMX, contains sufficient oxygen to propagate a steady back burn from the MSD unit to the HE reservoir without any additional oxygen. To reduce the

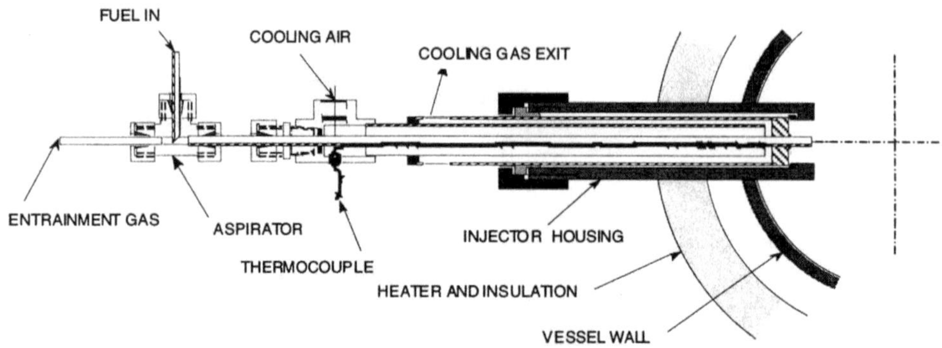

Figure 4. Details of the nozzle.

probability of this, we developed a special nozzle design, shown in Figure 4. The key features of the nozzle are side injection of the feed into the crucible, external cooling of the feed, and introduction of a large volume of inert gas, such as nitrogen, as a carrier for the feed. The side injection, coupled with external cooling, keep the feed temperature well below 100°C to minimize spontaneous decomposition of the HE. The large volume of inert carrier gas further dilutes the HE, provides a large heat capacity, and keeps the velocity of the feed inside the nozzle relatively high. As a result of all these conditions, the HE decomposes only inside the molten salt, and not in the feed nozzle. The high injection velocity also helps maintain a high degree of agitation and mixing inside the crucible.

Propellants react differently from high explosives when subjected to heat. Therefore, the problem of burn-back may be more severe in some case for propellants than explosives. To avoid the possibility of back-burn, we diluted the propellant with water in the ratio of 2 parts by weight of water per one part by weight of the propellant. This dilution renders the material nonexplosive and nonflammable. This diluted mixture can be safely injected into the molten salt bath without any back-propagation.

EXPERIMENTAL PROGRAM

We have built an experimental unit to investigate the applicability of the molten salt destruction process to the destruction of HE and HE-containing wastes. Figure 2 shows a schematic of the current experimental setup.

The crucible is made of stainless steel, which is stable in the presence of alkali carbonates. For wastes of high chlorine content, more corrosion resistant materials, such as Inconel, Hastalloy or ceramics may be necessary. The crucible has the dimensions 5.76″ inside diameter, 6.63″ outside diameter, and 24″ length. It is fitted with a flange on the top and a removable injection nozzle, described earlier, on the side. Thermocouples are placed at various locations, including the crucible, the feed nozzle, and the exhaust gas outlet. The coolant gas flow is controlled so as to maintain the feed nozzle temperature under 100°C. Gas sample bottles with manual and solenoid valves are attached to the exhaust line. The whole assembly is placed inside an explosion-proof cell, designed to withstand a detonation of up to 500 g HE (TNT equivalent). A remotely operated TV monitor allows us to monitor the experiment without entering the cell during the experiment. All the solenoid valves controlling the sample and the feed sequences are operated remotely. The experimental data are continuously logged on a computer.

To start an experiment, a measured amount of salt is introduced into the crucible from the top, and the top flange is secured. The sample bottles are evacuated, and the vacuum valves are closed off. The sample bottles are now ready to accept samples. The heaters are switched on, and data logging is initiated. A slurry of energetic material waste (with particle size of about 2 mm) and water of desired proportions (typically 30 wt % solids in water) is made and kept in suspension using air driven stirrers (see Figure 3). The nozzle coolant gas is turned on to keep the feed nozzle cool. Once the salt approaches its melting point, the carrier gas in the nozzle is turned on to prevent molten salt from entering the feed nozzle. After the salt is completely molten, the oxidant feed is initiated. When the temperature of the melt reaches a desired value (typically about 750°C), the system is ready for accepting the high explosive waste.

The waste to be treated is injected in to the crucible using a peristaltic pump through the side nozzle . Air is the driver gas for the feed slurry. Additional air, if needed, is introduced through the tube near the center of the crucible. The gaseous product of the crucible is sent to the vent as shown in Figure 3. A slurry of up to 500 g TNT equivalent HE in water is fed at rates varying between 300 g/hr to 1800 g/hr in each run. Gas samples were collected at various intervals after steady state is obtained. A salt sample is withdrawn from the crucible at the end of the final run, and analyzed for traces of HMX. The heaters are then shut off, and the unit is allowed to cool. The entrained gas, the coolant gas and the oxidant gas are continued until the salt freezes, at which time the gas flows are shut off.

EXPLOSION SAFETY

Any time a cold liquid such as water is contacted by a very hot liquid such as liquid metal or molten salt, the possibility of an explosion exists (5). We have considered two potential scenarios for explosion:

- If the outlet from the MSD reactor gets plugged or restricted, and if the liquid feed to the molten salt bath continues, the resultant pressure increase in the vessel may lead it its explosive rupture.
- The rapid heat transfer from the hot liquid to the cold liquid may give rise, temporarily, to a region of superheated liquid, which may spontaneously nucleate, resulting in an explosion (5).

The first scenario can be easily avoided by installing pressure sensors in the reactor, the lines and the auxiliary equipment, and monitoring the pressure rise continuously and carefully. Interlocks on the feed delivery system can be easily designed so that any excursion of pressure beyond the design limits stop the feed. Relief valves and rupture disks may also be used for additional safety.

The second scenario can be dealt with by choosing proper design and operating conditions. The nozzle has been designed to disperse the feed into a stream of rapidly moving gas, thereby eliminating or reducing the probability of large slugs of cold water contacting the hot molten salt. Second, the contents of the reactor are constantly agitated by the driver gas and the air needed for oxidation. This ensures that quiescent layers of hot and cold liquids do not form in the reactor. Our analyses show that this eliminates the possibility of steam explosion in the molten salt system. We have demonstrated the safety of our specific experimental set-up by repeated injection of pure water under the molten salt bed for prolonged duration without any adverse consequences.

Table 1.

Explosive	%C→CO	%N→NO_x	CO ppm	NO_x ppm	Throughput (kg/h)
RDX	0.035	0.05	110	332	1.8
HMX	0.035	0.11	90	528	1.4
Explosive D	0.028	0.18	60	240	0.49
LX-10	0.029	0.019	98	106	0.99
LX-16	0.064	0.14	231	340	1.3
LX-17	0.029	0.20	100	570	0.69
PBX 9404	0.09	0.25	124	543	0.3
LP XM-46	0.11	0.58	100	1900	2.6

RESULTS AND DISCUSSION

Gas samples were taken during the experiments, and analyzed using a gas chromatograph and mass spectrometer (GC/MS). Typical gas analyses for a number of pure component explosives and formulations are shown in Table 1.

Based on the gas analyses, the following observations can be made:

1. No HCN was found in the exhaust. Since HE combustion may be accompanied by evolution of some HCN, we may infer that the HCN, being an acid gas, was neutralized by the alkaline carbonate, producing NaCN. This is consistent with published data regarding the effectiveness of the molten salt as a scrubbing agent for acid gases (5,6).
2. The amount of NO_x and CO was relatively small. This indicates that the major combustion products are N2, CO_2 and H_2O. It may be possible to reduce further the NO_x emissions by changing the operating conditions of the MSD unit (6).

The salt samples are analyzed for traces of HE . All of the HE is expected to be fully decomposed in the molten salt. Fourier Transform-Infra Red (FT-IR) spectra of pure HEs and of molten salt into which the HE had been combusted in, respectively, show a complete absence of HE peaks in the cooled molten salt. Thus, within the detection limits of the method (5 ppm), no HE or propellant are detected in the salt, indicating complete oxidation. This is consistent with our expectations, given the ease of decomposition of the energetic materials at these temperatures.

CONCLUSIONS AND FUTURE WORK

We have demonstrated that HEs and liquid propellants can be safely and fully destroyed using the molten salt destruction process. We are currently working on a number of improvements to the process:

- We have designed and constructed a larger unit with advanced design features. We are currently installing the unit in our High Explosives Applications Facility at LLNL. After initial shekedown, the unit will be transferred to and demonstrated at a field site (Figure 5).
- We are implementing an advanced nozzle design for top injection of larger particles.
- We are defining operating envelopes for a number of high explosives and formulations.

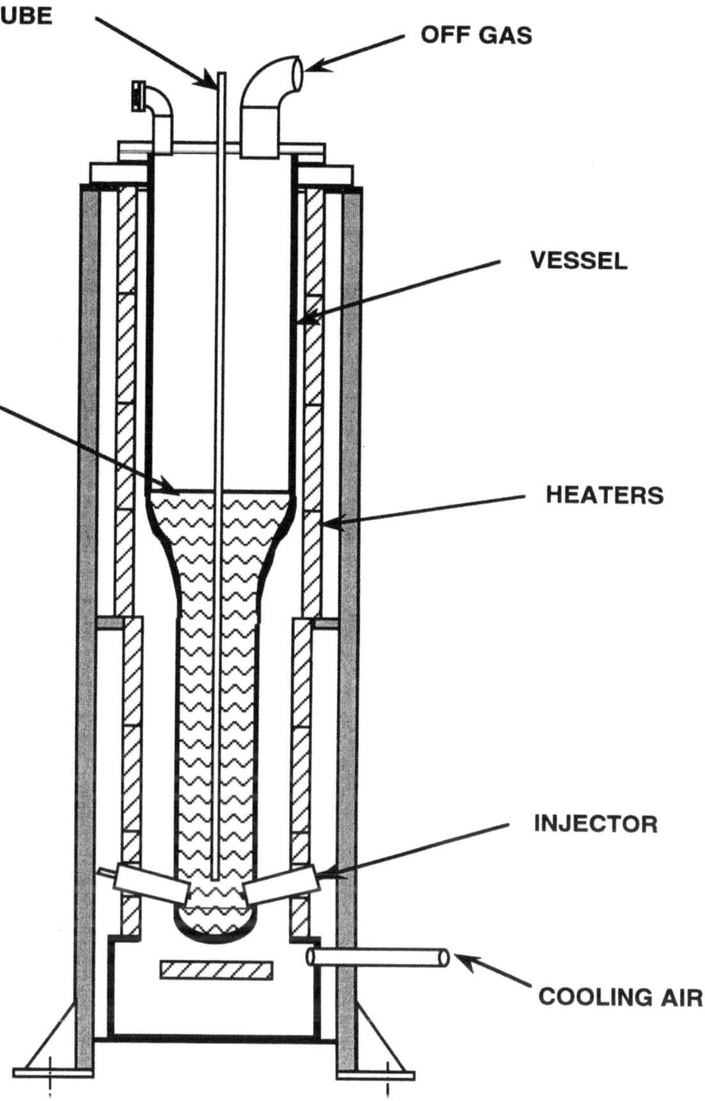

Figure 5. The advanced design unit under installation.

- We are developing models to study the temperature profile of a top-feed nozzle for feeding larger particles into the unit.

REFERENCES

1. Upadhye, R.S., W.A. Brummond, and C.O. Pruneda, "Destruction of High Explosives and Wastes Containing High Explosives using the Molten Salt Destruction Process", presented at the I&EC Special Symposium, American Chemical Society, Atlanta, Georgia, September 21–23, 1992.
2. Upadhye, R.S., W.A. Brummond, and C.O. Pruneda , Destruction of High Explosives and Wastes Containing High Explosives Using the Molten Salt Destruction Process", paper V-10, presented at the "waste Manage-

ment of Energetic Materials and Polymers", 23rd International Annual Conference of ICT, Karlsruhe, Germany, June 30–July 3, 1992.

3. Brummond, W.A., R.S. Upadhye and C.O. Pruneda, "Molten Salt Destruction of Energetic Materials", U. S. Patent 5,434,335, July 18, 1995.
4. Cooper, J.F., et. al., "Molten Salt Processing of Mixed Wastes with Off-gas Condensation", presented at the 1991 Incineration Conference, Knoxville, TN, 13 May 1991. Also, UCRL-JC-107288, Lawrence Livermore National Laboratory, Livermore, CA, 1991.
5. Henry, R.E., et. al., "Large Scale Vapor Explosions", Proceedings of the Fast Reactor Safety Meeting, April 2–4, 1974, Beverly Hills, CA (NTIS Ref. No. CONF-740401-P2).
6. Gay, R.L., et. al., "Treatment of Propellant and Explosive Wastes using Molten Salt Oxidation", presented at the I&EC Special Symposium, American Chemical Society, Atlanta, Georgia, September 21–23, 1992.

30

THE INFLUENCE OF MOLECULAR OXYGEN ON THE ADSORPTION OF ORGANIC POLLUTANTS

Piotr Anielak and Konrad Janio

Institute of General Chemistry
Technical University of Łódź

ABSTRACT

Influence of molecular oxygen on adsorptive capacity of the granular activated carbon was examined. Three experimental procedures for determining the adsorption isotherms were used. The novel "anaerobic" procedure where oxygen had been repeatedly removing from the test environment was applied. This procedure differs significantly from the widely used "aerobic" techniques which neglect the important influence of molecular oxygen on the adsorption phenomena. This study reveals that the presence of molecular oxygen has a significant impact on the adsorptive capacity of GAC for examined phenolic compounds. The study has shown that the increase in adsorptive capacity can be attributed not only to the polymerization of adsorbates on a carbon surface but also to their biological degradation.

INTRODUCTION

Granular activated carbon (GAC) is widely used in both water and wastewater treatment as an adsorbent for the removal of organic pollutants. The capacity of GAC for the retention of organic species is one of the key components needed for the scale-up of experimental data and for the prediction of adsorber performance. Procedures for the collection of capacity isotherm data differ in the carbon preparation, the particle size, volume of the bottles used in the experiment, volume of the adsorbate solution added, the buffer application and the equilibration time allowed. There are reports in the literature that molecular oxygen can affect the adsorptive capacity of GAC[1,2]. However, to the best of our knowledge there has been no systematic study designed to elucidate effects of the molecular oxygen on adsorption equilibrium so far.

Aim of this work was to estimate how the molecular oxygen influence adsorptive capacity of the granular activated carbon.

Chemistry for the Protection of the Environment 2, edited by Pawłowski et al.
Plenum Press, New York, 1996

EXPERIMENTAL

The granular activated carbon Carbopol N was used in this study. Prior to use, the GAC had been washing several times with deionized water until most of the fines had been removed. It was then dried in an oven at 110^0C for 2 days and stored in a desiccator until use.

All solutions were prepared with highly deionized water. The sorbates applied in this research were, pyridine, p-nitrophenol, tyrosine and nonionic surface active agent Rokafenol N-6. They were selected for this study because they are widely spread pollutants and can be easily analyzed using u.v. spectroscopy.

Measurements of concentrations of adsorbates were performed on the Carl-Zeiss VSU-2P UV/VIS spectrophotometer using 1 cm quartz cells. Wavelengths of 256 nm for pyridine, 318 nm for p-nitrophenol, 275 nm for tyrosine, and 276 nm for Rokafenol N-6 were used to determine their concentration in solution.

Three experimental procedures for determining the adsorption isotherms were applied. The procedures will be denoted subsequently as "aerobic I", "aerobic II" and "anaerobic".

Aerobic procedures designed by us are similar to procedures which are commonly used in the current research practice. Accurately weighed masses of the GAC areplaced in 100 ml glass bubbling washers. The bubblers are next filled with 100 ml of the adsorbate solution and sealed with the glass attachments. The samples were then saturated with atmospheric air until the equlibrium was achieved.

The described aerobic procedures facilitates interference of molecular oxygen arising from three sources: air entrapped within the pores of GAC particles, the dissolved oxygen present in the water used in preparing the solutions and the headspace in the bottles. Boiled water and standard dilution water for BOD were applied in the procedure aerobic I and aerobic II, respectively. Total carbon concentrations in the samples were measured using the Shimadzu TOC-5000 total organic carbon analyzer[3]. For each sample, inorganic carbon was measured to check for the biological activity. The organic carbon was calculated by subtracting the inorganic carbon from the total carbon. All aerobic measurements were carried out in device presented in Figure 1.

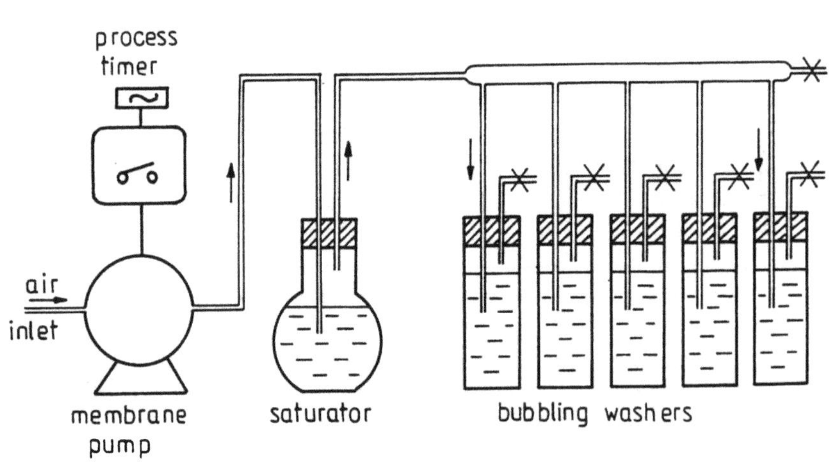

Figure 1. Schematic diagram of the '"aerobic" process.

The anaerobic procedure differs significantly from the commonly used aerobic techniques. The main feature of this procedure is a requirement for the exclusion of the interference of molecular oxygen. The isotherm data were collected using the "bottle-point" technique. After placing predetermined masses of GAC in the isotherm bottles, oxygen associated with air entrapped inside the carbon pores and reversibly adsorbed oxygen was removed from the test environment by boiling and purging with argon. Oxygen dissolved in the water used in preparing the adsorbate solution was elimented by boiling and stripping with argon prior to addition of the adsorbate. The bottles were subsequently filled with 100 ml of adsorbate solution and sealed with glass stoppersafter the air from the headspace was replaced by argon. Isotherm bottleswere then placed in a shaker and allowed to equilibrate for 2 days. Each series of isotherm experiments was accompanied by two blanks bottles without carbon.

Obtained isotherm data are presented in Figures 2–5.

RESULTS AND DISCUSSION

The results prove that the presence of molecular oxygen has a significant impact on adsorptive capacity of the GAC for examined phenolic compounds. Aerobic capacity of the GAC for the retention of tyrosine was increased twice or even more when compared to the anaerobic conditions. Similar but less pronounced tendency is observed for the adsorption of pyridine and Rokafenol N-6. However, presence of oxygen did not influence the adsorption capacity for the p-nitrophenol.

The increase in adsorption capacity of tyrosine under aerobic conditons can be attributed to biological degradation of the solute. The role of biological activity was confirmed by an increase in the concentration of inorganic carbon as well as a presence of

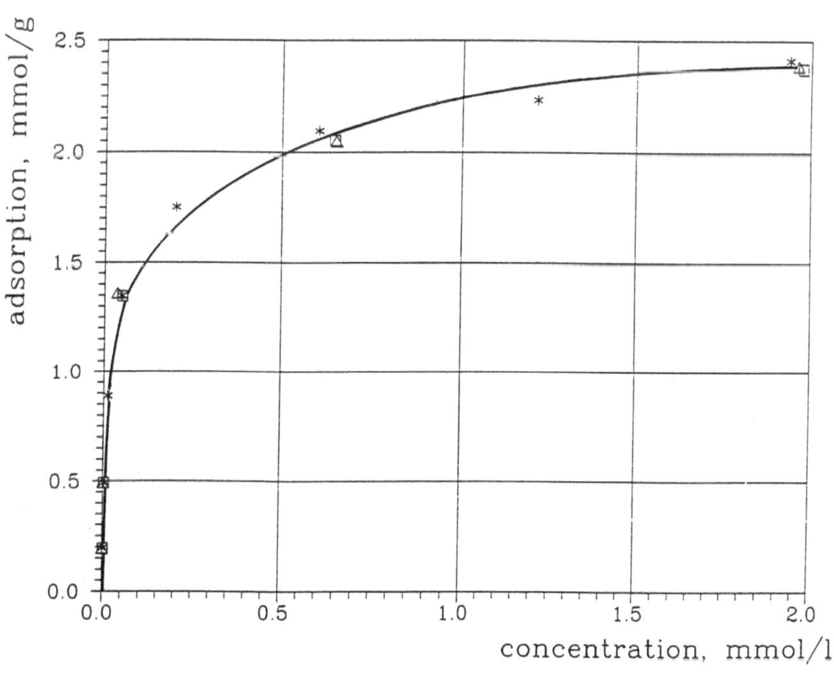

Figure 2. Adsorption isotherms for p-nitrophenol; procedures: □ aerobic I, Δ aerobic II, * anaerobic.

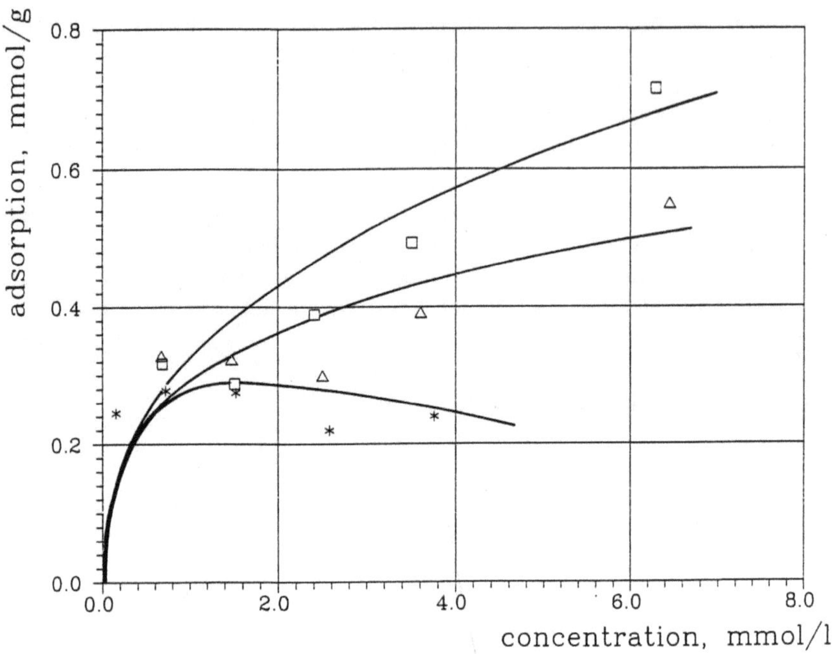

Figure 3. Adsorption isotherms for pyridine; procedures: □ aerobic I, Δ aerobic II, * anaerobic.

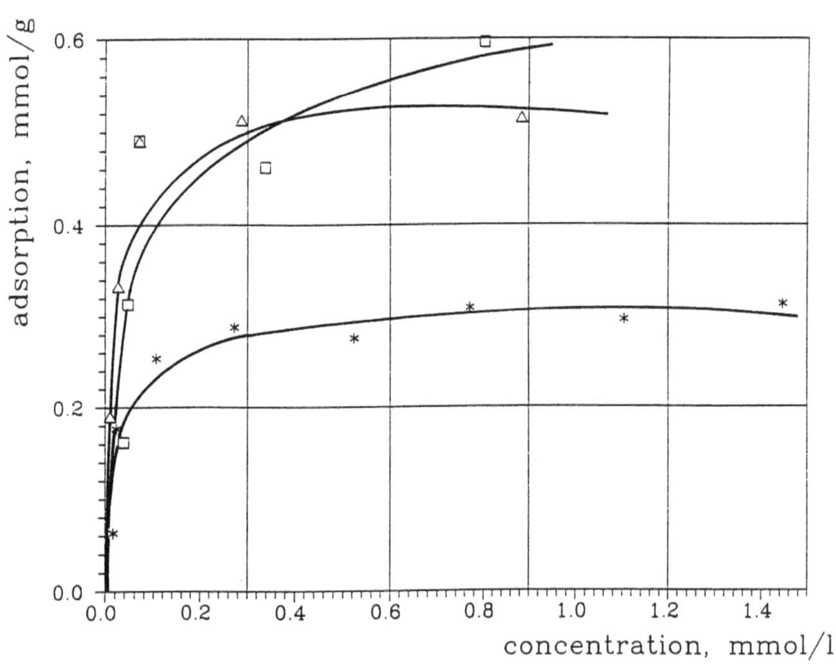

Figure 4. Adsorption isotherms for Rokafenol N-6; procedures: □ aerobic I, Δ aerobic II, * anaerobic.

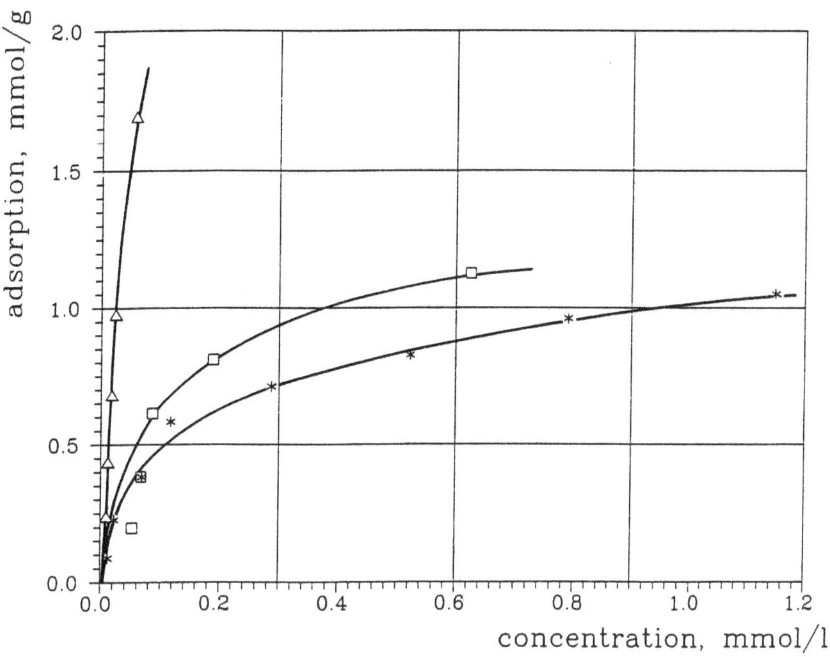

Figure 5. Adsorption isotherms for tyrosine; procedures: □ aerobic I, Δ aerobic II, * anaerobic.

some microorganisms capable of degrading tyrosine. The rise in carbon dioxide concentration and the amount of aerobes was especially high in aerobic procedure II.

On the other hand, no increase in the concentration of inorganic carbon under aerobic and anaerobic conditions was observed during isotherm tests of pyridine and Rokafenol N-6. Furthermore, the experimentally obtained contents of dissolved organic carbon are very closely related to the theoretically calculated values, based on concentrations of both adsorbates as measured by u.v. spectroscopy. Therefore, the increase in adsorptive capacity of pyridine and Rokafenol N-6 can be attributed to chemical reactions (polymerization) between the adsorbates and molecular oxygen which are catalyzed by the activated carbon surface. Mechanism of these reactions can be explained by the oxidative coupling of phenolic molecules[4].

However, substitution a nitro group on the parent phenol molecule suppressed polymerization reactions as well as its biodegradability. As a result, no increase in the GAC adsorptive capacity was observed under oxic conditions for p-nitrophenol.

The study has shown that the increase in adsorptive capacity can be attributed not only to the polymerization of adsorbates on a carbon surface but also to their biological degradation. The latter is also stimulated by the presence of molecular oxygen.

REFERENCES

1. Randtke S., Snoeyink W., JAWWA., **75**, 406 (1983).
2. Vidic R.D., Suidan M.T., Traegner U.K., Nakhla G.F., Water Res., **24** (10), 1187 (1990).
3. Carlson M.A., Heffernan K.M., Ziesemer C.C., Snyder E.G., JAWWA., **1994** (3), 91.
4. Warta C.L., Papaddimas S.P., Sorial G.A., Suidan M.T., Speth T.F., Water Res., **29** (2), 551 (1995).

ELECTROCHEMICAL OXIDATION OF CYANIDE COMPLEXES WITH COPPER AT CARBON FIBRE

A. Socha and E. Kuśmierek

Institute of General and Ecological Chemistry
Technical University of Łódź

1. INTRODUCTION

The oxidative degradation of cyanide wastes is very important in industrial waste management and for wastewater treatment. Electrochemical methods of waste treatment enable minimization of the need for chemical additives and sludge disposal. Anodic oxidation of cyanides can lead to formation of non toxic products such as N_2 and CO_2 or NO_3^- and CO_3^{2-}. A consideration of possible oxidation products in alkaline media, based on standard reduction potentials given below indicates that at low values of the applied potential, formation of the cyanates (CNO^-) is thermodynamically preferred[1].

$$CNO^- + H_2O + 2e \longrightarrow CN^- + 2OH^- \quad E° = -0.97\ V \tag{1}$$

$$CO_2 + 0.5\ N_2 + 2H^+ + 5e \longrightarrow CN^- + 2OH^- \quad E° = -0.49\ V \tag{2}$$

$$CO_3^{2-} + NO_3^- + 6H^+ + 10e \longrightarrow CN^- + 6OH^- \quad E° = +0.14\ V \tag{3}$$

However, cyanates are 1000 time less toxic than cyanides.

Anodic oxidation of concentrated aqueous cyanide wastes is an efficient alternative to their chemical oxidation. It is cost effective and produces no toxic products. Dilute aqueous cyanide solutions (<1 g/l) are not treated effectively by anodic oxidation because of the high overpotential due to the cyanide oxidation reaction. It leads to high power costs and low current efficiencies. The most common electrodes used in the process of cyanide electrooxidation are graphite, steel and titanium. They enable easy achievement of higher potential values in alkaline media. Electric energy consumption is about several dozen kWh/kg CN^-. A very good current efficiency in anodic oxidation of diluted cyanide solutions is obtained when using fluidal electrodes and electrodes made of strongly porous carbons with a developed surface (HSA reactors)[2]. Carbon fibre in the form of cloth, is

Chemistry for the Protection of the Environment 2, edited by Pawłowski et al.
Plenum Press, New York, 1996

significantly more stable and comfortable in application than other types of porous and powdered electrodes. This electrode material allows for filling up completely the anodic space of the electrolyzer and enables a uniform contact of the electrolyte with the electrode, through which it is pumped [3].

Certain metal ions, such as cuprous and nickelous catalyze the cyanide oxidation reaction by forming metallocyanide complexes that oxidize at less anodic potentials than free cyanide ions [4]. Copper oxide films have also been shown to catalyze the cyanide oxidation. Wels and Johnson [1] investigated the oxidation of cyanide on copper oxide surfaces in excess of hydroxide. However, these surfaces were not stable in moderately concentrated cyanide solutions (>1 mM).

The aim of this work was to increase the oxidation efficiency of diluted cyanide solutions and to accelerate this process by application of carbon fibre as an anode. The cyanide oxidation at carbon fibre modified with copper oxides was also investigated.

2. EXPERIMENTAL

The measurements were carried out in a flow electrolyzer (Fig. 1), in which anodic space was completely filled with carbon fibre. The volume of the anodic space was 22.4 cm^3. The real surface of the applied carbon fibre determined by the BET method was about 2.5 m^2/g, thus the surface of the electrode used was 6.43 m^2. In the cathodic spaces, copper or carbon fibre cathodes were placed. Electrode compartments were separated by a cationic membrane with $-SO_3H$ groups. The solution of cyanides was pumped through the anodic space at various flow rates in the range from 0.25 ml/min to 12 ml/min. Preparative oxidation of free and cyanide complexes with copper at carbon fibre and carbon fibre modified with copper oxides was carried out in an electrolyzer with non separated electrode compartments. The anode surface was about 0.1 m^2. In order to compare results obtained at the carbon fibre, a platinum sheet with comparative geometrical surface was applied as an anode. The change in cyanide concentration was controlled by recording square wave (SW) voltammograms at hanging mercury drop electrode (HMDE). The

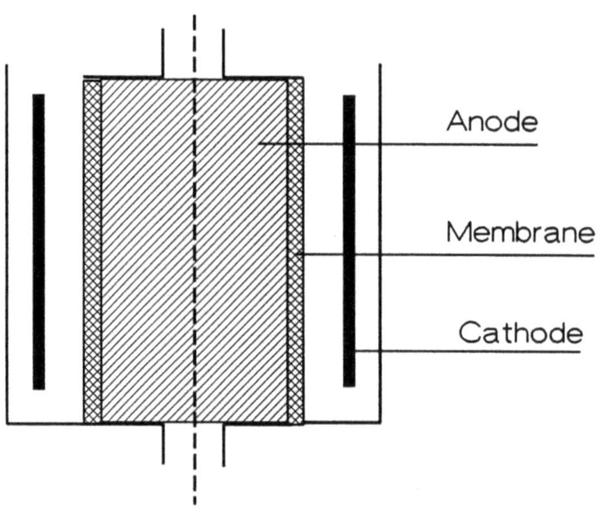

Figure 1. Flow electrolyzer with three electrode compartments.

change in concentration of copper ions was measured by recording SW voltammograms in sulphuric acid. Electrode potentials were measured vs. saturated calomel electrode. Three different solutions of copper complexes with two, three, and four ligands respectively were electrooxidized. Each of these solutions at concentration of $2.5*10^{-3}$ mole/l was prepared by mixing a stoichiometric amount of cuprous cyanide with potassium cyanide solved in 0.5 mole/l sodium hydroxide. In order to compare results obtained in complex solutions, the solution of potassium cyanide at concentration of 10 mmol/l was tested. Chemicals used were reagent grade.

3. RESULTS AND DISCUSSION

3.1. Electrooxidation of Free Cyanides at Carbon Fibre

Electrooxidation of cyanides at the carbon fibre causes a significant increase in the oxidation current and a decrease in oxidation potential in comparison with platinum anodes with comparable geometrical surface.Preparative investigations of free cyanide oxidation were carried out in the electrolyzer presented in Fig. 1. Primarily, the influence of electrode potential on the cyanide conversion degree was determined (Fig. 2). The electrolyses were carried out at the potentials in the range from 0.2 to 1.0 V up to the moment when a charge of 77.2 C passed through the system. Such a charge is theoretically sufficient to oxidize 80% of the initial cyanide amount. The electrolyte flowed in the electrolyzer in a closed cycle with a flow rate of 12 ml/min. Results dosplayed in Fig. 2, show that the highest cyanide conversion degree was obtained at the lowest potentials. However, the duration of the electrolyses was relatively large. Thus, the next experiments were

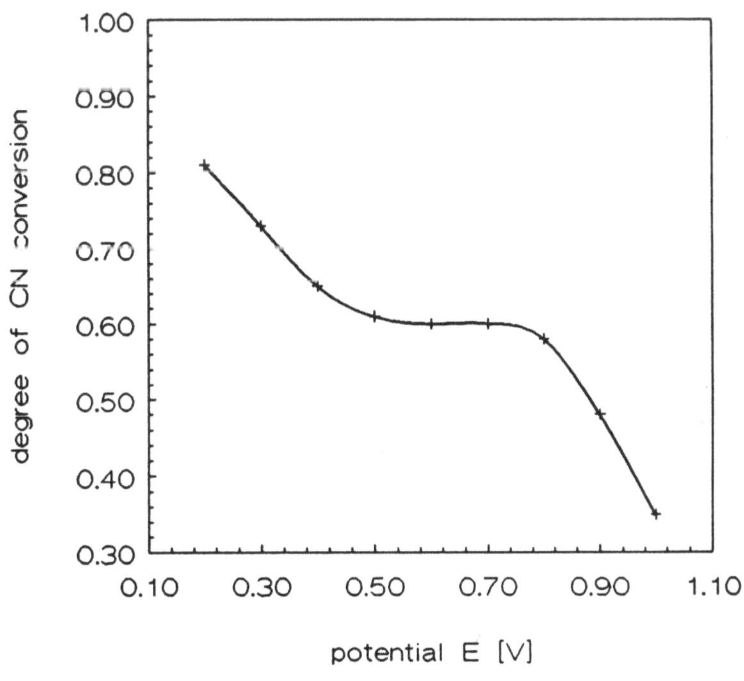

Figure 2. Dependence of CN^- conversion degree on electrode potential.

carried out at a potential of 0.6 V. The highest cyanide conversion degree was obtained at very low flow rates. This indicates that the process should be carried out under stationary conditions [3]. The investigation of the influence of initial cyanide concentration on the conversion degree showed that in the range of CN^- concentration from 1 to 20 mmol/l, the conversion degree is nearly constant (about 70%) and decreases with the increase in concentration.

3.2. Electrooxidation of Cyanide Complexes with Copper at Carbon Fibre

Cyclic voltammograms recorded at carbon fibre and platinum in stoichiometrically prepared solutions of copper complexes with different number of ligands are presented in Fig.3 and 4. A complex with four ligands is oxidized at a more positive potential than a complex with three and two ligands. All complexes are oxidized at less positive potential than potassium cyanide.

Standard reduction potentials for various complexes react according to the following general equation[5]:

$$Cu(CN)_n^{(n-1)-} + e = Cu + nCN$$

are as follows

for	n = 1	E°= 0.17 V	
	n = 2	E°= 0.38 V	
	n = 3	E°= -0.37 V	
	n = 4	E°= -0.48 V	vs. RHE in 1.0 mole/l NaOH, pH=13.6

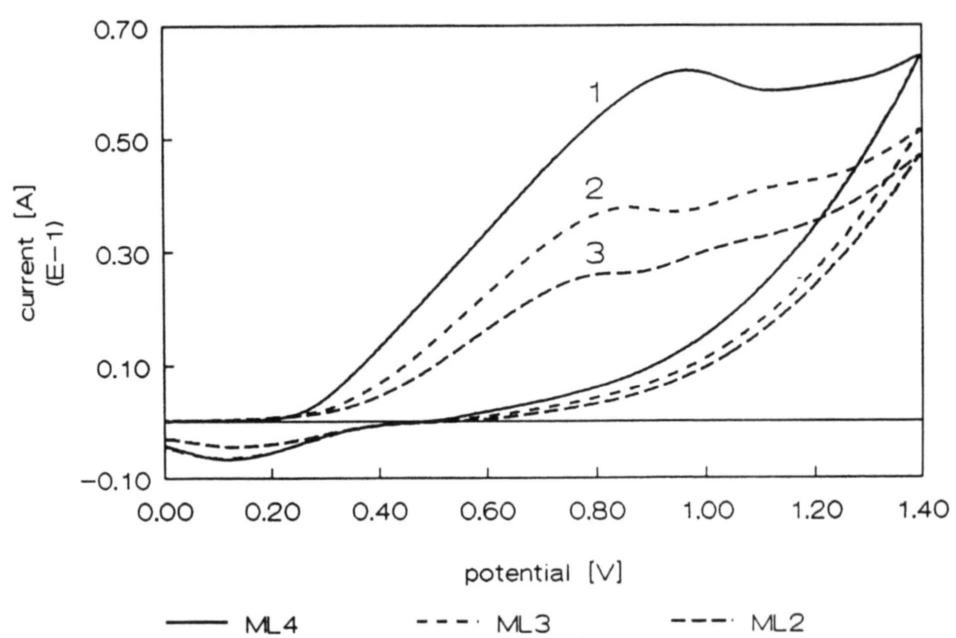

Figure 3. Cyclic voltammograms of cyanide oxidation at carbon fibre electrode; 1 - $K_3[Cu(CN)_4]$, 2.5 mmol/l; 2 - $K_2[Cu(CN)_3]$, 2.5 mmol/l; 3 - $K[Cu(CN)_2]$, 2.5 mmol/l.

As it can be predicted from the potentials presented above, oxidation of complexes should be in the reverse order.

The results of distribution calculations of cyanide complexes in Cu(I) - $(CN)_x$ and Cu(I) - $(OH)_y$ systems (Fig. 5) indicate that at least two complexes with different number of ligands are always prcscnt in thc solution within thc rangc of tcstcd ligand conccntration. Distribution of these complexes dictates their oxidation potentials and can explain rather small differences between them. Oxidation potentials for cyanide complexes increase with the number of ligands and with their unstability constants:

for	$[Cu(CN)_4]^{3-}$	pK =30.3
	$[Cu(CN)_3]^{2-}$	pK =28.6
	$[Cu(CN)_2]^{-}$	pK =24.0

According to these data, a complex with four ligands is more stable than others and can be more difficult to oxidize.

Preparative investigations of cyanide complex with copper (solution was stoichiometrically prepared) were carried out in electrolyzer with no separated electrode compartments. The change in substrate amount was controlled by recording SW voltammograms (Fig. 6) at HMDE in stoichiometrically prepared $K_3[Cu(CN)_4]^{3-}$ solution. Electrooxidation of Hg in cyanide complexes with copper is characterized by one wave (I) and two peaks (II, III) similar to those obtained in free cyanide solution. SW voltammograms recorded in solutions after electrolyses at different charges are presented in Fig. 7. As shown in Fig. 7, the charge of 7 C caused a decay of the I wave current. The charge of 14 C caused a decay of the II peak current. No peaks which correspond to the presence of cyanides were observed after the charge of about 40 C (the double theoretical charge enough to complete removal of cyanides) had passed through the electrolyzer. The charac-

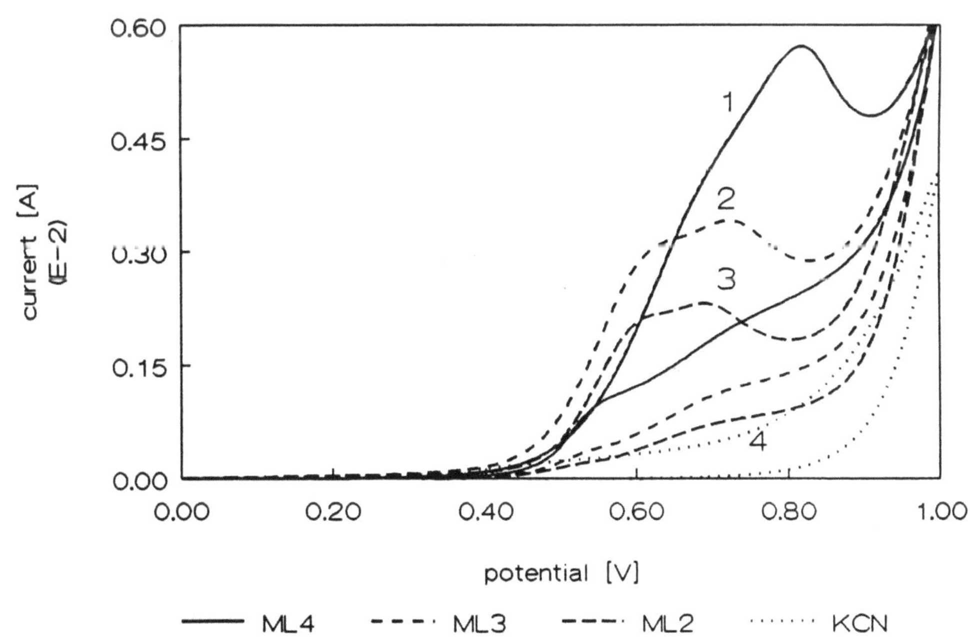

Figure 4. Cyclic voltammograms of cyanide oxidation at platinum electrode; 1 - $K_3[Cu(CN)_4]$, 2.5 mmol/l; 2 - $K_2[Cu(CN)_3]$, 2.5 mmol/l; 3 -$K[Cu(CN)_2]$, 2.5 mmol/l; 4 - KCN, 10 mmol/l.

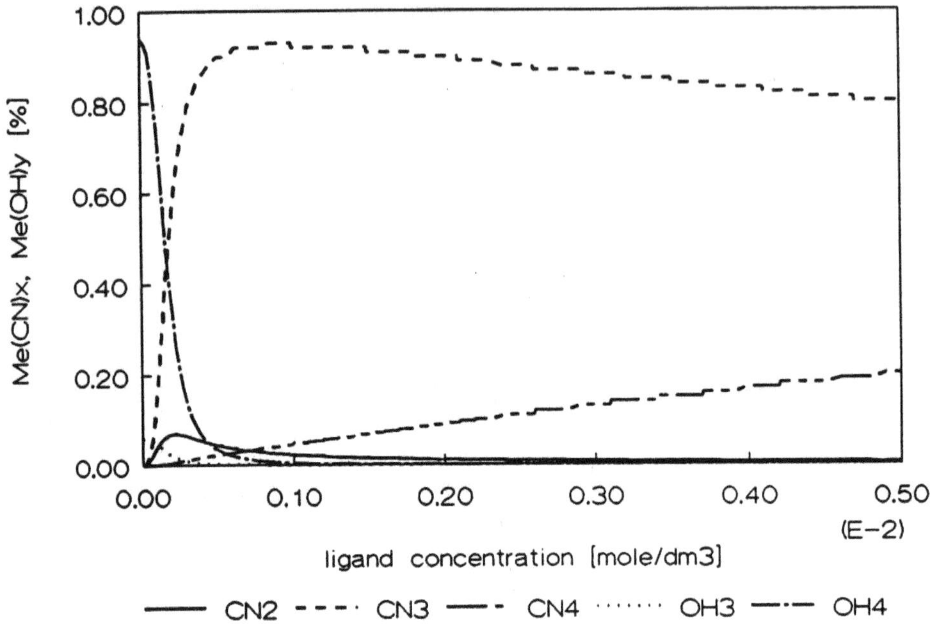

Figure 5. Distribution of cyanides in Cu(I) - $(CN)_x$ and Cu(I) - $(OH)_y$ systems at cyanide concentration up to 5 mmol/l and constant hydroxide concentration - 0.5 mol/l.

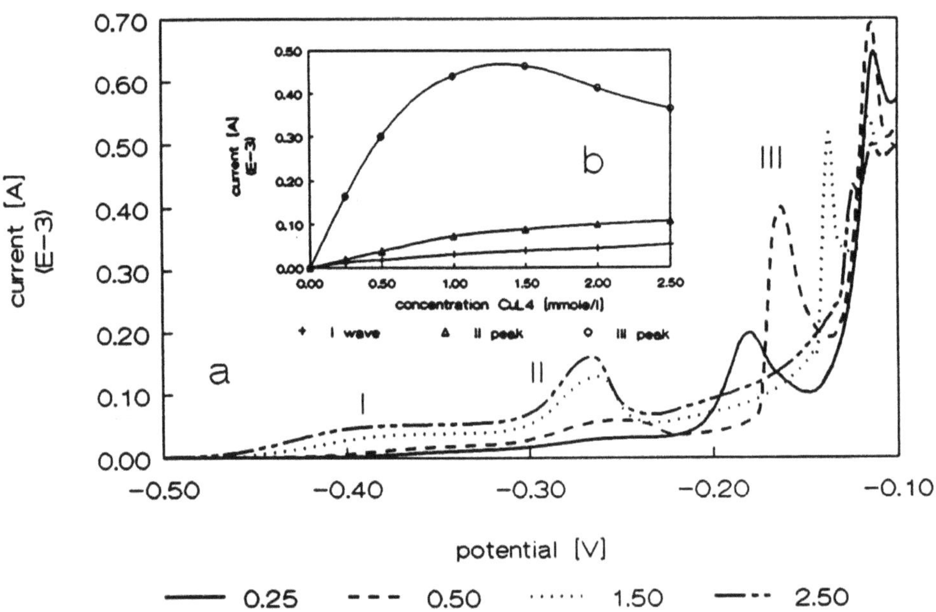

Figure 6. SW voltammograms recorded at HMDE in $K_3[Cu(CN)_4]$ solutions at different concentrations (a); the change of peak currents with $K_3[Cu(CN)_4]$ concentration (b).

ter of curves in Fig.7b is similar to curves presented in Fig. 6b. This indicates that the change in the content of cyanide complexes with copper which result from cyanide oxidation and copper reduction in the same way as while dissolution.

The electrolysis at a charge of about 10 C caused the change in colour of the solution. The solution became slightly blue. Simultaneously, two small peaks at the potentials -0.65 V and -0.58 V were observed in the voltammograms (Fig. 7a). The higher the charge the more intensive the colour and the higher the current at these potentials was observed. However, a decrease in colour and current was noticed at charges higher than 20 C. After electrolysis at 22 C, the colourless solution was obtained and no peaks were observed at a potential range from -0.8 to -0.55 V. Simultaneously with these analyses, the change in the amount of copper ions in the same solutions was determined. The influence of the charge passing through the system on concentrations of copper ions in solution and the copper amount deposited at the cathode, is presented in Fig. 8. The loss of copper ions from solutions higher than their amount deposited at the cathode was observed during electrolyses at charges up to 10 C. Such differences can be caused by the formation of copper oxides at the anode (electrode modification). A further increase of electrolysis charge up to 30 C results in the amount of copper deposited at the cathode to be higher than its loss from the solution. This can be caused by a dissolution of the copper oxide layer at anodes in cyanides[1]. The total amount of copper ions lost from the solution is totally deposited at the cathode at the charge of 40 C.

3.3. Electrooxidation of Cyanides at Carbon Fibre Modified with Copper Oxides

In order to increase the CN^- conversion degree and to decrease the electric energy consumption in the process of cyanide oxidation, the surface of carbon fibre was modified

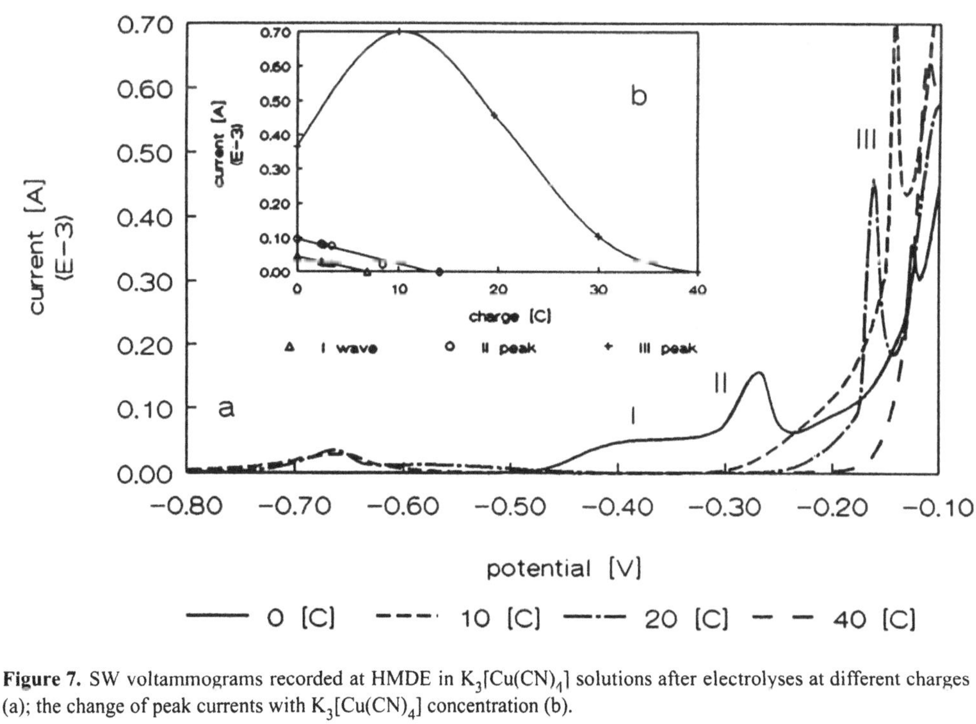

Figure 7. SW voltammograms recorded at HMDE in $K_3[Cu(CN)_4]$ solutions after electrolyses at different charges (a); the change of peak currents with $K_3[Cu(CN)_4]$ concentration (b).

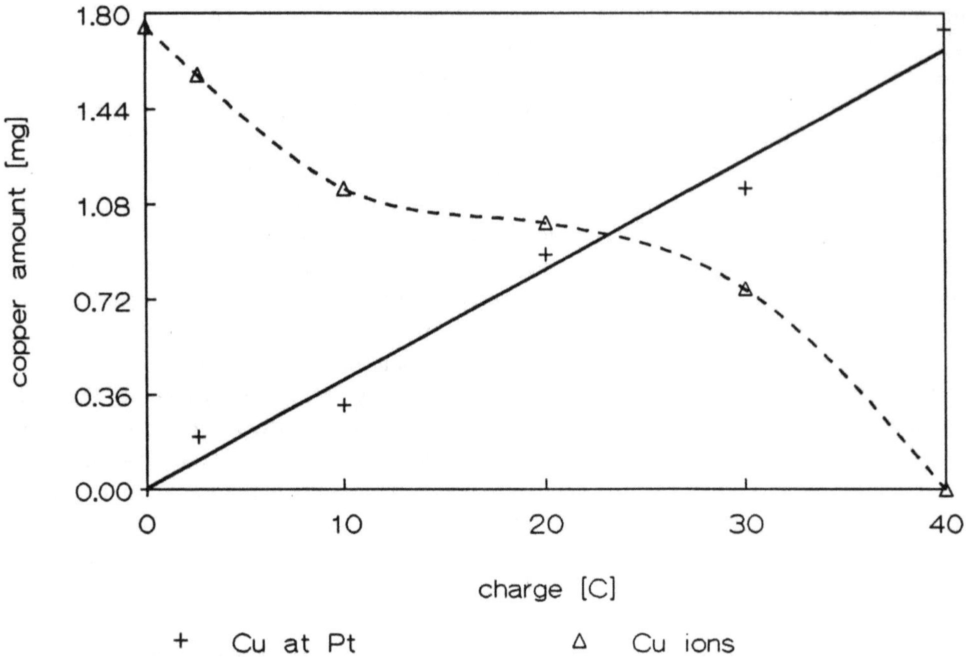

Figure 8. The change in copper amount deposited at cathode and copper ions removed from $K_3[Cu(CN)_4]$ solutions during electrolyses at the constant potential (0.6 V) and different charges.

by covering it with copper oxides. Modification was carried out in the $K_3[Cu(CN)_4]^{3-}$ solution (3 mmol/l) prepared by mixing stoichiometric amounts of KCN and CuCN at pH about 12 (0.01 mol/l NaOH). The electrolysis at the potential of 1.0 V lasted 30 minutes.

The electrode modification and its change during the electrolysis was determined by voltammetric curves recorded in KCN solution (10 mmol/l). Sufficient electrode modification was obtained when the current observed at the potential of 0.4 V was at least seven times larger than the current observed in the case of not modified electrode in the same solution. Typical voltammetric curves characterizing the investigated reaction and recorded at carbon fibre, modified carbon fibre and at modified electrode after cyanide oxidation, are presented in Fig. 9. The oxidation process at the carbon fibre starts at a potential of 0.44 V but at the modified carbon fibre at a potential of 0.24 V. Voltammetric curves recorded at modified and non modified platinum with comparable geometrical surface are presented in Fig. 10. In both cases, currents observed are significantly higher than at non modified electrodes.

Oxidation of free cyanides at modified carbon fibre causes the cyanide conversion degree to be higher than that resulting from the theoretical charge. All electrolyses were carried at a charge of 6.93 C which corresponded to a theoretical conversion of 50% of cyanides. An increase in anode potential from 0.2 up to 0.5 V causes less difference between theoretical and real CN^- conversion degree[3]. Simultaneously less destruction of the electrode surface (CuO - film) was observed. This can result from not only anodic polarization but also from a significantly shorter time of the electrode contact with KCN solution at higher anode potentials. The destruction of copper oxide films in the presence of high concentrations of CN^- was also reported by Wels and Johnson[1]. Cyclic voltammo-

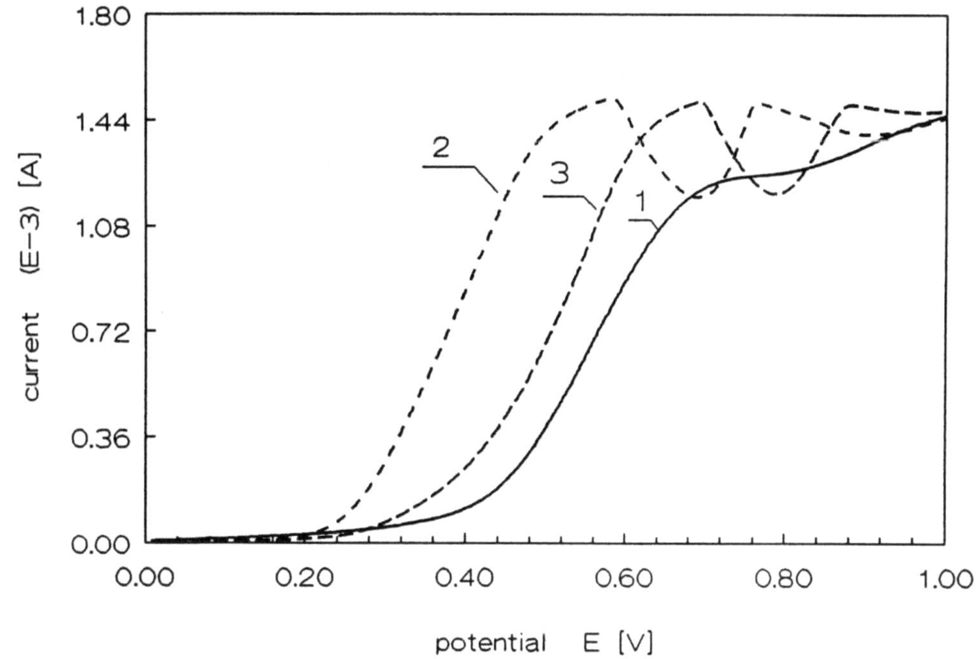

Figure 9. Voltammetric curves of 10 mmol/l KCN oxidation recorded at carbon fibre - curve 1, modified carbon fibre - 2 and modified carbon fibre after electrolysis - 3.

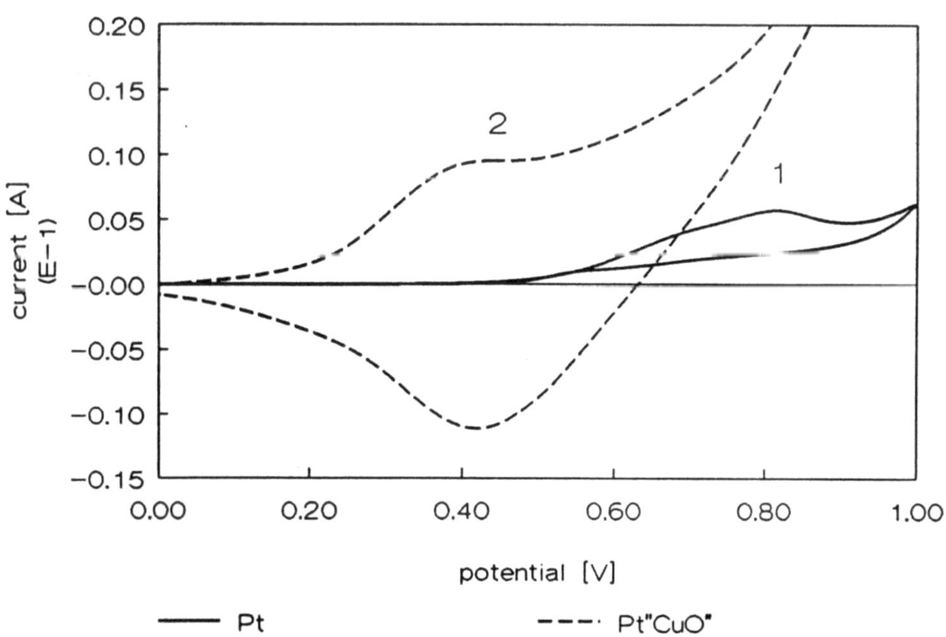

Figure 10. Cyclic voltammograms recorded in 2.5 mmol/l $K_3[Cu(CN)_4]$ at platinum electrode - 1 and platinum electrode modified with cupper oxides - 2.

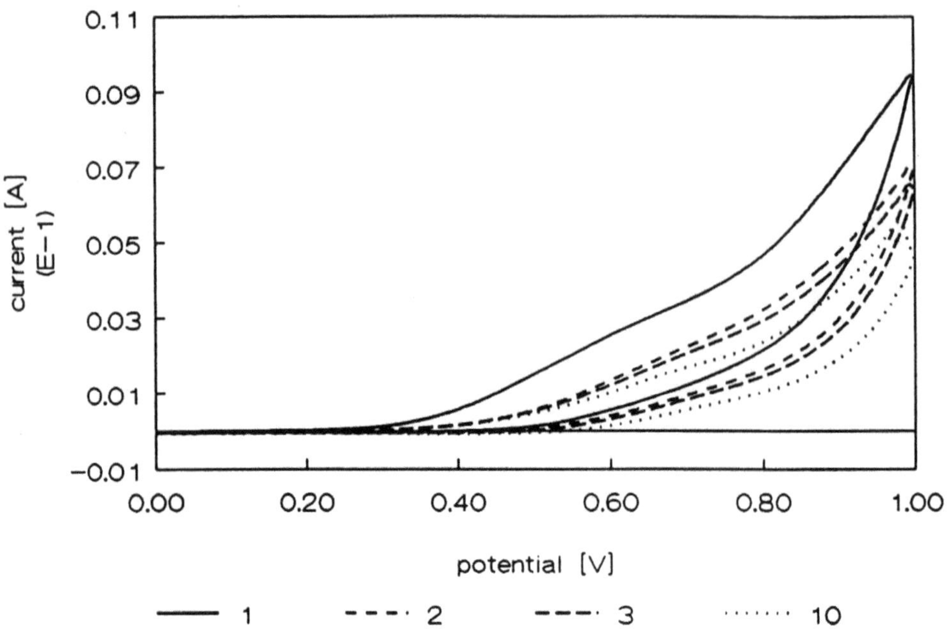

Figure 11. Cyclic voltammograms of KCN oxidation at 10 mmol/l concentration recorded at platinum electrode modified with copper oxides; 1, 2, 3, and 10th cycle, respectively.

grams recorded in KCN and $K_3[Cu(CN)_4]$ solution (the same concentration of cyanide ions) at carbon fibre modified with copper oxides are presented in Fig. 11 and 12. Modification of electrode was quickly destroyed in the case of free cyanides. Such destruction was not observed in the case of cyanide complexes with copper which can result from the possibility of electrode surface regeneration in cyanide complex solution. In the electrochemical aspect, the electrode modification is stable in the potential range from -0.2 up to 1.0 V.

4. CONCLUSIONS

Carbon fibre anodes enables the increase in efficiency of cyanide oxidation in comparison with other electrodes with comparable geometrical surfaces. The highest cyanide conversion degree was observed at a potential of 0.6 V and at very low flow rates. Thus, it seems to be convenient to lead the process under stationary conditions. Electrooxidation of cyanide complexes with copper is observed at less positive potentials than oxidation of free cyanides. In a charge range from 10 to 22 C, the formation of hydroxide complexes with copper was observed. The double theoretical charge was enough to complete destruction of cyanides and to removal of copper ions. Modification of anode surface with copper oxides causes the increase in reaction potential by 0.2 V and significant increase in oxidation current. Copper oxide films at the anode surface are quickly destroyed in free cyanide solution especially at concentrations higher than 1 mmol/l. Such destruction was not observed in solutions of cyanide complexes with copper which can result from the possibility of film regeneration.

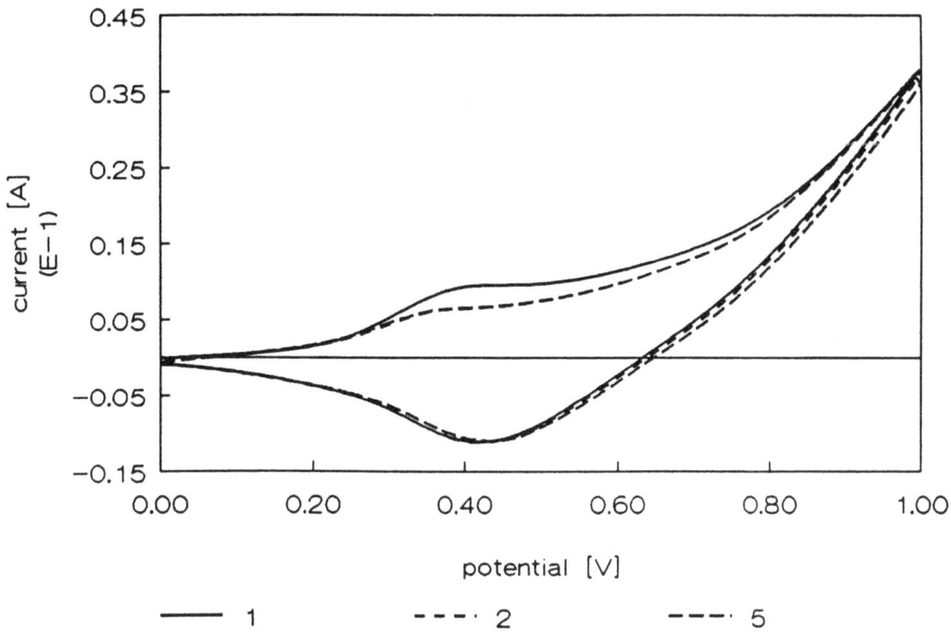

Figure 12. Cyclic voltammograms of $K_3[Cu(CN)_4]$ oxidation at 2.5 mmol/l concentration recorded at platinum electrode modified with copper oxides; 1, 2 and 5th cycle, respectively.

5. ACKNOWLEDGMENT

A grateful acknowledgement is made to the Polish State Committee for Scientific Research for the financial support of investigations - grant No 3 T09B 183 08

6. REFERENCES

1. Wels B., Johnson C., *J. Electrochem. Soc.*, **137 (9)**, 2785 (1990).
2. Poradnik Galwanotechnika, WNT Warzszawa (1985).
3. Socha A., Kuśmierek E., *Polish J. Chem.*, **67**, 517 (1993).
4. Hofseth C.S., Chapman T.W., *J. Electrochem. Soc.*, **138** ,8 (1991).
5. Burke L.D. Ryan T.G. *J.Electrochem. Soc.* **137**,1358 (1990)

32

THE EFFECT OF SULPHUR DIOXIDE ON REACTION OF TOTAL PHENOL OXIDATION AT COPPER-ZINC CATALYST

Marek Kaźmierczak

Institute of General and Ecological Chemistry
Technical University of Łódź
90-924 Łódź, ul. Żwirki 36, Poland

ABSTRACT

Copper-zinc catalyst was used for total phenol oxidation by pure air or air with sulphur dioxide. Aqueous solution of phenol was evaporated. Vapour with air was supplied to a tubular reactor with a catalyst. Change of catalyst activity as the result of SO_2 addition was determined by the composition of reaction products. High activity of the catalyst also in reaction medium which contains SO_2 at slight concentrations, was shown. However, temperatures higher than 600^0C should be applied. At high concentrations of SO_2, the catalyst loses its activity but it can be regenerated partly or totally by supply of pure air. Inactivation of the catalyst results in a decrease in decomposition rate of indirect product of p-benzoquinone oxidation.

1. INTRODUCTION

The development of utilization and treatment methods of wastes is necessary to meet requirements with regard to protection of the natural environment, human health and life conditions. Regulations governing the emission of these substances to the natural environment are more and more stringent but not all societies can meet them.

Toxic and hard to decompose organic substances are one of the main groups among wastes which are connected with human economic activity. Noxious industrial organic wastes are a mixture of compounds which often contains different types of phenol derivatives - high-molecular compounds. If the mixture contains also inorganic substances, then combustion is the most convenient and extreme method of treatment.

The installation for thermocatalytic treatment of liquid industrial wastes which contain toxic organic substances, is developed and improved in the Institute of General and

Ecological Chemistry in Technical University of Łódź 1. This system is designed as a section treatment plant which destroys wastes in the source of their formation , and protects biological treatment plants against irregular disposal of toxic compounds. In the installation, a copper-zinc catalyst is applied because of its low price and high activity in oxidation processes of organic substances. However, this catalyst can be sensitive to poisoning by sulphur compounds which often occur in industrial wastes.

Our experiments are part of the investigations on application of Cu-Zn catalyst for treatment of sulphur organic compounds.

2. EXPERIMENTAL

Investigations were carried out using the model for wastewater. It was an aqueous solution of phenol. This model for wastewater has been used in the process of total oxidation for a long time by investigators who are engaged in environmental protection[2–4].

Catalyst resistance to sulphur compounds was determined by the change of catalyst activity as the result of addition of sulphur dioxide to air used in the oxidation process.

Experiments were carried out in two tubular quartz reactors: one with a diameter of 4 mm filled with 5 mg of a catalyst and the second with a diameter of 10 mm filled with 20 mg of a catalyst. Some experiments were carried out with 100 mg of a catalyst. The copper-zinc catalyst TMC-2 produced by Zakłady Azotowe in Tarnów was comminuted to a grain size of 0.2 to 0.385 mm. It was placed on a packing of α-Al_2O_3 grains. In order to determine the effect of SO_2 on catalyst activity, pure air and air with SO_2 was used for oxidation. Some experiments were carried out with impulse or continuous addition of SO_2 to pumped air. Experiments were carried out with and without the catalyst.

A solution of phenol at concentration of 5 g/l was pumped to an evaporator at flow intensity in the range from 2.2 to 6 cm^3/h. Vapour from the evaporator and air was supplied to the reactor. The flow intensity of air was in the range from 4.5 to 40 dm^3/h. The contact time changed from 0.0012 to 0.0026 s.

Reaction products were cooled. In a condensate, the chemical oxygen demand (COD) or total organic carbon (TOC), UV-VIS spectra and pH were determined. The phenol concentration was also determined using gas chromatography and colorimetry[5]. The concentrations of SO_2 in dry combustion gases were determined using a colorimetric method[6]. Each experiment without changes of temperature and flow rates lasted 5–6 hours. Every hour samples for analysis were taken.

3. RESULTS AND DISCUSSION

The results of experiments are presented in Figures 1–7. Points at Figure 1, 2, 4 and 5 present mean values obtained for given parameters.

The dependence of TOC in a condensate after the oxidation process on temperature is presented in Figure 1. The TOC value (like COD value) characterizes the total amount of organic substances. It allows us to determine the course and efficiency of total oxidation of organic substances to CO_2 and H_2O. The upper curve was obtained for the process without the catalyst, the lower curve - for the process with 5 mg of the catalyst. The dashed line joins points obtained if the air necessary to oxidation contained the impulse doses of SO_2 in the range from 20 to 40 mg. These doses of SO_2 can be regarded as high in relation to the amount of the catalyst. As shown in Figure 1, the catalyst completely

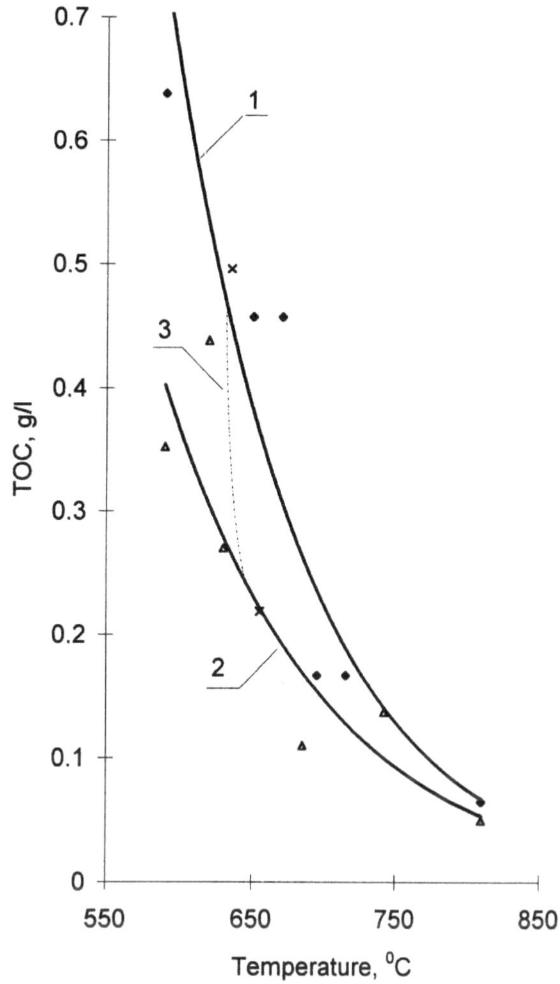

Figure 1. The dependence of TOC on temperature; 1 - without catalyst, 2 - Cu-Zn catalyst, 3 - Cu-Zn catalyst and SO_2.

loses its activity during total oxidation at temperatures lower than 625°C. At temperatures higher than 700°C the effect of SO_2 is slight.

The dependence of phenol concentration in condensate determined using chromatography on temperature is presented in Figure 2. The curves obtained for the catalytic process with and without SO_2 dosage, overlap. It proves that sulphur dioxide does not influence an increase in rate of catalytic decomposition of phenol to indirect products.

These investigations were carried out with impulse addition of SO_2 to a reactor. Therefore, catalyst activity changed after SO_2 addition with experiment duration. The example of dependence of organic substance concentration expressed as COD value on SO_2 dose and experiment duration expressed in days, is presented in Figure 3. As it results from Figure 3, the catalyst activity decreases after SO_2 addition. However, the change depends on sulphur dioxide dose. The catalyst activity is partially reproduced in a few hours if air supplied to a reactor is pure. This experiment presents the possibility of regeneration of the copper-zinc catalyst.

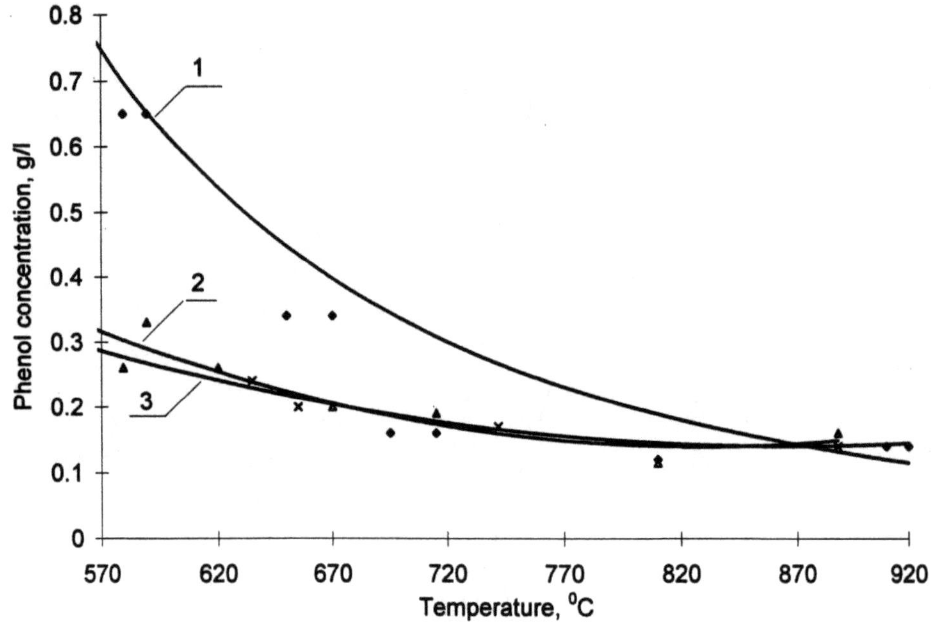

Figure 2. The dependence of phenol concentration on temperature; 1 - without catalyst, 2 - Cu-Zn catalyst, 3 - Cu-Zn catalyst and SO_2.

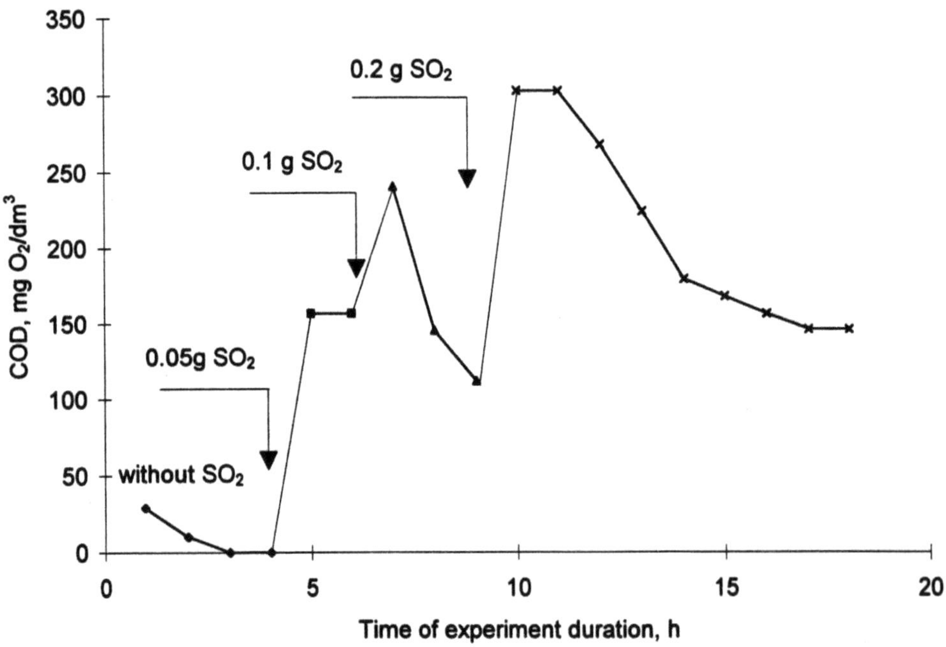

Figure 3. The dependence of COD on SO_2 dose and time of experiment duration. Temperature 500^0C.

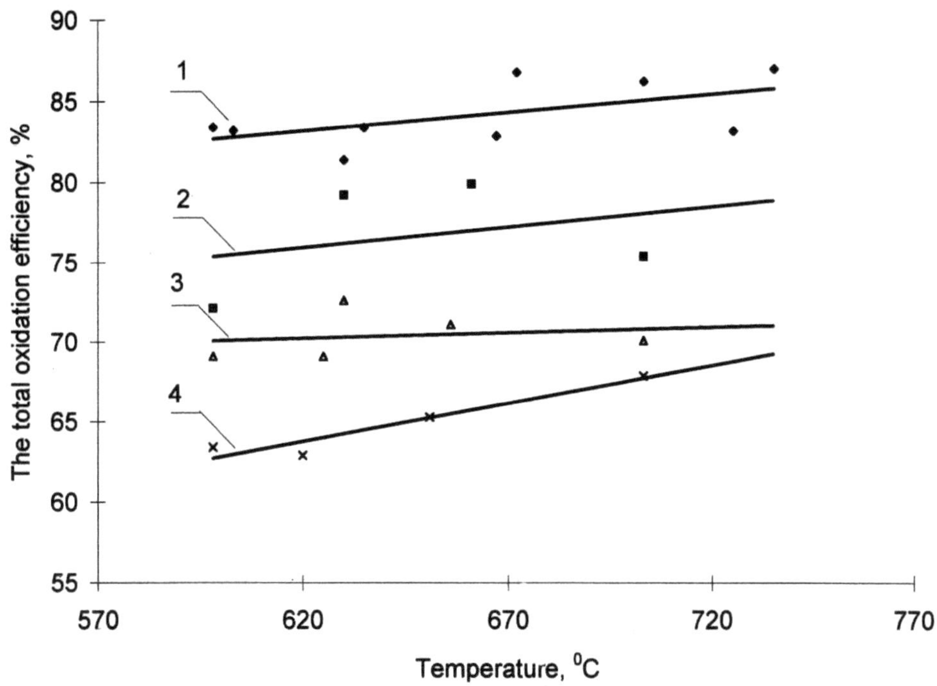

Figure 4. The dependence of total oxidation efficiency on temperature and SO_2 concentration; 1 - without SO_2, 2 - SO_2 concentration of 0.011 mg/dm^3, 3 - SO_2 concentration of 0.025 mg/dm^3, 4 - SO_2 concentration of 0.032 mg/dm^3. Copper-zinc catalyst.

The results of experiments show that at temperatures in the range from 625 to 730°C, qualitative change of the process may occur. Catalyst inactivation at temperatures below 600°C can be caused by the fact that inactive cupric sulphate is stable under conditions of, the cupric oxide reaction with SO_2.[7] Cupric oxide can be stable at temperatures higher than 730°C in a medium which contains SO_2. In the range of temperature from 600 to 730°C, Cu_2OSO_4 is stable. This compound is active during phenol oxidation according to experimental results. In this range of temperature, additional experiments with continuous dosage of SO_2 were carried out. The SO_2 concentration changed in the range from 0.011 to 0.032 mg/dm^3.

The dependence of oxidation efficiency (determined on the base of changes of the condensate COD values) on temperature and sulphur dioxide concentration is presented in Figure 4. The total content of organic substances in liquid reaction products depends clearly on SO_2 concentration. The efficiency of total phenol oxidation to CO_2 and H_2O distinctly decreases with the increase in SO_2 concentration in the reagent mixture.

The dependence of phenol oxidation efficiency, which results from this compound concentration in the condensate after the reaction, on temperature and sulphur dioxide concentration, is presented in Figure 5. In the presence of SO_2, the oxidation efficiency decreases slightly, especially at low temperatures. However, this efficiency slightly depends on this gas concentration in the tested range.

The dependence of catalytic reaction on temperature and SO_2 concentration obtained in experiments presented in Figures 1–2 and 4–5 is similar. TMC-2 catalyst, applied also at low concentrations of sulphur dioxide, is very active in the reaction of total oxidation of

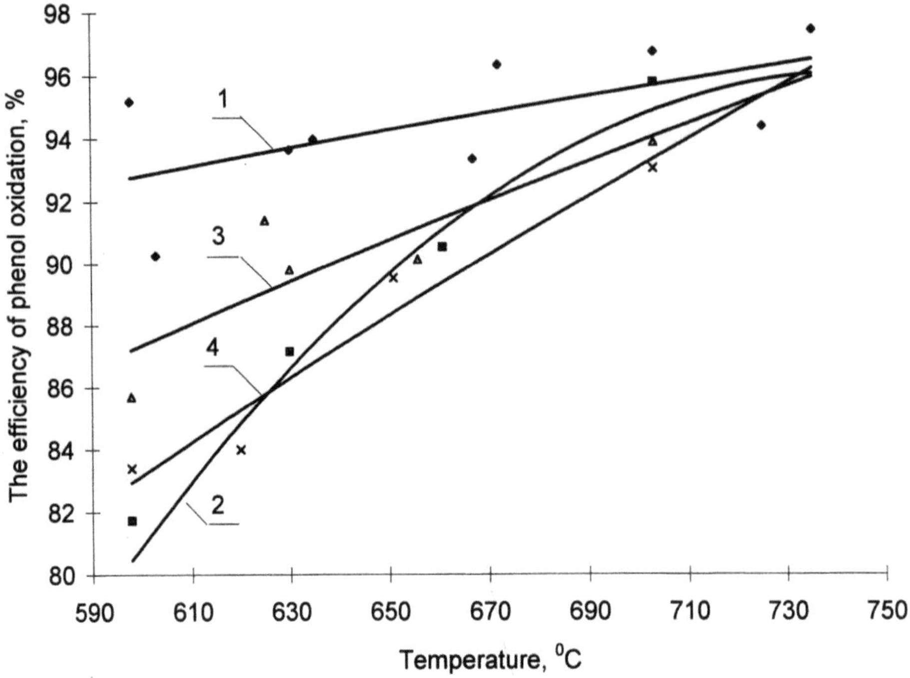

Figure 5. The dependence of phenol efficiency on temperature and SO_2 concentration; 1 - without SO_2, 2 - SO_2 concentration of 0.011 mg/dm^3, 3 - SO_2 concentration of 0.025 mg/dm^3, 4 - SO_2 concentration of 0.032 mg/dm^3. Copper-zinc catalyst.

hydrocarbons and their oxygen derivatives. At very low contact time about 0.002 s and at temperature 700^{0}C the efficiency of the total phenol oxidation to CO_2 and H_2O using pure air amounted to 84%. It decreased to 75% at SO_2 concentration of 0.011 mg/dm^3 and to 68% if SO_2 concentration was 0.032 mg/dm^3. The phenol concentration decreased by 95% at temperature of 700^{0}C if the mixture contained no SO_2. It decreased by 93–95% if the SO_2 concentration totalled 0.011–0.032 mg/dm^3. Significant differences between efficiencies of the first stage of the phenol reaction and the total oxidation shows that in the chain of reactions which leads to CO_2 and H_2O in the final stage, an indirect product which is difficult to oxidize can occur. Probably, p-benzoquinone is this indirect product.

UV spectra of liquid products of oxidation are presented in Figures 6 and 7. The application of the catalyst causes formation of p-benzoquinone. Its concentration increases with the increase in SO_2 concentration which is additionally confirmed by the yellow colour of the condensate solution.

The content of sulphur dioxide in the reaction mixture, caused a decrease in rate oxidation which occurs according to reaction:

$$\text{phenol} \longrightarrow \text{hydroquinon} \longrightarrow \text{p-benzoquinone} \longrightarrow \text{organic acid} \longrightarrow CO_2 + H_2O$$

The reaction rate of p-benzoquinone oxidation decreases in the presence of SO_2.

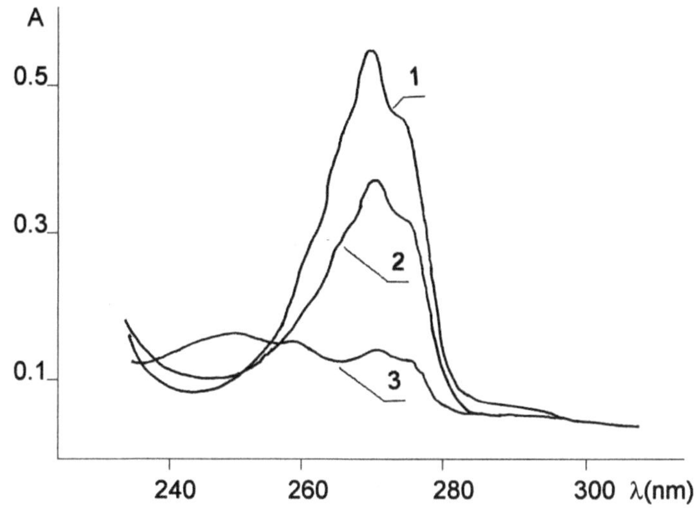

Figure 6. UV spectra of the condensate; 1 - without catalyst, temperature 598^{0}C, 2 - without catalyst, temperature 714^{0}C, 3 - Cu-Zn catalyst, temperature 609^{0}C.

4.CONCLUSIONS

Noxious industrial wastes are usually a mixture of compounds which can include toxic, hard to decompose organic substances. In many cases, their combustion is the most convenient method of their treatment. In processes of total oxidation of organic substances, copper-zinc catalyst is often used due to its low price and very high activity. However, this catalyst is sensitive to sulphur compounds.

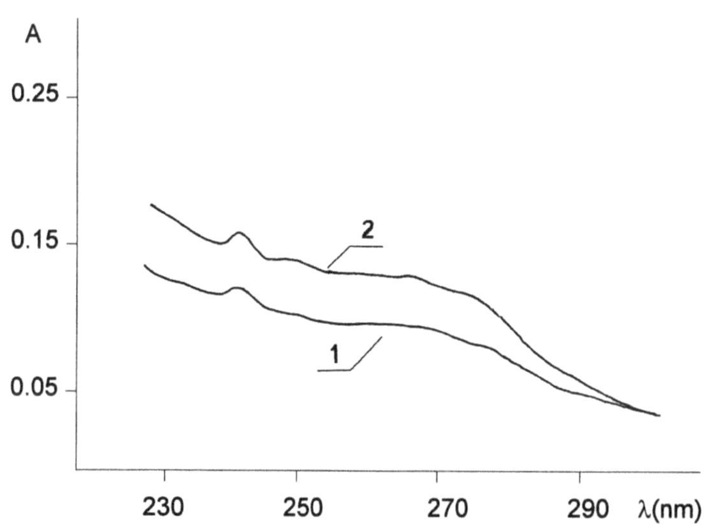

Figure 7. UV spectra of the condensate after catalytic reaction in the presence of SO_2; 1 - SO_2 concentration of 0.011 mg/dm^3 , 2 - SO_2 concentration of 0.032 mg/dm^3.

It can be stated on the basis of experimental results that the copper-zinc catalyst (TMC-2) used to total phenol oxidation in its aqueous solution is very active. This catalyst can also be used while sulphur concentration is low. If sulphur dioxide is present in the reaction medium, then, along with temperature, it has a principal effect on the process rate of total oxidation. The catalyst partly loses its activity at high concentrations of SO_2. However, it is possible to regenerate the catalyst by supply of pure air. In order to prevent the catalyst from becoming inactive while sulphur dioxide is present in the medium, it is necessary to apply temperatures above 600^0C. At higher temperatures, Cu_2OSO_4 compound is formed in reaction of copper oxide with sulphur dioxide. As it results from experiments, this compound is active in the reaction of organic substance oxidation. In order to avoid a decrease in mechanical resistance of the catalyst due to sulphate formation, temperatures higher than 730^0C should be applied.

Results of the investigations allow us to state that the decrease in activity of copper-zinc catalyst (TMC-2) in reaction of total phenol oxidation is not due to the rate change of the elementary stage of this compound decomposition. Probably, it results in a rate decrease of indirect product oxidation during p-benzoquinone reaction.

ACKNOWLEDGMENTS

The investigations were financially supported by grant KBN nr 0422/P4/94/06.

REFERENCES

1. Gorzka Z., Kaźmierczak M., Filipiak E., Chemistry for Protection of the Environment, Elsevier, Amsterdam, 1984, pp 437–443.
2. Borkowski B., Water Research 1(5), 367–85 (1967).
3. Walsh M.A., Katzer J.R., Ind. Eng. Chem. Proc. Des. Develop. 12,4,477–81 (1973).
4. Gorzka Z., Kaźmierczak M., Physiochemical Methods for Water and Wastewater Treatment, Pergamon Press, Oxford 1980, pp 175–83.
5. Polish Standard PN-72/C-04602.03
6. Polish Standard PN-68/Z-04039
7. Lorant B., Analitische Chemie, 219, 3, 256–71 (1966).

SORPTION OF BENZENE, TOLUENE, AND meta-XYLENE TO CONTAMINATED SITE SOILS

Christopher G. Uchrin, Koshy Koshy, and Izabela Wojtenko

Department of Environmental Sciences
Rutgers University
New Brunswick, New Jersey 08903

1. ABSTRACT

Results from adsorption studies are presented focusing on benzene, toluene and meta-xylene (BTX) as single components onto contaminated site soils obtained from an oil refinery site in New Jersey, USA. The soil material did not exhibit significant amounts of either benzene, toluene or xylene but was contaminated with other compounds. The organic matter content of these soil materials ranged from 0.14 to 2.91 percent. Results demonstrated that the sorption exhibited by the contaminated soils was not consistent with similar results obtained for "clean" reference soil systems. Results were unpredictable based on the organic carbon content of the soils.

2. INTRODUCTION

Biological treatment, as a method for remediating contaminated site soils, is becoming a more attractive technology, especially as other processes are limited in effectiveness (e.g., soil washing) or extremely costly (e.g., removal and treatment). A major limitation to the effectiveness of biological treatment, however, is related to the phase in which the contaminant resides. Contaminant molecules in saturated soils can exist in either the liquid (unadsorbed) or solid (adsorbed) phase. Within the solid phase, contaminant molecules can either be loosely associated with the solids or essentially contained or sequestered within a solid matrix. Biodegradation rates of sorbed chemicals are lower than in aqueous systems (Scow, et al., 1995). Thus, it is of paramount importance to accurately account for the sorption dynamics of contaminants in soil/water systems.

Most sorption studies focus on examining adsorption and desorption kinetics and equilibria of contaminants onto "clean" materials or reference materials. Some studies have attempted to examine the desorption dynamics of substances from contaminated

Chemistry for the Protection of the Environment 2, edited by Pawłowski et al.
Plenum Press, New York, 1996

soils. The research presented in this paper examined both adsorption and desorption dynamics from existing contaminated soils from a former oil refinery.

3. BACKGROUND

The soil materials used in this study were extracted from a former refinery site in New Jersey, USA. The site was used for the production of petroleum products for the greater part of the twentieth century. Over this period, numerous spills have occurred. A series of ground water samples were collected recently. Samples from 24 monitoring wells were analyzed for total petroleum hydrocarbons (TPHs), volatile organic compounds (VOCs), and semi-volatile organic compounds (SVOCs). TPHs were detected in 15 of the 24 wells ranging from 0.58 to 130 ppm. SVOCs were detected in 2 of the 24 wells and the VOC data showed quantities of benzene, chlorobenzene, 1,1,2,2 trichloroethane, methyl chloride, and xylenes exceeding the NJDEP class 11A groundwater quality criteria (greater than 3,000 ppm). The main petroleum hydrocarbon contamination appears to occur in shallow fill areas. Intermediate and deep monitoring wells also showed high metal and inorganic contaminant concentrations.

4. MATERIALS AND METHODS

4.1. Soils

Soil characteristics are listed in Table 1. All soils analyses were performed by the Soils Testing Laboratory, New Jersey Agricultural Experiment Station except for the chemical identification which was performed by the analytical section of the ARPA Project team (Arands, 1995). Organic matter content was determined by the Walkley and Black

modification of the rapid dichromate oxidation technique (Nelson and Somers, 1982) with organic carbon content defined as 59% of the organic matter content. Table 2 lists chemicals which were identified on the soils.

4.2. Experimental Protocol

A 15 g mass of soil was introduced into a series of 60-mL crimp-top vials. All glassware was washed carefully, rinsed with acetone, and baked at 225°C for 24 hours prior to use. The solids were then saturated with 40 mL of a 0.02 M solution of calcium chloride

Table 1. Soil characteristics

Characteristic	Site 1	Site 2	Site 3
Texture	sandy loam	sandy loam	loamy sand
Sand (%)	54	64	78
Silt (%)	26	24	8
Clay (%)	20	12	14
Organic Content (%)	1.34	2.91	0.14
CEC (meq/100 g)	7.69	7.76	1.83
pH	7.7	7.6	9.4

Table 2. Chemicals Identified in Soil Samples (Arands, 1995)

Site 1	Site 2	Site 3
3,4-dimethyl-1-decane	4-methyl-decane	2-cyclohexen-1-ol
2,3,7-trimethyl-octane	4,8-dimethyl-undecane	2-cyclohexen-1-one
2,6,10-trimethyl-dodecane	2,6,8-trimethyl-decane	4-chloro-,trans-cyclohexanol
5-ethyl-5-methyl-decane	2,6-dimethyl-octadecane	decahydro-2-methyl-naphthalene
sulfur	2,6,10-trimethyl-dodecane	sulfur
2-cyclohezen-1-ol	hexadecane	
4-chloro-,trans-cyclohexanol	sulfur	
5-methyl-decane		
decahydro-2-methyl-naphthalene		
2,6-dimethyl-undecane		
2-butyl-1,1,3-trimethyl-cyclohexane		
tetradecane		
2,6,11-trimethyl-dodecane		
4,6-dimethyl-undecane		

in water. Prior to use, all water was distilled, deionized, boiled and cooled. The vials with the soil/water slurry were then purged with nitrogen gas for seven minutes to reduced the oxygen level and minimize the possibility of aerobic degradation (Evans, et al., 1992; Wojtenko, et al., in press). Each vial was then sealed with a teflon septum and crimped with an aluminum seal. The septa and seals were autoclaved prior to use. The contaminant of interest was then introduced into the experimental vial by piercing the septum with a syringe and injecting a predetermined mass. Parallel control vials (without soil) were also set up to monitor for experimental losses. Solid phase concentration was computed from the differential between the liquid phase concentrations of the experimental vial and the parallel control vial. The vials were then agitated and stored in a 20°C room.

Prior studies (Wojtenko, et al., in press) had shown that adsorption equilibrium for similar systems had been reached within 96 hours. This was then determined to be the time-to-equilibrium for the current systems. The vials were placed in a water bath at 20°C 24 hours prior to analysis. Sampling for the contaminant of interest was performed by piercing the teflon cap with a syringe and drawing a liquid sample from the experimental vial. This sample was then injected into a Tekmar Model ALS 2016 Purge and Trap instrument and then analysized by a Hewlett Packard Model 5840 gas chromatograph equipped with a flame ionization detector.

5. RESULTS AND DISCUSSION

Results for single component toluene and meta-xylene equilibrium adsorption studies are shown in Figures 1 and 2, respectively. All seem to fit a linear isotherm fairly well and isotherm parameters are given in Tables 3 and 4 for toluene and meta-xylene, respectively. Indeed, it is somewhat surprising that a linear fit should be able to describe these data at the concentration ranges used. Earlier studies had shown the applicability of linear isotherms for toluene (Uchrin and Mangels, 1987) and meta-xylene (Uchrin, et al., 1986) onto "clean" reference soils but at concentration ranges several orders of magnitude lower than the relative solubility limits. Furthermore, the amount of adsorption exhibited by the Site 3 soil is truly astounding based on its organic carbon fraction. Table 5 lists calculated K_d values based on the literature relationship given by Karickhoff, et al. (1979):

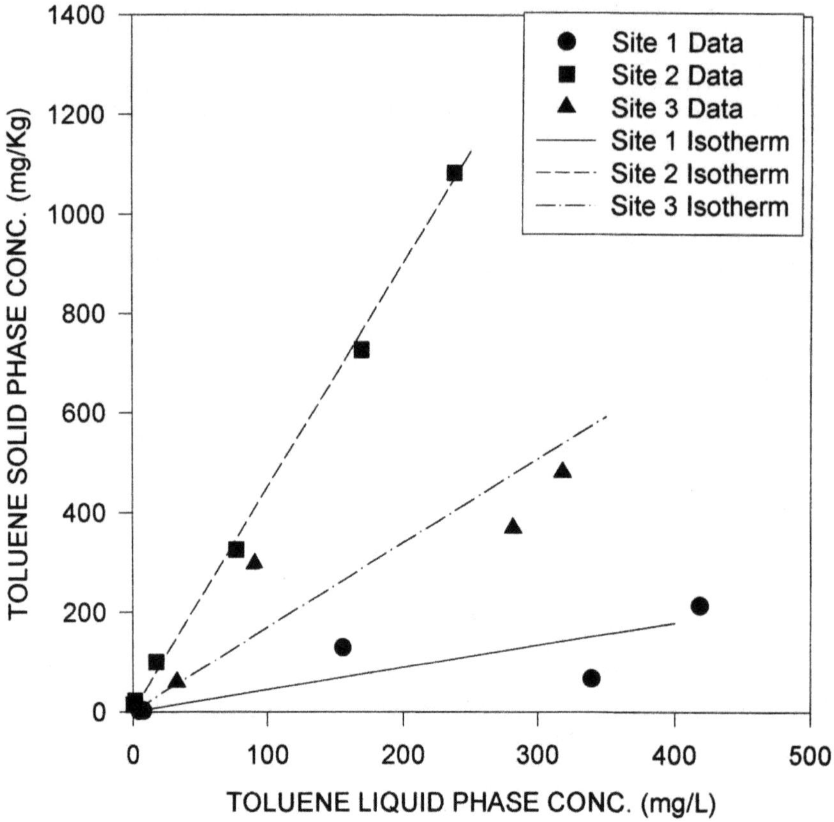

Figure 1. Toulene adsorption isotherms.

$$\log K_{oc} = \log K_{ow} - 0.21$$

where Koc is the organic carbon normalized partition coefficient:

$$K_{oc} = K_p/f_{oc}$$

Table 3. Toluene adsorption isotherm parameters

Parameter	Site 1	Site 2	Site 3
K_d	0.45	4.5	1.7
r^2	0.88	0.96	0.71

Table 4. meta-Xylene adsorption isotherm parameters

Parameter	Site 1	Site 2	Site 3
K_d	3.0	4.2	2.8
r^2	0.94	0.90	0.90

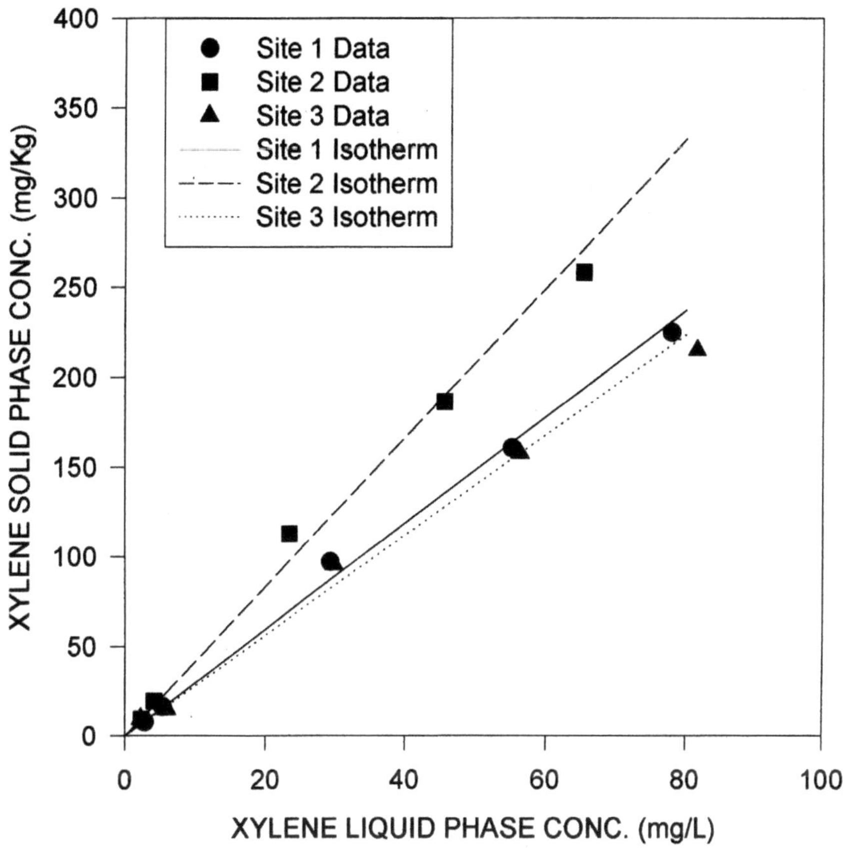

Figure 2. meta-Xylene adsorption isotherms.

where f_{oc} is the organic carbon fraction of the soil (which is defined as 59% of the organic matter. Indeed, for the toluene systems, the data suggest that B.E.T. type behavior, i.e., multi-layer adsorption, may exist for the Site 1 and 3 soils.

Results from the benzene equilibrium adsorption studies are shown in Figure 3. It is interesting to note that a Langmuir-type Isotherm is exhibited for benzene on the Site 2 soil. Only in this case was a definite adsorption maximum reached. The Site 2 soil did, however, consistently evidence the strongest adsorption for all three BTX compounds, however, a definite adsorption maximum was not reached for either the toluene or the meta-xylene. The benzene on the Site 1 and Site 3 soils, hoever do seem to show some effects of approaching the solubility limit of benzene in water (1780 mg/L @ 20°C; Ver-

Table 5. Literature based Kd values [log Kow values from Miller et al. (1985)]

Compound	log K_{ow}	K_d-Site 1	K_d-Site 2	K_d-Site 3
Toluene	2.65	2.2	4.7	0.23
meta-Xylene	3.20	7.7	17	0.81
Benzene	2.13	0.66	1.4	0.069

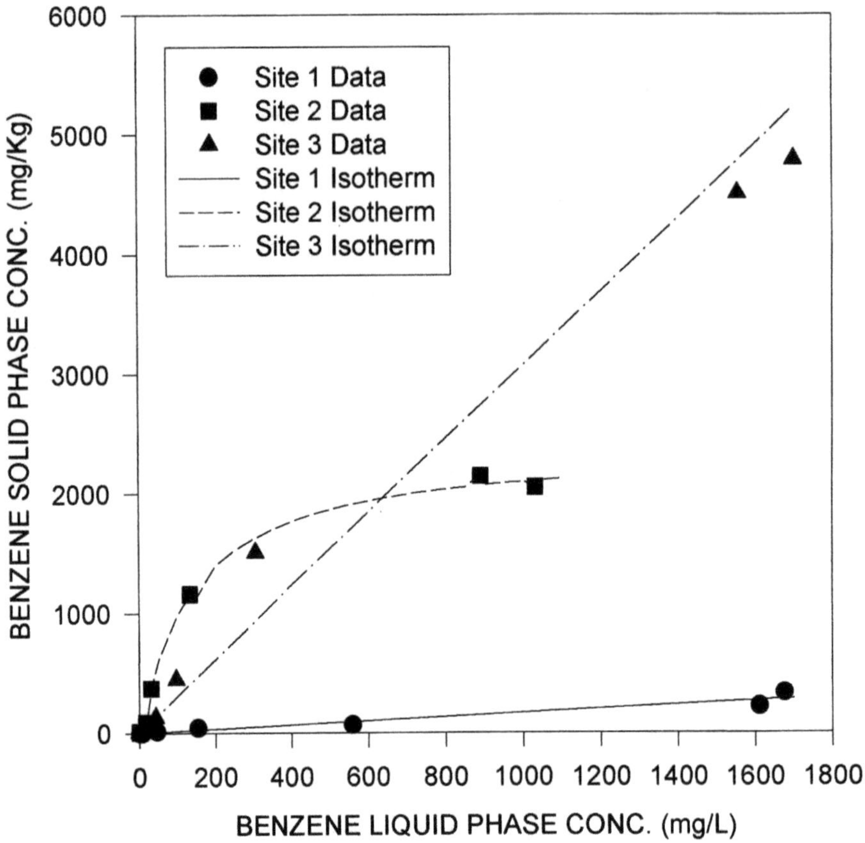

Figure 3. Benzene adosption isotherms.

schueren, 1983). Further experiments on the Site 2 soil are anticipated to push this system toward the solubility limit. Again, as with the toluene and meta-xylene systems, the large degree of adsorption evidenced by the Site 3 soil is unexpected based on organic matter content. It does appear that the presence of the organic contamination on the soils causes a more favorable condition for adsorption. Earlier studies (Uchrin and Katz, 1986) had shown that the presence of multiple components could cause an enhancement in adsorption. Isotherm parameters are given in Table 6.

Table 6. Benzene adsorption isotherm parameters

Parameter	Site 1	Site 2	Site 3
K_d	0.17	17*	3.1
r^2	0.95	xxx	0.83
Langmuir K (Kg/mg)	xxx	0.0070	xxx
Langmuir Q (mg/Kg)	xxx	2400	xxx
Langmuir r^2	xxx	0.94	xxx

*Kd estimated from linear portion of Langmuir Isotherm (K_d = KQ).

6. ACKNOWLEDGMENTS

This work was supported in part by the U.S. Department of Defense, Office of Naval Research and Advanced Research Projects Agency, University Research Initiative Program (Grant N00014–92-J-1888, R&T a40r41rui) and the New Jersey Agricultural Experiment Station, publication No D-07525–3–95.

7. REFERENCES

Arands, R., 1995. Personal Communication.

Evans, P.J., W. Ling, B. Goldschmidt, E.R. Ritter, and L.Y. Young, 1992. Metabolites Formed During Anaerobic Transformation of o-Xylene and Their Proposed Relationship to the Steps of Toluene Mineralization. *Applied and Environ. Microbiol.*, 58:496–501.

Karickhoff, S.W., Brown, D.S., and Scott, T.A. 1979. Sorption of Hydrophobic Pollutants in Sediment Suspensions. *Water Research*, 13:241–248.

Miller, M.M., Wasik, S.P., Huang, G.L., Shiu, W.Y., and Mackay, D. 1985. Relationships Between Octanol-Water Partition Coefficient and Aqueous Solubility. *Environ. Sci. Technol.*, 19:522–529.

Nelson, D.W., and L.E. Somers. 1982. Total Carbon, Organic Carbon, and Organic Matter. In: *Methods of Soil Analysis, Part 2: Chemical and Microbiological Properties, 2nd Edition* (Page, A.L., Miller, R.H., and Keeney, D.R., Eds.). Am. Sco. Agron. Inc. and Soil Sci. Soc. Amer. Inc., Madison, Wis., pp 565–571.

Scow, K.M., S. Fan, C. Johnson, and G.M. Ma, 1995. Biodegradation of Sorbed Chemicals in Soil. *Environ. Health Perspectives*, Vol. 103, Supplement 5, pp 93–95.

Uchrin, C.G., J.V. Hunter and P. Chan, 1986. *A Characterization of the Binding of Toxic and Hazardous Substances to Soil Particulates.* Report to the Industry/ University Center for Research in Hazardous and Toxic Substances, NJIT, Newark, NJ.

Uchrin, C.G., and J. Katz, 1986. Sorption Kinetics of Competing Organic Substances on New Jersey Coastal Plain Aquifer Solids, *Hazardous and Industrial Solid Waste Testing and Disposal: Sixth Volume. ASTM STP 933*, D. Lorenzen, R.A. Conway, L.P. Jackson, A Hamza, C.L. Perket, and W.J. Lacy, Eds., American Society for Testing and Materials, Philadelphia, pp 140–150.

Uchrin, C.G., and G. Mangels, 1987. Sorption Equilibria of Benzene and Toluene on Two New Jersey Coastal Plain Ground Water Aquifer Solids, *J. Environ. Sci. Health*, A22:743–758.

Verscheuren, K., 1983. Handbook of Environmental Data on Organic Chemicals, 2nd ed. Van Nostrand Reinhold Co., NY.

Wojtenko, I., K. Koshy, and C.G. Uchrin, in press. Long Term Sorption Studies of Benzene and Toluene onto Soils, submitted to *Bull. Environ. Contam. Toxicol.*

34

HEAVY METALS IN THE SOIL-WATER-PLANT SYSTEM

Importance of Interactions

Chen Huai-man and Zheng Chun-rong

Institute of Soil Science
Academia Sinica
Nanjing 210008, China

Heavy metals, like any other chemical element in a terrestrial ecosystem, may be dissolved in the soil solution, adsorbed onto colloidal surfaces, occluded into soil minerals, precipitated with other components in soil, or incorporated into biological materials. These processes are influenced by many factors, such as soil pH, Eh, temperature, organic matter content, mineralogical compositions as will as the type and concentration of other dissolved constituents. Furthermore, the chemical behavior of a heavy metal in the soil-water-plant system is affected by interactions with other elements and/or soil components and the environmental conditions. The knowledge of the interactions may lead to a better understanding of the relationship between crop yield and quality and the nutrient level of soils (Ohno and Grunes, 1985). Research on these interactions should attract more attention because it relates to ecosystem balance, environment qualities and human health (Chen and Zheng, 1992; 1994b). This, however, has not been the case until recently. To il lustrate the effects of interactions on chemical behavior of the elements in soil-water-plant system, some results obtained recently in laboratory and greenhouse studies by our research group were chosen (Zheng and Chen 1990; Chen *et al.*,1991a,b; Chen and Zheng, 1994a,b; Zang *et al.*, 1987, 1989).

1. INTERACTIONS BETWEEN Pb AND SOME ANIONS

1.1. Effect on Relative Extraction Rate

The incubation-extraction experiment of Pb in soil was carried out with calcic concretion black soil (CB, Vertisol), yellow brown soil (YB, Alfisol), red soil (R, Ustisol) and latosol (L, Oxisol) (Chen *et al.*, 1991a). The soils were incubated for 2, 4, 8 and 16 weeks after adding different Pb compounds at the rate of 2.41, 4.43, 9.65 and 19.31mmol/kg re-

Chemistry for the Protection of the Environment 2, edited by Pawłowski et al.
Plenum Press, New York, 1996

Table 1. Comparison of relative extraction rates for Pb(RERL) added as different compounds to soils (incubation time: 16 weeks; the highest rate is set to 100%) (Chen *et al.*, 1991a)

Concentration of Pb added (mmol/kg)	Compound	Soil			
		CB	YB	R	L
2.41	$PbCl_2$	100	100	100	100
	$Pb(NO_3)_2$	95.4	91.9	100	99.4
	$Pb(OAc)_2$	87.8	81.1	100	81.2
4.83	$PbCl_2$	88.7	100	100	99.8
	$Pb(NO_3)_2$	100	97.7	92.2	100
	$Pb(OAc)_2$	73.6	96.6	86.4	82.6
9.65	$PbCl_2$	99.5	65.6	98.8	100
	$Pb(NO_3)_2$	100	100	100	97.8
	$Pb(OAc)_2$	73.8	83.8	76.8	8
19.31	$PbCl_2$	92.2	88.2	100	100
	$Pb(NO_3)_2$	100	100	96.6	100
	$Pb(OAc)_2$	58.1	68.4	69.4	75.6

spectively. Each treatment had two replications. Then, the incubated soil samples were placed into centrifuge tubes, 1mol/l NH_4OAc (pH7) was added in the solution to soil ratio of 5 to 1, and the suspension was shaken for 1 hour at 25±1°C. Data presented in Table 1 show the average relative extraction rates of Pb added as $PbCl_2$, $Pb(NO_3)_2$ or $Pb(OAc)_2$. There were only small differences between the results for $PbCl_2$ and $PbNO_3$. In contrast the average relative rates for $Pb(OAc)_2$ were lower (80.1%). The reason for this difference has not been fully understood yet. However both Cl^- and NO_3^- are generally considered to be non-specifically adsorbed in soils, while specific adsorption may play a greater role for OAc^- (Mott,1981). As a result, the Pb sorption capacity of the soil was raised with the increase of negative charges on clay surfaces resulting in lower extractability of Pb added as $Pb(OAc)_2$.

1.2. Effect on Pb Uptake by Plants

Lead uptake by wetland rice when various Pb compounds were added to the soil is given in Table 2 (Chen *et al.*, 1991b). There were differences in Pb contents of root, straw and brown rice between treatments of $Pb(NO_3)_2$, $PbCl_2$ or $Pb(OAc)_2$ at 3000 and 4000 mgPb/kg, and the differences reached significant or very significant level in the one-way classification analysis of variance (Du *et al.*,1985). The average concentration of Pb in straw and brown rice in the $Pb(NO_3)_2$ treatment approached to about twice that in the treatment with $PbCl_2$ or $Pb(OAc)_2$. Lead content in all plant parts increased with increasing Pb addition except in the root in the treatment with 4000 mgPb/kg. The reason might mainly be the higher toxicity, therefore, the root weight and Pb uptake by roots decreased. However, Pb translocated from root to shoot in this treatment was higher than others and it may be related to the function of nitrogen. Because the uptake of Pb by plants belongs to passive absorption, for example, the more amount of $Pb(NO_3)_2$ applied in the soil, the more the growth of rice aerial parts and the more transpiration. In the result, the concentration of Pb in straw increased distinctly. Lead uptake by roots varied clearly between different Pb compounds, showing that Pb absorption by root and translocation of Pb from

Table 2. Effect of different Pb compounds on Pb uptake by wetland rice (average of 3 replicates) (Chen et al.,1991b)

Compound	Pb Conc. (mg/kg)	Root (mg/kg)	straw (mg/kg)	Ear stem (mg/kg)	Brown rice (mg/kg)
	0	27.2	15.7	1.9	0.04
$PbCl_2$	1000	672	44.6	8.4	0.36
	2000	1912	58.5	16.9	0.83
	3000	4652	77.9	21.2	1.04
	4000	6065	122	29.2	1.62
	Average	3327	75.8	18.9	0.96
$Pb(NO_3)_2$	1000	773	46.1	5.9	0.25
	2000	2909	69.5	8.6	0.60
	3000	3857	182	40.0	2.21
	4000	2492	324	98.3	4.91
	Average	2498	155.4	38.2	1.99
$Pb(AOc)_2$	1000	688	31.7	8.6	0.35
	2000	3031	49.6	15.9	0.70
	3000	6067	86.6	25.2	1.43
	4000	10437	134	32.1	2.05
	Average	5056	75.5	20.5	1.13

root to shoot not only depended on the concentration of Pb added in soil, but also on the accompanying anions.

1.3. Effect on Trace Elements Uptake by Plants

Pollutants reaching soils, may in addition to toxic effects, cause disturbances in micronutrient functions by antagonism or synergism, so that the balance of nutrients in the food chain is further affected, with possible adverse effects on humans and animals. Kannan and Kepple (cited by Koeppe, 1981) conducted a short-term experiment with pea (*pisum sativum*) seedlings and showed that Pb in the nutrient solution was highly inhibitory to the absorption of Zn^{2+}, Mn^{2+} and Fe^{2+} ions. The results in our study Chen *et al.*, 1991b)showed that the effects of various Pb compounds on the absorption of Zn, Mn, Fe and Cu by plants were different and varied between the parts of plant (Chen et al., 1991b). For example, $PbCl_2$ clearly reduced Zn concentration in roots was observed ($r = -0.88^{**}$), but the Zn concentration in brown rice increased with increasing Pb addition to the soil ($r = 0.49^{*}$); $Pb(NO_3)_2$ added to the soil caused a decreasing tendency of Zn in root (r = -0.43 NS), but higher Zn content of straw ($r = 0.54^{*}$) and brown rice ($r = 0.78^{**}$). $Pb(OAc)_2$ additions restrained the zinc absorption by roots ($r = -0.73^{**}$) and brown rice (r = -0.23NS). All Pb treatments caused increased Mn concentrations in straw and roots (r values for $PbCl_2$, $Pb(NO_3)_2$ and $Pb(OAc)_2$ were 0.80^{**},0.24NS and 0.65^{*} respectively), but the changes in brown rice were not obvious in the $Pb(NO_3)_2$ and $Pb(OAc)_2$ treatments. Effects of Pb added as $Pb(OAc)_2$ on Fe concentration in roots and grains showed a striking contrast, the former increased and the later decreased markedly with Pb increasing in soil with r values of 0.92^{**} and -0.92^{**}, respectively. The concentration of root Fe increased at first and then decreased with increasing the amounts of Pb addition in soil. This difference between Fe in straw and brown rice was not found with addition of $Pb(NO_3)_2$. The iron content of root increased significantly ($r = 0.88^{*}$) by addition of Pb as $PbCl_2$, but clear effects of $PbCl_2$ on Fe in straw and brown rice were not observed. Amounts of Cu absorbed by all parts of rice plants significantly increased with amount of Pb added in all treatments. Kannan and Keppel (cited by Koeppe,1981) considered that inhibition of Pb on absorption of Fe^{2+}, Mn^{2+} and Zn^{2+} by pea was mainly physical in nature, i.e. by blocking the entry or the binding sites of the ion-carriers. The results of the present experiments show that: (1)effects of Pb on absorption of Zn, Mn and Fe by different plant were not identical; and (2)various Pb compounds have different effects. It seems that the effects of Pb on uptake of microelements by wetland rice are depending not only a simple physical process but also on complex physiological and biochemical processes. Further research on the mechanisms is warranted.

2. INTERACTION AMONG CATIONS

2.1. Interaction between Pb and Cd

The uptake of Pb and Cd by wetland rice was affected by the interaction of Pb and Cd added to soils, and the effect on the uptake of Cd was more marked. The uptake of Cd by rice grown in red soil was accelerated by the addition of 250 mgPb/kg when the concentrations of Cd added were 1, 5, 10 mg Cd/kg (added as $CdCl_2$). Stepwise multiple regression analysis showed that the relationship between Cd content (Y, mg/kg) in brown rice and Pb and Cd added in soil (mg/kg) is

$$Y = 0.170 + 0.143\ (Cd) + 0.00044\ (CdxPb)$$

with $R^2 = 0.80$ ($P<0.001$). The stimulation of Cd uptake by Pb addition may be explained by the adsorption kinetics. Fig. 1 shows variations in Cd in effluence in leaching experiments with and without Pb addition under the same conditions as the adsorption experiments. Cadmium in effluence reached the concentration in the leaching solution more rapidly with Pb

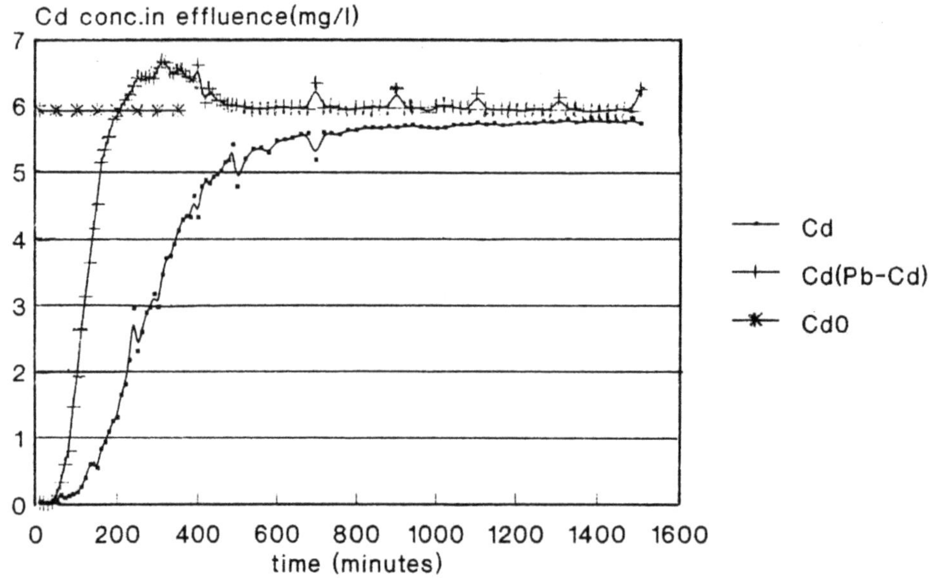

Figure 1. Effect of interaction between Pb and Cd on kinetics of Cd adsorption by soil.

addition than without, The concentration maximum in effluence exceeded that in the leaching solution. This can be explained by replacement of initially adsorbed Cd by Pb at a later time. More details about the experiment will be reported elsewhere.

2.2. Interaction among Pb, Cd, Cu and Zn (Zheng and Chen, 1990)

The concentrations of Pb, Cd, Cu, and Zn in brown rice in relation to metal addition in soil are given in Table 3. Stepwise multiple regression analysis of metal concentration in brown rice reveals a significant effect of added amounts of Pb, Cd, Cu and Zn on the metal contents. The best-fit equations with and without interaction terms are given in equations (1)–(8). Where Y (mg/kg) is the metal content of brown rice, and concentrations of added metals is in mg/kg soil:

$$Y_{Pb} = -0.41 - 0.0011\,(Pb) + 0.11\,(Cd) + 0.0086\,(Cu) + 0.0047\,(Zn) \quad (1)$$
$$R^2 = 0.71\ (p<0.001)$$

or

$$Y_{Pb} = 0.15 + 0.000012\,(CdxCuxZn) + 0.000033\,(CuxZn) \quad (2)$$
$$R^2 = 0.88\ (p<0.001)$$

$$Y_{Cd} = -3.91 - 0.0085\,(Pb) + 1.19\,(Cd) + 0.044\,(Cu) + 0.028\,(Zn) \quad (3)$$
$$R^2 = 0.76\ (p<0.001)$$

or

$$Y_{Cd} = -0.32 - 0.000041\,(PbxZn) + 0.0084\,(CdxZn) + 0.00025\,(CuxZn) \quad (4)$$
$$R^2 = 0.86\ (p<0.001)$$

$$Y_{Cu} = 7.67 + 0.085\,(Cu) + 0.035\,(Zn) \quad (5)$$
$$R^2 = 0.68\ (p<0.001)$$

Table 3. Concentrations of Pb, Cd, Cu, and Zn in rice grain under various treatments with three replicates (Zheng et al., 1990)

	Average concentration (mg/kg)			
Treatment no.	Pb	Cd	Cu	Zn
1	0.34	0.26	9.76	35.31
2	0.35	2.20	18.18	45.50
3	2.05	12.15	22.84	81.98
4	0.64	1.90	20.39	57.10
5	0.45	1.76	16.85	43.05
6	0.52	3.89	15.44	46.04
7	0.62	1.07	19.84	43.71
8	0.47	1.73	13.94	41.71
9	0.30	1.43	13.14	38.62

or

$$Y_{Cu} = 4.36 + 2.07\,(Cd) + 0.050\,(Cu) + 0.043\,(Zn) + 0.00014\,(PbxZn) - 0.00007\,(PbxCdxZn) \quad (6)$$

$$R^2 = 0.81\ (p<0.001)$$

$$Y_{Zn} = 21.93 - 0.034\,(Pb) + 2.54\,(Cd) + 0.20\,(Cu) + 0.14\,(Zn) \quad (7)$$

$$R^2 = 0.76\ (p<0.001)$$

or

$$Y_{Zn} = 35.93 + 0.0028\,(CuxZn) - 3.67 \times 10^{-6}\,(PbxCuxZn) \quad (8)$$

$$R^2 = 0.81\ (p<0.001)$$

In order to stress the importance of combination and interaction among metals, simple regression analysis was carried out and results can be expressed as follows:

$$Y_{Pb} = 0.95 - 1.08 \times 10^{-3}\,(Pb)\ r = 0.32\ (p>0.05) \quad (9)$$

$$Y_{Cd} = 0.63 + 1.19\,(Cd)\ r = 0.54\ (p<0.01) \quad (10)$$

$$Y_{Cu} = 11.74 + 0.085\,(Cu)\ r = 0.65\ (p<0.01) \quad (11)$$

$$Y_{Zn} = 31.40 + 0.14\,(Zn)\ r = 0.62\ (p<0.01) \quad (12)$$

Equations (1)–(12) show very clearly that the values calculated by models including interaction of metals fit very satisfactorily with measurements and were much better than results obtained by simple regression equations. This mean that the equations (2), (4), (6) and (8), which include interactions, can closely predict the actual contents of heavy metals in brown rice. From the view of environment protection, these results are of practical significance because metals such as Pb and Cd are dangerous environmental pollutants that cause several diseases in man and animals. As an example, the values of Cd calculated with equation (10), (3) and (4) are 42.8, 89.2 and 99.5% of the measurements, respectively. Obviously, countermeasures for controlling Cd pollution based on the simple regression equations may be inefficient.

3. INTERACTION BETWEEN HEAVY METAL IONS AND SOIL COMPONENTS

3.1. Effect of Clay Mineral and Oxide on Hydrolysis of Pb

The characteristics of Pb hydrolysis are affected by adsorption of clay minerals and metal oxides. The hydrolysis of Pb can be expressed as follows (Lindsay, 1979):

$$Pb^{2+} + H_2O = Pb(OH)^+ + H^+ \qquad pk_1 = 7.7 \quad (13)$$

There is a good linear relationship between $\log R_d$, and pH, where the distribution ratio (R_d) of logarithm and R_d is equal to [%sorbed/(100-%sorbed)]. If the slope of the linear relationship was changes around pH 7.7, it may easily be understood due to the change of ionic valence. However, experimental results showed an inflecting point around pH5.8 when Pb was adsorbed to montmorillonite, and around pH4.3 when Pb adsorbed to hematite. Thus the hydrolysis character of Pb was markedly affected by hematite and montmorillonite.

3.2. Effect of Active Si, Fe, Al, Mn and Available Si on Behavior of Cd

Active Si, Fe, Al, Mn in soil (extracted by $H_2C_2O_4$-$(NH_4)_2C_2O_4$, pH3.2) influence composition of soil solution, leaching of soil nutrients, plant growth and behavior of pollutant directly and indirectly. Experimental results showed that high contents of active Fe, Al, Mn significantly decreased the absorption of Cd by soil and thereby increased the concentration of Cd in soil solution and the uptake of Cd by plants. In contrast, active and available Si in soil (extracted by 0.5 mol/l HCl) increased the adsorption of Cd by soil and thus decreased Cd content in soil solution and plant. Furthermore, experimental results showed the Cd content in wheat grain, wheat straw, and Chinese cabbage has a relationship with available Si content in soil as follows (Zang et al., 1987; 1989):

$$Y_1 = a_1 X^{b1} \quad r = -0.77^{*} \tag{14}$$

where Y_1 is the Cd content (mg/kg) in wheat grain, X is the available Si (mg/100g soil) content in soil, $\log a_1 = 8.41$, $b_1 = -1.22$.

$$1/Y_2 = a_2 + b_2 X \quad r = 0.77^{**} \tag{15}$$

where Y_2 is the Cd content in wheat straw, X is the available Si content in soil, $a_2 = 8.41$, $b_2 = -1.22$.

$$1/Y_3 = a_3 + b_3 X \quad r = 0.91^{**} \tag{16}$$

where Y_3 is the Cd content in Chinese cabbage, X is the available Si content in soil, $a_3 = 0.026$, $b_3 = 5.47 \times 10^{-4}$. Equations (14)–(16) indicate that increase of available Si decreases the Cd concentration in plant significantly.

4. CONCLUSION

The chemical behavior of heavy metals in the soil- water-plant system is affected significantly by the interactions between elements and soil components. Unfortunately, the effects of interactions of heavy metals on the criterion estimates have not been fully considered. Results of the present research suggest that the interaction of heavy metals added to soil must be considered in the estimation of environmental standards.

ACKNOWLEDGMENTS

This study was supported by the National Natural Science Foundation of China and Laboratory of Material Cycling in Pedosphere, Institute of Soil Science Academia Sinica.

REFERENCES

Chen, H.M., C. R. Zheng and X. H. Sun, 1991a, Effect of anions on adsorbality and extractability of lead added in soil. Pedosphere. 1:51–62.

Chen,H.M., C. R. Zheng and X. H. Sun,1991b, Effects of different lead compounds on growth and heavy metal uptake of wetland rice. 1:253–264.

Chen,H.M. and C.R.Zheng, 1992, Discussion about loading capacity of soil for pollutants. Acta Pedologica Sinica (in Chinese). 29:219–225.

Chen,H.M. and C.R.Zheng, 1994a, Effect of interaction among Pb,Cd, Cu and Zn on yield and metal uptake of rice. PEDOSPHERE (edited by Zhao Qiguo), pp273–281, Nanjing University Press.

Chen, H. M. and C. R. Zheng, 1994b, Effect of interaction on plant growth and elements cycling. Progress in Soil Science (in Chinese), 22:47–49.

Du, R. J. 1985, Statistical Analysis for Biology (in Chinese). pp164–178, Higher Education Press, Beijing.

Koeppe, D. E., 1981, Lead: understanding the minimal toxic of lead in plants. In L. W. Lepp (ed.) Effect of Heavy Metal Pollution on Plants. Vol.I. App. Sci. Pub., pp55–66,London and New Jersey.

Lindsay, W. L. 1979, Chemical Equlibria in Soil. pp331, John Wiley & Sons, New York, Chichester, Brisbane, Toronto.

Mott, C.J.B., 1981, Anion and ligand exchange. In Greenland D.J. and Hayes M. H. B., (eds.) Chemistry of Soil Process. pp179–219, John wiley and sons.

Ohno, T. and D. L. Grunes, 1985, Potassium - Magnesium interactions affecting nutrient uptake by wheat forage. Soil Sci. Soc. Am. J. 49:685–690.

Zang, H. L., C. R. Zheng and H. M. Chen, 1987, Study on controlling cadmium absorption by crops and cadmium-cotamina tion soil. Agro-Environmental Protection (in Chinese),6(3): 28–29, 27.

Zang, H. L., C. R. Zheng and H. M. Chen, 1989, Study on the inhibition of cadmium absorption by crops in cadmium-contaminated soil (II). Agro-Environmental Protection (in Chinese) 8(1): 33–34.

Zheng,C.R. and H. M. Chen, 1990, Transfer of heavy metals through soil-plant system and its influence on the growth of rice. Acta Scientiae Circumstantiae (in Chinese). 10:146–152.

35

OBJECT-ORIENTED SIMULATION OF PATHWAYS OF CHEMICALS IN NATURE

Barbara Zwerger, Mahshid Sotoudeh, and Andreas Windsperger

Research Institute of Chemistry and Environment
Getreidemarkt 9/191, A-1060 Wien, Austria

ABSTRACT

In this study a basic modular (object-oriented) model for the dispersion of chlorinated substances in the environment was simulated with the OOSimulation tool named ASPECT. The model was tested by simulating a substance input with varying amounts of different substances and varying the properties of the environmental compartments. The simulations proved the model to be valid and showed that it can help to a better understanding of the pathways of substances in the environment.

1. INTRODUCTION

Environmental protection only can be practised if we understand the processes governing the pollution of the environment. To have a successful pollution control it is very important to know where the chemicals accumulate. The environment is a very complex system with a variety of different subsystems. For studying the pathways of chemicals in the environment the use of models and simulation-programs can help to understand the complex processes. A computer-program can support the modelling. But the commonly used simulation-programs are seldom flexible enough to respond to new scientific results in the field of impact assessment. It is very young and still developing, and marked by the difficulty in understanding natural systems as they are not comparable to technical processes because each process influences the other. A computing tool must be able to manage user-defined models and offer the user structural help for his research. A computing tool can be used to test new ideas at the first stages of research and develop the models after evaluating them step by step. The software-package ASPECT is an object-oriented simulation tool which has been designed at our institute to satisfy these requirements.

Chemistry for the Protection of the Environment 2, edited by Pawłowski et al.
Plenum Press, New York, 1996

2. MATERIAL AND METHODS

2.1 Analysis and Simulation of Processes with an Efficient Computing Tool—ASPECT

The software ASPECT was developed at our institute for the modelling and simulation of environmental and industrial processes, as well as for substance flow analysis so that it can be considered as a tool for the teaching and understanding of these processes [SOTOUDEH, 1994], [SOTOUDEH, 1995]. With the help of ASPECT we studied a basic object-oriented model for the dispersion of chlorinated substances in the environment.

2.2 Modelling

The main purpose of modelling the dispersion of chlorinated substances in the environment in this study was to compare the behaviour of these substances related to their physical and chemical properties. To design a basic model we chose some environmental compartments as model-elements, namely Atmosphere, River, Groundwater, Lake, Soil1, Soil2, Sediment1 and Sediment2. For starting we considered the dispersion of substances by water and vapor through the various compartments. The substances chosen were 2,4-di-chlorophenoxy-acetic acid (2,4-D), which is a solid at 25°C, chloro-ethene (or vinyl-chlorine VC), which is a gas and finally tetrachloro-ethene (PER), which is a liquid.

2.3 Description of the Model

See Figure 1.

2.4 Assumptions (in This Phase of Modelling)

- The model is studied in a continuous state. A constant input of chemicals is assumed.
- Before the application of the substance there is no substance present in the systems.
- We neglected particle and aerosol presence of the compounds in the atmosphere.
- Except in the atmosphere all systems are studied under stationary conditions.
- Microbial degradation only takes place in some systems, namely Groundwater, Lake, Soil1, Sediment1 and Sediment2, and follows first order kinetics.
- Adsorption occurs in Soil and Sediment irreversibly (there is no desorption).
- Microbial degradation and adsorption in the Soil and Sediment are parallel processes.
- The reference system for the data is Austria.

2.5 Structure and Behaviour of the Model

2.5.1. Atmosphere. We assumed a direct substance input into Atmosphere. A part of the substance leaves the system by rain. From the compartments, which are adjacent to the Atmosphere, there can occur a backflow of substance due to evaporation.

We described the atmosphere in this first phase as a totally mixed system with no microbiological degradation and no chemical or photochemical reactions. This mixed behaviour is valid regarding the lower layers of atmosphere due to the thermal conditions and the wind.

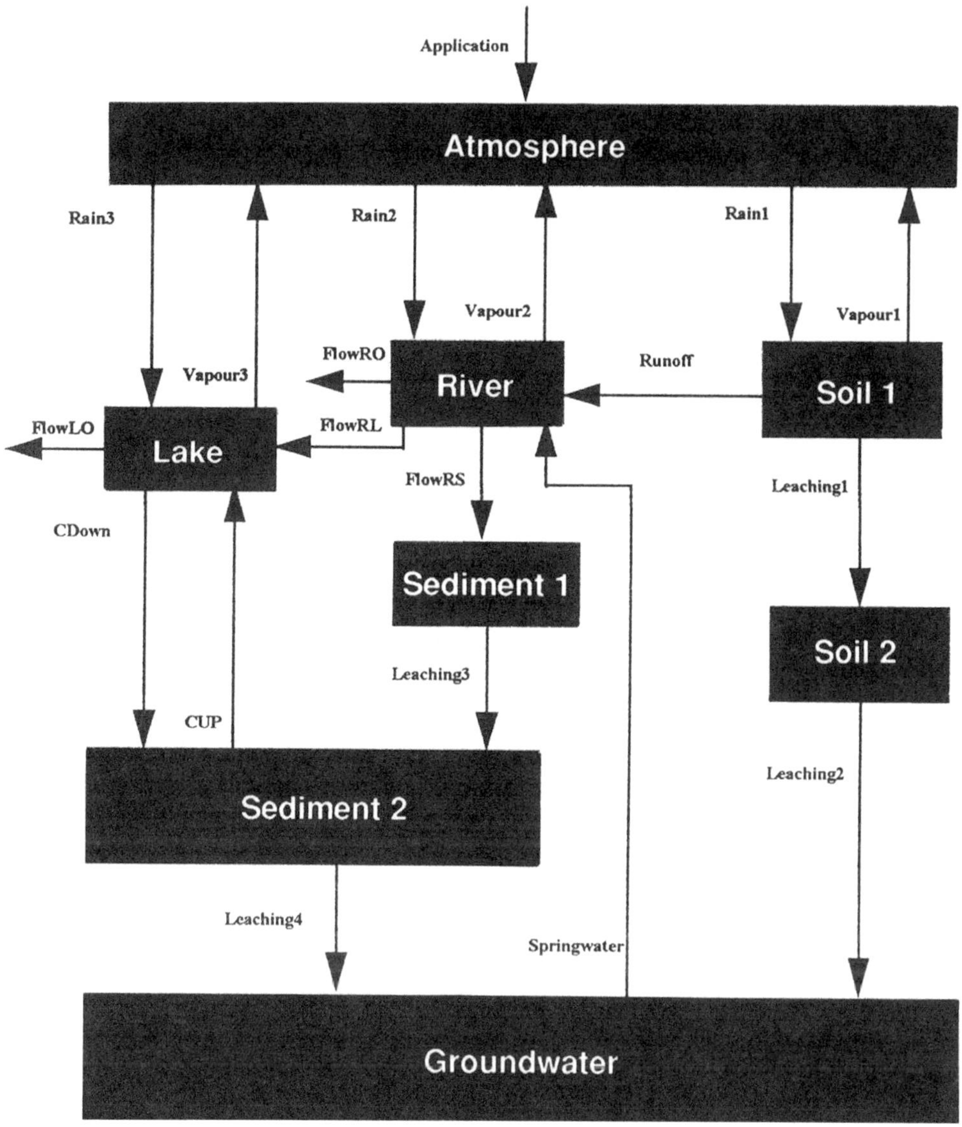

CDown: Downside Circulation
CUP : Upside Circulation
FlowLO : Flow from Lake to Surroundings
FlowRL : Flow from River to Lake
FlowRO : Flow from River to Surroundings
FlowRS : Flow from River to Sediment1

Figure 1. Structure of the Model [Zwerger, 1993]. The pathways of pollutants in the environment.

Table 1. Properties of the substances used in our model

	2,4-D	VC	PER
Solubility in the liquid phase [kg/m^3]	0.68	1.6	0.14
Vapor pressure of the pure substance [atm]	$1.38*10^{-5}$	0.714	0.0227
Molecular weight [g/mole]	221	62.5	165.83

The dominating process is the transfer of the substance from the gas phase into the liquid phase, which we described with Henry's law and the ideal gas equation. As we considered a direct application into the atmosphere and as Henry's law is only valid for dilute solutions (till saturation is reached) we had to consider two cases. In case one the input is higher than the maximum output, when the rain is saturated with the respective substances. In this case the concentration of the substance in the gas phase is independent of the concentration in the rain. The rain is saturated with the substance and therefore the atmosphere output is constant. It is only related to the vapor pressure and solubility of the pure substance. The concentration in the gas phase is described according to the ideal gas equation. The other case is, when the input is lower than the maximum output, where a distribution according to Henry's law was used.

2.5.2. River. The substance leaves the Atmosphere with the rain, which evaporates partly at the surface of the system. The rest of the rain enters the River and increases the concentration of the substance. Additionally to the rain there is a substance input from the Groundwaterflow, which we called Springwater, and a substance input from Runoff from the surface of the soil. The output from the River directly enters the Lake. A vertical flow goes to the Groundwater through the Sediment layers. To keep the volume of the system constant, even by high amounts of rain, we added a flow which ends openly (enters no system in our model).

The compartment River represents only the upper part of a river, where we consider turbulent conditions and for this reason we modelled it as an ideal stirred tank reactor without microbial reaction. The lower parts and the laminar region at the side, which are

Table 2. List of systems

Atmosphere	*input streams:* Application, Vapor1, Vapor2, Vapor3 *output streams:* Rain1, Rain2, Rain3 *behaviour:* totally mixed, no microbial activity
River	*input streams:* Rain2, Runoff, Springwater *output streams:* Vapor2, FlowRO (Flow from River to Surroundings), FlowRL (Flow from River to Lake), FlowRS (Flow from River to Sediment1) *behaviour:* totally mixed, no microbial activity
Groundwater	*input streams:* Leaching2, Leaching4 *output streams:* Springwater *behaviour:* totally mixed, microbial activity
Lake	*input streams:* Rain3, FlowRL (Flow from River to Lake), CUP (Upside Circulation) *output streams:* Vapor3, FlowLO (Flow from Lake to Surroundings), CDown (Downside Circulation) *behaviour:* totally mixed, microbial activity, aerob
Soil1	*input streams:* Rain1 *output streams:* Vapor1, Runoff, Leaching1 *behaviour:* thin layer, plug flow characteristics, adsorption, microbial activity, aerob
Soil2	*input streams:* Leaching1 *output streams:* Leaching2 *behaviour:* plug flow characteristics, adsorption
Sediment1	*input streams:* FlowRS *output streams:* Leaching3 *behaviour:* thin layer, plug flow characteristics, adsorption, microbial activity, aerob
Sediment2	*input streams:* Leaching3, CDown (Downside Circulation) *output streams:* Leaching4, CUP (Upside Circulation) *behaviour:* plug flow characteristics, adsorption, microbial activity, anaerob

not very well mixed, were considered as the real active section, due to high cell density and low residence time (Sediment1). Compared with that activity we neglected microbial reactions in the mixed regions. The concentration of the substance in the vapor from the surface of the system River was described by Henry's law.

2.5.3. Groundwater. The water leaving the Soil layers and the Sediments enters the Groundwater. There is an output which directly enters the River. We modelled the system Groundwater as a mixed system. Although the microbial density should be extremely low, we assumed microbial activity according to the very long residence time.

2.5.4. Lake. In analogy to the River we divided the Lake compartment into a well and poorly mixed region. The system Lake represents the upper well mixed region and can be modelled as a stirred tank reactor. Additionally a distinct aerobic microbial activity was assumed.

The substance enters the system Lake through the rain from the Atmosphere and through River. One part of the rain evaporates at the surface of the Lake before entering the system. One part of the water of the Lake will enter systems out of our balance frame (open flow) and can be considered as an overflow to keep a constant system volume, another part will enter the deeper poorly mixed zones (Sediment2). In Sediment2 only anaerobic reaction was assumed (same modelling as anaerobic zones in river). As there is a close relation between the upper, well mixed parts of a lake and the lower parts, the sediments, we took into account a backflow from the sediments. This backflow is the part of the flow which does not enter the deeper parts of the sediments.

2.5.5. Soil 1 and Soil2. For modelling we split the soil in two parts with different properties. Due to its structure the soil behaves as a plug flow reactor. The substance enters the upper soil layer by the rain. At the soil surface there occurs on one hand an evaporation of the rain water, on the other hand a run off to River. The water entering the system will also pass deeper layers till it reaches the Groundwaterflow.

The adsorption and microbial degradation in the upper soil layer (Soil1) is not so easy to handle as these are competing processes. We considered adsorption and microbial reaction as parallel processes, as particles adsorbed by soil components cannot undergo microbial degradation. For describing the adsorption we used the Langmuirn isotherm. We neglected the desorption and considered the number of already occupied adsorbent sites as very small, so that we could set it equal to zero.

2.5.6. Sediment1. Sediment1 represents the microbial active part of a river, these are the lower layers and the side regions together with its sediments. It was modelled according to Soil1. Sediment1 is only a very thin layer and therefore microbial degradation is considered to be aerobic . Only water from the River (FlowRS) enters this system. The flow which leaves this system enters Sediment2.

2.5.7. Sediment2. Sediment2 combines anaerobic reactions with simultaneous adsorption. It represents the deepest layers of the sediments of River and Lake, although these systems are locally separated. A flow from the River, which has already passed aerobic degradation in Sediment1, and a flow from the Lake enter this system. The water which passes this system enters the Groundwater. In Sediment2 an adsorption and an anaerobic degradation of the substance occurs. The modelling is analogous to Soil1 and Sediment1.

3. RESULTS OF SIMULATION

For simulation we first challenged the behaviour of the single systems. Each system was tested separately by varying first the input concentrations of a model substrate (2,4-D) and afterwards some system parameters (as the density or thickness of the soil).

At the end, the systems were assembled together and the whole process was tested with different chlorinated substances (2,4-D, VC and PER).

3.1 Atmosphere

If the input is higher than the maximum output (case1) the substance concentration in the gas phase is increasing linearly with time, whereas in the liquid phase the concentration of the substance has already reached its saturation. With respect to the gas phase the system is instable, but with respect to the liquid phase the Atmosphere is stable and we obtain a constant substance input to the systems below.

If the input is lower than the maximum output (case2) the system reaches a final stable value in the gas phase as well as in the liquid phase. This final value can be smaller than the maximum solubility or can be the maximum solubility of the substance in the liquid phase. As we had obtained with the time varying input into the systems below we considered the condition after an indefinitly long time where saturation is reached and a stable input in the other systems is guaranteed.

A substance with good solubility and a very low vapor pressure (2,4-D) prefers to remain in the liquid phase. Relatively highly volatile substances (such as VC and PER) re-

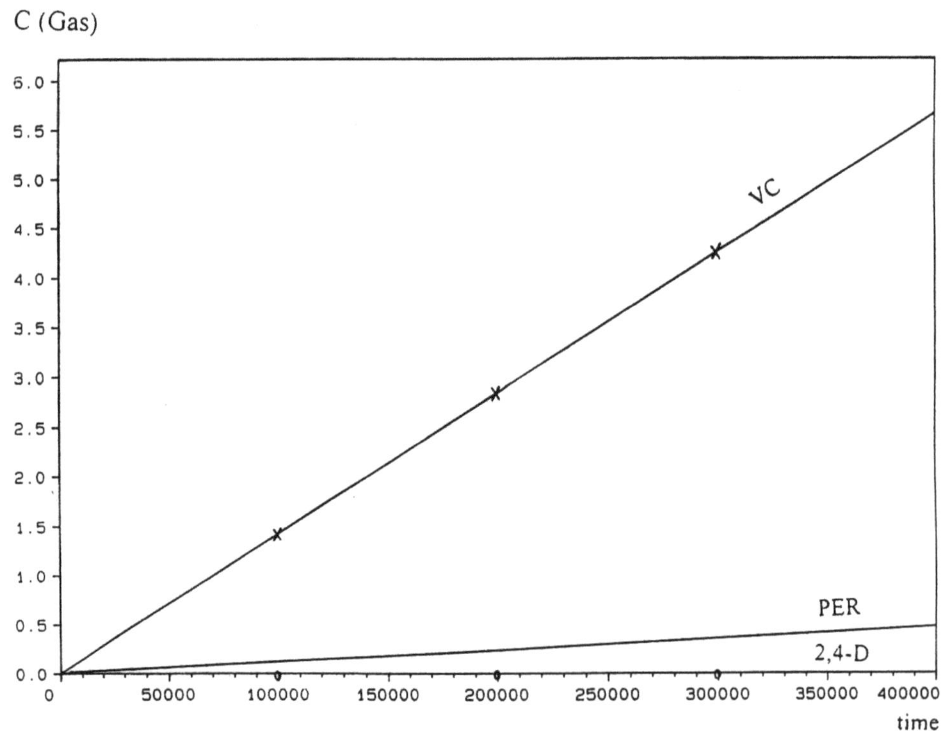

Figure 2. Behaviour of 2,4-D, VC and PER in the gas phase.

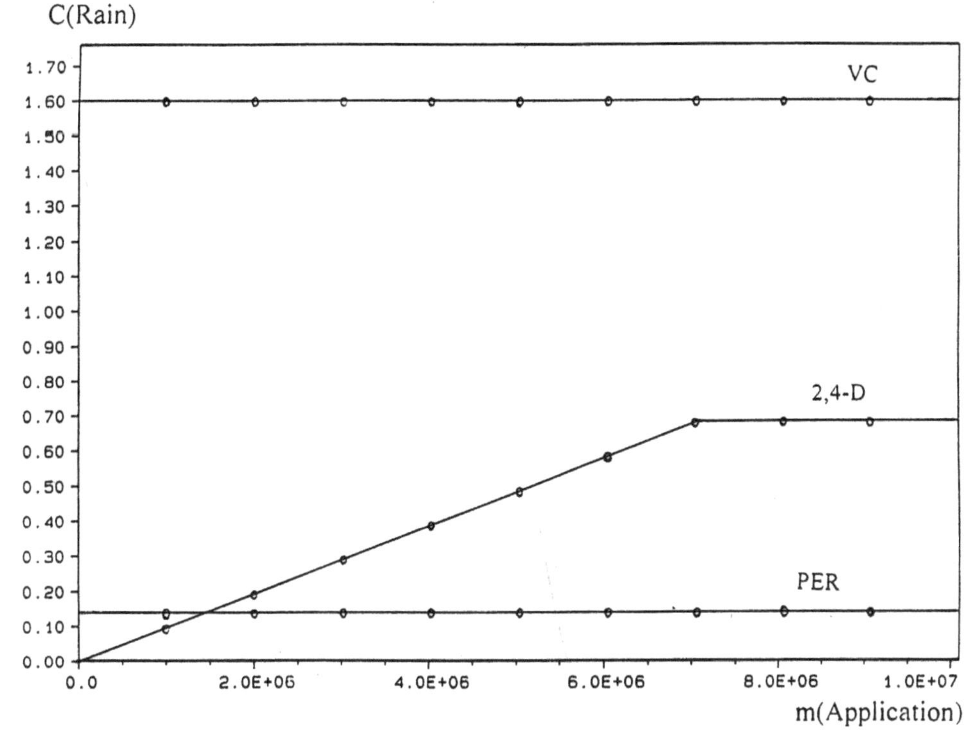

Figure 3. Behaviour of 2,4-D, VC and PER in the rain.

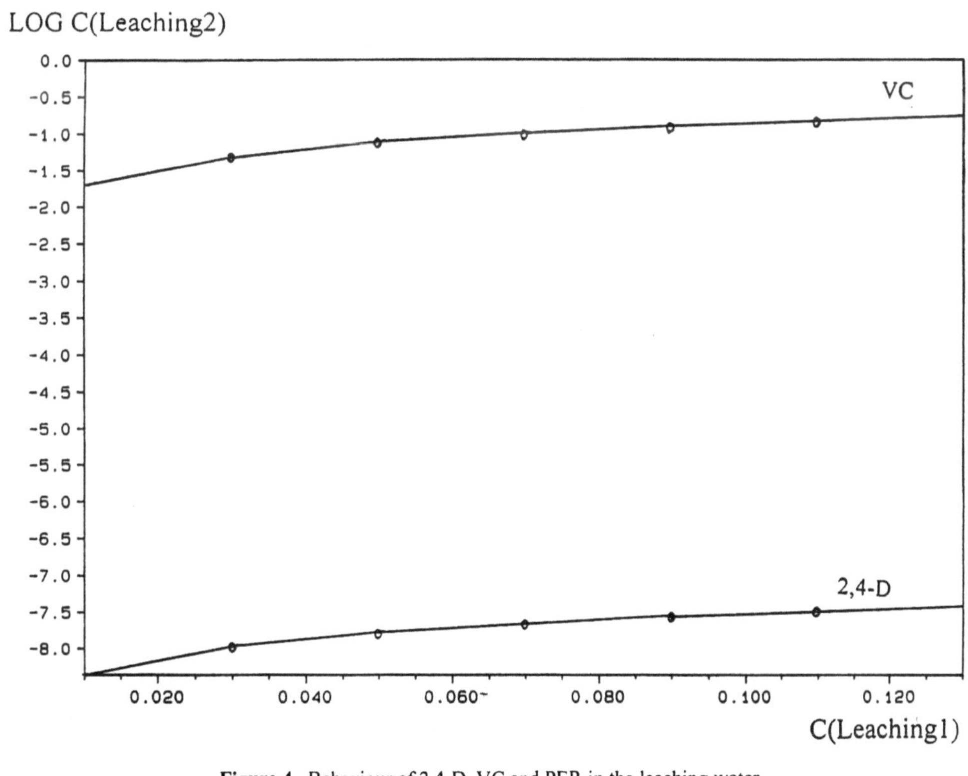

Figure 4. Behaviour of 2,4-D, VC and PER in the leaching water.

main preferably in the gas phase at nearly the same amount as which leaves the system with rain and will be recirculated with the evaporation. Due to its high vapor pressure VC reached high concentrations in the gas phase quicker than PER, although the solubility of VC in the liquid phase is much higher than the solubility of PER. This showed the influence of the relation between the vapor pressure and the solubility.

3.2 Soil2

The challenging of the systems showed expected results. An increase in thickness of the soil increased the accumulation. An increasing water content increased the wash out of substances and decreased the adsorption. Increasing the substance charge led to a minimisation of the effects of the soil parameters. Substances with low adsorption constants and low degradation constants ,like VC , passed the soil column and reached the Groundwater. The higher the adsorption constant the higher the accumulation in the system.

3.3 Combination of the Systems Lake, River, Sediment1, Sediment2 and Groundwater

By combining the systems we found that we cannot determine the substance concentration in the system groundwater with the mass balances which we had already written. The reason was a closed loop between the systems River, Lake, Sediment1, Sediment2 and Groundwater. Therefore the concentration of the substance for the system Groundwater must be calculated by a TRY AND ERROR method.

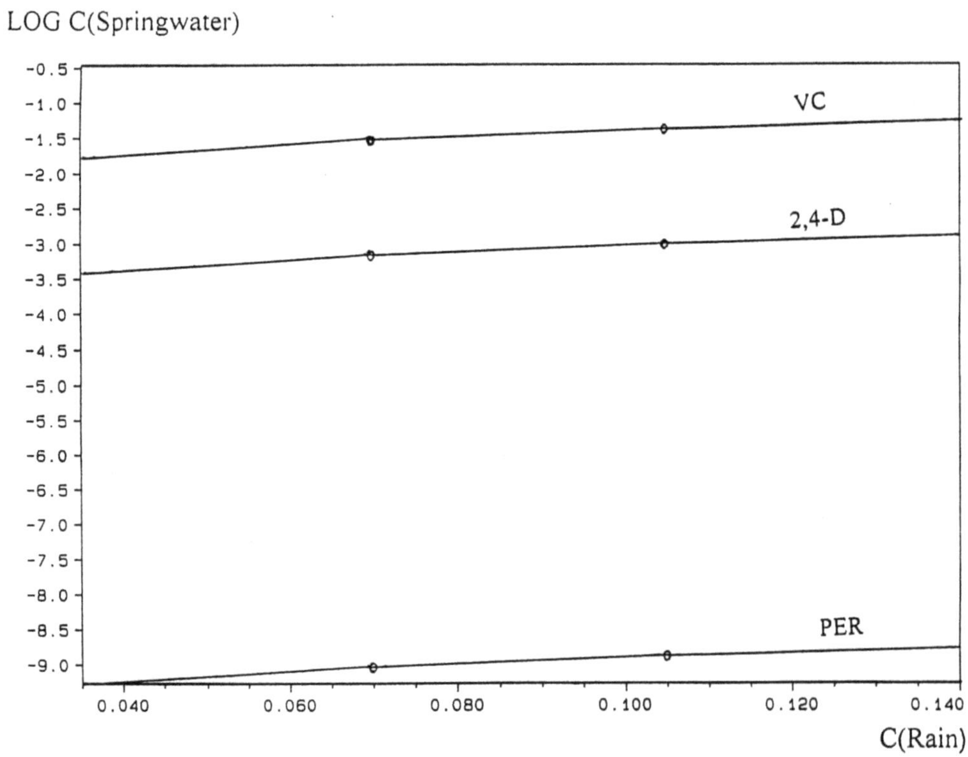

Figure 5. Behaviour of 2,4-D, VC and PER in the Springwater.

Regarding the adsorption potential and the degradation rate of the substance, the sediment layers had a greater influence on the concentration of the substance in the Groundwater than the systems River and Lake. Substances with low adsorption constants and low degradation rates could be found therefore preferably in Groundwater.

3.4 Simulation of the Whole Process (Combination of All Environmental Compartments)

The system combination was only tested with the substance 2,4-D, as for an application of VC and PER the model must be modified considering an application directly into the water and into the soil. For the combination of the systems we studied the behaviour of the systems after the concentration in the Atmosphere had reached stable values so that we had a stable substance input into the systems Lake, River, Soil, Sediment and Groundwater.

In the Lake the concentration of the substance did not reach the threshold values, but did it in the system River. This is due to our modelling, where we considered a direct microbial degradation in the system Lake, but described an additional system (Sediment1) for the microbial activity of the system River. The Groundwater showed due to the high accumulation potential of the soil column and the sediments low substance concentrations, which laid below the limited value for this substance in drinking water.

4. DISCUSSION AND CONCLUSIONS

By applying different substances it is possible to discover the influence of physical and chemical parameters on the dispersion of substances in the environment. Vapor pressure and solubility predict where a substance reaches high concentrations, as could be seen in the system Atmosphere.

Comparing the distribution of a substance in the different environmental compartments indicates where a substances will accumulate (for example more soluble substances will accumulate in the liquid phases and will be found mainly in the water basins , like Lake and River). It helps us to find the threshold input for substances with defined physical chemical properties that drive at least one compartment to instability. This shows us at the same time the danger potential of the substances. As it could be seen for VC and PER, highly volatile chlorinated substances lead at low application rates to instability in the Atmosphere.

Additionally the model allows simulation of different environmental conditions and their effects. One can vary the volumes of the systems, the height of the soil columns and follow the effect of change in one compartment on the concentration of the substance in the other compartments.

We are now improving the model regarding also photochemical reactions in the gas phase and evaporation not only from the surface of the soil but also from the inside layers. Additionally we are also considering the application of the model for other substances.

REFERENCES

[SOTOUDEH, 1994]: Sotoudeh M., Windsperger A., Object-Oriented Simulation of Chemical Processes, ACHEMA 94, Abstracts of the lecture groups Process Engineering, Thermal Process Engineering

[SOTOUDEH, 1995]: Sotoudeh M., Windsperger A., Object-Oriented Modelling and Simulation in Environmental Engineering, Proceedings of the Workshop New Trends in Environmental Education and Training, May 15–17,1995, Vienna, Austria

[ZWERGER, 1993]: Zwerger B., Bestandsaufnahme Stofflicher Wechselwirkungen zwischen Industrie und Umwelt an Beispielen, Master Thesis TU Wien, Institute of Chemical Engineering

ADSORPTION AND DESORPTION OF CADMIUM, COPPER, AND LEAD IN POLISH MINERAL SOILS

Ryszard Gąszczyk and Tadeusz Paszko

Department of General Chemistry
University of Agriculture
20–033 Lublin, Akademicka 15, Poland

1. INTRODUCTION

Soils with different physicochemical properties will adsorb/desorb metals to different degrees. Investigation of desorption of metals in soils with different properties may give useful information for assessing consequences of heavy metal contamination.

Cations possesing specific energies for entering the soil sorption complex (SSC) are combined with it to some extents. These depend, among other things, on the ion radius, chemical affinity, kind of the deposition ion, soil pH, content of organic substance, and the structure of soil minerals. It was, therefore, possible to determine the susceptibility of soil to poisoning with heavy metals by studying the course of desorption proccsscs in soil of know physico-chemical composition using exchangeable cations.

The purpose of this research was to investigate the chemical affinity of the three heavy metals: copper, cadmium, and lead introduced together as a chloride solution into the soil material as well as to determine how these metals are washed out by a solution of by barium cations.

2. MATERIALS AND METHODS

Three kinds of mineral soils were used for these studies: grey-brown podzolic soil formed from a weakly loamy sand (Podzol Luvisols), grey-brown podzolic soil formed from loess (Orthic Luvisols), and brown soil formed from heavy loam (EutricCambisols); they had different chemical properties and granulometric compositions[1,2]—Table 1.

To find the level of heavy metals (Cd, Cu, Pb) before the experiment soil samples were dissolved in a mixture containing spectrally pure, concentrated hydrofluoric, nitric (V), and chloric (VII) acids in the ratio 4:3:1. Solid rests were dissolved in the diluted

Chemistry for the Protection of the Environment 2, edited by Pawłowski et al.
Plenum Press, New York, 1996

Table 1. Some properties of soils used in the experiments

Soil	Percent of soil size fractions (mm)				$N_{tot.}$ (%)	$C_{org.}$ (%)	pH		Specific surface area (m^2/g)	Available forms (mg/kg soil)		
	1.0–0.1	0.1–0.02	0.02–0.002	<0.002			H_2O	KCl		P_2O_5	K_2O	Mg
Grey-brown podzolic soil from weakly loamy sand	80	14	2	4	0.04	0.50	5.5	4.7	13	12.8	10.7	1.0
Grey-brown soil from loess	1	63	27	9	0.10	2.13	4.8	4.3	24	40.0	22.7	4.1
Brown soil from heavy loam	29	19	34	18	0.12	1.70	6.2	5.9	59	11.8	6.7	3.4

(1:1) ultra-pure hydrochloric acid. Analytical measurements were carried out by AAS. The total contents of studied metals were the following:

- in sandy soil: Cd - 0.3, Cu - 3.0, Pb - 15.0 mg/kg dry matter,
- in loess soil: Cd - 0.6, Cu - 6.0, Pb - 65.0 mg/kg dry matter,
- in loamy soil: Cd - 1.5, Cu - 19.0, Pb - 20.0 mg/kg dry matter.

The sorptive–desorptive processes of Cd^{2+}, Cu^{2+}, and Pb^{2+} were carried out in columns made of plexiglass in which there was 100 g soil material dried in air; it was saturated with deionized water (ionite–Amberlit, reversed osmosis) with specific conductance 10 $\mu S\ cm^{-1}$ by the means of capillary rise. Then at the top of the column there was introduced the three-component solution: $CdCl_2$, $CuCl_2$, and $Pb(NO_3)_2$ at concentrations $1.67 \cdot 10^{-2}$ M Me^{2+} till sorptive saturation occurring in dynamic conditions[3]. The next operation was washing the soil columns by the water till the heavy metals in the effluent disappeared. Then their desorption from the soil by 0.05 M $BaCl_2$ solution was studied. The effluent from the columns was collected in 10 cm^3 aliquots in which the contents of cadmium, copper, and lead were measured by of AAS at following wavelengths: for Cd - 228.8 nm, for Cu - 324.7 nm, and for Pb - 217.0 nm.

3. RESULTS

After passing 140 cm^3 of the three-component solution $CdCl_2$-$CuCl_2$-$Pb(NO_3)_2$ through the soil columns the following amounts (per cent of input) were sorbed: the sandy soil Pb^{2+} - 46.9%, Cu^{2+} - 35.4%, and Cd^{2+} - 30.4%; the loess soil: Pb^{2+} - 88.1%, Cu^{2+} - 54.9%, and Cd^{2+} - 38.1%, and the loamy soil: Pb^{2+} - 81.6%, Cu^{2+} - 71.2%, and Cd^{2+} - 54.9%. In Figure 1 the area A illustrating the contents of metals in effluent.

The binding of cadmium, copper, and lead depends on the specific surface area of soils. In all the soils the amounts adsorbed decreased in the order:

$$Pb > Cu > Cd.$$

Passing the 140 cm^3 of water through the columns caused washing out of heavy metals in interstitial water and the forms least strongly bond in the soils. From the sandy and loamy soils of cadmium was washed out mosteasily (27.9% and 13.7%, respectively), and from the loess soil—copper (32.4%). These values are presented in Figure 1 as area B.

The desorption process of previously bound heavy metals occurred during passing of 160 cm^3 of barium chloride solution through the columns. The effluent from loamy soil contained 21.2% Cd and 18.5% Cu of the amount added to the columns. In effluents from loess and sandy soils these amounts were less: 15.4% Cu and 2.4% Cd (loess) respectively 4.8% Cu and 0.8% Cd (sand). Lead was washed out in the highest amount from the loess soil (34.6%) and from the loamy and sandy soils 27.7% and 16.4%, respectively. These values are presented by the area C in Figure 1.

The distribution of heavy metals contents in particular portions of effluent is shown in Figure 2: Ba^{2+} cation displaced cadmium in the least amount in sandy soil (area C in Figure 1).

All the above results suggest the contribution of chemical sorption in sandy soil and ion exchange in loamy soil as well as in loess soil for lead and copper[4,5]. Thus, ion-exchange sorptive processes can be connected with large specific surface area of loamy soil and relatively high contents of organic matter in loess soil.

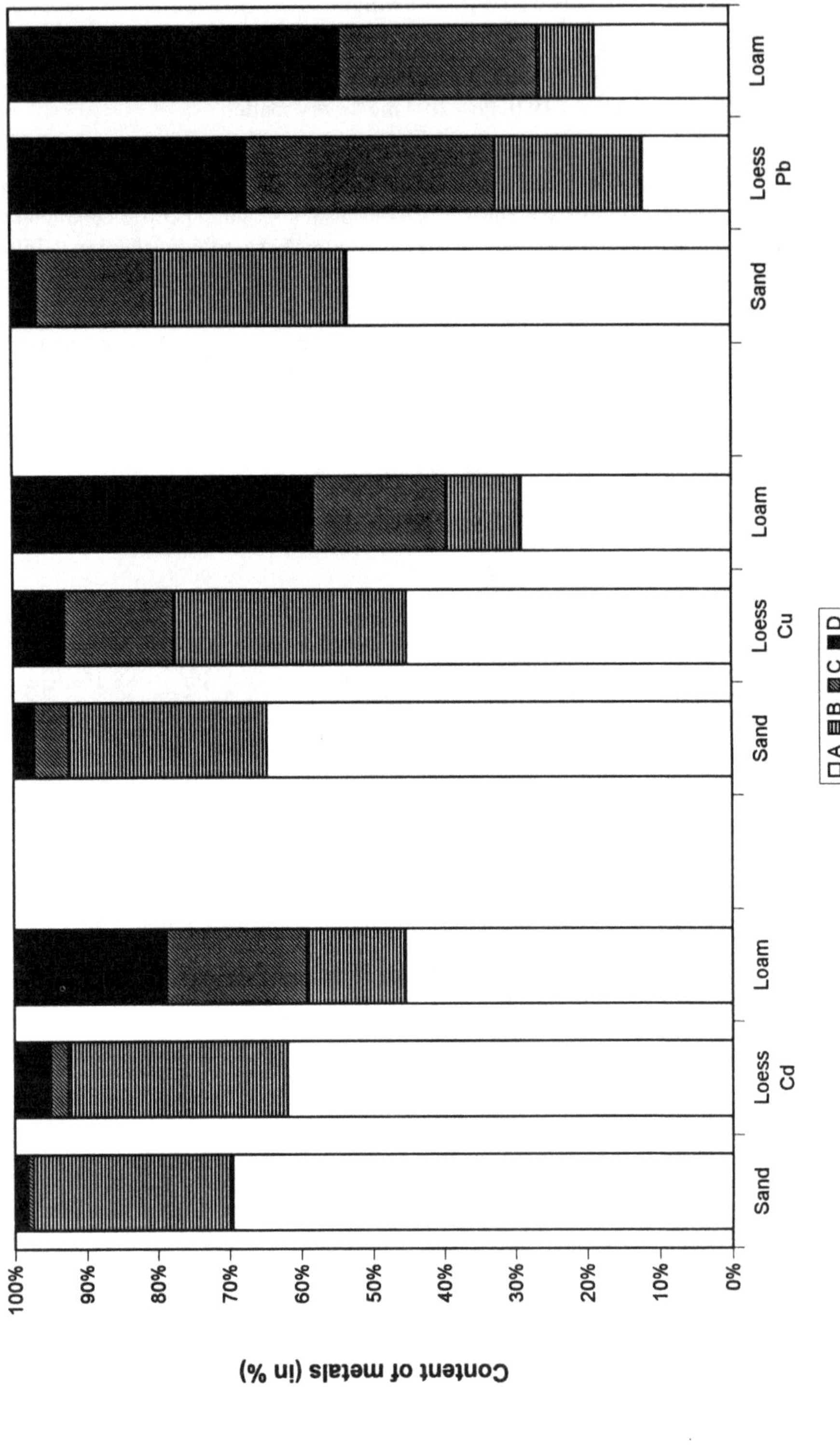

Figure 1. The contents of cadmium, copper, and lead in effluents from the soil columns after the process of: A—saturation, B—washing out by the water, C—desorption by the $BaCl_2$ solution; D—the rest in the soil.

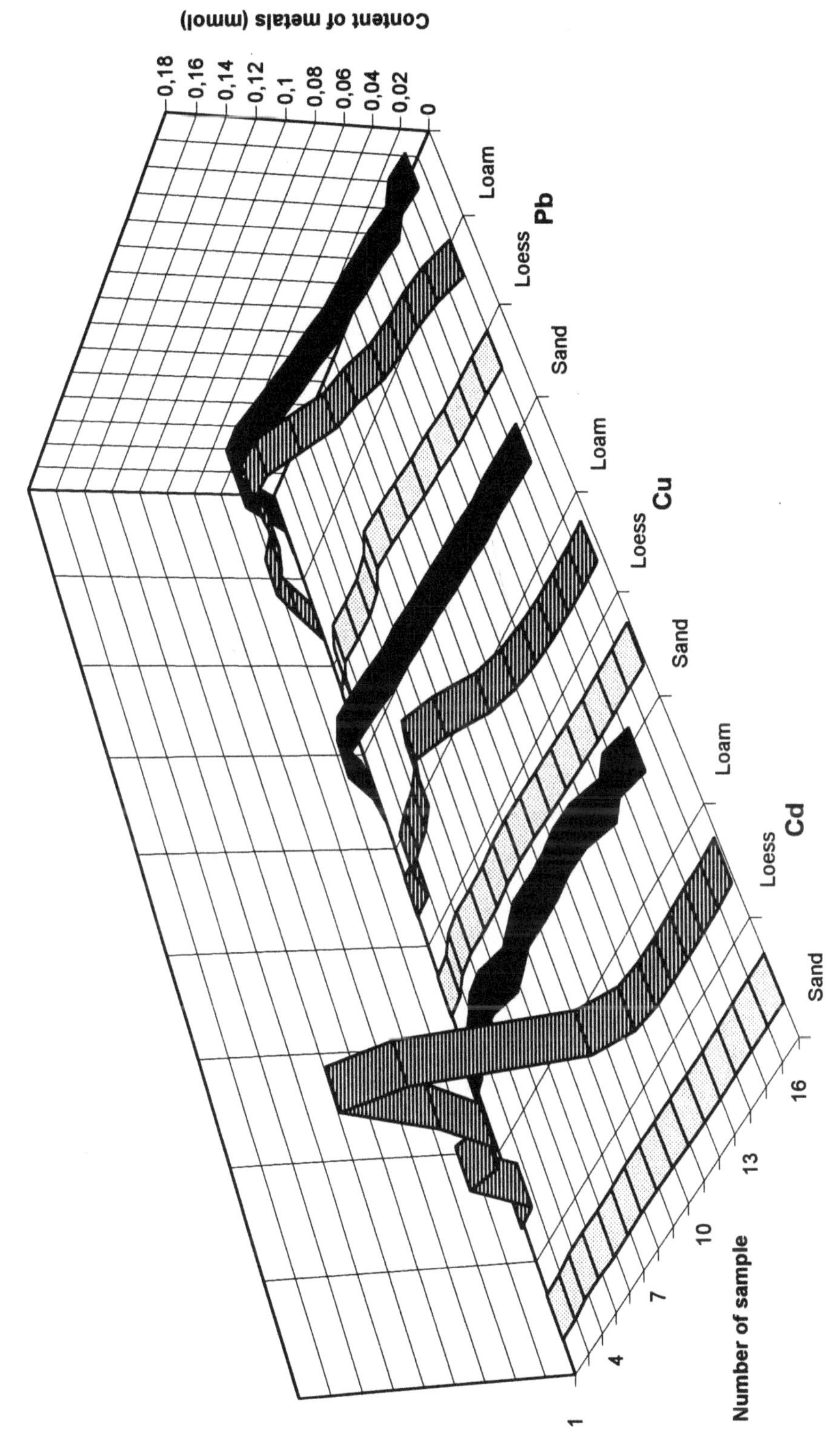

Figure 2. The course of desorption of cadmium, copper (II), and lead (II) cations from the soils as a function of the effluent volume.

The analysis of the curves of leaching of heavy metal cations with a barium ion indicates the participation of forces characteristic to ion-exchange sorption in the case of loessial soil. It can be proved by a large content of cadmium, copper and lead in a 50–90 cm^3 volume of the effluent. In loamy soil the displacement was not that rapid but it included all of the metals.

A part of heavy metals was not removed from the studied soils by the applied desorption methods. It mainly refers to the lead in loamy soil (46.0% in relation to the amount given into the column) and in loess (32.8%) as well as to copper and cadmium in loamy soil (42.1% and 21.2%, respectively). In the other cases persistent sorption was from 1.8% (cadmium in sandy soil) to 7.1% (copper in loess soil)—area D in Figure 1.

On the base of presented results the investigated heavy metals can be lined up into the series due to the influence of washing them out by the barium cation:

- in sandy soil: Pb > Cu > Cd,
- in loess soil: Cu > Pb > Cd,
- in loamy soil: Cd > Cu > Pb.

These series display the interactions of heavy metals taking place in sorptive complexes of investigated soils due to the outer factor—barium cation. The same sequences were obtained by the authors using calcium cation for heavy metals displacement[1].

The elimination of heavy metal cations by exchangeable cations in the above mentioned conditions refers to ion exchangeable sorption which explains their desorpion degrees from soil. However, we shouldn't forget about the part irreversibly adsorbed by SSC which results from such processes as complex formation, hydroxide precipitation, as well as from the phenomena of occlusion and hydrolysis.

The change of physical and chemical conditions in the soil environment (eg soil acidification by roots of plants) may change the balance in the sorptive complex and cause the transfer of heavy metal cations to the soil solution making them available for plants.

The problems discussed in this study belong, according to the criteria applied in surface physicochemistry, to the adsorption with soil sorbents from multicomponent solutions. This approach to the problem, especially in a dynamic aspect, is not frequent in the publications on this subject. Gorlach[4] showed that simultaneous application of heavy metals from solutions undergoing the process of sorption, decreases the process of their desorption, compared to the application of single metals. The established series of inputs is as follows: Pb > Cu > Cd. Basta[5] achieved a similar series when both single metals and multicomponent solutions underwent sorption.

The easiness of cadmium desorption points to the high probability of its filtering into the soil solution whereit may be taken up by plants[6–8]. Another aspect of this mobility in the soils is the possibility of its penetration to the lower soil layers if added to the surface by atmospheric deposition.

4. CONCLUSIONS

Based on the above observations we draw the following conclusions:

1. The investigated soils sorbed in the widest range the Pb^{2+} cation. The level of this sorption depended on the organic carbon contents.
2. The levels of sorption of Cd^{2+} and Cu^{2+} cations depended on the soils' specific surface areas.

3. Barium cation displaced in the widest range the Cd^{2+} and Cu^{2+} cations in the loamy soil, and Pb^{2+} cation in the loess soil.
4. The largest contribution of chemisorption occurred for Pb^{2+} cation in loamy and loess soils, and for Cu^{2+} cation in loamy soil.

5. REFERENCES

1. Gąszczyk R., Paszko T.: Studies on interactions between exchangeable cations and cadmium, copper, and lead in mineral soils. Part I: Effect of calcium cation. Pol. J. Soil Sci., 28, 2(1995), 35–39.
2. Gąszczyk R., Bubicz M., Jackowska I., Paszko T.: Sorptive-desorptive processes of zinc in the soils. Annales UMCS, Sec. E, 50(1995), 191–195.
3. Gąszczyk R., Paszko T.: The investigations of magnesium interaction with cadmium, copper and lead in mineral soils on the dynamic conditions. Biul. Magn., 4(1994), 62–65.
4. Gorlach E., Gambuś F.: A study of the effect of sorption and desorption of selected heavy metals in soils on their uptake by plants. Zesz. Probl. Post. Nauk Rol., 398(1992) 47–52.
5. Basta N.T., Tabatabai M.A.: Effect of cropping system on adsorption of metals by soils. II. Effect of pH. Soil Sci., 153, 3(1992), 195–204.
6. Konig N., Baccini P., Ulrich B.: Effect of organic matter on the translocation of metals between soil and soil solution in an acid forest soil. Z. Pflanzenern. Bodenk., 149, 1(1986), 68–82.
7. King L.D.: Retention of metals by several soils of the Southeastern United States. J. Environ. Qual., 17, 2(1988), 239–246.
8. Ahumada I.A., Schalscha E.B., Pratt P.F., Mattigod S.V.: Extraction by barium salts of heavy metals from two Chilean soils. Soil Sci. Soc. Ann. J., 48, 6(1984), 1240–1243.

37

MOBILIZATION OF HEAVY METALS FROM SOILS BY ACIDIFICATION

Izabella Jackowska

Department of General Chemistry
University of Agriculture
20–033 Lublin, Akademicka 15, Poland

1. INTRODUCTION

Soil is the basic natural environment element and its properties (physical, chemical and biological ones) being shaped due to the soil-forming process are in the dynamic equilibrium. Throwing out of this balance causes the negative consequences for environment.

For many years about 60% of the soils in Poland have been clasified as acidic or very acidic, although in recent years the fraction of very acidic soils has increased[1,5,6,7]. Among the industrial pollutions emitted into the atmosphere, and thus into the soils, the acidic components are in the majority. Many chemical reactions are accompanied the change of soils pH values[2,3,4]. The heavy metals often occur in soils as slightly soluble compounds that, depending on the conditions, can be mobilized from these soils (the mean contents of metals in Polish arable land amount respectively: Cd—0.22, Cu—6.7, Ni—5.5, Pb—13.8, Zn—33.2 mg dry soil)[8].

Since metal pollution is a problem in many areas of Poland and acidification tends to mobilize metals from soils, we have carried out the laboratory experiments to study the mobilization of Cu, Cd, Cr, Co, Fe, Mn, Ni, Pb, and Zn under various conditions.

2. MATERIAL AND METHODS

The studies were carried out using three typical Polish soils taken from the A_1 horizon:

- brown-grey developed from loamy sand,
- brown-grey developed from loess,
- brown developed from heavy loam.

The sandy soil was characterized by low content of clayes fraction (6%), low humus content (0.5%) as well as small specific surface area (13 m^2/g). The loess soil contained

Chemistry for the Protection of the Environment 2, edited by Pawłowski et al.
Plenum Press, New York, 1996

Table 1. Total content of metals in soils [mg / kg dry mass]

Soil	Fe [%]	Mn	Ni	Co	Cu	Cr	Zn	Pb	Cd
Brown-grey developed from loamy sand	0.35	183	10	1.6	3.0	10	16	15	0.3
Brown-grey developed from loess	0.70	300	16	8.1	6.0	32	35	65	0.6
Brown developed from heavy loam	2.90	226	48	16.0	19.0	68	227	20	14
The lower detection limit(μg/ml)	0.5	0.2	0.2	0.2	0.1	0.2	0.05	0.1	0.1

much more clayes fraction (36%) and humus (2.1%). Its specific surface area was 24 m^2/g. In loamy soil, the clayes fraction amounted 52%, humus 1.7% and specific surface area was 59 m^2/g. All these soils were of acidic reaction.

The mineralogical compositions of the three soils were significantly different: quartz content was from 90% in sandy soil through 72% in loess soil to 48% in loamy soil; the loamy minerals contents were 4, 17, and 46%, respectively.

Soil samples, dried in air and sieved through a riddle with a 1 mm mesh, were transferred to containers made of polypropylene. 0.05 kg soil samples were suspended in: a) redistilled water, b) 0.1 M HCl, c) 1 M HCl, in the mass ratio 1:1. The tightly closed samples were incubated at 293K for 24 days. After incubation was completed, metals in solutions were determined in filtrates with the AAS technique. The obtained results were converted into mg metals / kg dry soil. The pH values of the soils were determined before and after the experiments. Total content of metals in soil samples was determined using the emission spectrography (Table 1).

3. DISCUSSION

The pH values measured in the various suspensions are given in Table 2.

The amounts of mobilized metals under the experimental conditions are shown in Figures 1–9.

Treating the samples of soil with the redistilled water caused Mn mobilization only from the loess soil (Fig. 1). This kind of soil was the richest in Mn as well as it was of the highest pH value[6, 1]. Despite of the fact that pH values of sandy and loess soils were similar after treatment by the 0.1 M HCl, the highest amount of Mn was mobilized from the sandy soil. This soil is of the smallest specific surface area and of the least organic matter content. 1 M HCl extracted over 80% Mn from all investigated soils. The mobilization of manganese and its compounds in soils is mainly under influence of pH conditions.

Table 2. pH in suspension of soil samples in:
a) redistilled water, b) 0.1 M HCl, c) 1 M HCl

	pH		
Sample	a	b	c
Sandy soil	6.6	2.6	0.67
Loess soil	6.1	2.8	0.72
Loamy soil	7.1	4.1	1.44

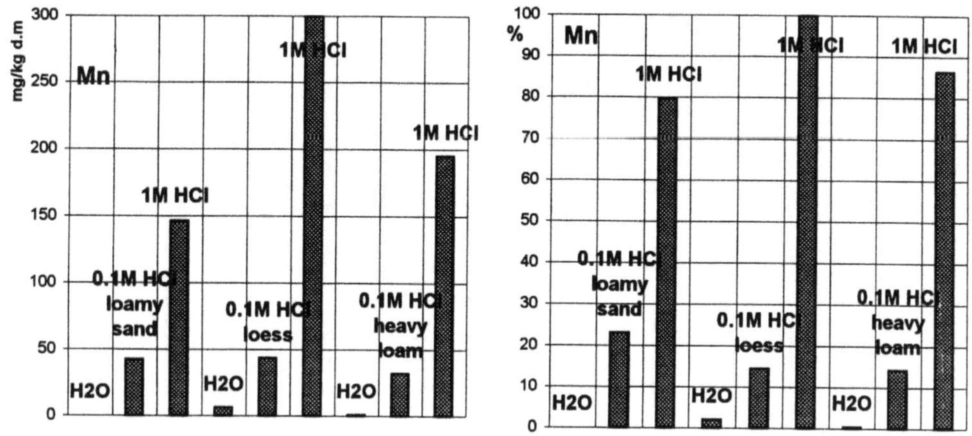

Figure 1. The amounts of Mn mobilized from the soils.

Cobalt was mobilized from studied soils by the redistilled water in amounts less than 1.5% (Fig. 2). The reason of such weak Co mobilization in these conditions might have been its sorption by the soil organic matter. Then 0.1 M HCl mobilized the most of Co from soil of pH 2.6–2.8. There was mobilized only 2.35% Co from the richest loamy soil but it was of pH 4.1. The analogous dependency occurred with mobilization of Co by the 1 M HCl. Here pH value of the soil also appeared to be the most fundamental factor of metal mobilization.

Cadmium appeared to be weak mobile in soils of pH 6.1–7.1. Both redistilled water and 0.1 M HCl mobilized only 0.29% of total content of Cd from the loamy soil (Fig. 3) which points that the metal is immobilized at pH higher than 4.1.

Cd was the easiest mobilized from sandy soil what, due to the similar pH values of sand and loess after treatment of the soil samples with the 0.1 M HCl (2.6 and 2.8, respectively) as well as 1 M HCl (0.67 and 0.72, respectively), indicates to the organic matter content as an essential factor for mobilization of this metal from soils.

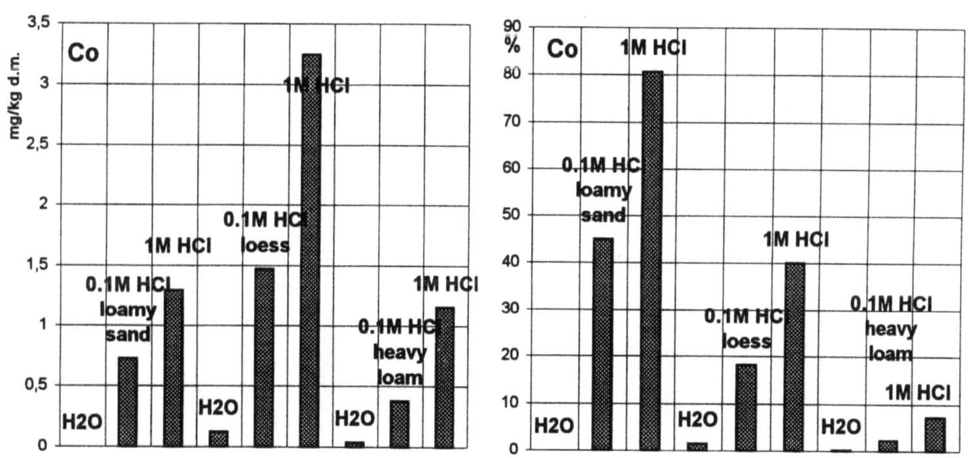

Figure 2. The amounts of Co mobilized from the soils.

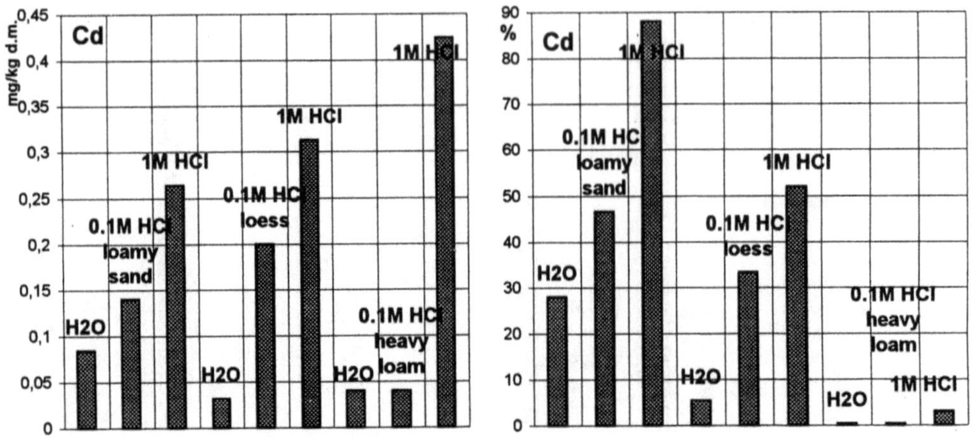

Figure 3. The amounts of Cd mobilized from the soils.

Chromium—metal of weak migration activity—has not been mobilized from investigated soil samples treating with redistilled water or 0.1 M HCl (Fig. 4). 1 M HCl used as an eluent mobilized the least percentage of Cr from the loamy soil. It could be caused by the highest pH value of the soil (1.44). Among very acidic soils: sandy (pH 0.67) and loess (pH 0.72) almost two times more Cr was mobilized from the sandy soil that is of the least specific surface area and of the least total chromium content. This results from the fact that Cr has exceptional tendency towards complex compounds forming both organic and inorganic ones. Thus chromium cycle in the soil is very complicated.

Copper was not mobilized from investigated soils by the redistilled water (Fig. 5). This indicates to strong bounding of copper with organic matter and loamy minerals of the soil. Along with soil acidification the solubility of copper increased. Such increase was proportional to the total copper content in soils as well as to specific surface area of the soil. Only 0.1 M HCl treatment of the loess soil did not confirm that dependency. But this soil is of the highest organic matter content (2.13%).

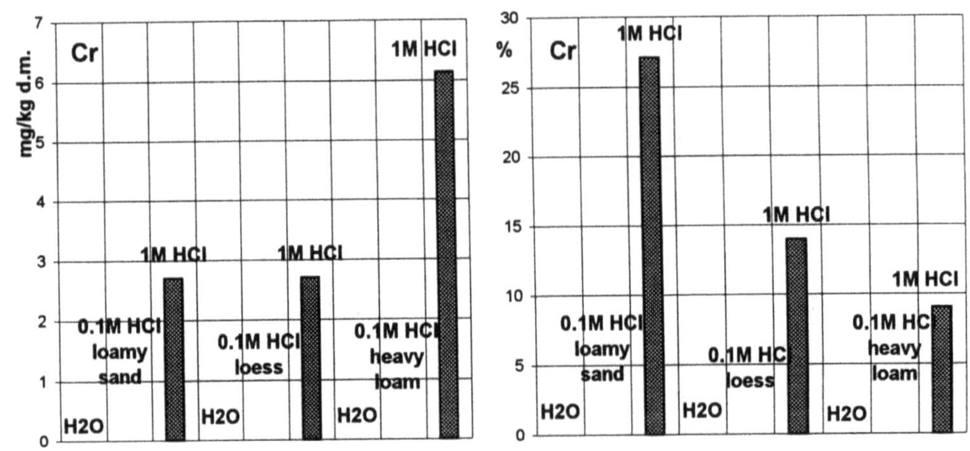

Figure 4. The amounts of Cr mobilized from the soils.

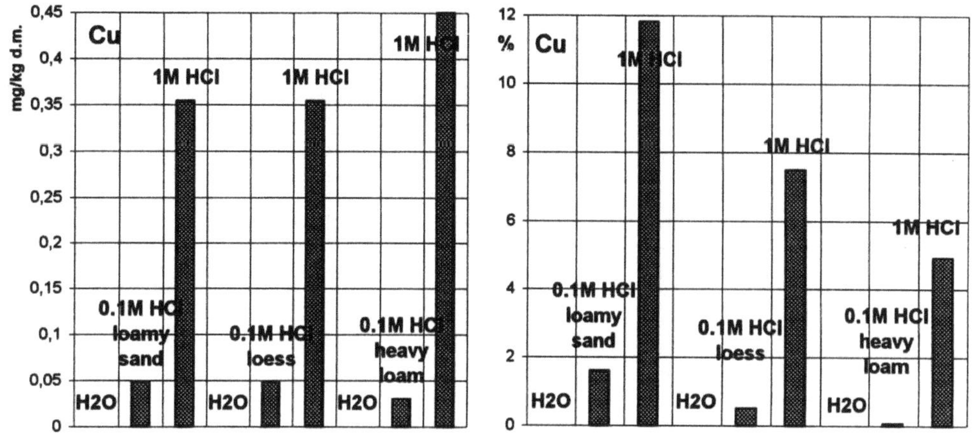

Figure 5. The amounts of Cu mobilized from the soils.

Under conditions of the experiment, redistilled water (when pH of the soil was 6.1 to 7.1) mobilized small amounts of lead (Fig. 6), however clear dependency between these amounts and organic matter contents occurred. Study results concerning Pb displayed that decrease of pH below 4.1 caused significant increase of its solubility, especially from the loamy soil treated with 1 M HCl. It might have been caused by the elution of organic complexing compounds.

There was slight mobilization of Zn from the soils at pH > 6.1 (Fig. 7). In acidic soils the mobilization of zinc was connected to specific surface area. Increasing of zinc mobilization from the soils at pH < 3 may be result of zinc sorptive reduction by humus acids.

Nickel was mobilized by redistilled water in amounts below 1% from investigated soils (Fig. 8). Small amounts of Ni were also mobilized from the soils at pH < 3. In soils treated with 1 M HCl, the nickel mobilization was connected to organic matter content.

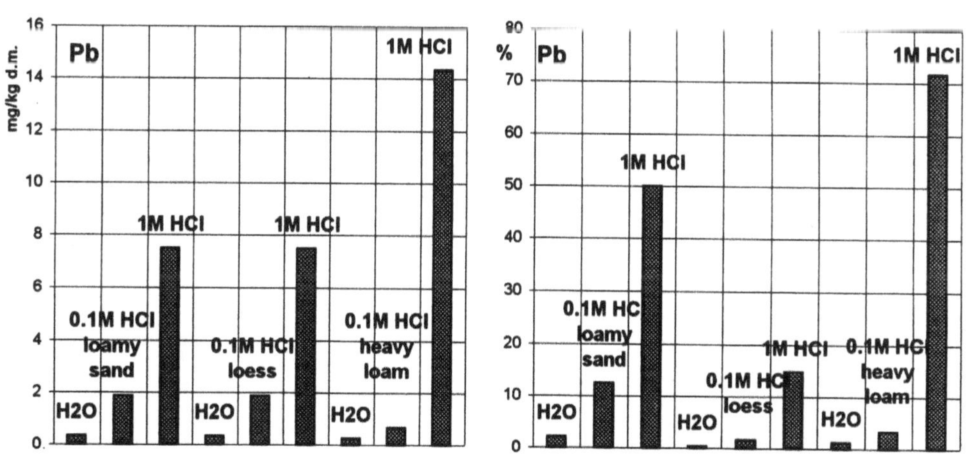

Figure 6. The amounts of Pb mobilized from the soils.

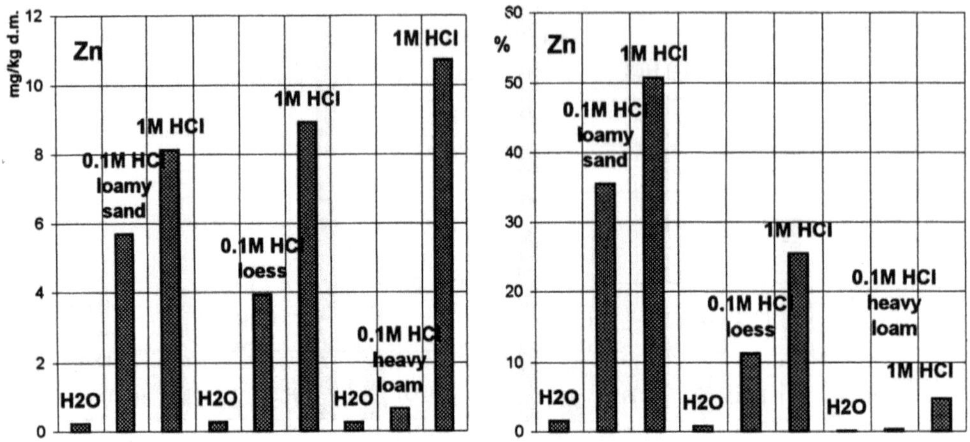

Figure 7. The amounts of Zn mobilized from the soils.

The most of Fe was mobilized from the loess soil—the richest in organic matter—by the redistilled water (Fig. 9). Mobilization of iron by HCl was connected to specific surface area. The iron solubility increased proportionally to the soil acidification level.

4. CONCLUSIONS

1. In studied soils the connexion between total contents of Mn as well as Pb and organic matter content occurred.
2. Processes of treating the soils with redistilled water may lead to partial mobilization of metals from investigated soils (except Cr and Pb).
3. Mobilization of heavy metals from soils by 0.1 M HCl is reciprocally correlated to specific surface area of the soil (except Cr).
4. Mobilization of Mn and Ni by 1 M HCl is positively correlated to organic matter content in soils.

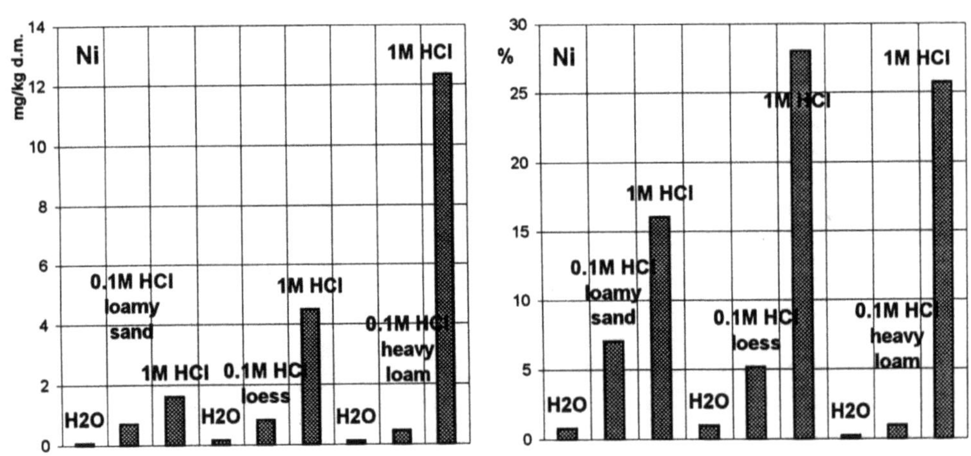

Figure 8. The amounts of Ni mobilized from the soils.

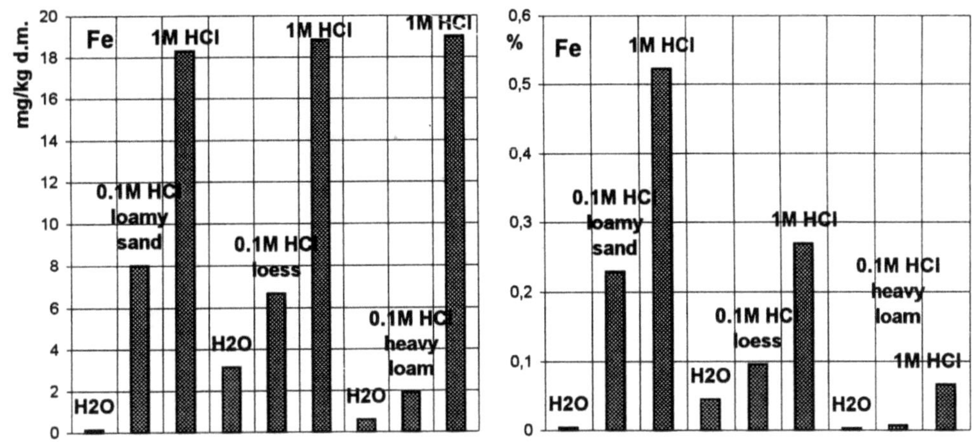

Figure 9. The amounts of Fe mobilized from the soils.

5. REFERENCES

1. Gliński J. 1987: Problems of soil degradation in Poland. Zesz. Probl. Post. Nauk Rol., 344, 7–15.
2. Gliński J, Turski R. 1993: Heavy metal pollution of Polish soils. Zesz. Probl. Post. Nauk Rol., 400, 1–5.
3. Gorlach E., Gambuś F. 1991: Desorpcja i fitotoksyczność metali ciężkich zależnie od właściwości gleby. Rocz. Glebozn., 42, 3/4, 207–215.
4. Gąszczyk R., Jackowska I. 1994: Copper (II), lead (II) and chromium (III) displacement from soils of various reaction. Zesz. Probl. Post. Nauk Rol., 413, 115–118.
5. Motowicka-Terelak T., Terelak H. 1987: Acidity indices of Polish soils. Zesz. Probl. Post. Nauk Rol., 344, 65–85.
6. Muranyi A. 1987: Soil Acidification and the Soil Properties. Zesz. Probl. Post. Nauk Rol., 344, 123–126.
7. Sinta J. 1993: Przemysłowe czynniki kwasowej degradacji pokrywy glebowo-roślinnej. Konf. nauk. "Przyrodnicze i antropogeniczne przyczyny oraz skutki zakwaszenia gleb". Lublin, 27–43.
8. Terelak H., Piotrowska M., Motowicka-Terelak T., Stuczyński T., Budzyńska K. 1995: Zawartość metali ciężkich i siarki w glebach użytków rolnych Polski oraz ich zanieczyszczenie tymi składnikami. Zesz. Probl. Post. Nauk Rol., 418, 45–60.

REDOX RESISTANCE AS A FEATURE DETERMINING FATE AND TRANSPORT OF POLLUTANTS IN SOILS USING THE EXAMPLE OF MINERAL SOILS OF POLAND

Z. Stępniewska,[1] W. Stępniewski,[1,2] J. Gliński,[1] and J. Ostrowski[3]

[1] Institute of Agrophysics PAN
P.O. Box 121, 20–236 Lublin, Poland
[2] Technical University of Lublin
Department of Environmental Engineering and Protection
Nadbystrzycka 40, 20–618 Lublin, Poland
[3] Institute of Reclamation and Grassland
Falenty, 00–550 Raszyn

ABSTRACT

Soil is an absorber of different pollutants from water and the atmosphere. Their fate and migration depend, besides the external conditions, also on the conditions and the properties of the soil itself. One of the most important soil features in this context is soil redox status which decidedly influences solubility and the biotransformations of pollutants. The dynamics of the variation of the oxidation - reduction conditions depend on redox buffering capacity being an intrinsic property of a given soil.

This new soil feature is a resultant effect of soil microbial activity, carbon availability, temperature, and of the pool of oxidized forms of nitrogen, manganese, and iron. It can be characterized on a chemical or biological basis. In the latter context it is called soil redox resistance and is defined as the time needed to decrease soil redox potential under flood conditions to 400 mV corresponding to nitrate decomposition (t_{400}), or to 300 mV corresponding to reduction of iron (t_{300}). The studies have been performed with the use of 953 soil profiles (2777 soil horizons investigated) representing 30 soil units from the entire territory of Poland.

It was found that the values of t_{300} at temperature 15°C varied from 0.7 to 6 days in Ap horizons, and 1.2–17 days in the subsoil. Nitrate half decomposition time was usually within 6–12 days at 20°C and the related Eh was within 140–350 mV. An unique map of soil redox resistance (t_{300}), for the temperature of 15 °C is presented.

Determination of soil redox resistance makes it possible, on the basis of climate data (temperatures, precipitation), to make a prognosis of transformations and leaching of nu-

Chemistry for the Protection of the Environment 2, edited by Pawłowski et al.
Plenum Press, New York, 1996

trients in soil. This is of particular importance for rational management of nitrogen fertilizers (nitrates) and for assessement of acceptable saturation time for drainage design.

1. INTRODUCTION AND DEFINITION

Soil is a sink of diverse pollutants from air and water. Their fate and migration depend, besides the external factors also on the conditions of the soil itself and in particular on the oxidation - reduction status.

The awareness of the importance of soil redox conditions has extensively increased since the first measurement of redox potential in soil performed in 1892 i.e. more than a century ago by Bancroft(after Hesse[8]). According to present day knowledge, the soil redox status is a key factor related to a majority of soil properties[7]. Biodegradability of organic polutants[2] and the availability of heavy metals[3] is strongly dependent on soil redox conditions.

Redox properties of a given soil can be characterized by its redox intensity i.e. by the value of redox potential itself, and by soil redox buffering capacity preventing the soil from sudden changes of its redox potential. Both these characteristics are necessary for full description of the soil redox status.

For the first time the notion of "poising" used in the sense of redox buffering in soil was introduced by Jeffery[9]. Kaurichev et al.[10] used the term redox buffering defined as the soil's ability to prevent changes of redox potential after addition of an oxidizer or reducer. They determined this feature by chemical titration of the soil with some oxidizing or reducing agents eg. ferrocyanide and ferricyanide. A modification of this chemical approach is a polarographic method of determination of reducing substances in soil used by Yu[14] to characterize instantaneous content of these forms rather than to characterize a given soil unit.

An alternative approach has been presented by Gliński and Stępniewska[6] and Stępniewska[12], who proposed to characterize soil redox buffering capacity with the preservation of the biological component of the soil. In this approach soil redox resistance is defined as the time during which soil redox potential under flood conditions drops to the value of 400 mV corresponding to the beginning of nitrate decomposition (t_{400}), or to 300 mV corresponding to the initiation of the reduction of manganese and iron (t_{300}). Thus the value obtained is a resultant effect of soil microbial activity, carbon availability, temperature, and of the pool of oxidized forms of nitrogen and manganese. These soil characteristics are of particular importance for predicting stability of many nutrients and in particular of nitrates in soil[5,11,13]. It was also found[1] that the plant response to reduced conditions in soil was related to the number of days with redox potential < 300 mV rather than to the mean value of redox potential itself.

The aim of the study was to determine redox resistance of cultivated soils of Poland. The final purpose was to elaborate maps of t_{300} and t_{400} with the consideration of the effect of temperature in order to make it possible—in the prospective—to evaluate losses of the incorporated nitrates and of the plant yield. In this paper a map of soil redox resistance t_{300} for humus (Ap) horizons at a temperature of 15 ^{o}C is presented.

2. MATERIALS

The studies were performed with the use of Ap horizons of 953 profiles of arable soils from 25 soil units comprised from the entire territory of Poland. Each soil unit was

Table 1. Basic soil units of Poland included in the study and their t_{300} values at a temperature of 15°C for particular horizons: Ap—plow horizon, B—second genetically defined soil horizon under the Ap horizon, C—subsoil i.e. third genetically defined horizon in a given profile

No.	Soil unit	Area thous. ha	Number of profiles	t_{300} 15°C Ap	B	C
1	Rendzinas "proper".	190	20	0.7	1.3	1.3
2	Rendzinas "mixed".	45	22	0.2	1.2	1.3
3	Chernozems.	236	21	1.1	2.1	2.6
4	Podz. brown, dev. from loose sands.	4098	182	5.0	9.8	14.9
5	Podz. brown, dev. from light loamy sands.	163	28	5.2	8.1	15.0
6	Podz. brown, dev. from loamy sands.	605	15	4.6	1.3	9.8
7	Podz. brown, dev. from loamy sands over heavier materials.	1858	96	5.3	8.3	8.3
8	Podz. brown, dev. from light loams.	1897	61	5.2	6.9	11.8
9	Podz. brown, dev. from medium loams.	937	62	4.3	7.8	10.5
10	Podz. brown, dev. from heavy loams.	121	25	3.5	6.4	12.7
11	Podz. brown, dev. from non-uniform loams.	570	18	2.1	3.9	10.8
12	Podz. brown, dev. from gravel.	88	23	1.7	3.1	6.0
13	Podz. brown, dev. from silts of water origin.	739	30	3.8	7.2	9.1
14	Podz. brown, dev. from loess.	1056	50	1.7	2.2	4.5
15	Podz. brown, dev. from clay	50	15	2.4	3.3	11.4
16	Podz. brown, dev. from loam and skeleton loam.	168	22	1.6	2.4	3.1
17	Podz. brown, dev. from loam.	192	24	2.5	5.3	6.4
18	Podz. brown, dev. from argillaceous slates.	38	14	1.4	2.3	2.3
19	Podz. brown, dev. from parent material of silt.	201	45	1.8	2.3	5.2
20	Heavy alluvial soils.	505	34	2.8	3.8	5.9
21	Light alluvial soils.	211	29	3.9	4.6	7.0
22	Medium alluvial soils.	70	24	2.8	3.9	7.3
23	Mollic gleysols - dev. from loams and silts.	660	45	2.1	3.2	4.8
24	Mollic gleysols - dev. from sands.	394	28	2.4	2.6	6.4
25	Histosols.	114	22	3.1	6.9	14.9

represented by 12 to 182 profiles. One soil profile corresponded to 2–40 thousand hectares. The samples were collected during 1988 and kept air dry at ambient temperatures.

Specification of the soil units is presented in Table 1 together with the areas represented by each of them.

3. METHODS

The soil samples (30 g) were incubated under flood conditions (soil : water = 1:1) at constant temperatures (20, 15, 10 and 4 °C). During the incubation, which lasted up to 40 days measurements of redox potential (Eh) were performed with the use of platinum electrodes v.s. saturated calomel electrode were performed. From the patterns of the Eh changes in time the values of t_{300} were determined. The illustration of the definition of t_{300} is presented in Fig. 1. showing an example of changes in soil redox potential under flood conditions.

On the basis of the data gathered for each of the horizons mean values for particular soil units have been calculated. The units have been grouped according to their t_{300} values into five categories. At 15 °C for t_{300} the intervals <1, 1–2, 2–3, 3–5, and > 5 days have

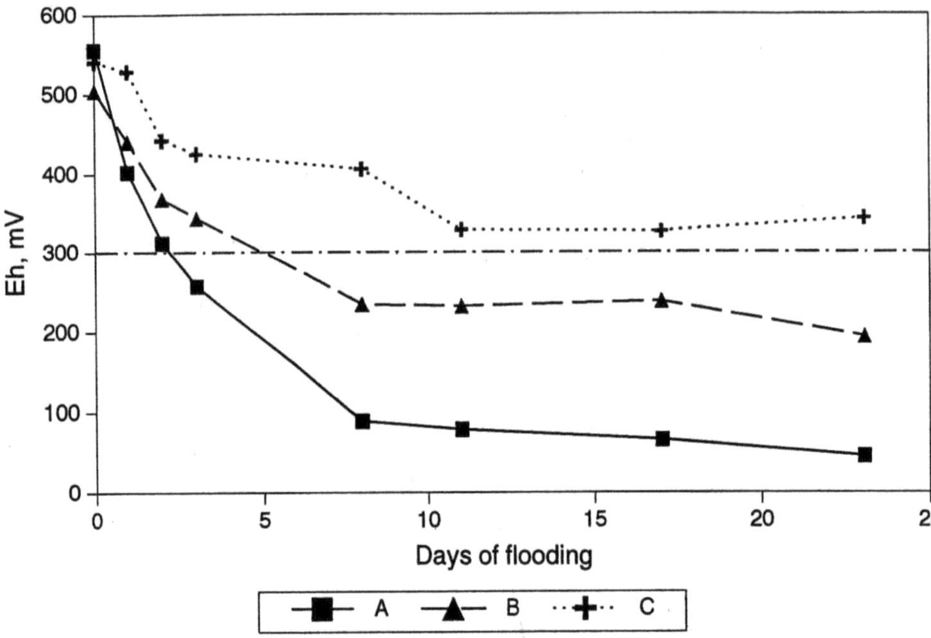

Figure 1. Example pattern of redox potential changes during flood incubation for one soil profile and related t_{300} values (profile no. 370, a—Ap horizon, sand loam; b—eluvial horizon, silty sand loam; c - brown horizon, medium loam).

been distinguished. Taking into account the distinguished categories a map of soil redox resistance for the entire territory of Poland was prepared.

4. RESULTS

The values of t_{300} for the three horizons of particular soil units at 15 °C are presented in Table 1.

As it can be seen, the values of t_{300} at 15°C were in the interval 0.3–5.3 days on the Ap horizon, 1.2–10 days in the second horizon, and they ranged from 1.3 to 15 days in the subsoil. As stated previously[12] the value of t_{300} increases at lower temperatures as the rate of biochemical processes generally decreases under such conditions.

Spatial distribution of the values of t_{300} for the entire territory of Poland at 15°C (mean soil temperature in the surface horizon at the end of May and at the beginning of September) is presented in the Map. It can be noticed that South and South East parts of the country plus alluvial soils along a majority of rivers are characterized by lower redox resistances, while the remaining part, covered mainly by light soils, is distinguished by higher redox resistances.

Of the investigated soils the most susceptible to reduction processes, (i.e. characterized by the values of t_{300} <2 days) turned out to be rendzinas ("proper" and "mixed"), chernozems, as well as podzolic and brown soils developed from gravels, silts of non water origin, skeleton loams and argillaceous slates.

The most resistant to reduction processes (with the highest redox resistance values) were podzolic and brown soils developed from loose sands, light loamy sands, loamy sands, light loams and medium loams.

5. DISCUSSION

Previous attempts to characterize soil redox buffering capacity treated the soil as a reserve of oxidized or reduced chemical compounds only. The approach presented here is an attempt to approximate the situation occurring under field conditions by preserving the biological side of the processes.

It is known[5] that reduction processes in soil are connected with a certain sequence of the redox reactions. Each soil redox couple is characterized by its specific redox potential at which the reduction process begins, and by a certain number of oxygen equivalents involved in the corresponding reaction. The abundance of a given form in the soil prevents the reduction of the next one. Thus the manganese and iron oxides constitute a buffer protecting the soil from reduction of sulphates. The redox buffering indicator considered here (i.e. t_{300}) combines this chemical aspect with the amount and decomposibility of soil organic matter as well as the rate of the biochemical processes performed with the help of microbes.

The results obtained are of importance both from theoretical and practical point of view. The theoretical importance consists in elaboration of the spatial distribution of the new soil feature. This will enable us to have a better understanding of the interrelations between physical, chemical, physicochemical and biological soil properties and components.

Practical aspects are connected with the following potential applications:

- assessment of degree of threath and agricultural yield losses due to conditions of temporal overmoistening in different parts of the country.
- evaluation of ecological damages connected with the losses of nitrogen fertilizer due to denitrification, and in consequence—an increase of the emission of nitrous oxide (a gas highly active in the green—house effect).
- predicting negative ecological and agricultural effects connected with the perspective of climate and soil temperature changes.
- practical application of the values of t_{300} during drainage and irrigation design.

6. CONCLUSIONS

Soil redox resistance, a new soil feature determining the fate and the transformations of polutants in soil, depends on the type of the soil material. At 15 °C it varies from 0.2 to 6 days in the Ap horizon, from 1.2 to 10 days in the subhumus horizon and from 1.2 to 15 days in the subsoil. This soil parameter is of particular importance for rational management of nitrates in soil and for assessment of acceptable saturation time for drainage design.

ACKNOWLEDGMENT

The project was supported by grant PB05.0268.91.01 of the Polish Committee for Scientific Research (KBN).

REFERENCES

1. Carter C.E. 1980. Redox potential and sugarcane yield relationship. Trans. A.S.A.E. 23, 924–927

2. Doelmann P, 1994. Microbiology of soil and sediments. In Biogeodynamics of pollutants in soils and sediments: Risk assessment of delayed and non linear responses. Synthesis repport for the Foundation for Ecodevelpoment "Mondiaal Alternatief" Hoofddorp, The Netherlands. (Eds -Salomons and Stigliani), 1994.
3. Gambrell R.P. 1994. Toxic and trace metals in wetlands—a review. J. Environ. Qual. 23, 833–891.
4. Gliński K., Ostrowski J., Stępniewska Z., Stępniewski W. 1992: Bank of soil samples representing mineral soils of Poland. (In Polish). Problemy Agrofizyki 66, Ossolineum.
5. Gliński J., Stahr K., Stępniewska Z., Brzezińska M. 1992: Changes of redox and pH conditions in a flooded soil amended with glucose and nitrate under laboratory conditions. Z. Pflanzenernahr. Bodenk., 155, 13–17.
6. Gliński K., Stępniewska Z. 1986: An evaluation of soil resistance to reduction processes. Pol. J. Soil Soc. 19, 15–19.
7. Gliński J. Stępniewski W. Soil Aeration and its Role for Plants. CRC Press, Boca Raton, Florida, 1985.
8. Hesse P.R. A Textbook of Soil Chemical Analysis. J. Murray, London, 1971, 520 pp.
9. Jeffery J.W.O. 1960. Iron and the Eh of waterlogged soils with particular reference to paddy. J.of Soil Sci. 11, 1, 140- 148.
10. Kaurichev I.S., Latfulina G.G. Sawicz W.I. 1974. (Oxidation - reduction buffering in soil)—in Russian. Izv. Timiriaz. Sel'.-khoz. Akad. 2, 75–85, Moscow.
11. Patrick W.H.Jr, Jugsujinda A. 1992: Sequential reduction and oxidation of inorganic nitrogen, manganese and iron in flooded soil. Soil Sci. Soc. Am. J. 56, 1071–1073.
12. Stępniewska Z. 1988: Redox Properties of Mineral Soils of Poland (In Polish). Problemy Agrofizyki 56, Ossolineum,
13. Stępniewska Z., Stępniewski W., Włodarczyk T. 1993. Agrophysical factors determining nitrate and heavy metal transformations in soils. Zesz. Probl. Post. Nauk. Roln. 400, 23–32.
14. Yu Tian-Ren 1985: Physical Chemistry of Paddy Soils. Science Press. Beijng, Springer-Verlag, Berlin Heidelberg, New York, Tokyo.

SORPTION CHARACTERISTICS OF ORGANIC COMPOUNDS IN LANDFILL LEACHATE BY MODIFIED CLAYS

Jiunn-Fwu Lee,[1*] Pao-Mei Liaw,[1] Chi-Chan Hwan,[1] and E. E. Chang[2]

[1] Graduate Institute of Environ. Eng.
Natl. Central University
Chung Li, Taiwan
R.O.C.
[2] Dept. Analytical Chem.
Taipei Medical College
Taipei, Taiwan
R.O.C.

ABSTRACT

Natural clays were modified by substituting inorganic ions with the organic cations hexadecyltrimethylammonium (HDTMA), dodecyltrimethylammonium (DDTMA) and Tetramethylammonium (TMA). This resulted in the clays having significantly higher organic matter contents in proportion to the amount of carbon in each organic cation. The sorption behavior of different molecular-weight organic compounds from leachate by different sorbents was also investigated. The batch experiments indicated that the sorption of modified clays is more sensitive to pH than granular activated carbon (GAC). In general, the sorption capacities of the modified clays increased with increasing sorbate molecular weights. In contrast, the activated carbon was found to be an effective sorbent rather than modified clays for organic compounds with small molecular weights. Moreover, the surface characteristics of sorbents with their sorption behaviors indicated that the sorption mechanism of modified clays is different from that in activated carbon. The uptake of organic compounds in landfill leachates by modified clays depends on the organic contents of the modified clays, and the sorption mechanism of activated carbon is surface adsorption. Experimental results indicated that GAC combined with a modified clay effectively reduces leachate Chemical Oxygen Demand (COD), and might be further used to enhance the containment capabilities of clay landfill liners.

* Corresponding author.

Chemistry for the Protection of the Environment 2, edited by Pawłowski et al.
Plenum Press, New York, 1996

INTRODUCTION

The sorption of nonionic organic compounds (NOCs) in soil water systems is attributed to two mechanisms. First, organic chemical sorption may include ion exchange, protonation, hydrogen bonding and coordination/ion-dipole reactions with clays[1,2]. These sorptive mechanisms account for the uptake of organic cations and some polar organic molecules on soil. The second major mechanism of organic chemical-soil interaction involves nonionic organic compounds (NOCs), the largest and most problematic group of organic substrates in leachate. Chiou et al.[4,5] described this mechanism as a partition process involving nonpolar interactions between soil organic matter and NOCs. Sorption of this type depends almost entirely on the solute water solubility[4,5] and the soil organic matter content[6–9].

A typical lay mineral has metal cations occupying cation-exchange sites; its surface is hydrophilic because of the water molecules in the hydration shell solvating the cation. As a result, NOC sorption by clays is suppressed in water because relatively nonpolar organic chemicals cannot effectively compete with highly polar water for adsorption sites on the clay surface.

The inability of clays to remove substantial amounts of NOCs from aqueous solution can be resolved by substituting natural metal cations with large organic cations through ion-exchange reactions. Lee et al.[10–13] examined this relatively simple modification to improve the retardation capabilities of low organic matter soils and clays. They used hexadecyltrimethylammonium (HDTMA) ions to replace naturally occurring metal ions, thereby resulting in a significantly higher organic matter content and significant enhancement in sorptive capabilities for NOCs. Moreover, previous investigations have demonstrated the organic phase derived from HDTMA exchanged on soils[14] and on smectite clays[10,15] to be a powerful partition medium in removing NOCs from water. Furthermore, such modified soils and clays may be used to (a) improve the retardation capabilities of low organic matter soils and aquifer materials and (b) enhance the containment capabilities of clay landfill liners and slurry walls.

In this article, studies involving soil modification for attenuating organic contaminant mobility have been expanded to organic moiety in a landfill leachate. The landfill leachates studied here are divided into five fractions, depending on the different molecular weights. Also, three organic cations of the general form $[(CH_3)_3NR]^+$, HDTMA ($R=C_{16}H_{33}$), dodecyltrimethylammonium (DDTMA) ($R=C_{12}H_{25}$), and tetramethylammonium (TMA) ($R=CH_3$), were used to prepare clay-organic complexes. The different behaviors of modified clays and activated carbons in relation to the sorption of organic compounds from landfill leachate are elucidated. Finally, the sorptive properties of granular activated carbon (GAC) is compared with those of modified clays to evaluate its potential application as an active sorbent in the sorption of NOCs.

MATERIALS AND METHODS

Sorbents

Clays. The clay used in this work was Wyoming Bentonite (American Colloid Company), whose cation-exchange capacity (CEC) is 90 meq/100g. Organochemical-clay complexes were prepared by reacting 10g of clay with HDTMA, DDTMA, or TMA chloride salts. The organic cation salts were dissolved in water and added in an amount equal to the CEC of the clay. After thorough mixing, the organochemical-clay complexes were dialyzed against distilled water until a negative chloride test was obtained. The exchanged clays were than freeze-dried and stored in refrigerator.

Activated Carbons. Seven activated carbon adsorbents were used in this study. Their sources and surface properties are listed in Table 1.

Sorbates

Leachates were collected from a solid waste landfill located in the northern part of Taiwan. The chemical characteristics of the leachate samples are displayed in Table 2. Before batch sorption experiments, the leachate samples were pretreated by filtration through a 0.45 μm cellulose nitrate filter to remove particular matter. Next, a comprehensive fractionation scheme was developed to divide the organic molecules of the leachate into five groups on the basis of their molecular weight (MW) using High Performance Gel Permeation Chromatography (HPGPC).

Apparent Equilibrium Sorption

A 0.5 gram mass of either the organochemical-clay complexes, or activated carbon, and 20 mL of leachate were placed in Corex glass tubes. The mixed suspension was shaken for 24 hr at 20°C in a reciprocating shaker. At the end of the equilibration period, the solid and liquid phases were separated by centrifugation at 7000 rpm for 12 min. in a Sorvall RC-5C (DU PONT) centrifuge equipped with an SS-34 rotor. The residual organic matter in the aqueous phase was immediately characterized by measuring the chemical oxygen demand (COD), Total Organic Carbon (TOC) or UV adsorption.

Analytical Procedures

HPGPC was carried out at ambient temperatures on an ALCOTT HPLC system consisting of a Model U6K injector, a Linear UVIS 200 absorbance detector operating at 254 nm, and an Advantecsf-2120 fraction collector. The column used was a 25 × 280 nm Ultropac column packed with TSK HW-55(F) gel. The void volume and the total permeation volume were determined with Blue Dextran and water, respectively.

Organic compound recovery was determined by comparing total UV peak areas for identical samples passing through and bypassing the column. Separation of humic substances was performed with a 0.02M KH_2PO_4 buffer at a flow rate of 1.0 mL/min. Injection volumes ranged from 50 μL to 1,000 μL for preparative separations from which fractions were collected. Finally, carbon yields throughout each fraction were monitored by measuring Total Organic Carbon (TOC) using a TOC Analyzer.

RESULTS AND DISCUSSION

The leachate used was obtained from a municipal disposal site in the northern part of Taiwan. Its chemical composition is summarized in Table 2. The majority of leachate molecules had a molecular weight between 3500 and 15000. High molecular weight compounds, normally consisting of fulvic and humic-like compounds, were found in a rela-

Table 1. The sources and surface properties of different sorbents

Property	Treated clays			PAC	A	GAC	D	C-5260	C-2889	C-2764
	HDTMA	DDTMA	TMA							
Total intrusion volume (mg/g)	0.92	2.17	6.36	6.38	1.75	0.48	1.51	2.13	1.31	1.02
Total pore area (m^2/g)	205.85	49.00	108.39	157.55	127.56	113.22	180.36	707.08	134.51	158.08
Median pore diameter (volume) (mm)	0.14	21.68	7.41	8.60	3.28	3.03	101.60	0.04	1.59	0.80
Median pore diameter (area) (nm)	0.58	0.72	0.59	0.73	0.45	1.09	0.76	1.07	0.52	0.47
Average pore diameter (4V/A) (nm)	1.78	17.69	23.45	16.19	5.48	1.73	3.35	1.20	3.90	2.59
BET surface area (m^2/g)	880.60	183.52	46.43	22.07	329.59	1218.27	33.84	610.69	720.52	692.19
Langmuir surface area (m^2/g)	1191.02	250.23			423.57	1606.55	68.19	873.63	957.42	927.90

PAC: power activated carbon

GAC: granular activated carbon

A: power activated carbon (Taipei Chemical Industry)

D: power activated carbon (Aldrich)

C-5260: Charcoal Activated, Power 100-400 mesh, Sigma Chemical Co.

C-2889: Charcoal Activated, granular 8-20 mesh, Sigma Chemical Co.

C-2764: Charcoal Activated, granular 4-14 mesh, Sigma Chemical Co.

Table 2. The chemical characteristics of the leachate samples

	Original leachate	Fraction I	Fraction II	Fraction III	Fraction IV	Fraction V
Organics (mg/L)						
M.W. range		>250,000	60,000-250,000	15,000-60,000	3,500-15,000	1,000-3,500
TOC	1,462	22.9%	17.7%	15.2%	40.1%	4%
BOD	1,683					
BOD_S*	1,540					
COD	3,836	24.5%	15.1%	12.9%	43.8%	3.6%
COD_S*	3,503					
UV peak area		18.4%	16.9%	11.3%	47.9%	5.4%
Conductivity (μmhos/cm)	13,310					
pH	8.5					
Solids (mg/L)						
TS	7615					
SS	265					
TVS	2,640					
Nutrients (mg/L)						
NH_3-N	1,173					
Organic-N	42					
T-P	41					

*Filtrated by Whatman GF/C

tively low concentration. The gross properties analysis of each fraction revealed that the Chemical Oxygen Demand (COD) ranged from 14 mg/L (fraction V) to 1680 mg/L (fraction IV). Results obtained from the supernatant of a leachate collected from the influent of a treatment plant, indicated that 56% of the original TOC derived from organic moiety had an apparent molecular weight larger than 15,000 Daltons. A relatively higher percentage (75%) of the original TOC was observed in an effluent derived from same size of organic compounds.

The physical and chemical properties of untreated Wyoming Bentonite and the Wyoming Bentonite treated with three different organic cations are listed in Table 3. The three different types of organic cations used were all of the form $[(CH_3)_3NR]^+$, where R varied in size from a C_{16} (HDTMA) to a C_{12} (DDTMA) to a C_1 (TMA) unbranched hydro-

Table 3. Properties of the untreated and organic-cation treated clays

	Wyoming Bentonite			
Property	Untreated	HDTMA-treated	DDTMA-treated	TMA-treated
CEC (meq/100g)	90			
Organic carbon (%)	0.9	19.8	13.7	4.10
Organic matter (%)				
Soil organic matter	1.8	1.8	1.8	1.80
Cation organic matter	0	23.5	16.2	3.55
Total organic matter	1.8	25.3	18.0	5.35

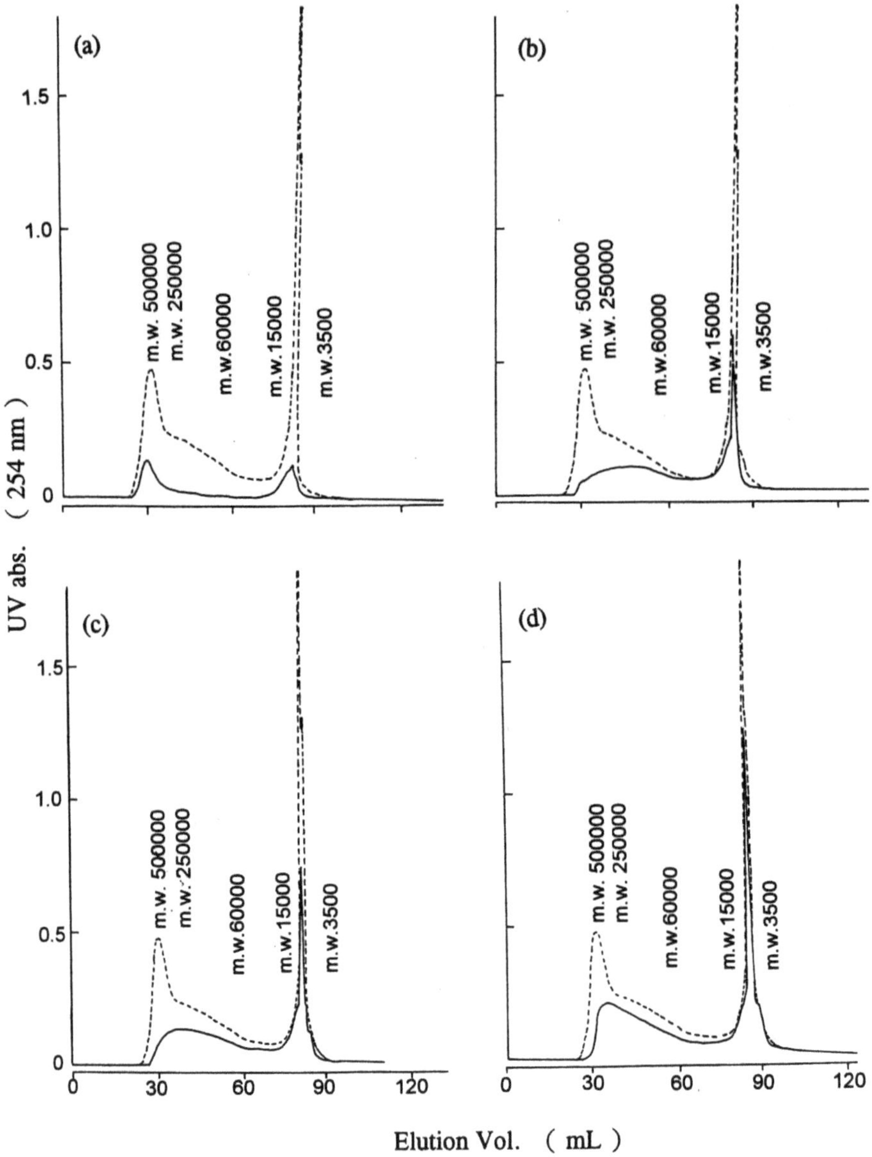

Figure 1. GPC chromatogram of leachate before (dashed line) and after (solid line) the adsorption experiment by using (a) GPC, (b) HDTMA-Bentonite, (c) DDTMA-Bentonite, and (d) TMA-Bentonite as the adsorbents.

carbon. When Wyoming Bentonite was treated by HDTMA, DDTMA, and TMA ions, the organic carbon (OC) contents increased significantly. As expected, the larger organic cation (HDTMA) more effectively contributed to OM than the smaller organic cations (DDTMA and TMA).

In order to evaluate the differences in uptake of organic moiety from leachate by different sorbents, the elution patterns of leachate (Fig. 1) are expressed in terms of reduced absorbance. Figure 1 plots UV absorbance (254 nm) against elution volume. A comparison of the elution patterns reveals that the uptake of a larger organic moiety from leachate by HDTMA-Bentonite is greater than that of the solutes by activated carbon. Also, the uptake of low MW organic moiety from leachate by organic clays decreased in the order: HDTMA > DDTMA > TMA; however, the overall differences were relatively small. In contrast to the lack of sorption by organo-clays, an extremely high uptake of small organic moiety was found on activated carbon. This is reflected in the significant decrease in the UV peak area (Fig. 1).

A closer examination of the sorption of organic moiety from leachate on granular activated carbon (GAC), HDTMA-, DDTMA-, and TMA-Bentonite was made by utilizing the separated leachate. Figure 2 shows a sorption tendency of organic compounds from fraction I of the leachate onto several modified clays and GAC. As shown in Table 2, fraction I contained a relatively large amount of high molecular weight compounds, i.e., about 23% of the TOC with a molecular weight higher than 250,000. Bentonite containing HDTMA as the exchanged cation showed evidence of being a very powerful adsorbent to remove larger organic moiety from the leachate. This phenomenon appears in Fig. 2(b), which shows the disappearance of the UV absorption peak of the leachate on the chromatogram. In contrast, bentonite saturated with TMA cations showed very weak sorptive characteristics for larger organic compounds. The GAC showed somewhat stronger affinities for larger organic compounds present in leachate.

Whether a low molecular weight of organic moiety would have a different sorption behavior from larger organic compounds was next examined. The chromatograms obtained for the small organic compounds before and after sorption experiment are displayed in Fig. 3. A pronounced difference was noted in the UV peak area of the elution patterns shown in Figure 3, indicating progressively weaker sorption capacities for small organic moiety (fraction IV) in the order: GACHDTMA-clay > DDTMA-clay > TMA-clay. From the change in the UV peak area, both the large and small molecular organic compounds clearly indicated weaker interactions in an aqueous environment with TMA-Bentonite. However, the smaller organic moiety uptake from leachate was markedly higher for the GAC than for the organo-clays (Fig. 3). This obviously contrasts with the uptake of larger organic compounds by modified clays, in which the highest uptake was by HDTMA-Bentonite.

The data presented here demonstrate the wide diversity of clay-organic complexes for the sorption of organic moiety from leachate. In general, the more hydrophobicity of the group associated with the quaternary ammonium entity implies the greater the sorption of organic moiety in leachate. The sorption process for the extremely hydrophobic (DDTMA- and HDTMA-) clays can be represented as a partitioning of organic moiety between leachate and the hydrophobic environment provided by the alkyl groups; the organic phase formed by the alkyl groups bears some resemblance to a bulk organic solvent such as hexane or octanol. For these complexes, the mineral surface is obscured by the large organic cations. Thus, their sorption behavior is characteristic of an organic sorbent, such as soil organic matter. In this study, the clay-organic complex containing smaller organic cation (TMA^+) behave as a conventional mineral sorbent. The organic phase of this complex is comprised of separate small organic cations; this phase does not act as an effective partition medium, despite a significant carbon content. The mineral surface in this complex is exposed and occupied largely by water.

On the other hand, the adsorption mechanism of activated carbon for the uptake of organic compounds from leachate shows non-selective and sorptive competition in a multisolute system. In the multisolute system, activated carbon is not allowed to accommo-

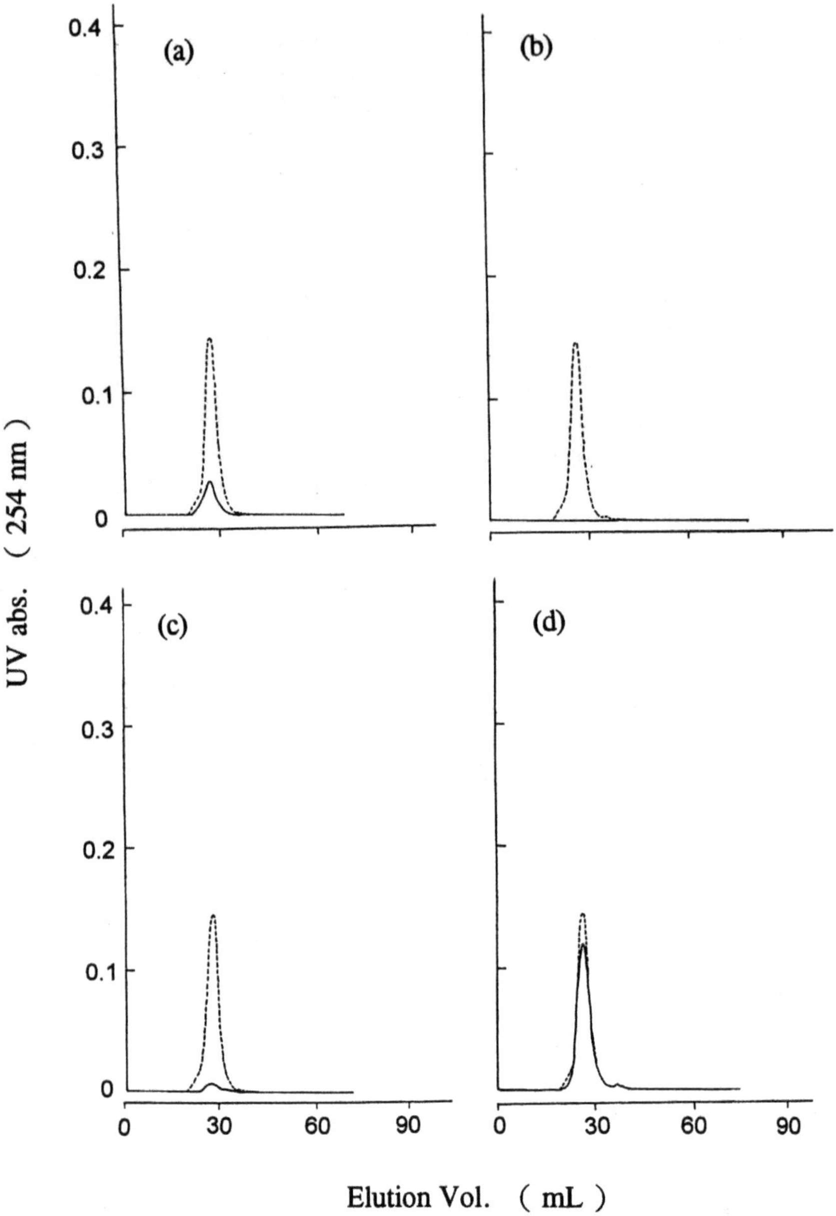

Figure 2. UV adsorption peaks of Fraction I leachate before (dashed line) and after (solid line) adsorption by different adsorbents. Plot (a) is GAC, (b) is HDTMA-Bentonite, (c) is DDTMA-Bentonite, and (d) is TMA-Bentonite.

date relatively larger molecules. In this work, the primary factor affecting the degree of adsorption of organic moiety by activated carbon is the molecular size and sorption mechanism. Results in this study demonstrate the potential versatility of modified clays for the sorption of different molecular sizes of organic compounds from leachates. A practical application of this simple reaction may be the *in situ* formation of an organo-clay sorption phase for the removal of the larger organic compounds that resist biodegradation.

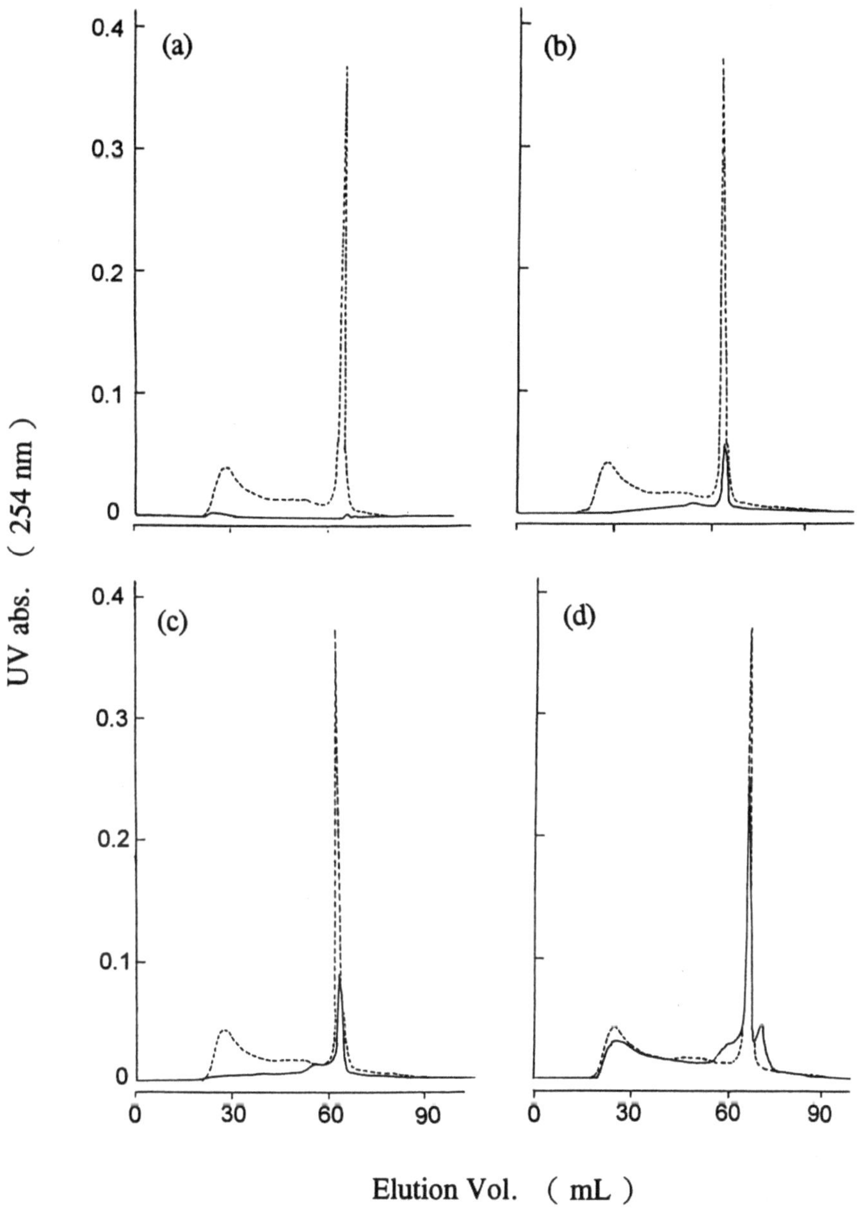

Figure 3. UV adsorption peaks of Fraction IV leachate before (dashed line) and after (solid line) adsorption by different adsorbents. Plot (a) is GAC, (b) is HDTMA-Bentonite, (c) is DDTMA-Bentonite, and (d) is TMA-Bentonite.

REFERENCES

1. Mortland, M. M. *Adv. Agron.*, 22 (1972), 75–117.
2. Theng,, B. K. G. *The Chemistry of Clay-Organic Reactions*; Wiley: New York, (1974), 343.
3. Parris, G. E. *Residue Rev.*, 76 (1980), 1–30.
4. Chiou, C. T., Peter, L. J., Frees, V. H. *Science*, 206 (1979), 831–832
5. Chiou, C. T.; Porter, P. E. Schemdding, D. W. *Environ. Sci. Technol.*, 17 (1983), 227–231

6. Briggs, G. G. *Nature*, 223 (1969), 1288.
7. Karickhoff, S. W.; Brown, D. S.; Scott, T. A. *Water Res.*, 13 (1979), 241–248
8. Garbarini, D. R.; Lion, L. W. *Environ. Sci. Technol.*, 20 (1980), 1263–1269
9. Hassett, J. J.; Means, J. C.; Banwart, W. J.; Wood, S. A.; Khan, A. J. *Environ. Qual.*, 9 (1980), 184–186
10. Boyd, S. A.; Lee, J. F.; Mortland, M. M.; *Nature*, 333 (1988), 345–347
11. Lee , J. F.; Crum , J. R.; Boyd, S. A.;*Environ. Sci. Technol.*, 23 (1989), 1365–1372
12. Lee , J. F.; Mortland, M. M.; Boyd, S. A.; Chiou, C. T.; *J. Chem. Soc.,Faraday Trans. 1*, , 85 (1989), 2953–2962
13. Lee , J. F.; Mortland, M. M.; Chiou, C. T.; Kile, D. E.; Boyd, S. A.; *Clays Clay Miner.*, 38 (1990), 113–120
14. Boyd, S. A.; Mortland, M. M.; Chiou, C. T.; *Soil Sci. Soc. Am. J.*, 52 (1988), 652–657
15. Mortland, M. M.; Shaobai, S.; Boyd, S. A.; *Clays Clay Miner.*, 34 (1986), 581–588

ASSESSMENT OF THE EFFECTS OF HYPOLIMNION WATER REMOVAL FROM THE KORTOWSKIE LAKE

C. Mientki and M. Teodorowicz

Faculty of Water Protection and Inland Fishery
University of Agriculture and Technology in Olsztyn
ul. Prawocheńskiego 1, 10–957 Olsztyn-Kortowo, Poland

ABSTRACT

By comparison to the values in the northern part, the influence was shown of the hypolimnion waters removal on thermal and oxygen structure, visibility, and phosphorus and nitrogen content in the southern part of the Kortowskie Lake, which has been rejuvenated (reclaimed) since 1956. The tendency of the parameters to change is also presented in this paper.

1. INTRODUCTION

The increasing input of nutrients to lakes, mainly of phosphorus and nitrogen, accelerates their eutrophication. One of the results of this process is alteration in oxygen conditions in a reservoir. In eutrophic lakes the consequences of an intensive photosynthesis are: oxygen supersaturated surface layers, considerable oxygen deficits in metalimnion and predominant lack of oxygen and presence of hydrogen sulphide in hypolimnion. Therefore, the balance between bioproduction and degradation of the produced organic matter is upset - the amount of the organic matter surpasses the capability to break it down in aerobic conditions.

Deficit of oxygen in hypolimnion waters and on the surface of bottom deposits does not mean only the elimination of a large water volume and bottom surface from fisheries. Only below 0.1 mg $O_2 \cdot dm^{-3}$ the redox potential on the surface of bottom sediments reaches the value equal to, or less than 200 mV [1]. Under reducing conditions nutrients stored in bottom deposits are released—the eutrophication runs in an avalanche mode ("rasante Seeneutrophierung"[2]). In consequence the circulation of substances in a lake is substantially accelerated, especially of phosphorus.

Chemistry for the Protection of the Environment 2, edited by Pawłowski et al.
Plenum Press, New York, 1996

Elimination or reduction of the external sources of pollution in many cases does not restore the primary condition of a lake. In order to allow for the restoration process, or to accelerate it, it has to be reclaimed.

One of the first methods of reclamation employed in the world was the removal of hypolimnion water to an outlet. It was introduced by Olszewski in the Kortowskie Lake in 1956,[3] and has been continued to the present.

This reclamation method is based on selective exchange of water. Instead of surface layers, which are rich in oxygen and relatively poor in nutrients, water above the bottom, devoid of oxygen, containing hydro sulphide, and mineral and organic substances, is removed. In this way not only the results of eutrophication are eliminated through the oxygen balance improvement but, what is of great importance, is that the degree of trophy is diminished by the transfer of nutrients outside the ecosystem. Thus, this method is considered, apart from bottom deposits removal, to be the most efficient. Its advantages are simplicity and operation without energy expenditure.

Many lakes in the world are reclaimed by the application of the „kortowska" method and Olszewski's pipeline,[4] and since 1982 also the Wielkie Rudnickie Lake in Grudzi•dz, Poland.

2. DESCRIPTION OF THE KORTOWSKIE LAKE

The Kortowskie Lake is located in Poland, in the Mazurian Lake District, within the administrative borders of the city Olsztyn.

The main morphometrical data are[5]: surface of the water mirror 90 ha, maximal length 1660 m, maximal width 715 m, length of the shoreline 4800 m, maximal depth 17.2 m, mean depth 5.9 m, relative depth 0.02, volume $5.323 \cdot 10^6$ m^3.

The lake bowl has two overdeepenings (Fig. 1): in the northern part with the max. depth 15.7 m (st. N) and in the southern part with the max. depth 17.2 m (st. S). The pipeline for the hypolimnion waters removal is laid down on the bottom of the latter. The shallowness in the central part of the lake (6 m deep), allows for the comparison of the effects acquired at the st. S (experimental station) and the st. N (reference station).

The main inflow and outflow of the Kortowskie Lake is the Kortówka river. Its catchment is 38 m^2, where forests constitute 32% and agricultural area about 40%. The Leœny stream plays very important role, in regard to the volume of the inflowing water and nutrients input. The other tributaries (Fig. 1), although seriously polluted, do not represent an actual danger, since their flows are very small. In the direct catchment of the lake, which is 102 ha in area, forests comprise 23.5% and the area used for agricultural purposes 76.5%.

3. HISTORY OF THE "KORTOWSKI" EXPERIMENT

The limnological studies carried out in the years 1952–1953 revealed a notable degradation of the reservoir.[3] In both periods of stagnation the water layers above bottom deposits were devoid of oxygen and hydrogen sulphide was present at 5 m beneath the water surface.

Olszewski had employed an original method for the lake's recovery. In winter 1956 in the southern part of the lake a pipeline was set down. It was constructed from planks. Its dimensions were 0.4 x 0.5 m, length 178 m, maximal water flow 0.187 $m^3 \cdot s^{-1}$, and the inlet was at 13 m (Fig. 1).

In 1956–1967 waters from above the bottom were carried away only in summer, and since 1968 also in winter. In 1973 due to the deterioration of the technical condition of the

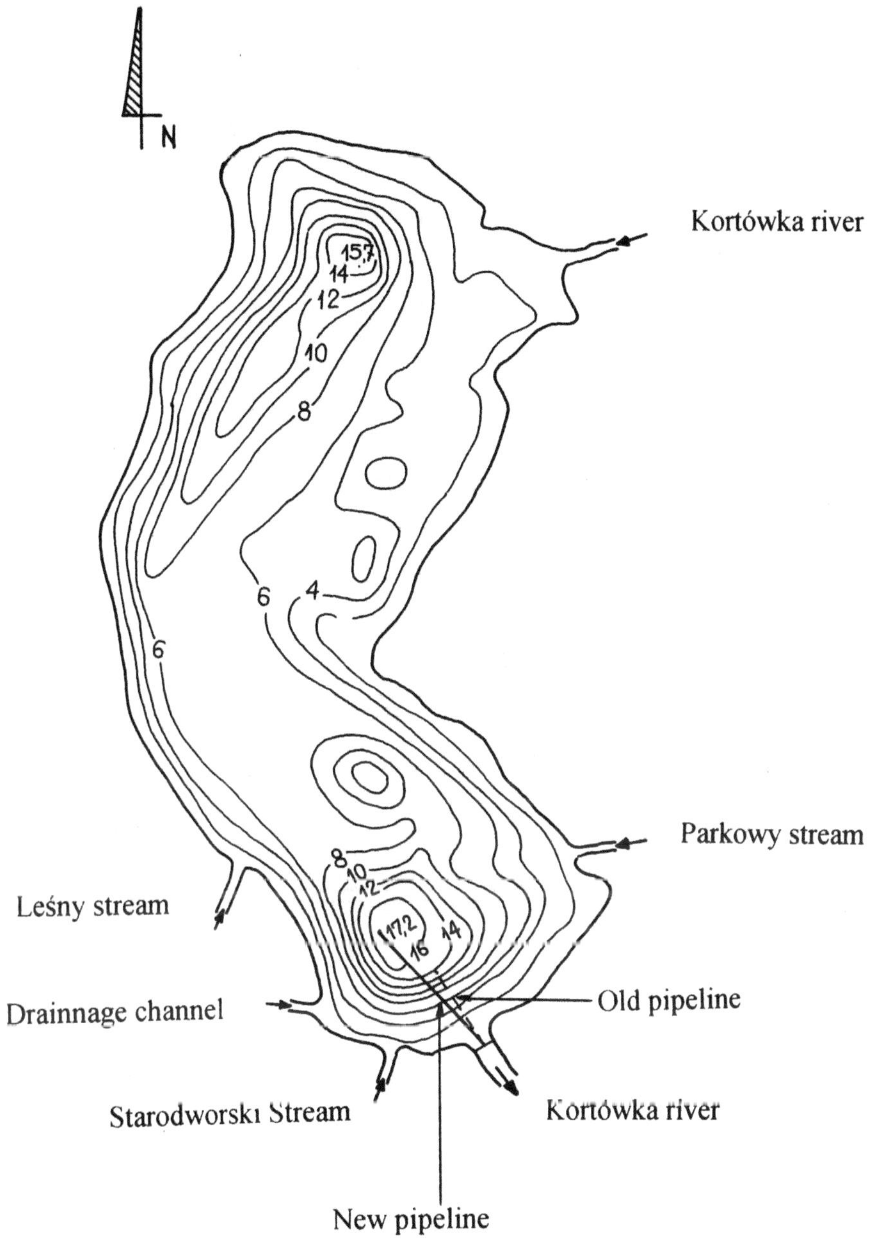

Figure 1. Bathymetric map of Kortowskie Lake.

appliances, the experiment based on the wooden pipe was closed down. Since then the years 1956–1973 are regarded as the Ist stage of the Lake Kortowskie reclamation. The hydrochemical and hydrobiological effects obtained in that period were summarised by, among others, Olszewski[6] and Sikorowa et al.[7]

In 1974–1975 the new devices were constructed for the experiment. The pipeline with diameter 600 mm, made of glass fibre and polyester resins is 250 m long, with the maximal discharge of 0.25 $m^3 \cdot s^{-1}$. Its inlet is situated at the deepest spot in the southern part of the lake

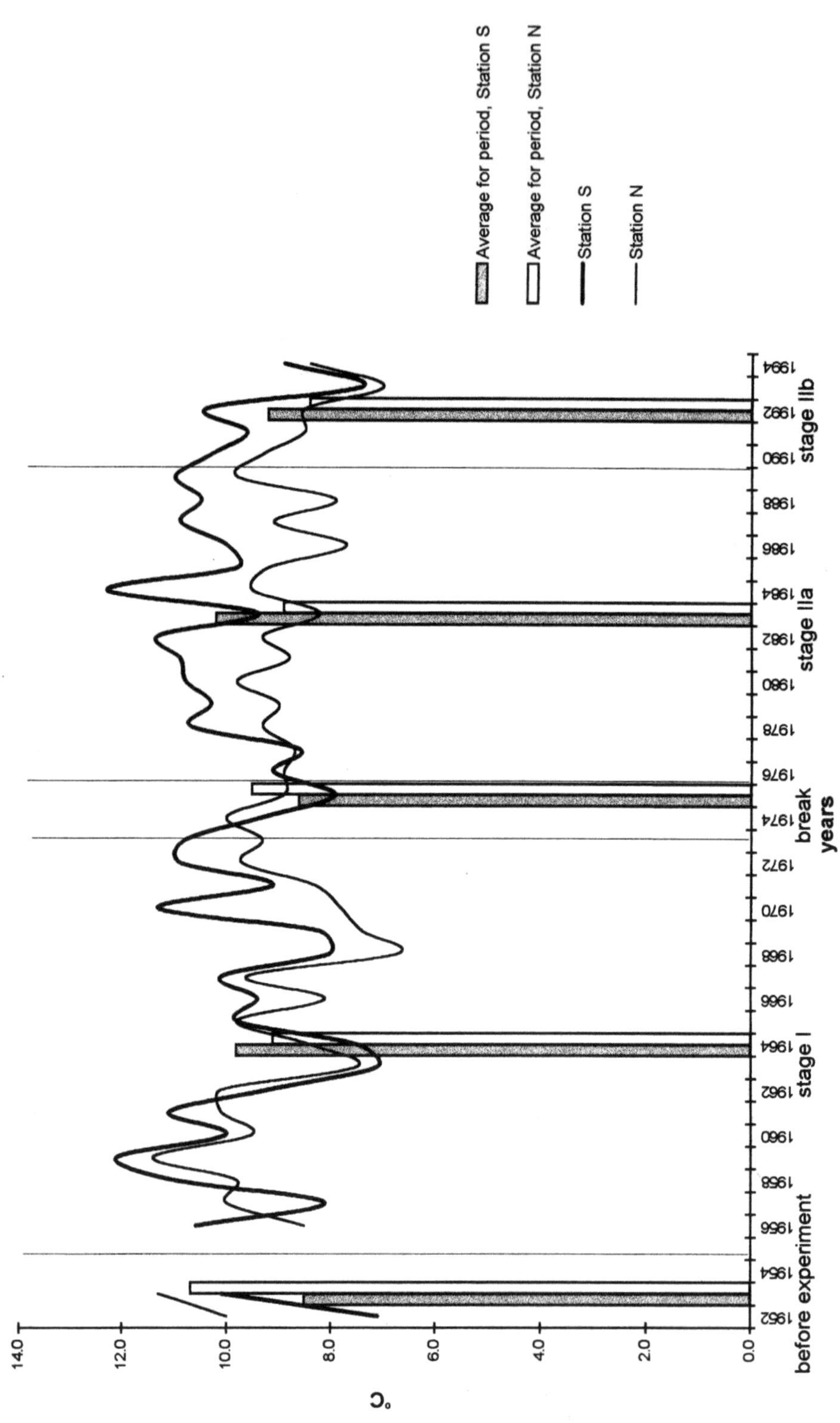

Figure 2. Average summer temperatures in layer below 13 m in years 1952–1994.

(Fig. 1). In 1976 the IInd stage of the lake's rejuvenation was begun. A comprehensive summary of the effects obtained in this stage, till 1982, is provided in Mientki's work.[8]

Till the end of the winter stagnation in 1990 the near-bottom waters were removed at the maximal flow rate (phase II a), and since the summer in 1990 till the end of the winter stagnation in 1995 (phase II b) it was reduced to 0.05 $m^3 \cdot s^{-1}$, in order to examine the effects of the work of the new PVC pipes, with the diameter 200–300 mm. From the beginning of the summer stagnation in 1995 water has been again carried away in the amount 0.25 $m^3 \cdot s^{-1}$.

4. DIRECT EFFECTS OBTAINED FROM THE PIPELINE'S OPERATION AND TENDENCY OF CHANGES IN THE KORTOWSKIE LAKE

The pipeline lays on the bottom in the southern part of the lake and does not have a direct effect on the physico-chemical structures in the northern part. The surface layers are not affected by the hypolimnion waters removal either, nor is the entire water volume during circulation.

4.1 Thermal Stratification

The experimental section of the lake freezes predominantly 2 to 3 days earlier and melts 2 to 3 days later than the northern part. This results from better exposure of the latter to wind action.

Spring circulation in the Kortowskie Lake lasts usually from 4 to 6 weeks. The thermal stratification sets up in both parts of the lake, in general, at the same time. The summer stagnation in 1952–1953 (before reclamation), in 1974–1975 (interval after the Ist stage) and in 1990–1994 (phase II b) lasted 1–4 weeks longer in the southern part, subject to the meteorological conditions. In the Ist stage (1956–1973) the summer stagnation terminated generally in the whole reservoir at once. In phase II a (1976–1989) it was shorter in the experimental section by 2 to 9 weeks. This considerable shortening of the summer stagnation results directly from the hypolimnion waters removal and is connected with the turbulence in the thermal structures caused at the big water flows in the pipeline (0.25 $m^3 \cdot s^{-1}$).

This process is accompanied by temperature rise in the bottom layers of water which were always higher in the southern part (Fig. 2).

In the peak of the summer stagnation, in the second half of August, the temperature at the bottom in the northern part was between 6.3 and 11.5°C, and in the southern part between 6.6 and 16.6°C. The highest temperatures of this water layer were recorded in 1976–1989. Smaller temperature difference between the surface and the bottom waters influenced the maximal thermal gradients in thermocline in the southern section (maximal thermal gradient 4.9°C $\cdot$ m^{-1}). They were lower compared to the north (7.1°C $\cdot$ m^{-1}).

The autumn circulation in the experimental part started at a much higher water temperature compared to the reference section. At the end of this period, just before freezing, the water temperature in the whole lake was equalised.

4.2 Oxygen Stratification

The approximate indicator of a lake's eutrophication degree is the oxygen depletion in deeper water layers. It is mainly determined by the amount of organic matter, auto- and allochtonous, the hypolimnion volume and its temperature.

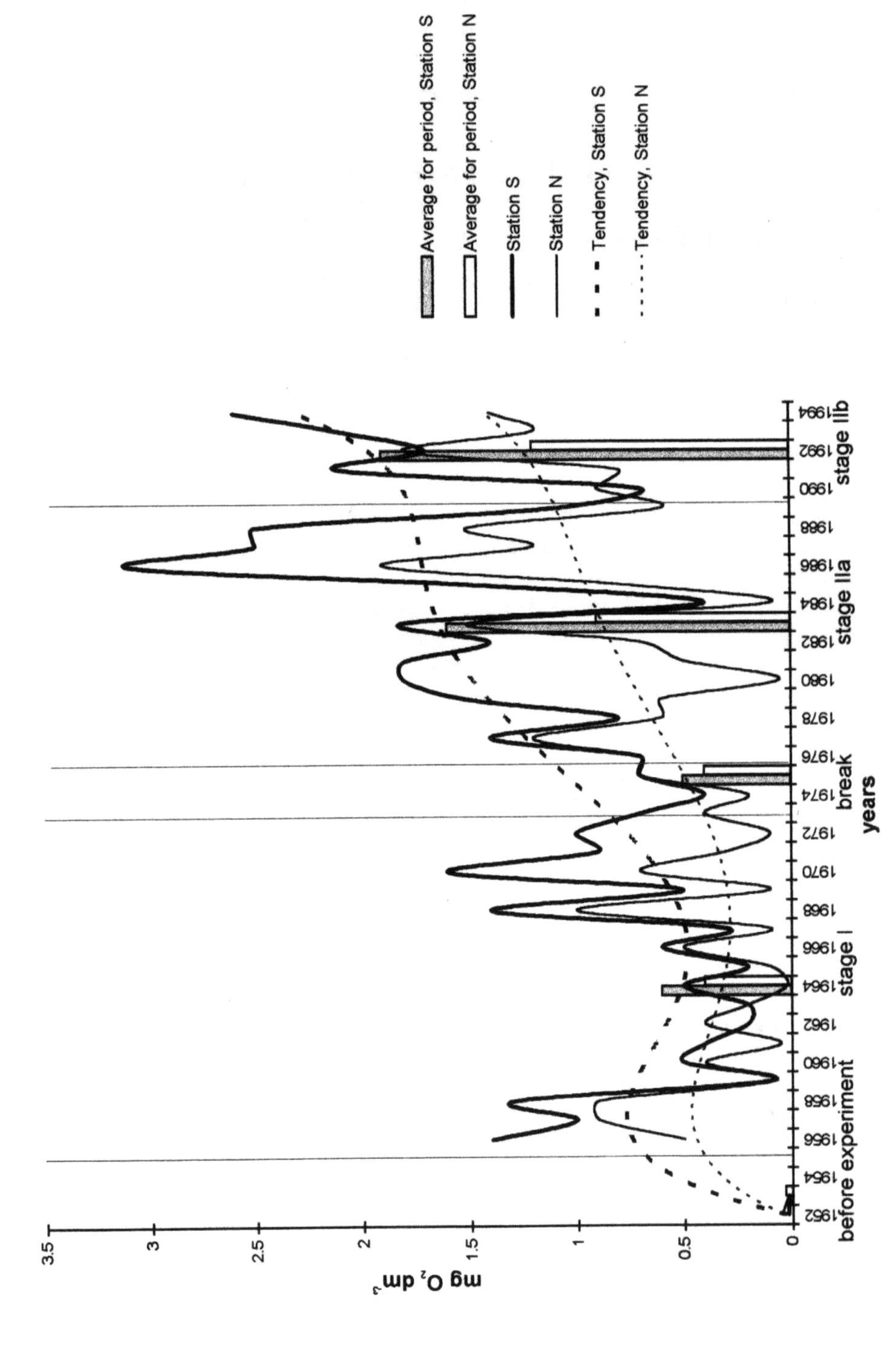

Figure 3. Average summer concentrations of dissolved oxygen in the layer below 13 m in years 1952–1994.

In the early 1950's the oxygen conditions in the Kortowskie Lake were catastrophic. Both in winter and in summer the near-bottom water was devoid of oxygen. Whereas during the winter stagnation oxygen deficits were reported in the water layer up to 2–3 m above the bottom, in the summer stagnation water without oxygen reached from the bottom to 5 m below the surface. At the same time hydrogen sulphide was present in the entire volume.

The hypolimnion waters displacement, since 1956 in summer and since 1968 also in winter, has been causing a gradual improvement of the oxygen conditions in this water layer. The improvement however, did not reach the extend expected by Olszewski,[3] mainly due to: the considerable and increasing (until 1990) lake pollution, not fully efficient pipeline's inlet at 13 m in 1956–1973 (Ist stage) and finally, substantial hypolimnion water warming in the summer months.

The changes observed in the lake were directed towards: decreasing the number of days when the near-bottom water was devoid of oxygen, lowering the vertical stretch of the anaerobic conditions and increasing the concentrations of oxygen in the near-bottom layers. In the years 1952–1953 at the S station the anaerobic conditions lasted maximally 198 days, and at the N station—183 days. During the few years of the IInd phase of reclamation, although in vestigial amounts, oxygen in the near-bottom layers in the experimental section was present over the whole year. Parallel improvement of the aerobic conditions was taking place in the whole lake (Fig. 3). Whereas the anaerobic conditions during the summer stagnation were, on the average, two weeks longer in the southern part before reclamation, in the IInd stage it was, on the average, four weeks shorter here, compared to the north.

The vertical stretch of the anaerobic water also lowered. In the IInd phase, in the southern part it reached as much as 15 m on the average (max. depth 17.2 m) and 11 m in the northern part (max. depth 15.6 m). This means that in 1952–1953, in the summer stagnation 40% of water in the whole lake was deprived of oxygen, but in the IInd stage only 1% at the st. S and 4% at the st. N.

Despite the clear tendency of the aerobic conditions improvement in the near-bottom layers during the summer stagnation (Fig. 3), in hypolimnion the oxygen concentrations are still small. A considerable increase has occurred during the winter stagnation, especially in the southern part of the lake (Tab. 1). In this season mineralization of organic substance, excluding its limited input from the catchment, runs at a much lower water temperature (mean water temp. below 7 m is 2.5°C) than in summer (mean temp.

Table 1. Average oxygen concentration (mg $O_2 \cdot dm^{-3}$) in the deeper layer of water

Period of study	Depth [m]	Years									
		1952 - 1953 1		1956 - 1973 2		1974 - 1975 3		1976 - 1989 4		1990 - 1994 5	
		Station									
		S	N	S	N	S	N	S	N	S	N
Winter stagnation	8 - 12	4.0	2.2	5.1	4.0	6.9	5.7	7.2	5.3	10.8	11.3
	13 - bottom	1.5	0.4	2.8	2.2	4.2	2.6	6.1	2.7	9.6	8.8
	8 - bottom	3.5	2.0	4.7	3.9	6.4	5.5	7.0	5.1	10.6	11.1
Summer stagnation	8 - 12	1.9	2.4	2.3	1.7	1.7	1.9	2.3	2.6	3.5	2.8
	13 - bottom	0.1	0.1	1.0	0.4	0.5	0.4	1.6	0.9	1.9	1.2
	8 - bottom	1.5	2.2	2.0	1.6	1.4	1.8	2.2	2.4	3.2	2.7

1 - before the experiment, 2 - I stage, 3 - break, 4 - IIa stage, 5 - IIb stage

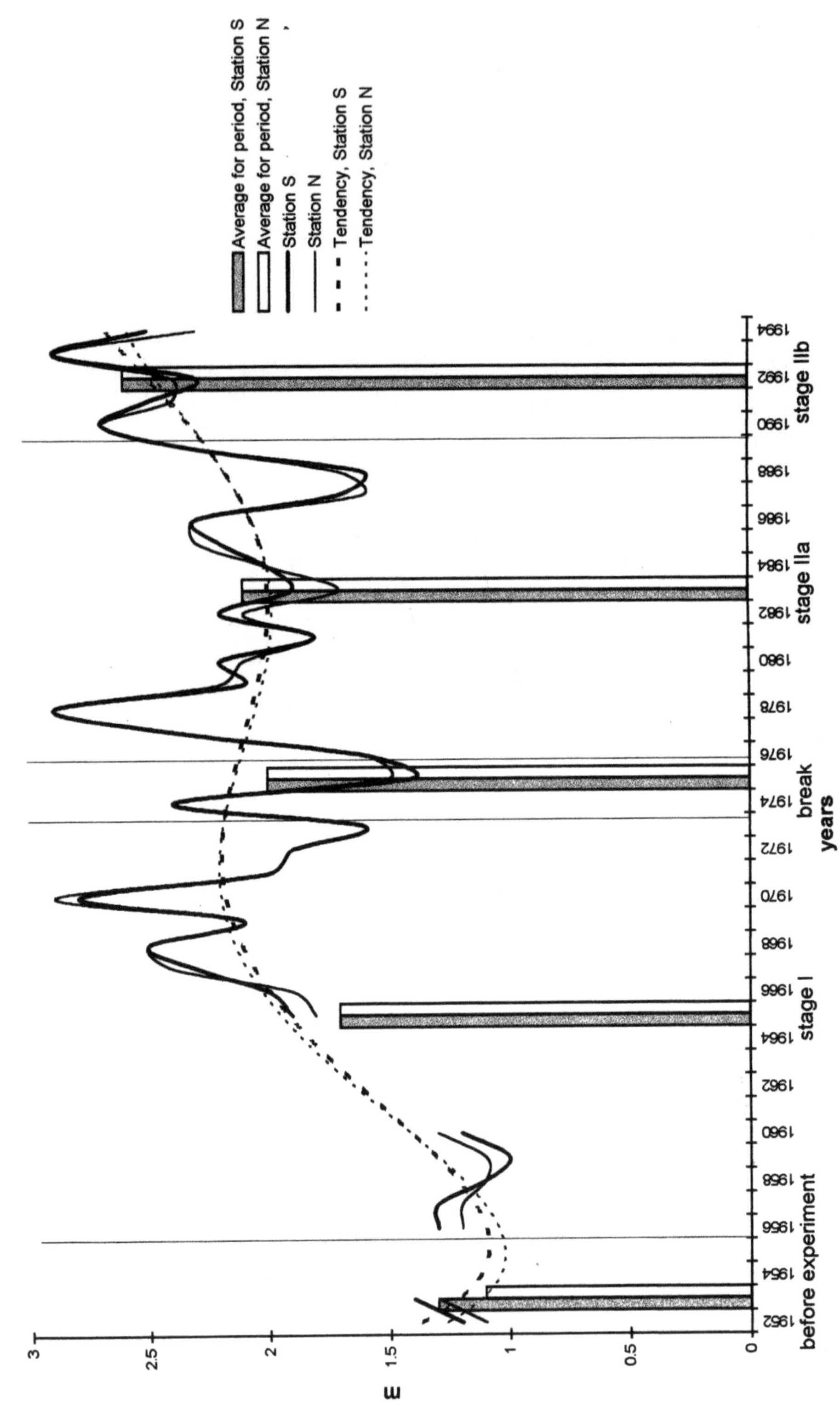

Figure 4. Visibility of Secchi disk in years 1952–1994 (average spring and summer).

11–12°C, and even higher in the IInd phase at the st. S)—water oxygen demand is a few times higher in summer than in winter.

A very important achievement of the experiment in the Kortowskie Lake is the hydrogen sulphide elimination. Its absence, proved by means of the commonly applied chemical analysis methods, has been reported since the early 1970's.

4.3 Visibility

The indirect indicator of phytoplankton concentration in a lake is Secchi disk visibility. In the past 40 years the visibility has considerably increased which, along with other indications, demonstrates the effectiveness of the reclamation processes in the Kortowskie Lake.

As it was earlier emphasised, hypolimnion water removal does not directly affect the surface water layers. Therefore, the displacement of the near-bottom water influences indirectly phytoplankton growth, mainly by limiting the role of the bottom sediments as the source of nutrients and increasing their function as nutrient trap (mainly phosphorus).

In the early 1950's visibility during the summer stagnation was restricted to a value of less than 1 m (min. 0.5 m). It was low in the beginning of the Ist reclamation stage, too. Over the years it has been steadily increasing. Since the surface water layers are mixed freely in the entire water body, the visibility improvement was occurring parallel in both parts of the lake (Fig. 4).

A significant visibility increase has taken place after the year 1968, when the near-bottom waters were also removed in winter, and after 1990 when the main pollution sources of the lake were cut off. This increase reached on the average 0.5 m in relation to the previous periods. Similar changes have been noted not only in the summer stagnation, but also during the autumn circulation. In the latter season visibility in the years 1990–1994 gets close to 4 m on the average, whereas in 1952–1953 it was lower than 3 m.

Complete chemical analyses of the Kortowskie Lake waters, including also phosphorus and nitrogen compounds, have been executed since 1970. The hypolimnion removal has its best effects during summer stagnation. In this season inorganic phosphorus dominates in the total phosphorus in the near-bottom layer, and ammonia in the total nitrogen. Therefore, these have been applied to assess the efficiency of the reclamation method in the Kortowskie Lake.

4.4 Inorganic Phosphorus

The maximal concentrations of inorganic phosphorus in the near-bottom water layers (st. S 15 m and st. N 13 m) ranged from 0.33 to 1.45 mg in the southern part and from 0.53 to 2.48 mg $PO_4 \cdot dm^{-3}$ in the northern. They were recorded always in August and September, with a strong tendency in the northern part to be present in September. The maximal values were predominantly higher at the reference station (st. N), and only in the years 1977, 1984 and 1988 slightly higher (max. by 0.1 mg), and much higher in 1974 (by 0.27 mg $PO_4 \cdot dm^{-3}$) at the experimental section (st. S).

The differences in the maximal concentrations were smaller in both parts of the lake in the upper hypolimnion (10 m), and alike in the near-bottom layers. They were higher in the northern part. In this water layer they ranged between 0.20 and 0.85 mg at the st. S and from 0.22 to 2.14 mg $PO_4 \cdot dm^{-3}$ at the st. N, with the domination of the concentrations up to 0.5 mg in the south and from 0.5 to 1.0 mg $PO_4 \cdot dm^{-3}$ in the north.

Also the mean concentrations of inorganic phosphorus in the near bottom waters calculated for the summer stagnation period were always higher at the st. N. Exclusively

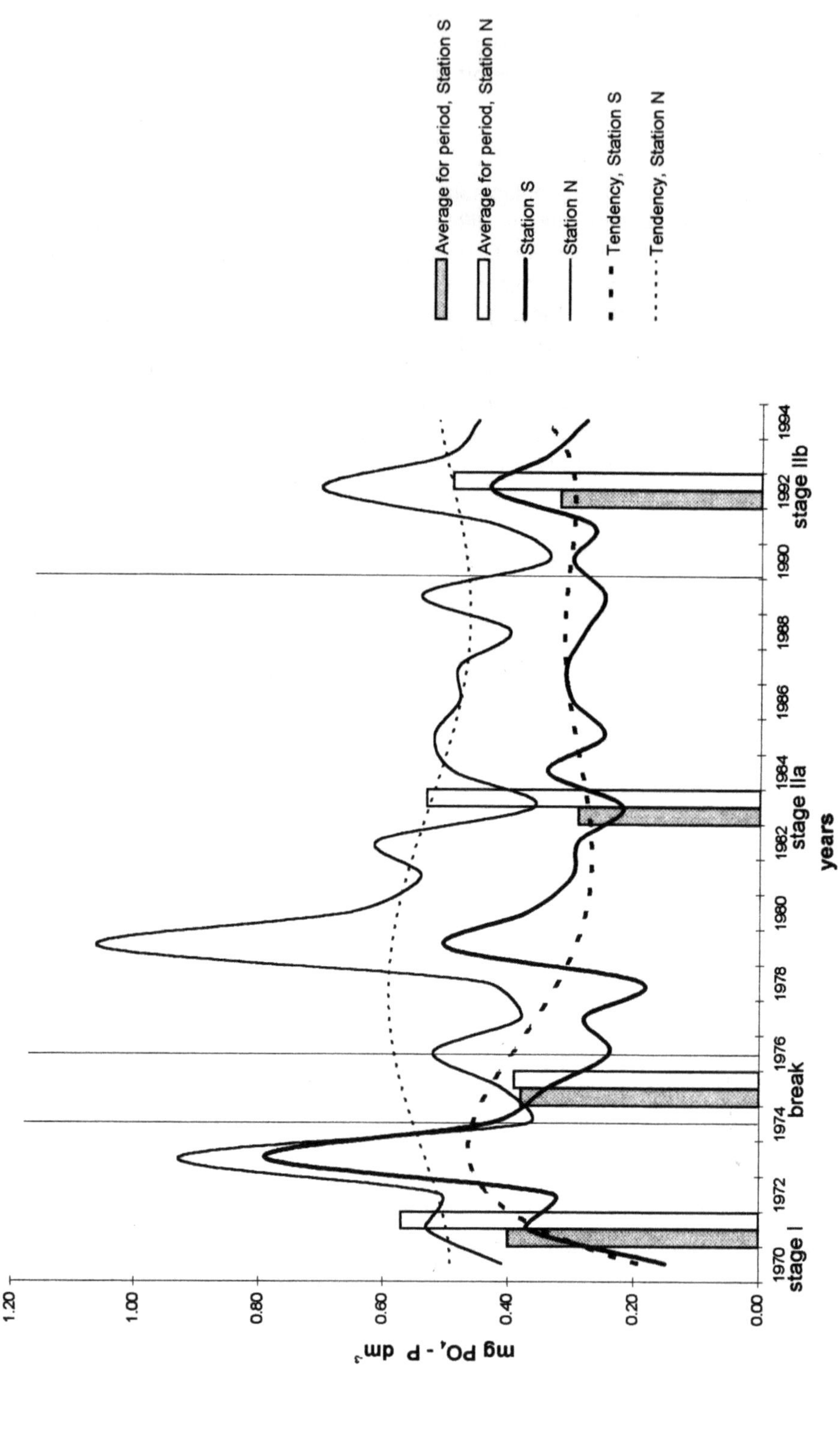

Figure 5. Average concentrations of mineral phosphorus in near-bottom water layer of Kortowskie Lake during summer stagnation.

in 1974, when the pipeline was out of order, the relations were inverse (Fig. 5). Lower values reported in the experimental section were resulting directly from the hypolimnion waters removal to the outflow by means of the pipeline.

The tendency observed over the years of the research of diminishing concentrations, together with the decreased input from the catchment since 1990, results in an important reduction of the eutrophying role of phosphorus and, moreover, in a more radical improvement of the lake's trophic state.

4.5 Ammonia Nitrogen

The differentiation in the ammonia nitrogen concentrations in the near-bottom waters was similar to that of the inorganic phosphorus. However, the maximal concentrations of this form of nitrogen in both the bottom-adjacent waters and in the upper hypolimnion (10 m) were considerably higher. Invariably in all years the concentrations were higher in the north and the difference in values in both parts of the lake was much bigger than in the case of phosphate phosphorus.

At 10 m the maximal concentrations ranged from 0.25 to 2.60 mg with the domination of up to 1.0 mg NH_4-N · dm^{-3} in the south, and from 0.53 to 3.35 mg with the prevalence of up to 2.0 mg in the northern part. In the layers near the bottom the values were higher and ranged accordingly from 0.54 to 4.40 mg and from 1.44 to 4.65 mg with the domination accordingly of up to 2.0 mg and from 2.0 to 3.0 mg NH_4-N · dm^{-3}.

The mean values of the ammonia nitrogen during summer stagnation in the near-bottom waters were also considerably lower at the experimental section in comparison to the reference station, and have shown the tendency to decrease in the following years of the study (Fig. 6).

The lower concentrations of the ammonia nitrogen in the southern part are undoubtedly an effect of the hypolimnion waters displacement through the pipeline. The cut-off of the main pollution sources in the late 1980's and consequently, further improvement of the oxygen conditions in the bottom-adjacent waters, will lead to the redox potential alteration on the sediments surface towards oxidation conditions. The tendency to diminish the concentrations of inorganic phosphorus and ammonia nitrogen in the deeper waters in both parts of the lake, confirms the above-mentioned predicaments. The increase in the redox potential on the bottom deposits surface will eventually lessen internal fertilisation with nutrients, and will further diminish the lake's trophy.

4.6 Balance of Total Phosphorus and Total Nitrogen

Since 1973 on all tributaries and on the outflow from the lake, the water flows have been systematically monitored. Together with the results of the water chemical analyses they are used to calculate the in- and outflowing nutrients' loads. The balance includes also the loads coming from the direct catchment and the atmospheric deposition. In the years 1973–1989 the latter comprised about 10 % of the total load supplying the lake. The reduction of loads from the external sources after 1990 increased the contribution of the direct catchment in the total load to over 16% in II b phase.

The balance of the total phosphorus and total nitrogen for 1974–1994, indicating the reclamation phases, is presented in Tab. 2.

Interpretation of the data obtained is very difficult and requires further and detailed study.

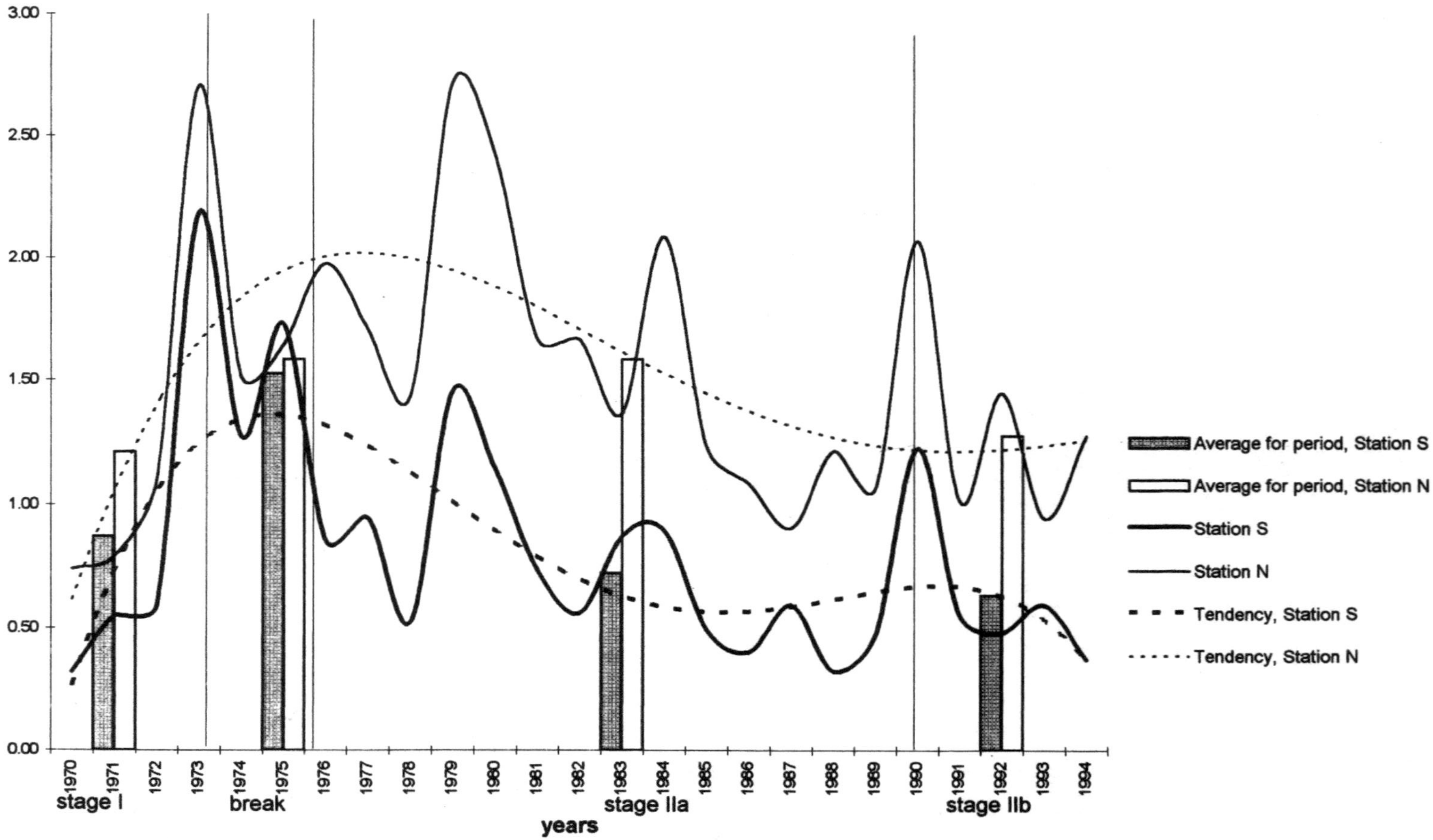

Figure 6. Average concentrations of ammonia nitrogen in the near-bottom water layer of Kortowskie Lake during summer stagnation.

Table 2. Balance of total nitrogen and total phosphorus in Kortowskie Lake

Years	Nutrients	Inflow kg	Outflow kg	Difference kg	Difference %
1974-1975	N_{tot}	25 675	13 245	12 430	48.4
	P_{tot}	2 410	845	1 565	64.9
1976-1989	N_{tot}	255 005	150 835	104 170	40.9
	P_{tot}	25 035	19 330	5 705	22.8
1990-1994	N_{tot}	49 600	29 045	20 555	41.4
	P_{tot}	5 260	4 050	1 210	23.0
1974-1994	N_{tot}	330 280	193 125	137 155	41.5
	P_{tot}	32 705	24 225	8 480	25.9

At the start of the balance study period, the main load of the two nutrients was coming with the Kortówka river to the northern part of the lake. The loads supplying the lake in the south from the LeÑny stream were, in the early 1970's, negligible. The surface outflow and the pipeline's outlet are both situated in the south. After 1990 the loads contributed by the Kortówka river have considerably diminished, but the situation with the LeÑny stream is exactly opposite.

One of the factors determining the balance of nutrients is water balance in the lake. In the average year, in regard to precipitation, there is not enough water to be discharged at the maximal capacity. Usually in the second half of summer stagnation the efficiency of the pipeline declines and consequently, the amount of the removed nutrients declines.

For the years 1973–1982[8] it has been calculated that by discharging exactly the same amount of water through the surface outflow only, the load of the removed total nitrogen is increased by about 3%. At the same time though, the load of the total phosphorus would be less by about 11%. Such calculations, however, can mislead. Lack of swelling devices on the lake causes a rapid discharge of the water surplus in spring, and in summer lack of surface outflow can be expected.

In 1981 the amount of water in the lake guaranteed the maximal flow in the pipeline over the whole summer stagnation with parallel surface discharge. The balance of the total nitrogen was then 21%, and of the total phosphorus negative 28.4%, which means that more phosphorus was carried away from the lake, than carried in. From the above, it can be concluded that the best effects are obtained at the maximal flow in the pipeline together with the surface outflow, during the whole summer stagnation. The results prove also that removal of the hypolimnion waters to outflow is an effective method of lake rejuvenation based on reduction of the phosphorus content.

5. CONCLUSIONS

Removal of the hypolimnion waters directly increases oxygen concentrations and diminishes concentrations of inorganic nitrogen and phosphorus in the southern part of the lake. In spite of the growing pollution, visibility, oxygen and nutrients content show a positive tendency. The balance of nutrients in the Kortowskie Lake proves that the hypolimnion waters removal is an efficient reclamation method, mainly due to elimination of phosphorus from the lake's ecosystem.

ACKNOWLEDGMENTS

The equipment for the research and the study were financed by the Institute of Ecology of the Polish Academy of Sciences, the University of Agriculture and Technology in Olsztyn and by the National Committee for the Scientific Research in Warsaw.

REFERENCES

1. Boström, B., Jansson, M., and Forsberg, C., 1982, Phosphorus release from lake sediments. Arch. Hydrobiol. Beih. Ergebn. Limnol., 18, 5–59.
2. Ohle, W., 1953, Der Vorgang rasanter Seenalterung in Holstein. Die Naturwiss., 40, 153–162.
3. Olszewski, P., 1961, Versuch einer Ableitung des hypolimnischen Wassers aus einem See. Verh. Internat. Verein. Limnol., 14, 855–861.
4. Dunst, R. C., Born, S. M., Uttormark, P. D., Smith, S. A., Nichols, A. S., Peterson, J. O., Knauer, D. R., Serus, S. L., Winter, D. R., and Wirth, T. L., 1974, Survey of lake rehabilitation techniques and experiences. Department of Natural Resources. Tech. Bull. 75. Madison, Wisconsin.
5. Synowiec, A., 1965, Morphology of Kortowo Lake. Zesz. nauk. WSR Olszt., 19, 3–17 (in Polish, Engl. summ.).
6. Olszewski, P., 1973, Fünfzehn Jahre Experiment auf dem Kortowo See. Verh. Internat. Verein. Limnol., 18, 1792–1797.
7. Sikorowa, A., Mientki, C., Chudybowa, D., Widuto, J., Zmys³owska, I., and Sobierajska, M., 1975, Experiment on the rejuvenation of Kortowskie Lake. Preliminary results of the years 1971–1073 and some earlier results. Pol. Arch. Hydrobiol., 22, 251–269.
8. Mientki, C., 1986, The effect of removing hypolimnion waters on thermal and oxygen conditions, and content of nitrogen and phosphorus in water of Lake Kortowskie. Acta Acad. Agricult. Techn. Olszt., Protectio Aquarum et Piscatoria. 14, 1–53 (in Polish, Engl. summ.).

41

COMPLEXATION OF COPPER (II), CADMIUM (II) AND LEAD (II) WITH HUMIC AND FULVIC ACIDS OF YAMUNA RIVER SEDIMENTS

Satyabrata Sahu and D. K. Banerjee

School of Environmental Sciences
Jawaharlal Nehru University
New Delhi 110 067, India

ABSTRACT

The complexation of humic and fulvic acids extracted from Yamuna river sediments, collected in summer and winter months with Cu^{+2}, Cd^{+2} and Pb^{+2} were investigated at different pH of the medium employing ion selective electrode potentiometry. The different complexation parameters were computed from Double Surface Langmuir, Scatchard and Hill plots. The degree of complexation was found to increase with increase in pH and decrease with rise in total metal concentration. Maximum binding abilities of humic acids were higher than those of fulvic acids. As regards the metal ions, the conditional stability constants were found to increase with increase in pH and follow the order: Cu-L > Pb-L > Cd-L. Metal-humates were found to be more stable than metal-fulvates. The ligands extracted from the sediments collected in summer showed stronger metal binding ability than those extracted from winter samples. The cooperative binding ability of the ligands did not indicate any trend of variation and thus was found to be an unreliable parameter to assess the complexation behaviour of the poly-functional humic and fulvic acids towards metals.

INTRODUCTION

Humic and fulvic acids are a group of naturally occurring organic ligands, widely distributed in terrestrial and aquatic environments, formed from the chemical and microbial degradation of dead organic matter. Their molecular weights vary from few hundreds to several thousands. They widely interact with metal ions through a range of ionizable functional groups of differential acidities (1) with important ecological implications. According to Saar and Weber (2) metal–humus complexation controls the mobility, stability and toxicity of metals in aquatic ecosystems. Hering and Morel (3) opined that the speciation and bio-availability of transition metals in aquatic environment are intimately related

Chemistry for the Protection of the Environment 2, edited by Pawłowski et al.
Plenum Press, New York, 1996

to the metal-humus complexation. Of the several heavy metals, copper, cadmium and lead assume special importance as regards their interaction with humic substances on account of their (i) role as micro-nutrients when present in trace amounts, (ii) toxicity when present in excess, and (iii) comparatively high release into the environment.

The present study has investigated the complexation of humic and fulvic acids extracted from Yamuna river sediments in Delhi with copper, cadmium and lead in their divalent states in terms of the complexation parameters, such as degree of complexation, maximum binding ability, conditional stability constant, macroscopic stability constants and cooperative binding ability. The study has, in particular, investigated the influence of the pH and redox status of the sampling sites on the complexation parameters. In addition, the dependence of complexation on the primary source of the humic substances was also assessed.

There has been some quantitative models based on Adair's equation, normally used to examine the ligand-binding properties of proteins and other macro-molecules. The graphical approaches developed on those models, such as Scatchard, Double Surface Langmuir and Hill plots, have been employed in the present work.

Several workers have used Double Surface Langmuir plot for obtaining the maximum binding ability of humus as well as some other macro-molecules (4, 5). Scatchard plot was used by Mantoura and Riley (6) and Bresnahan et al (7) for calculating the conditional stability constants of the complexes. The same plotting method is also used for distinguishing between different classes of binding sites. Humic substances have a range of acidic functional groups, situated in different geometrical locations, which participate in complex formation and which may influence one another in their metal binding ability. Thus, to substantiate the intricate process of metal–humus complexation several workers (8, 9, 5) have used Hill plot to investigate cooperative binding. The same plotting method is also used to determine macroscopic binding constants.

METHODOLOGY

The humic and fulvic acids were extracted from the bottom sediments of four well defined sites of Yamuna river flowing through Delhi (i) Wazirabad (WAZ)—where the river enters Delhi, (ii) Old Yamuna Barage (OYB), (iii) Nizamuddin Bridge (NZB) and (iv) Okhla (OKH)—where the river leaves the city (Figure 1). The river near Wazirabad does not receive any major effluent discharge from the city. The main source of humic substances at this site is the *in situ* decomposition and humification of the local aquatic vegetation and organic matter that comes along with the upstream river water. At Old Yamuna Bridge, the decay of domestic organic wastes contributes to the formation of the bulk of the humic substances. The river receives a bulk of organic input from the main effluent discharge drain of Delhi, the Najafgarh drain, right below the Wazirabad reservoir. At Nizamuddin Bridge, the decay of natural vegetation and agricultural wastes contribute to the humus formation. By the time the river reaches Okhla, it has received effluent discharges from several additional sources, both industrial and domestic. At this site, the sewage treatment works of the city further adds organic-loaded effluents to the river. In addition, Okhla abounds in natural vegetation, especially *Eichornia crassipes*—a submerged aquatic weed. The decay of all this organic matter accounts for a significant fraction of humic and fulvic acids.

Humification is a function of time, temperature, redox status etc. at the location. Delhi experiences two extremes of the ambient temperature variation. Therefore, the sediments were collected from the respective locations once in summer (June) and once in winter (December).

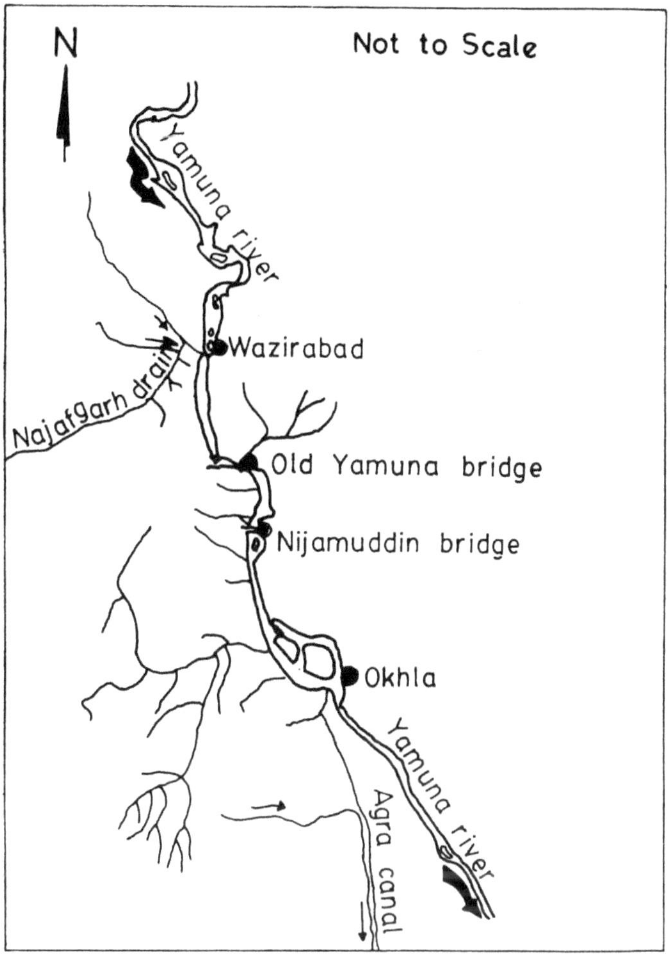

Figure 1. River Yamuna around Delhi showing the sampling sites.

The extraction of the humic and fulvic acids from the sediment samples was done following the method by Holtzclaw et al (10). The complexation of purified humic and fulvic acid extracts with bivalent Cu, Cd and Pb, in aqueous phase, was studied by ion selective electrode potentiometry (11, 5) using Orion Ionalyser (model 901) and Orion specific ion selective electrodes (Cu:94–82; Cd:94–48 and Pb:94–29). Titrations of humic and fulvic acid solutions (30 mg/l) with 10^{-3} molar solutions of Cu^{+2} ($CuNO_3$), Cd^{+2} ($CdNO_3$) and Pb^{+2} ($PbNO_3$) were performed at pH 6.0, 7.0 and 8.0 for ionic strength 0.1. At each step of a titration 0.2 ml of the respective metal ion solution was added to 100 ml of aliquot and the corresponding electrode potential was recorded from the ionalyser. The blank titrations were also performed simultaneously.

The degree of complexation (α) in each step of titration was calculated from the ratio $[M_t]/[M]$ (11); where $[M_t]$ stands for total metal ion concentration and [M] for the free metal ion concentration. In terms of electrode potential, α can be expressed as:

$$\alpha = [M_t] \times 10^{(E_0 - E)/P}$$

where E_0, E and P stand for the standard electrode potential, the recorded electrode potential at some particular step and the slope of the electrode respectively.

The maximum binding abilities (MBA) were determined from the Y-intercepts at higher free metal ion concentration of Double Surface Langmuir Plots (5) drawn by plotting [M] along the X- axis and $[M_b]/[M]$ along the Y- axis, where $[M_b]$ is the bound metal concentration.

Conditional stability constants were obtained by Scatchard's method of plotting $\Theta/[M]$ along Y- axis and Θ along X- axis (5), where Θ stands for the fraction of the reactive sites bound. The stability constants K_1 and K_n were obtained by extrapolating the linear segments of the plots and the subsequent Y- intercepts corresponding to the lower and higher metal ion concentrations.

Hill plots were drawn by plotting log [M] along the X- axis and log $[\Theta/(1-\Theta)]$ along the Y- axis for determining macroscopic binding constants (K_*) and the total number of binding sites (n). K_* was obtained by extrapolating the straight line segment corresponding to higher metal ion concentration range and subsequently finding out the Y- intercept. `n' was calculated from the slope of the line in the vicinity of log $[\Theta/(1-\Theta)]$ approaching 0.

RESULTS AND DISCUSSION

The variation of the degree of complexation (α) with the total metal ion concentration are shown in Figure 2. Log α values were found to vary from near zero to even more than two. Some common trends seen in the variation of log α are the following: (i) log α decreases with increase in metal ion concentration, (ii) it increases with increase in pH, (iii) humic acid complexes show higher values of log α than fulvic acid complexes, (iv) summer samples show higher values of log α than the winter ones.

The Double Surface Langmuir plots shown in Figure 3 are non-linear and concave up. The same nature of the plot has also been reported by earlier workers (4, 5). The concave shape of the plots apparently indicate two major classes of binding sites in the humic or fulvic acid macro-molecules. According to Zunino and Martin (9) 'group a' sites involve strong and weak acidic functional groups capable of forming a chelate configuration involving high level of bonding energy. On the other hand `group b' binding sites would have functional groups either capable of forming open complexes, involving intermediate level of bonding energy or active sites buried within the macro-molecule, in which case they are less accessible to the metal ions and therefore, associated with low bonding energy. Further, the assumption of two centres of pK_a distribution in humic acid molecules has been confirmed by acid–base titration data (12).

The MBA values (*Table 1a,b*) for humic acids were found to remain around 2.5 x 10^{-5}M. Higher MBA values were obtained as the medium became alkaline. These values were commonly found to be less than 1.0 x 10^{-5}M and in some instances even less than 0.5 x 10^{-5}M for fulvic acids but not for humic acids, which showed comparatively higher values. Normally, fulvic acids have substantially higher total acidity and -COOH group contents as compared to humic acids, therefore, the former are expected to exhibit higher metal binding capacity; which is not in accordance with the experimental results. This contradiction might be resolved by arguing that for humic acids physical adsorption could play a significant role in metal binding along with complexation. This is because relatively more extensive aromatic condensation occurs in humic acids which along with the greater surface area and consequent augmented surface activity of humic acids is responsible for their higher MBAs. The fulvic acids, on the other hand contain more of aliphatic structures, present mostly as side chains.

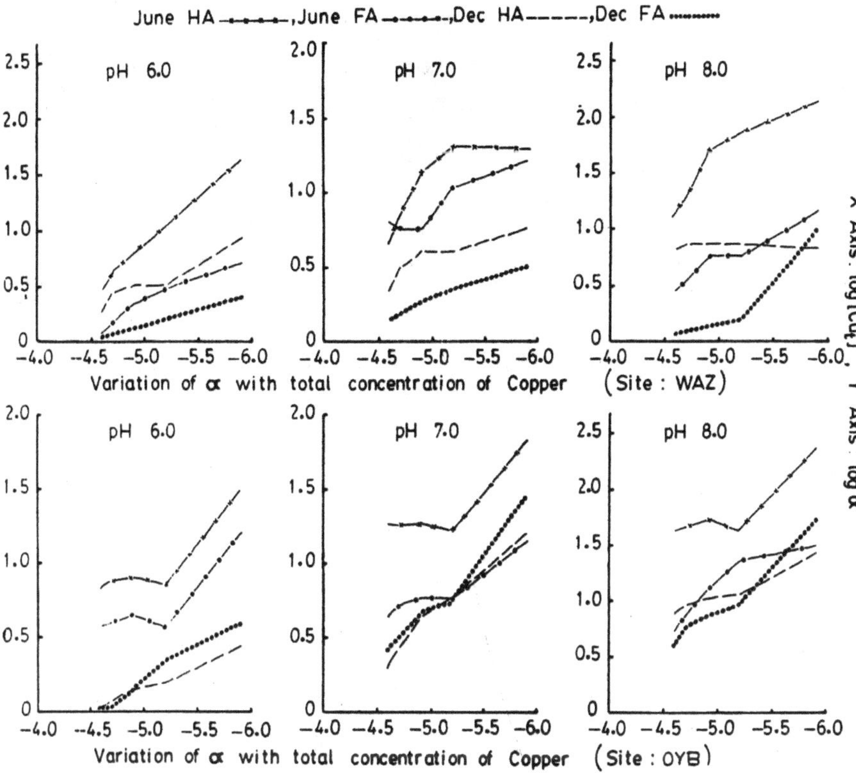

Figure 2. Variation of α with total concentration of copper (site: OYB).

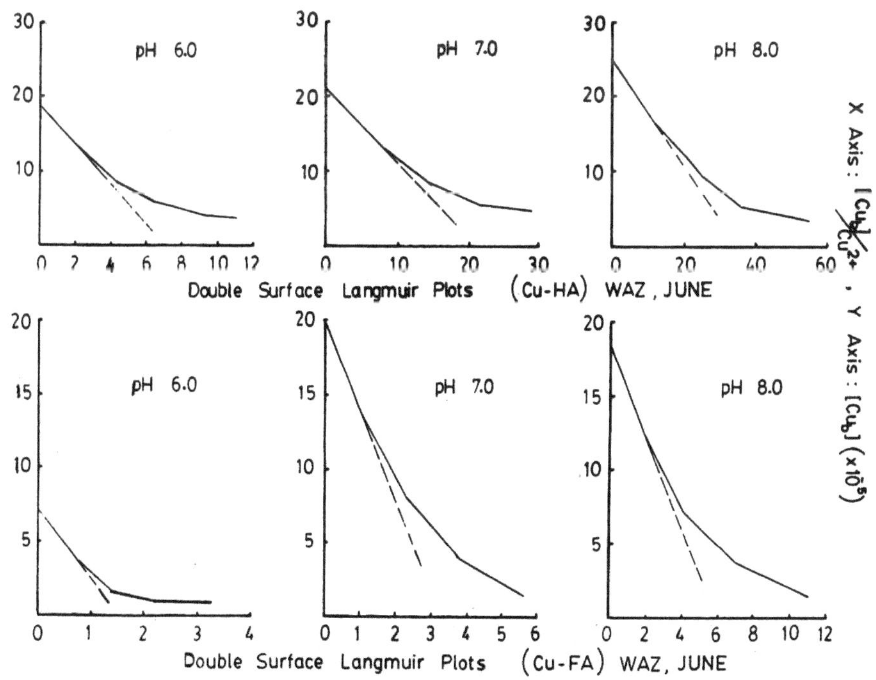

Figure 3. Double surface Langmuir plots (Cu-FA) Waz, June.

Table 1. MBA (x 10^{-5}) of humic and fulvic acids for bivalent Cu, Cd and Pb at ionic strength 0.1

	pH	WAZ June	WAZ Dec	OYB June	OYB Dec	NIZ June	NIZ Dec	OKH June	OKH Dec
(a) MBA of humic acids									
Cu									
	6.0	1.91	2.09	2.17	0.44	2.21	1.70	2.57	2.35
	7.0	2.10	2.17	2.39	1.57	2.47	1.94	2.76	2.39
	8.0	2.47	2.83	2.42	2.37	2.59	2.16	2.49	2.35
Cd									
	6.0	1.95	1.31	1.97	0.86	2.00	1.65	0.30	1.47
	7.0	2.34	2.08	2.49	2.10	2.13	1.92	0.36	2.23
	8.0	2.09	1.76	2.52	2.35	2.22	1.90	0.92	2.29
Pb									
	6.0	2.28	1.53	2.08	1.05	2.22	1.99	2.13	2.22
	7.0	2.43	2.09	2.33	1.56	2.61	2.04	2.34	2.51
	8.0	2.44	2.20	2.40	2.04	2.51	1.97	2.62	2.46
(b) MBA of fulvic acids									
Cu									
	6.0	0.71	0.42	1.81	0.55	2.24	1.94	2.15	0.33
	7.0	2.00	0.65	2.09	1.75	1.33	0.62	2.66	1.53
	8.0	1.80	0.40	2.16	1.96	1.42	0.66	2.34	1.89
Cd									
	6.0	1.33	0.66	0.64	0.35	0.75	0.63	0.25	0.30
	7.0	1.90	0.47	2.33	0.31	0.92	0.70	1.51	0.37
	8.0	2.15	1.72	0.91	1.23	0.96	0.83	1.89	0.77
Pb									
	6.0	1.80	1.23	1.29	0.96	0.54	0.30	1.99	0.77
	7.0	1.92	1.74	2.04	1.45	1.16	0.67	1.80	0.94
	8.0	2.22	1.82	2.40	1.66	1.39	0.77	2.30	1.06

MBA = Maximum Binding Ability, WAZ = Wazirabad, OYB = Old Yamuna Bridge, NZB = Nizamuddin Bridge, OKH = Okhla, Dec = December

Most of the functional groups would be attached to these side chains than on aromatic rings. This would contribute mostly to complexation rather than to adsorption.

The Scatchard plots (Figure 4) drawn for obtaining conditional stability constants were found to be non-linear; and the nature of the curvilinearity indicated a continuum of binding sites in the macro-molecules. However, the plots could be divided into two apparent parts, indicating two distinct classes of binding sites. The non-linearity of the Scatchard plots indicated the formation of 2:1 complexes with metal ion as the central species (5, 11). The formation of 2:1 metal–humus complexes along with 1:1 complexes of has also been suggested (13). However, 2:1 complexes are not likely to be formed within a single aggregate; rather the metals would serve to establish a link between different humus macro-molecules to form extended chain like structures (13). Tao et al. (14) reported the formation of 1:1 and 1:2 complexes of humic and fulvic acids at different concentration ratios of metal to the humic substances.

The conditional stability constants (log K_1 and log K_n, given in *Table 2a,b,c,d*) showed higher values at lower metal ion concentrations. This is explained by the preferential binding of divalent metal cations to the functional groups of stronger binding strengths at low metal ion concentration and to the weaker functional groups at high metal ion concentration (15).

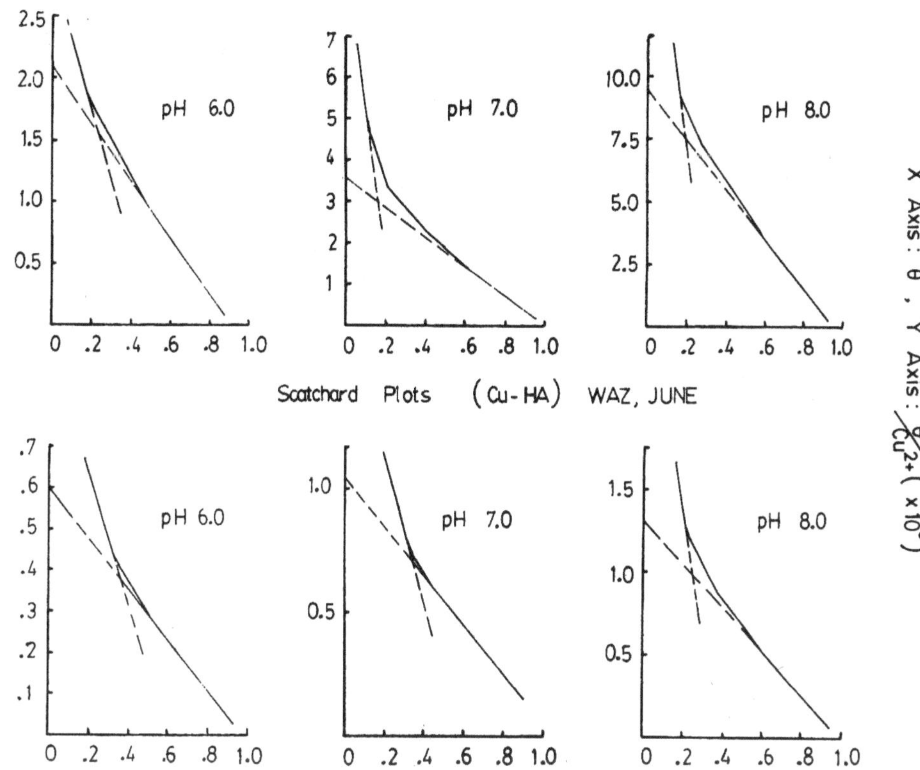

Figure 4. Scatchard plots (Cu-FA) Waz, June.

The log K_n values were found to rise with increase in pH of the medium. The dependence of the conditional stability on pH of the medium may be explained (15) in the following ways: (i) in higher pH range metal cations would be bound to only those ionized functional groups which have potentials of forming strong complexes; in lower pH, however, such functional groups may be partly occupied by hydrogen ions, and (ii) charge neutralization and the steric stabilities achieved with chelation would facilitate complex formation at higher pH. Town et al. (16) have shown a positive correlation between pH and the percentage of carboxyl groups involved in copper complexation. Unlike the log K_n values, which showed a uniform variation with pH, the log K_1 in some instances showed an irregular trend. These deviations could be accounted for by the low sensitivity of the electrodes at low metal ion concentrations.

With respect to metals the stability of the humic and fulvic acid complexes followed the order: Cu-L > Pb-L > Cd-L. This observation corroborates with the observations of the earlier workers (13, 15). Moreover, this is in well accord with the empirical order of stability of complexes, formed between metals and ligands with O-O linkages, proposed by Perrin (17). The ionic potential of Cu^{+2}, Cd^{+2} and Pb^{+2} are 2.74, 2.17 and 1.69 respectively. This would suggest a maximum covalent character in M-L bonds for copper complexes followed by the cadmium and the lead ones. However, Pb^{+2} is an intermediate Lewis acid and is likely to have higher affinity for intermediate donors of humic or fulvic acid moiety, as compared to Cd^{+2}, which is a moderately soft Lewis acid with fairly high affinity for sulfhydryl groups rather than for the oxygen-carrying donors of humic acids.

Table 2. Conditional stability constants (Log K_1 and K_n) of metal–humic and metal–fulvic acid complexes at ionic strength 0.1

	pH	WAZ June	WAZ Dec	OYB June	OYB Dec	NIZ June	NIZ Dec	OKH June	OKH Dec
(a) Log K_1 constants of metal–humic acid complexes									
Cu									
	6.0	6.91	6.18	7.38	7.18	7.33	7.18	8.05	6.88
	7.0	7.57	6.54	7.60	6.88	7.27	7.12	7.27	6.86
	8.0	7.94	6.27	7.90	7.15	7.80	7.44	8.48	7.65
Cd									
	6.0	6.88	5.78	6.07	5.85	5.90	5.83	6.68	5.65
	7.0	6.90	5.90	7.04	6.12	6.03	5.85	6.26	5.92
	8.0	6.62	6.94	6.42	6.53	6.03	6.19	6.42	6.92
Pb									
	6.0	6.81	5.86	6.20	5.77	7.41	6.15	6.90	6.85
	7.0	6.88	6.10	6.58	6.16	7.88	6.60	7.20	7.70
	8.0	7.60	6.64	7.48	6.19	7.67	7.12	7.85	7.70
(b) Log K_1 constants of metal–fulvic acid complexes									
Cu									
	6.0	6.24	6.32	6.80	5.40	5.81	5.10	6.90	6.06
	7.0	6.44	6.54	7.04	6.24	6.51	6.14	6.99	5.97
	8.0	6.88	6.51	7.35	5.27	6.39	6.48	7.17	6.07
Cd									
	6.0	5.87	5.49	5.57	5.90	5.22	5.13	5.45	4.85
	7.0	6.54	6.48	5.85	5.26	5.34	5.45	5.48	5.18
	8.0	6.10	6.60	5.51	5.41	5.30	5.48	5.85	5.32
Pb									
	6.0	6.22	5.91	6.02	4.80	6.52	5.45	6.24	5.78
	7.0	6.23	5.91	6.25	4.90	5.81	5.60	7.00	5.98
	8.0	6.41	5.84	6.40	5.13	6.43	5.60	7.03	6.14
(c) Log K_n constants of metal–humic acid complexes									
Cu									
	6.0	6.23	5.57	6.01	5.36	6.34	5.76	6.48	6.08
	7.0	6.56	5.78	6.34	5.78	6.98	6.03	6.78	6.26
	8.0	6.98	5.89	6.78	6.02	7.02	6.27	7.02	6.46
Cd									
	6.0	5.66	5.23	5.62	5.20	5.46	5.26	5.52	5.30
	7.0	5.85	5.45	6.02	5.72	5.58	5.46	5.90	5.68
	8.0	6.08	5.98	6.18	6.13	5.72	5.53	6.18	6.01
Pb									
	6.0	6.24	5.45	5.90	5.30	6.27	5.70	6.44	6.05
	7.0	6.46	5.69	6.23	5.69	6.85	5.98	6.73	6.18
	8.0	6.81	5.81	6.54	5.88	6.97	6.10	6.98	6.36
(d) Log K_n constants of metal–fulvic acid complexes									
Cu	6.0	5.78	5.30	5.60	4.48	5.00	4.78	5.90	5.04
	7.0	6.02	5.46	5.84	4.54	5.36	5.18	6.11	5.41
	8.0	6.12	5.51	6.02	5.08	5.62	5.36	6.34	5.66
Cd									
	6.0	5.20	4.98	4.78	4.38	4.75	4.41	4.98	4.68
	7.0	5.54	5.13	4.90	4.89	5.03	5.89	5.20	4.95
	8.0	5.76	5.38	5.18	5.11	5.18	5.01	5.53	5.20
Pb									
	6.0	5.65	5.20	5.45	4.36	4.93	4.71	5.80	4.99
	7.0	5.91	5.40	5.72	4.42	5.30	4.95	6.07	5.34
	8.0	5.99	5.45	5.95	4.69	5.58	5.25	6.29	5.64

WAZ = Wazirabad, OYB = Old Yamuna Bridge, NZB = Nizamuddin Bridge, OKH = Okhla, Dec = December

The conditional stability constants of the metal-humic acid complexes were found to be higher than those of the metal-fulvic acid complexes, despite the fact that the fulvic acids are normally found to have a greater abundance of acidic functional groups as compared to humic acids. This apparent anomaly could be partially explained in terms of chelate effect. The fulvic acid macro-molecules are largely composed of carbohydrate like structures; whereas in humic acids aromatic components predominate and a large proportion of the carboxyl groups are bound to aromatic skeletons (18). Therefore, in humic acid an optimal disposition of active sights might occur in favour of chelation, thus, enabling them to form stronger complexes as compared to fulvic acids. An extensive aromatic condensation in humic acids also might lead to added stability to the complexes through adsorptive interaction.

Another important observation was the summer samples exhibiting higher values of stability constants than their winter counterparts. For underwater soils (recent river sediments), the stability constants depend on the redox status of the samples (19). An intensive oxidizing environment in summer might have a significant impact on the physico-chemical characteristics of humic substances. It may so happen that the prevailing oxidizing environment in summer chemically modulates the humic substances, by extensively oxidizing the functional groups in favour of complexation. Johnsen et al. (20) studied the seasonal variation in composition and complexation properties of aquatic humic substances and found that the spring and summer humus showed substantially higher complexation or adsorption capacity as compared to the others.

Although the stability data of complexes from the four different locations did not differ much, there were some perceptable trends. The log K_n for Okhla and Wazirabad samples were relatively higher than those from Old Yamuna Bridge and Nizamuddin Bridge. The four different sources differed from one another with respect to the organic waste inputs, vegetation and also in the prevailing microbial and enviromental conditions. Thus humic substances extracted from these would differ in the extent of humification and degradation. In addition, at Wazirabad and Okhla the water flow and turbulence of the stream are greatly controlled by a reservoir and a barrage, thereby minimizing the disturbance of sediments and in turn, enhancing the humification. This promotes their complexation capacity. Of the Okhla and Wazirabad samples, the former ones, have generally shown higher conditional stability constants. The river in Okhla region is widely infested with water hyacinth, the dominant form of aquatic vegetation there. On decomposition this produces humic substances capable of forming complexes of higher stability (21).

The Hill plots (Figure 5) were found to be perceptively curved (in some instances straight lines were also obtained). The macroscopic stability constants obtained from the Hill plots (*Table 3a,b*) were generally lower than the conditional stability constants obtained from Scatchard plots. This may be quite expected as the stability data obtained from Scatchard plots designate the intrinsic or microscopic stability constants. However, in some instances the macroscopic stability constants obtained from Hill plots unexpectedly exceeded the conditional stability constants obtained from Scatchard plots. It has to be mentioned that the stability data obtained from Hill plots are considered valid provided the slopes of the asymptotes are unity, which was rarely found in the present study. Moreover, the stability data from Hill plots are acceptable if the binding at one site enhances the binding at the subsequent sites (slope of the line in the vicinity of log $[\Theta/(1-\Theta)]$ approaching 0 is > 1). This is referred to as `positive cooperative binding', the basic assumption of Hill plots. However, the slopes of these lines obtained in the present study were not always > 1 (*Table 4a,b*). Thus, the intensive use of Hill plots, initially developed to study the metal–protein interactions, in the study of metal–humus interactions is open to ques-

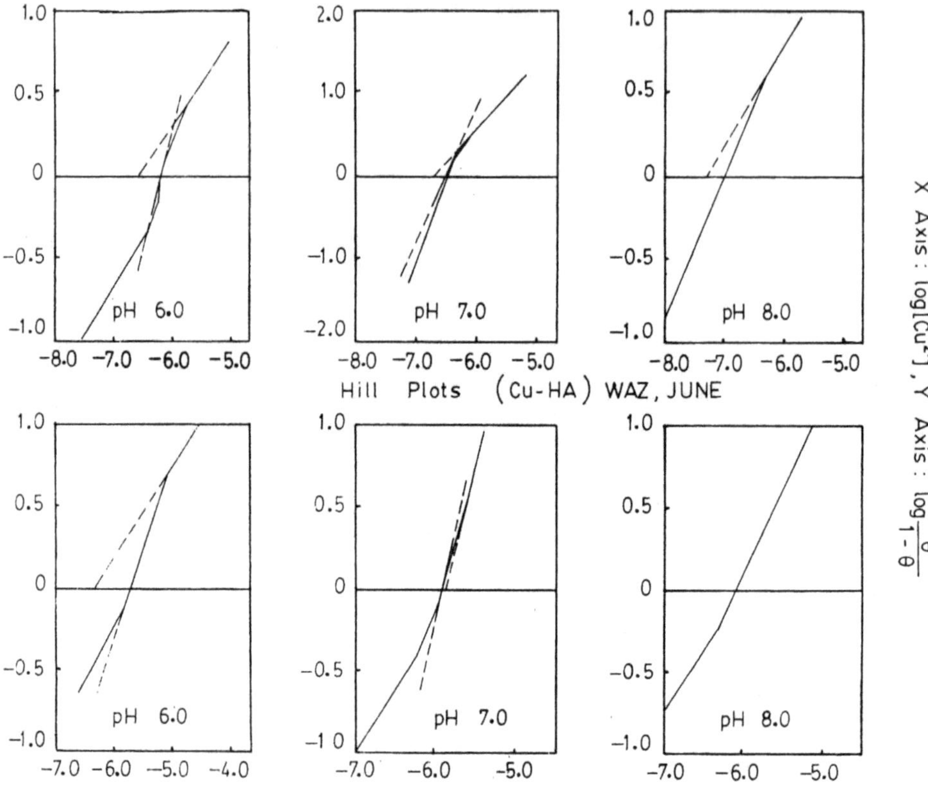

Figure 5. Hill plots (Cu-FA) Waz, June.

Table 3. Macroscopic binding constants (Log K*) of metal–humic and metal–fulvic acid complexes at ionic strength 0.1

		WAZ		OYB		NIZ		OKH	
	pH	June	Dec	June	Dec	June	Dec	June	Dec
(a) Constants of metal–humic acid									
Cu									
	6.0	6.53	5.00	5.87	5.23	6.27	5.33	6.53	6.10
	7.0	6.67	5.80	6.00	5.80	7.00	5.60	6.47	6.33
	8.0	7.27	5.90	6.73	6.00	6.87	6.27	7.00	6.47
Cd									
	6.0	5.40	5.27	5.53	5.07	5.47	5.33	5.53	5.40
	7.0	5.87	5.53	5.93	5.67	5.93	5.47	5.87	5.73
	8.0	6.07	6.00	6.33	6.07	5.33	4.47	5.93	5.60
Pb									
	6.0	5.93	5.47	5.73	4.93	6.07	5.60	6.27	5.73
	7.0	6.33	5.47	5.93	5.33	6.47	6.00	6.27	6.00
	8.0	6.73	5.80	6.53	5.73	6.67	6.07	7.00	6.13

Table 3. (Continued)

	pH	WAZ		OYB		NIZ		OKH	
		June	Dec	June	Dec	June	Dec	June	Dec
(b) Constants of metal–fulvic acid									
Cu									
	6.0	6.33	5.13	5.53	4.47	4.87	4.40	6.00	5.07
	7.0	5.85	5.33	5.67	4.73	5.47	4.87	6.13	5.27
	8.0	6.08	5.53	5.86	4.93	5.20	5.17	6.40	5.80
Cd									
	6.0	5.13	4.93	4.80	4.65	4.80	4.40	4.87	4.47
	7.0	5.40	4.87	4.80	4.87	4.73	4.87	5.27	5.00
	8.0	5.67	5.33	5.07	4.93	5.00	4.93	5.33	5.00
Pb									
	6.0	5.73	5.00	5.07	4.50	4.73	4.53	5.67	4.80
	7.0	5.67	5.40	5.74	4.87	5.27	4.87	5.67	5.12
	8.0	5.70	5.33	5.93	5.15	5.40	5.13	6.07	5.13

WAZ = Wazirabad, OYB = Old Yamuna Bridge, NZB = Nizamuddin Bridge, OKH = Okhla, Dec = December

Table 4. Number of cooperative binding sites (n) of humic and fulvic acids for bivalent Cu, Cd and Pb at ionic strength 0.1

	pH	WAZ		OYB		NIZ		OKH	
		June	Dec	June	Dec	June	Dec	June	Dec
(a) Number of sites for humic acids									
Cu									
	6.0	1.25	0.61	0.88	0.54	1.03	1.00	1.13	1.13
	7.0	1.50	0.88	0.94	0.94	1.00	0.90	1.05	1.38
	8.0	0.80	0.92	2.00	0.96	1.04	0.70	0.56	1.63
Cd									
	6.0	1.25	1.25	1.31	0.94	1.50	1.19	1.18	1.50
	7.0	0.70	1.95	1.05	1.38	1.25	1.41	1.13	1.42
	8.0	1.75	1.16	1.80	1.50	1.33	1.38	1.15	0.75
Pb									
	6.0	1.36	0.93	1.18	0.90	1.29	0.81	0.61	1.02
	7.0	1.50	1.05	1.00	0.83	0.83	0.75	0.88	0.90
	8.0	1.20	1.03	1.25	0.94	1.18	0.89	1.26	1.05
(b) Number of sites for fulvic acids									
Cu									
	6.0	1.39	0.60	1.08	0.83	0.80	0.58	1.02	0.75
	7.0	2.25	0.58	1.06	1.22	1.19	1.07	1.18	0.98
	8.0	1.06	0.54	1.13	1.23	0.83	0.61	0.96	1.50
Cd									
	6.0	1.00	1.13	0.75	1.67	0.64	0.69	0.90	1.04
	7.0	1.80	1.43	1.31	1.29	1.22	1.29	1.29	1.13
	8.0	1.97	1.41	0.94	1.05	1.13	1.59	1.07	1.28
Pb									
	6.0	1.39	0.83	0.75	1.74	0.75	0.69	0.81	0.82
	7.0	1.13	1.82	1.05	1.67	0.83	0.75	0.82	0.95
	8.0	0.92	1.06	1.39	1.66	0.94	0.94	0.75	1.20

WAZ = Wazirabad, OYB = Old Yamuna Bridge, NZB = Nizamuddin Bridge, OKH = Okhla, Dec = December tion. They may not be ideal to explain the intricate interaction of poly-disperse, oligo-electrolyte humus macro-molecules with metals.

As an outcome of the sustained investigations over the past few decades the major contours of how humic substances play a functional role in natural environments, such as controlling the bio-availability of micro-nutrients, binding the inorganic and organic chemicals, particularly the fertilizers, pesticides etc., have been broadly delineated. However, the queries, like what will happen when the adsorption or retention capacities of humic materials are exhausted, have to be duly addressed to. A better understanding of their role in the environment is contingent upon the elucidation of their chemical structure. The recent developments in instrumentation have helped in obtaining valuable information on the `building blocks' that constitute the humus macro-molecules. However, little is known about how they align and the subsequent structural arrangement. Further, there is need to ascertain how the information generated from the metal–humus complexation studies could be possibly used towards the pollution abatement of certain eco-toxic metals.

ACKNOWLEDGMENT

The authors are thankful to Rajkumar Ghosh for concisely editing and updating the manuscript.

REFERENCES

1. Reuter, J. H. and E. M. Perdue, Importance of Heavy Metal Organic Matter Interactions in Natural Waters. *Geochim. Cosmochim. Acta.* (1977) 41:325–34.
2. Saar, R. A. and J. H. Weber, Comparison of Spectrofluorometry and Ion-Selective Electrode Potentiometry for Determination of Complexes between Fulvic Acid and Heavy-Metal Ions. *Anal. Chem.* (1980) 52:2095–2100.
3. Hering, J. G. and F. M. M. Moral, Humic Acid Complexation of Calcium and Coppper. *Env. Sci. Tech.* (1988) 22:1234–37.
4. Sposito, G., On the use of Langmuir Equation in the Interpretation of Adsorption Phenomenon II. The Two-Surface Langmuir Equation. *Soil Sci. Soc. Am. J.* (1982) 46:1147–52.
5. Fitch, A. and F. J. Stevenson, Comparison of Models for Determining Stability Constants of Metal Complexes with Humic Substances. *Soil Sci. Soc. Am. J.* (1984) 48:1044–49.
6. Mantoura, R. F. C. and J. P. Riley, The Use of Gel Filtration in the Study of Metal Binding by Humic Acids and Related Compounds. *Anal. Chem. Acta.* (1975) 78:193–200.
7. Bresnahan, W. T., C. L. Grant and J. H. Weber, Stability Constants for Complexation of Copper (II) Ions with Water and Soil Fulvic Acids Measured by an Ion Selective Electrode. *Anal. Chem.* (1978) 50:1675–79.
8. Cornish-Bowden, A., and D. E. Koshland, A General Method for the Quantitative Determination of Saturation Curves for Multisubmit Proteins. *Biochemistry.* (1970) 9(17):3325–3336.
9. Zunino, H., and J. P. Martin, Metal Binding Organic Macromolecule in Soil; 2. Characterization of the Maximum Binding Ability of the Macromolecules. *Soil Sci.* (1977) 123(3):188–202.
10. Holtzclaw, K. M., G. Sposito and G. R. Bradford, Analytical Properties of the Soluble Metal Complexing Fractions in Sludge-Soil Mixtures. I. Extraction and Purification of Fulvic Acid. *Soil Sci. Soc. Am. J.* (1976) 40:254–258.
11. Buffle, J., F. L. Greter and W. Haerdi, Measurement of Complexation Properties of Humic and Fulvic Acids in Natural Waters with Lead and Copper Ion Selective Electrodes. *Anal. Chem.* (1977) 49:216–22.
12. Tanaka, S., K. Hasebe, M. Taga, and H. Nakamura, Interpretation of the Acid–Base Equilibrium of Humic Hcid by a Continuous pK Distribution and Electrostatic Model. *Anal. Chem.* (1995) 302(2–3):365–373.
13. Stevenson, F. J., Nature of Divalent Transition Metal Complexes of Humic Acids as Revealedv by a Modified Potentiometric Titration Method. *Soil Sci.* (1977) 123(1):10–17.
14. Tao, Z., and H. Gao, Use of Ion Exchange Method for the Determination of Stability Constants of Thorium with Humic and Fulvic Acids. *Radiochim. Acta.* (1994) 65(2):121–123.

15. Takamatsu, T. and T. Yoshida, Determination of Stability Constants of Metal Humic Acid Complexes by Potentiometric Titrations and Ion Selective Electrodes. *Soil Sci.* (1978) 125:377–86.
16. Town, R. M., H. Powell and J. Kipton, Ion-Selective Electrode Potentiometric Studies on the Complexation of Copper (II) by Soil derived Humic and Fulvic Acids. *Anal. Chim. Acta.* (1993) 279(2):221–233.
17. Perrin, O. D., Organic Complexing Reagents. *Chemical Analysis Series, Vol. XVIII,* Interscience, New York, (1964) pp. 34–53.
18. Frund, R., H. D. Ludemann, F. J. Gozalez Villa, G. Almendros, J. C. del Rio and F. Martin, Structural Differences between Humic Fractions from Different Soil Types as Determined by FT-IR and *13*C-NMR Studies. *Sci. Tot. Environ.* (1989) 81/82:187–194.
19. Fischer, W. R., Complexation of Heavy Metals by Water Soluble Humic Substances. S. E. Lindberg and T. C. Hutchinson (Eds.) *International Conference on Heavy Metals in the Environment.* New Orleans, E P Consultants, Edinburg, (1987) pp. 304–306.
20. Johnsen, S., K. Martinsen, G. E. Carlberg, E. T. Gjessing, G. Becher and M. Laegreid, Seasonal Variation in Composition and Properties of Aquatic Humic Substances. *Sci. Tot. Environ.* (1987) 62:13–25.
21. Sahu, S., and D. K. Banerjee, Complexation Properties of Typical Soil and Peat Humic Acids with Copper (II) and Cadmium (II). *Int. J. Environ. Anal. Chem.* (1990) 42(1–4):35–44.

APPLICATION OF A PASSIVE SOIL VAPOR SURVEY AT A FORMER MANUFACTURED GAS PLANT

Bruce A. Scamoffa,[1] John P. Cusick,[2] and Richard Leckenwalter[3]

[1] South Jersey Gas Company
Folsom, New Jersey
[2] W.L. Gore & Associates
Elkton, Maryland
[3] W.L. Gore & Associates GmbH
Putzbrunn, Germany

ABSTRACT

Soil gas screening methods have long been utilized as a means for reducing overall site investigation costs by minimizing the number of soil borings and monitoring well samples required to suitably characterize a site. While active soil gas methods are capable of detecting volatile organics, they are not applicable in polycyclic aromatic hydrocarbon (PAHs) investigations due to the lower vapor pressure of these target compounds. Some passive soil gas methods are capable of detecting low vapor compounds but , until now, have shown limited success on manufactured gas plant (MGP) sites (1).

Advances in passive soil gas technology are discussed here along with an exemplary application of this new technology on a former MGP site in the Northeast United States.

SCREENING TECHNOLOGIES AT MGP SITES

Screening level methods employed at MGP sites range from immunoassay test kits to active and passive soil gas techniques. While immunoassays can provide quick, on-site answers for a preselected range of hydrocarbons, they are limited to discretely collected soil samples only. One advantage of some soil gas methods as compared to immunoassays is the ability to indirectly detect subsurface contaminants at greater depths than the sampling point—in some cases detecting contaminants in the saturated zone twenty-seven (27) feet below the sampling depth (2).

Chemistry for the Protection of the Environment 2, edited by Pawłowski et al.
Plenum Press, New York, 1996

Active soil gas methods can provide relatively quick answers also but are traditionally limited to volatile organic compounds and to highly permeable sites. These factors significantly limit their applicability in typical MGP site investigations where low vapor pressure compounds are of primary interest and shallow ground water is not uncommon.

Passive soil gas techniques have shown some utility on MGP sites by detecting low vapor pressure, high molecular weight organics due to their time-integrated sampling approach. However, collector design and analytical techniques have limited more widespread successful applications (1). Recent advances address some of these limitations of early passive technologies.

NOVEL PASSIVE COLLECTOR DESIGN

A unique approach to passive soil gas sampling was introduced two years ago by W.L. Gore & Associates, Inc., Elkton, MD, USA which takes advantage of the unique properties of a fluoropolymer membrane. Expanded polytetrafluoroethylene (ePTFE) membranes are hydrophobic, microporous structures commonly marketed as GORE-TEX®* membranes or laminates. The incorporation of this polymer into a passive soil gas sampler design allows for high vapor transfer but preserves sample integrity by preventing direct contact with water and soil particles.

GORE-SORBER® Modules are patented, sorbent-based passive collector apparatus which overcome the limitations of traditional collectors. The sorbent containers are housed inside insertion/retrieval cords each of which is constructed solely of GORE-TEX® ePTFE allowing for vapor transfer across the cords as well as the entire surface area of the sorbent containers. The entire apparatus resembles a "shoestring" (Figure 1) and is typically 1 meter long. This design also allows for flexibility as to the choice of adsorbents utilized inside the collectors.

Installation is achieved by creating a small pilot hole (18 mm diameter) in the ground with hand tools and inserting the module with the aid of a simple insertion tool placed in a specially designed insertion pocket. After the recommended exposure time to the subsurface (typically 14 days) modules are retrieved by hand and shipped in coolers back to the laboratory for analysis. Adsorbed hydrocarbons are typically thermally desorbed from the sorbent containers and analyzed via GC/MS.

SITE INFORMATION & SURVEY DESIGN

A former manufactured gas plant, located in the Northeast United States and in operation from 1910–1925, was chosen for this study. The survey area is characterized by relatively permeable mixed soils consisting of about 1 meter of fill material overlying silty sands, clay sands/sandy clays, silts and sands. Depth to ground water averages 1–1.5 meters on the 1.0 acre parcel that was surveyed. The study area also included nineteen (19) soil borings and six (6) monitoring wells from previous site investigation work. Analytical results from sampling at these locations were used for comparison to the passive soil gas survey results.

---

* GORE-SORBER Module is a registered trademark of W.L. Gore & Associates, Inc.
GORE-TEX is a registered trademark of W.L. Gore & Associates, Inc.

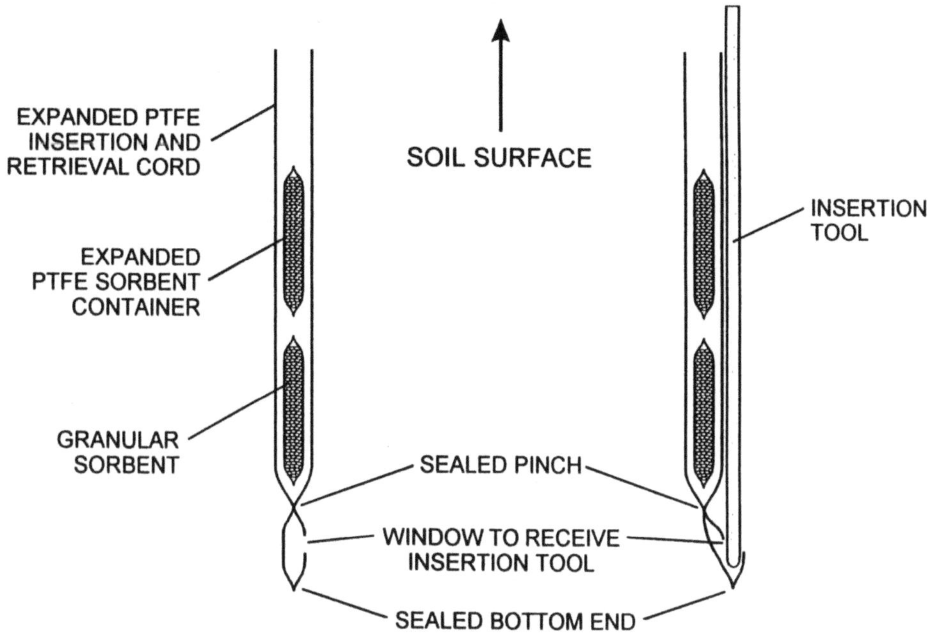

Figure 1. GORE-SORBER® screening module.

Sixty-six (66) GORE-SORBER Modules were deployed throughout the site to an average depth of 1.0 meter. Spacing was approximately ten (10) meters on-center (Figure 2). Modules were also placed directly in ground water in a clustered well triplet in the central portion of the site. The objective of this part of the project was to determine the effectiveness of this passive technique when applied to screening for low vapor pressure compounds directly in the saturated zone.

After exposure to the subsurface for 21 days the modules were retrieved and returned to Gore's lab for analysis using thermal desorption and GC/MSD analysis. Several modules were also solvent extracted in an attempt to identify heavier polycyclic aromatic hydrocarbon compounds (PAHs) that may not have been extracted via thermal desorption.

FIELD DEMONSTRATION RESULTS

Comparisons of GORE-SORBER Module analytical results for five PAHs (naphthalene, 2-methyl naphthalene, fluorene, acenaphthene and acenaphthylene) were made with soil sampling data previously collected between 0.75–1.75 meters depth and with ground water quality results from well sampling events.

Soil sample data for these five PAHs collected from 0.75–1.75 feet is presented as isoline contours in Figure 3. These data compared well with the soil gas results presented in Figure 4. When making these comparisons it is important to note that the soil vapor modules collect hydrocarbon vapors that may originate from both soil-adsorbed or dissolved-phase constituents.

Ground water quality data (obtained from previous sampling events) and soil gas results also showed a good relationship as shown on Figure 4. Comparison of analytical results for modules submerged directly in ground water in the well triplet to water quality

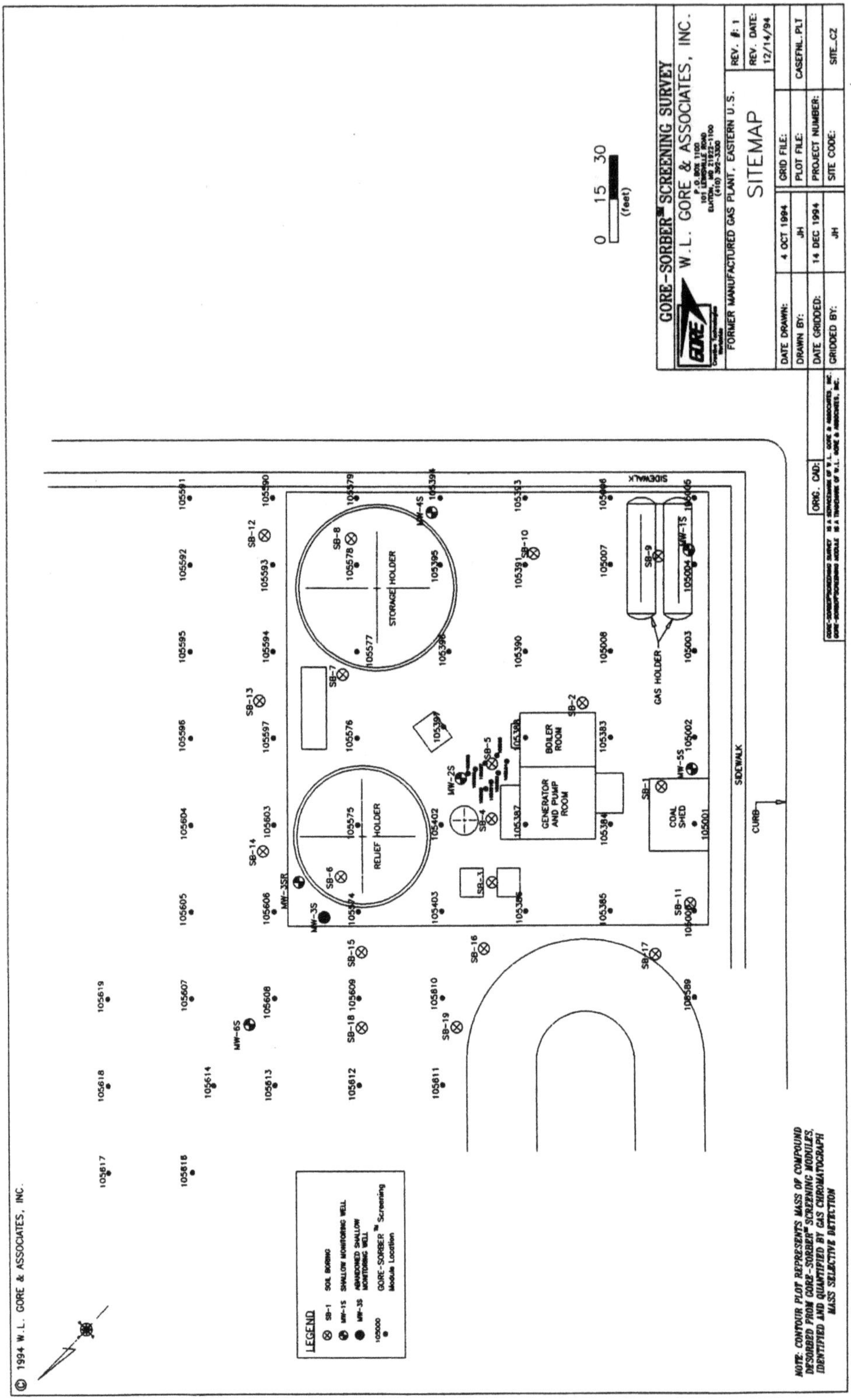
GORE-SORBER SCREENING SURVEY
W.L. GORE & ASSOCIATES, INC.
FORMER MANUFACTURED GAS PLANT, EASTERN U.S.
SITEMAP
DATE DRAWN: 4 OCT 1994
DRAWN BY: JH
DATE GRIDDED: 14 DEC 1994
GRIDDED BY: JH
GRID FILE:
PLOT FILE: CASEFNL.PLT
PROJECT NUMBER:
SITE CODE: SITE_CZ
REV. #: 1
REV. DATE: 12/14/94
0 15 30
(feet)
© 1994 W.L. GORE & ASSOCIATES, INC.
LEGEND
SB-1 SOIL BORING
MW-1S SHALLOW MONITORING WELL
MW-3S ABANDONED SHALLOW MONITORING WELL
105000 GORE-SORBER Screening Module Location
STORAGE HOLDER
RELIEF HOLDER
GAS HOLDER
BOILER ROOM
GENERATOR AND PUMP ROOM
COAL SHED
SIDEWALK
CURB
NOTE: CONTOUR PLOT REPRESENTS MASS OF COMPOUND DESORBED FROM GORE-SORBER SCREENING MODULES, IDENTIFIED AND QUANTIFIED BY GAS CHROMATOGRAPH MASS SELECTIVE DETECTION

Figure 2

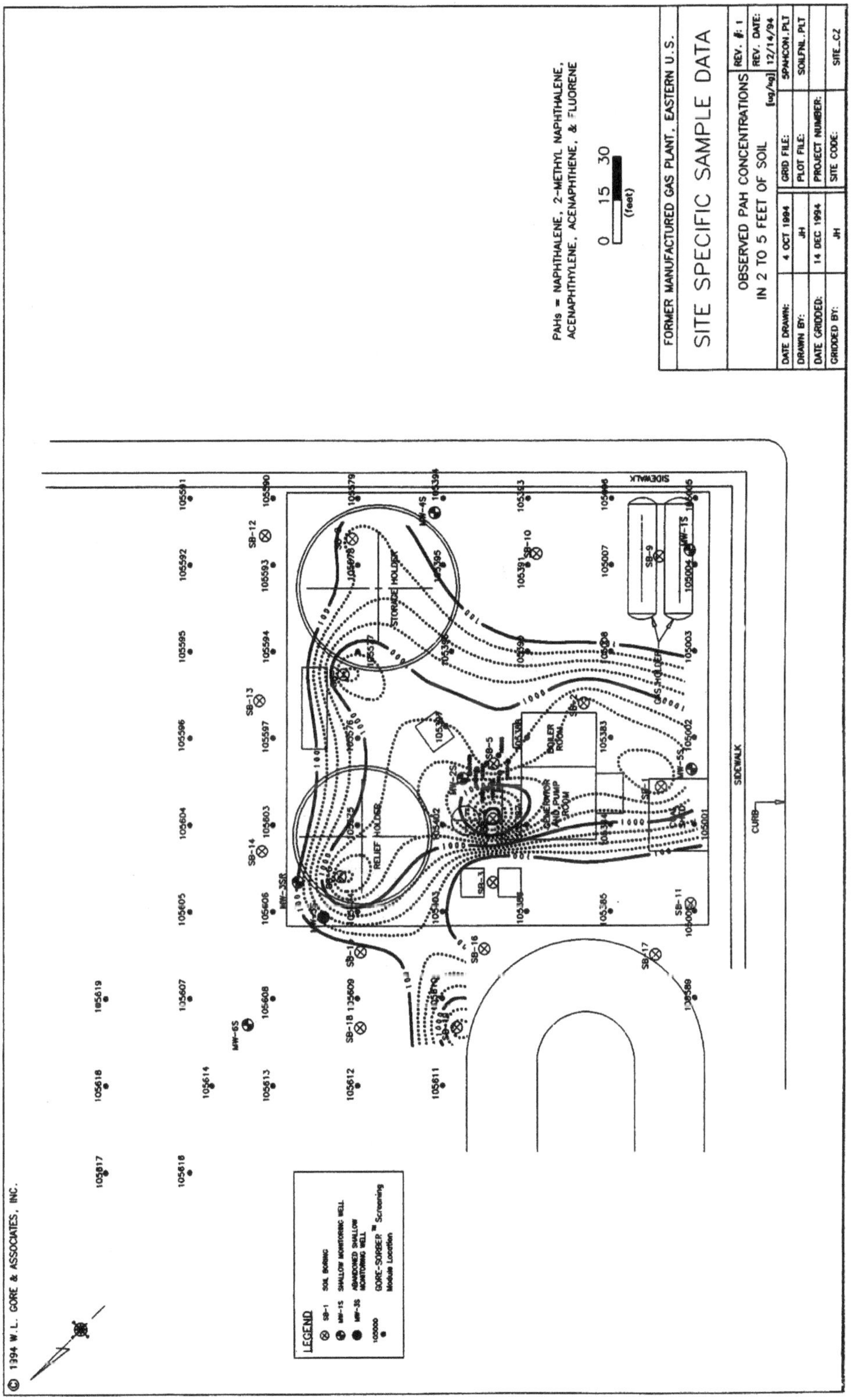
FORMER MANUFACTURED GAS PLANT, EASTERN U.S.
SITE SPECIFIC SAMPLE DATA
OBSERVED PAH CONCENTRATIONS IN 2 TO 5 FEET OF SOIL [ug/kg]
PAHs = NAPHTHALENE, 2-METHYL NAPHTHALENE, ACENAPHTHYLENE, ACENAPHTHENE, & FLUORENE
0 15 30 (feet)
DATE DRAWN: 4 OCT 1994
DRAWN BY: JH
DATE GRIDDED: 14 DEC 1994
GRIDDED BY: JH
GRID FILE: SPAHCON.PLT
PLOT FILE: SOILFNL.PLT
PROJECT NUMBER:
SITE CODE: SITE_CZ
REV. #: 1
REV. DATE: 12/14/94
© 1994 W.L. GORE & ASSOCIATES, INC.
LEGEND
SB-1 SOIL BORING
MW-1S SHALLOW MONITORING WELL
MW-3S ABANDONED SHALLOW MONITORING WELL
105000 GORE-SORBER™ Screening Module Location
STORAGE HOLDER
RELIEF HOLDER
GAS HOLDER
BOILER ROOM
GENERATOR AND PUMP ROOM
SIDEWALK
CURB

Figure 3

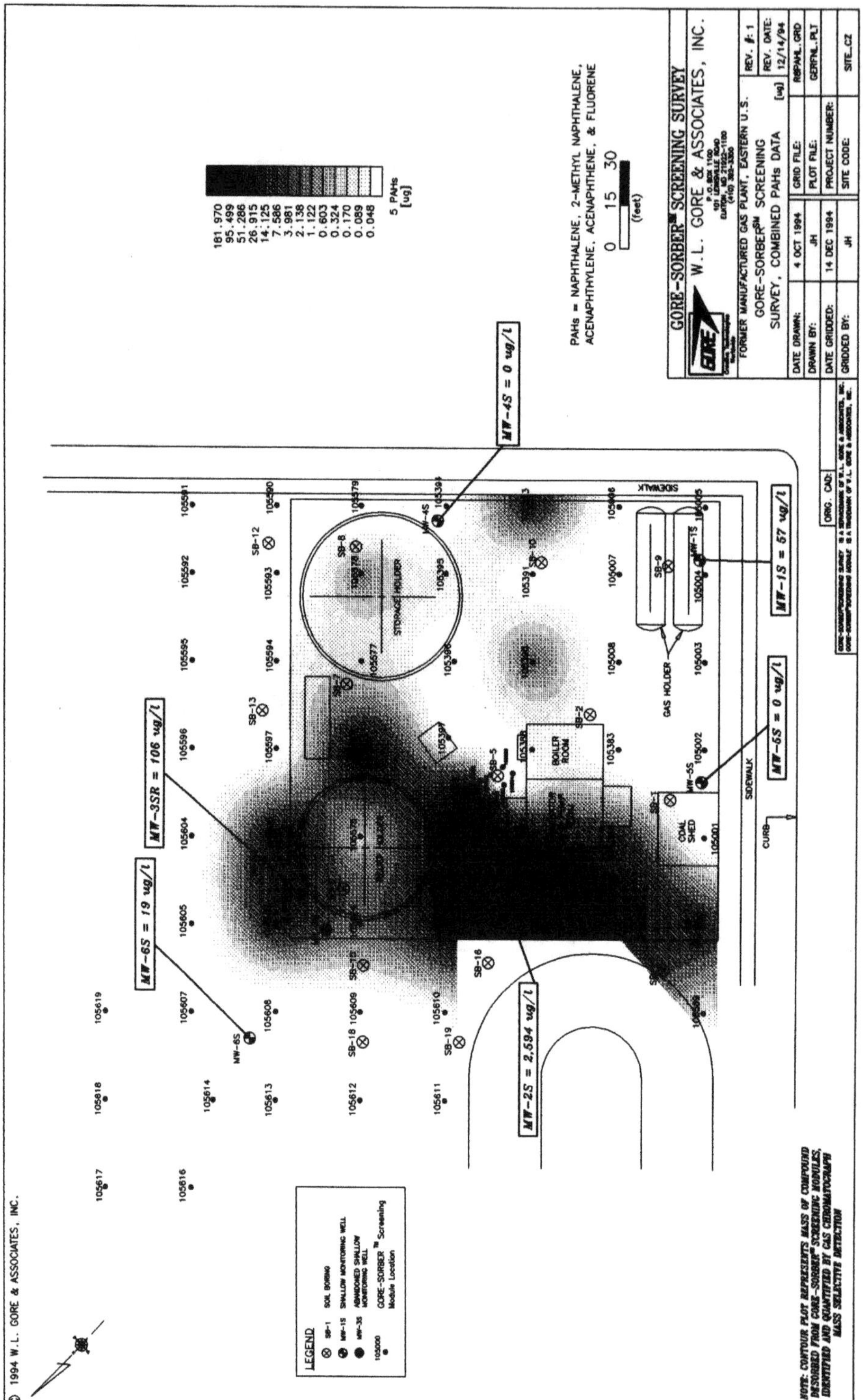
GORE-SORBER SCREENING SURVEY
W.L. GORE & ASSOCIATES, INC.
FORMER MANUFACTURED GAS PLANT, EASTERN U.S.
GORE-SORBER SCREENING SURVEY, COMBINED PAHs DATA
PAHs = NAPHTHALENE, 2-METHYL NAPHTHALENE, ACENAPHTHYLENE, ACENAPHTHENE, & FLUORENE
5 PAHs [ug]
181.970
95.499
51.286
26.915
14.125
7.586
3.981
2.138
1.122
0.603
0.324
0.170
0.089
0.048
0 15 30
(feet)
MW-4S = 0 ug/l
MW-1S = 57 ug/l
MW-5S = 0 ug/l
MW-3SR = 106 ug/l
MW-6S = 19 ug/l
MW-2S = 2,594 ug/l
STORAGE HOLDER
GAS HOLDER
BOILER ROOM
COAL SHED
SIDEWALK
CURB
LEGEND
© 1994 W.L. GORE & ASSOCIATES, INC.

Figure 4

Well Identification	Water Quality 5 PAHs (µg/l)	GORE-SORBER 5 PAHs (µg)
MW-2S	2,594	180
MW-2I	8,229	561
MW-2D	nd	<0.01

Note: "5 PAHs" = naphthalene, 2-methyl naphthalene, fluorene, acenaphthene and acenaphthylene

Figure 5. Water quality data versus GORE-SORBER®. Results for monitoring well cluster #2. The five PAHs are naphthalene, 2-methyl naphthalene, fluorene, acenaphthene, and acenaphthylene.

data is shown in Figure 5. The results of solvent extraction (using methylene chloride) and subsequent analysis via GC/MS of replicate sorbent containers yielded detection of heavier PAH compounds up to and including pyrene. Although heavier PAHs were detected by previous matrix sampling events they were not detected in this soil gas survey. This result confirms other findings from field work with this passive technology that pyrene may be the heaviest PAH compound detectable in the vapor phase. The following laboratory experiment was aimed at further examining this area.

PAH VAPOR ADSORPTION BENCH SCALE EXPERIMENT

A bench scale experiment was designed with the primary objective of better understanding the effects of time and vapor pressure on adsorption of PAH vapors onto passive soil gas modules.

GORE-SORBER Modules were exposed to samples of a commercially available PAH soil matrix from Resource Technology Corporation, Wyoming, USA. Each sample jar contained a small module with replicate sorbent containers ("sorbers") covered with thirty (30) grams of PAH soil. Blank control modules were placed in sample jars with no soil and the entire experiment was run in duplicate (Figure 6). Exposure times were one, two and three weeks after which sorbers were removed and analyzed via thermal desorption followed by GC/MSD. Modifications to the thermal desorption process at Gore's laboratory allow for desorption of up to six ring PAHs if they are present on the adsorbent being analyzed. For this reason solvent extraction was not employed as part of this experiment.

Results are presented in graphical format on Figures 7–9 and include Henry's Law constants and vapor pressure for each of the analytes (3). Although the PAH soil matrix contained less volatile compounds than pyrene, none were detected in appreciable quantities by the soil gas modules. From review of Figure 7 it is apparent that the more volatile PAHs are more readily adsorbed (for a given amount in the soil). Although this result was expected, it is interesting to note the transition that occurs near phenanthrene in the relationship between the mass adsorbed and the concentration in the soil. From review of Figure 8 it can be seen that this change is largely due to significant reduction in vapor pressure as the size of these compounds increases.

These results also indicate that exposure time may not have a significant impact on mass of analyte adsorbed in the vapor phase. However, it should be noted that this experiment is analogous to modules exposed directly to "source material" and it may not be appropriate to extrapolate these conclusions to field surveys where the source material may

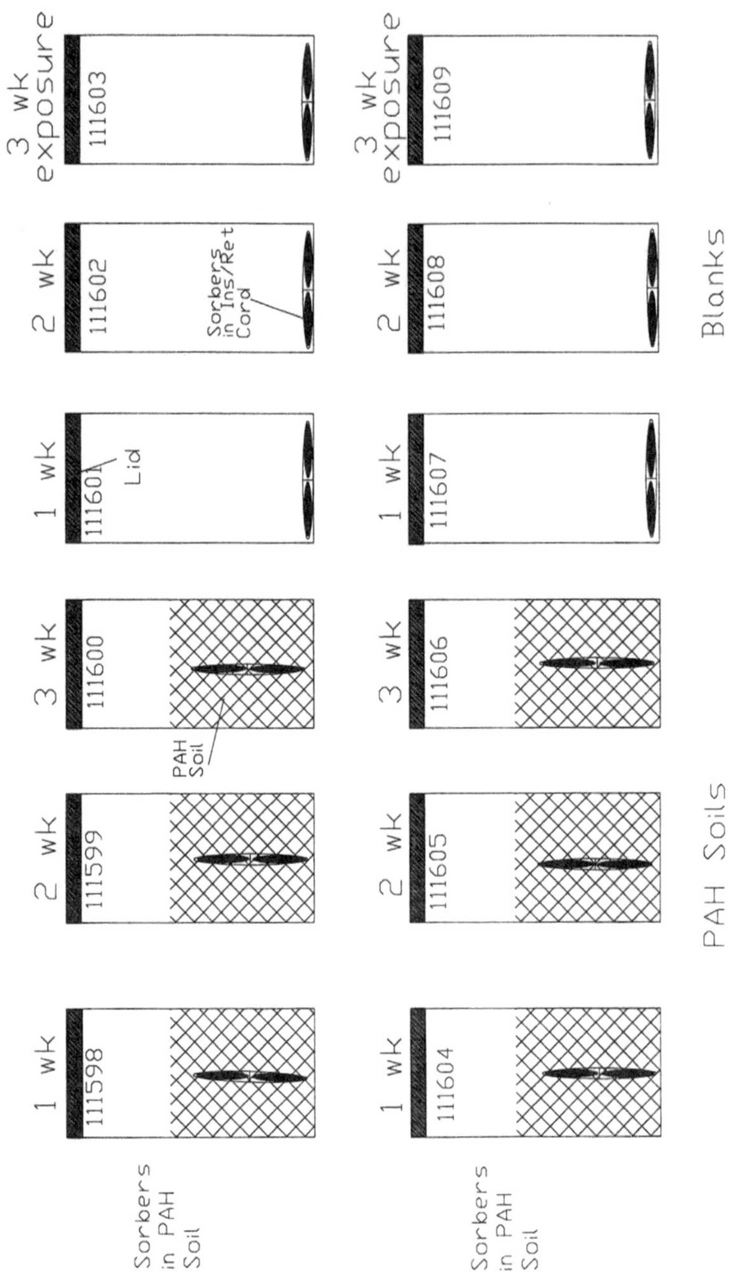

Figure 6. PAH vapor adsorption experiment design.

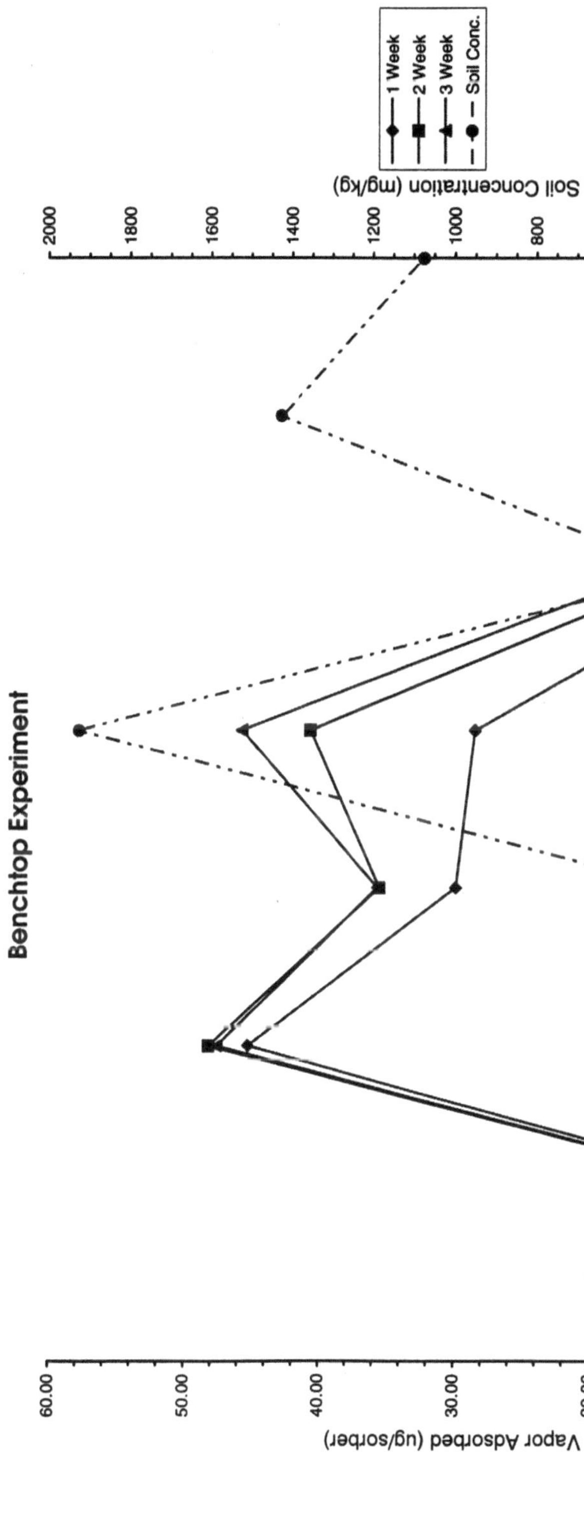

Figure 7. Effect of exposure time on PAH vapor adsorption: Benchtop experiment.

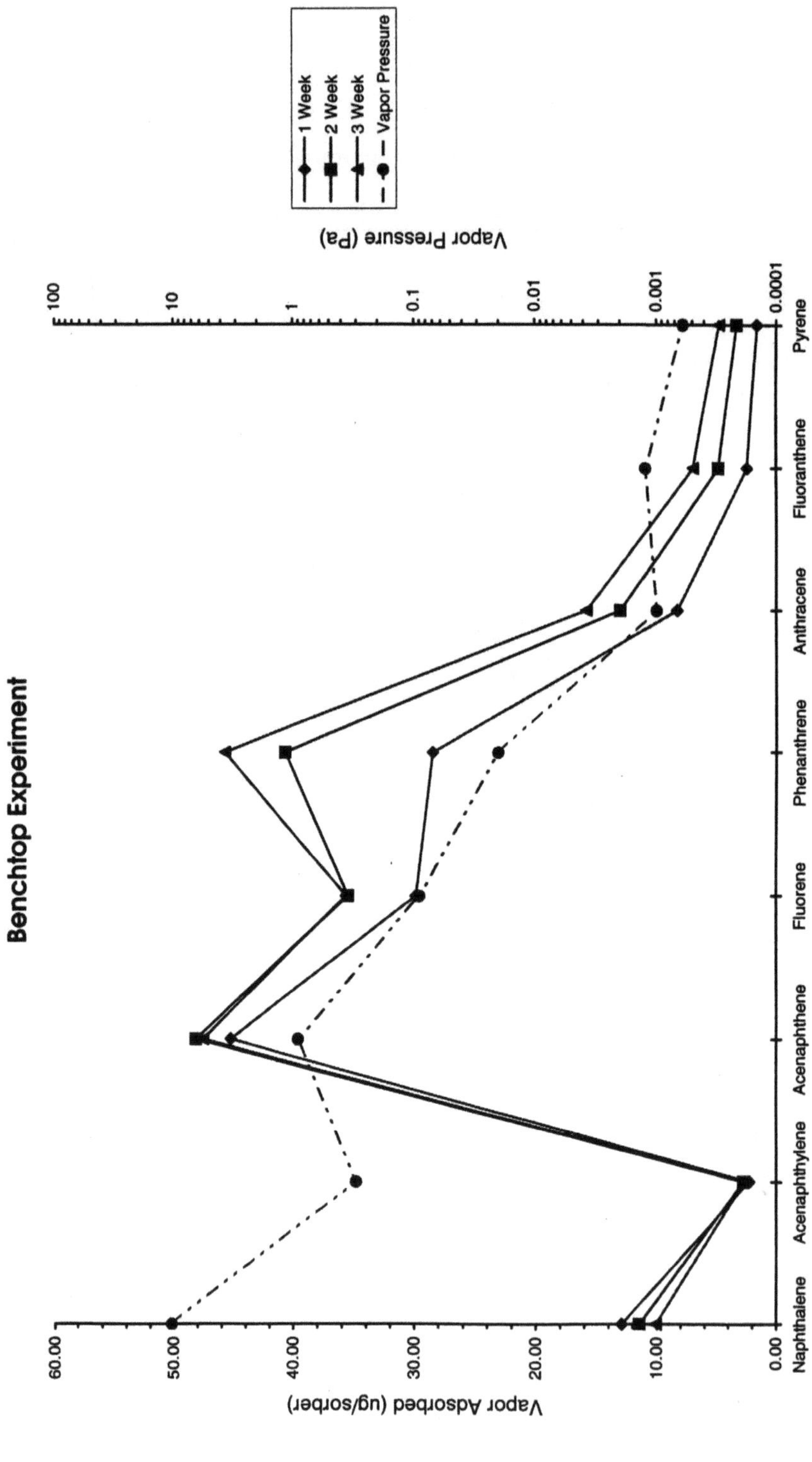

Figure 8. Effect of exposure time and vapor pressure on PAH vapor adsorption: Benchtop experiment.

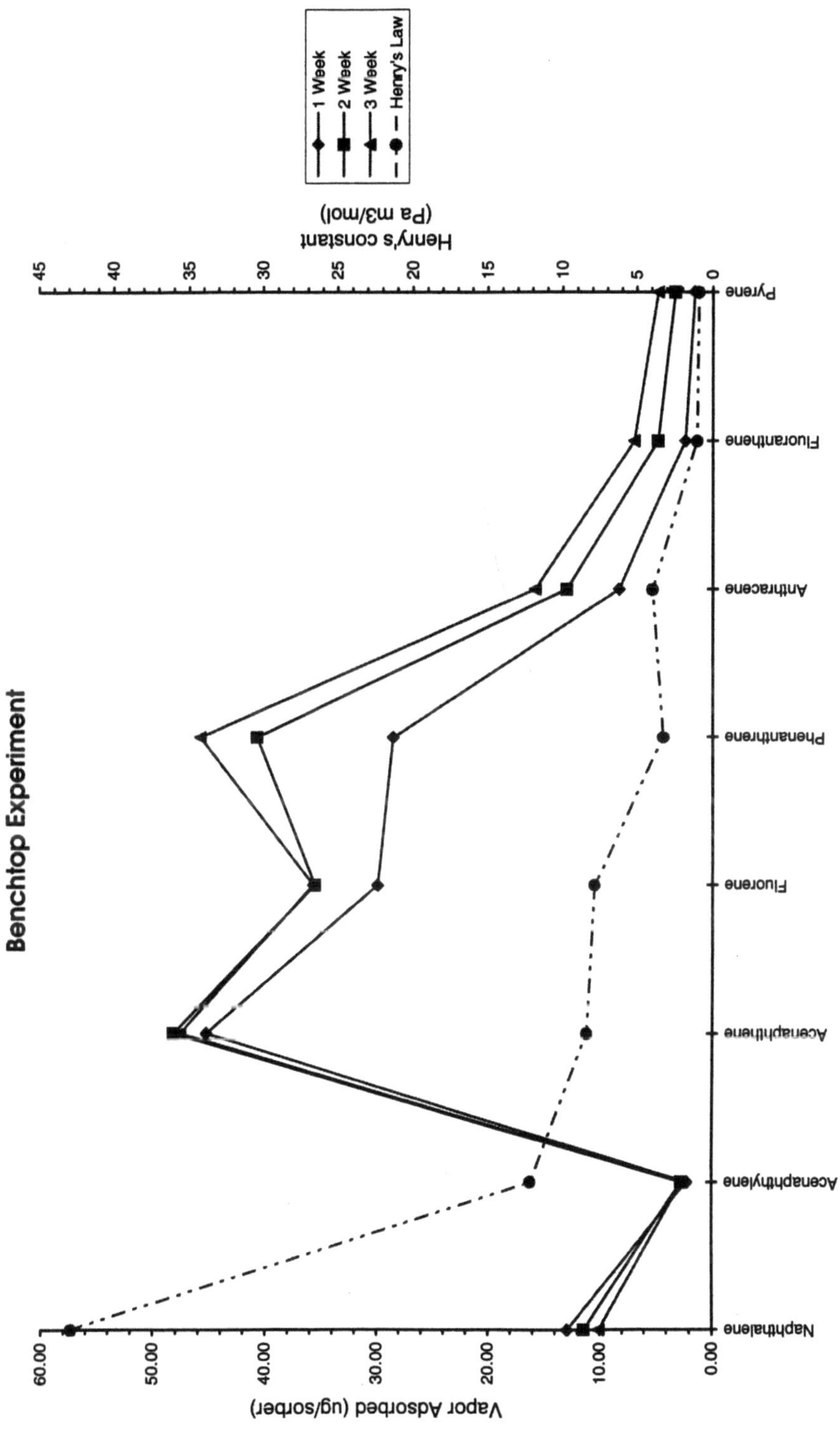

Figure 9. Effect of exposure time and Henry's constant on PAH vapor adsorption: Benchtop experiment.

be located several meters deep in the soil column or in the saturated zone. Field studies have shown exposure time effects to be more significant in these type of applications.

REFERENCES

1. Gas Research Institute (GRI). 1992. "Evaluation of Three Soil Gas Techniques at an MGP Site", Topical Report, September 1990-April, 1992
2. Stutman, Mark B. 1993. "A Novel Passive Sorptive Method for Site Screening of VOCs and SVOCs in Soil and Ground Water", *Proceedings of the 8th Annual Conference on Contaminated Soils*, Amherst, MA September,1993
3. Mackay, Shiu, Ma, 1992. "Illustrated Handbook of Physical-Chemical Properties and Environmental Fate for Organic Chemicals, Volume II", Lewis Publishers, Ann Arbor, MI.

43

USE OF A PORTABLE, FIBER-OPTICS, CCD SPECTROPHOTOMETER TO MEASURE FRIEDEL-CRAFTS PRODUCTS IN THE DETECTION OF CRUDE OIL, FUEL, AND SOLVENT CONTAMINATION OF SOIL

John David Hanby

Hanby Environmental Laboratory Procedures, Inc.
501 Sandy Point Road, Wimberley, Texas 78676

ABSTRACT

The utilization of a test kit employing Friedel-Crafts alkylation reactions to produce intensely colored products of aromatic compounds in the analytes (typically crude oils and fuels) which are caused to react with solvent constituents has facilitated removal of contaminated soils and provided an extremely accurate and rapid analysis of remediation processes. The extraction/colorimetric method has employed visual comparison of results with photograph standards. Testing of a new, portable spectrophotometric read-out device has been completed on a selected group of crude oils, fuels, and solvents. This paper describes results of the use of the device in determining concentration of a typical West Texas crude oil, a gasoline, and a diesel fuel in soil. The extremely small size (5″ x 7″ x 3″) of the device is made possible by the use of a recently-developed single-fiber optic/CCD spectrometer "bench". The instrument is interfaced to a 486 SX "notebook" PC. An algorithm for software development using color values developed by the International Commission on Illumination (CIE) was incorporated to provide quantitative analytical data.

INTRODUCTION

Since the discovery, in 1986, of the technique of extracting soil and water samples with various solvents and then causing the extracts to undergo Friedel-Crafts (F-C) reactions (Fig. 1) by the addition of stoichiometrically great excess (>100x) amounts of appropriate Lewis acid catalysts, the procedure has been utilized as a field

Chemistry for the Protection of the Environment 2, edited by Pawłowski et al.
Plenum Press, New York, 1996

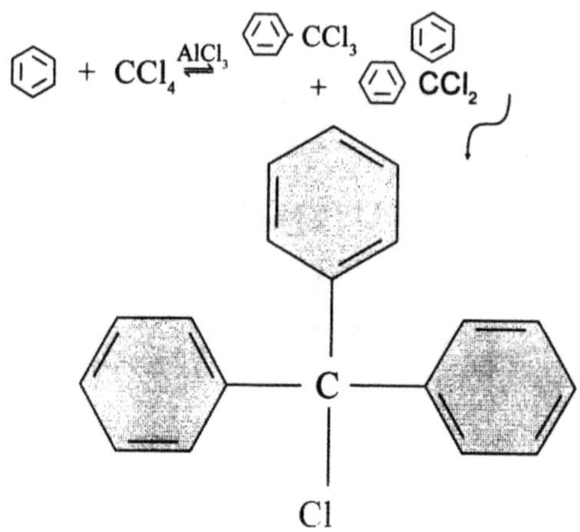

Figure 1. Friedel-Crafts alkylation reactions: Formation of mono-, di-, and tri-arylalkylhalide structures, intensely colored, UV-unstable.

method to provide extremely accurate quantitative analyses of these substances on site in thousands of cleanup and remediation projects around the world. Optimization of this procedure in order to maximize the visual detection sensitivity has typically involved the use of various amounts of an alkyl halide, carbon tetrachloride, which is, although an extremely good Friedel-Crafts reactant, high on the list of chemical "betes noirs". The subsequent search for solvents which would serve to provide sufficient reactivity and color via this method, coincident with a general focus by the EPA and other regulatory agencies on the larger scale environmental problems such as leaking underground fuel tanks, crude oil production, storage and processing areas, pipelines, etc., led to the realization that these substances of concern were generally composed of the requisite chemical species, e.g. aromatics, alkenes, ketones, to allow higher level (ca. 100 PPM) detection without the use of an alkyl halide solvent. This, however, further spurred efforts to develop a spectrophotometric read-out device which would alleviate the problems of subjective interpretation of the colors obtained in the precipitates which are the determining parameters in the method. That is, in comparing the colored products of the FC reaction caused by the sample extract in the test tube to the photograph standard, visual acuity, lighting and other uncontrollable factors played a part in determining the result.

Two relatively important technical considerations had stymied successful development of a suitable, field-portable, spectrophotometric instrument for this method: 1) an appropriately sized and focused optical viewing device, and 2) an appropriately sized and powered detector. The first technical problem centered on the fact of the powdered reflective surface composed of the excess catalyst and the F-C precipitates. Numerous researches on these reflection/scattering phenomena have pointed to the considerable effects of parameters such as particle size, packing, interstitial fluid, etc. A solution to many of these problems seemed to be offered in the technique understood by printers and Post-Impressionist painters for many years: optical integration of a large area. Just as a too-close inspection of a Seurat painting or a photograph reveals a confusing jumble of dots, a microscopic look at the color in the Hanby method test tubes showed a wide variation in the color of the catalyst/ precipitate catalyst/precipitate mixture. The unreacted catalyst particles ($AlCl_3$) and hydrated catalyst ($AlCl_3 \cdot 6H_2O$) were typically white while those particles of the F-C reaction, or adsorbed product were colored to an extent indicative of the relative amounts of the reactants (analytes).

The solution to the second technical problem, i.e., size and power requirements, lay in the utilization of a charge-transfer device of some type. In the fall of 1994 the author was introduced to a group, Ocean Optics, Inc. who had developed a technology in 1992 which seemed ideally suited to the needs of the method. In the development of optical technologies diversely used for pH and spectrophotometric applications, this group had produced an extremely small optical detector employing charge coupled devices and single fiber optic transmission. Essentially all that was necessary in the development of the present device was the design and manufacture of the appropriate optical cell (test tube) holder. This was accomplished by the author after experiments with various materials indicated that, probably due to the rather inconsistent reflection characteristics of relatively inexpensive, commercially-available test tubes, a non-reflective material should be utilized. After the completion of the test tube / fiber optic probe module (Fig. 2), a series of tests were run to determine the correct distance the probe should be placed from the sample tube. This would optimize the view area and the signal strength. In effect, probe distance would primarily determine the signal to noise ratio (SNR).

For the initial trials of the instrument a particularly appropriate algorithm was available for the conversion of input to output, i.e., the tri-stimulus values established by the International Commission on Illumination (CIE). Essentially, the defined wavelength/color relationships of Red (700nm), Green (546.1), and Blue (435.8) are used in the computation of values in the CIE-derived colour space such as: L*, a*, b*, xtri, ytri,and ztri. As the primary indicator of quantitative analytical results with the method, as heretofore used as a visual method, had been lightness/darkness of precipitated color, it was assumed that the L* (color intensity from white to total saturation) value would be the more indicative parameter, however the data was to prove otherwise. In more spectrophotometrically familiar (to a chemist) terms this was of benefit as the chromophoric effect various functional groups have is a long-established body of chemical knowledge, and using other parameters, namely the tristimulus values: xtri, ytri, and ztri, a simpler translation to wavelength/absorption numbers would be available. Hence, the expansion of the method in terms of a qualitative technique for identification of substances would be enhanced.

EXPERIMENTAL

Selection of substances to be utilized in the initial trials of the instrument was prioritized roughly by production and environmental importance. Thus, crude oil (Fig. 3), gasoline (Fig. 4), diesel fuel (Fig. 5), and toluene were chosen. Of course, an extremely wide range of crude oils exists, and the definition of a "standard" gasoline or diesel fuel is not chemically available. Hence the use of the various terms to describe these substances, e.g., "heavy, medium, light" or, "high or low" octane or cetane numbers, etc. This lack of exact chemical definition, of course, is understandable and perhaps has given rise to the often denigrated term "Total Petroleum Hydrocarbon". Given the site-

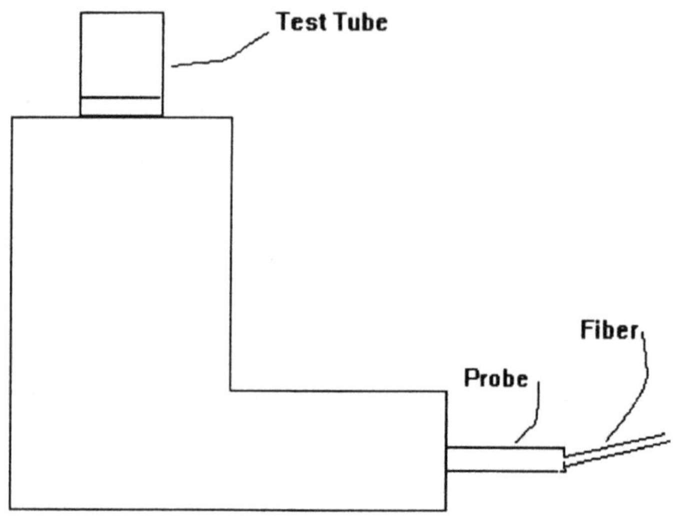

Test Tube / Fiber Optic Probe Module

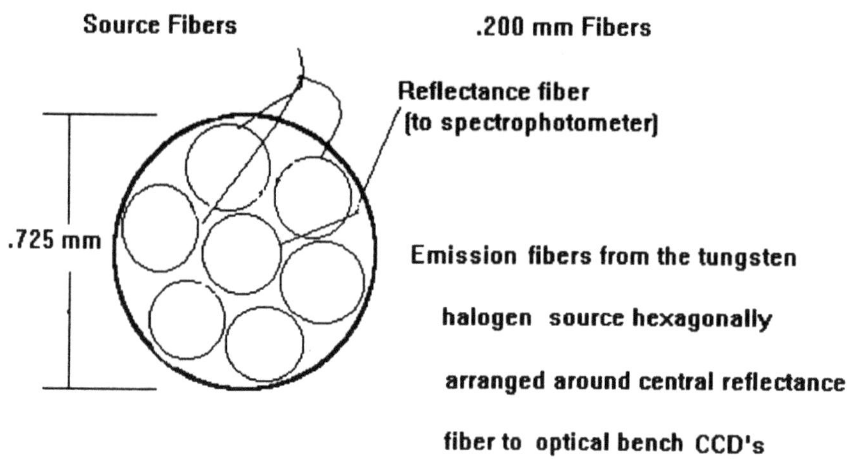

Fiber Optic Probe End Window

Figure 2

specific, and, often, substance-specific uses this method has found it was appropriate to prepare exact mass/solvent concentration standards of various "typical" samples of crude oils, gasolines, and diesel fuels which were ampoulized as reference materials for the procedure. All standards were prepared using HPLC grade n-Heptane or a 20% (v/v) solution of carbon tetrachloride in heptane (Fisher Scientific). Two ranges were established according to the solvent selected: 1. (CCl_4/Heptane--low range--) 2, 10, 25, 50, 75, 100, 200, 500, 750, 1000 mg/Kg and, 2. (Heptane --high range--) 500, 2500, 10,000, 25,000, and 50,000 mg/Kg.

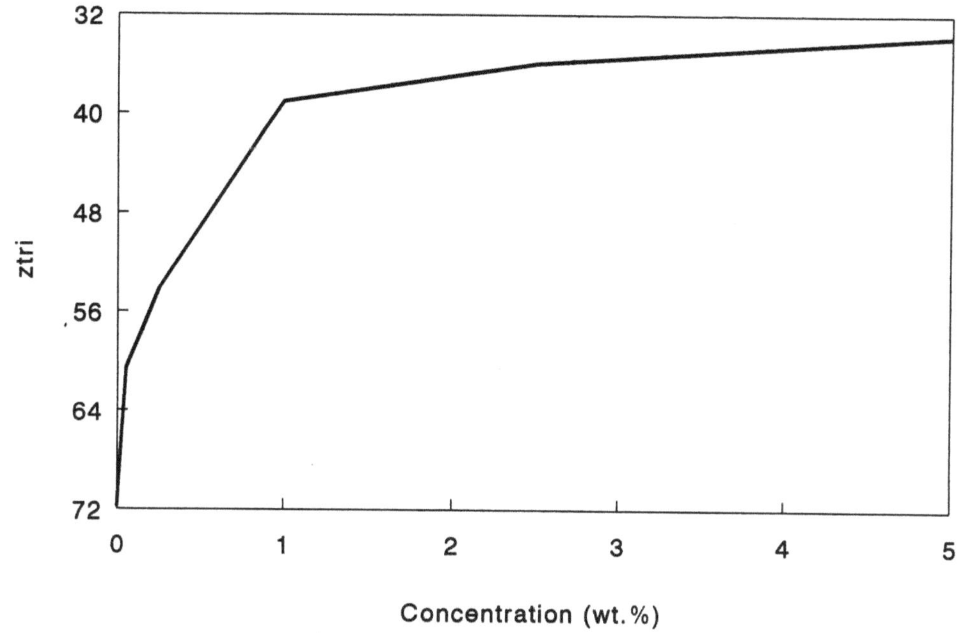

Figure 3. West Texas crude oil in soil.

INSTRUMENTAL

The L-shaped test tube/probe holder was fabricated from black Delrin to configure the fiber optic probe orthogonally to the test tube at a distance of 7.0 mm. This resulted in a focused viewing area of 3.4 mm in diameter. The reflectance probe/fiber optic used was

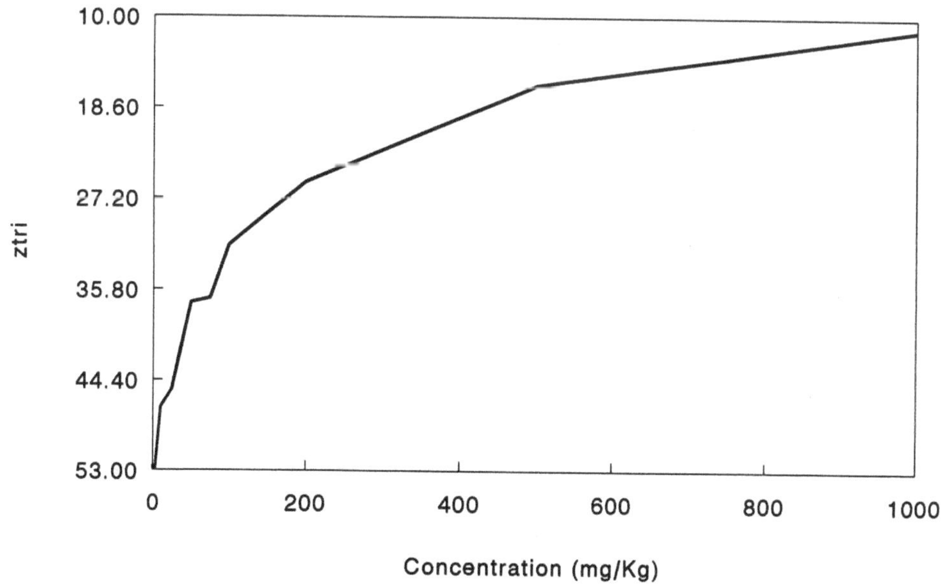

Figure 4. Gasoline in soil.

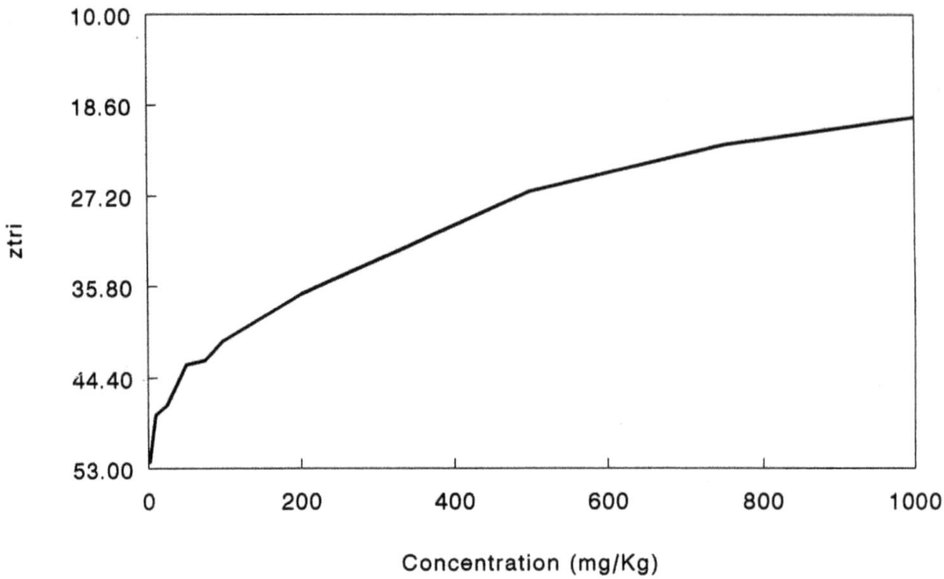

Figure 5. Diesel in soil.

the R-200–7-LR; tungsten-halogen source, LS-1; spectrometer optical bench, PS1000 (Ocean Optics, Inc. Dunedin, Fl). An aluminum housing for the complete assembly was manufactured by Preferred Stampings of Texas, Inc., Round Rock, TX. The spectrometer was interfaced via a ribbon cable/A/D card (DAQCard-700, National Instruments) to an AST 486 SX/25 notebook PC. Data and graphs were printed on a portable printer (HP DeskJet 310). The portability of the complete system: Field test kit, spectrophotometer, computer, and printer is such that it can easily be carried on-site and operated, and complete reports can be generated by the analyst.

EXPERIMENTAL PROTOCOL

Solutions were prepared in the ranges listed above corresponding to the amount of analyte extracted from a 5.0 gram soil sample using 10 ml (C_7/CCl_4) or 20 ml (C_7). One gram catalyst amounts were added to 4.2 ml aliquots of these solutions in the standard 100 mm x 15 mm test tubes according to the Hanby Field Test Kit protocol, and the solutions were hand shaken intermittently for 4 minutes, allowed to settle for 1 minute, and then read in the instrument. Four readings (with ca. 30 degree rotation of the tube between each) were taken for each concentration. The tristimulus value ztri was found to correlate extremely well with concentration for each set of readings. Virtually all readings were found to lie within five percent of the mean of the four readings. This verified the fact of the optimization of the tube/probe module configuration, and corresponded well with the typical test kit weighing and extraction error of ± 5%.

DISCUSSION

The primary aim of this research was to test the application of the new instrument to this field test kit method of analysis. As illustrated by the data, confirmation of the applicability is proven. In the course of the experimental work, another desirable feature of the method was demonstrated. The utilization of this application of Friedel-Craft chemistry now extends to the "other" branch of this time-honored discovery, i.e., acylation. That is, previous employment of the method primarily exploited the intense colors produced by strong alkylation reactions promoted by use of very high ratios (>100:1) of the Lewis acid catalyst employed, and the use of the very reactive F-C solvent, carbon tetrachloride. Acylation reactions as well as reactions in which other alkylating agents, e.g., alkenes, are available in substances such as crude oils and fuels. Again, the fundamental principle of the Hanby method, i.e., the use of stoichiometrically very large proportions of the aluminum chloride catalyst which serve to dehydrate the extract and enhance the Friedel-Crafts reaction, is certainly key to the successful use of this procedure for the high concentration (ca. 50,000 PPM) ranges that are being allowed as interim soil contamination levels at designated sites. Implementation of these regulatory limits has been carefully considered by a number of oil-producing states and was recently effected by the Texas Railroad Commission. As stated in a classic text of analytical chemistry, "In the broadest sense, an instrument for chemical analysis converts an analytical signal that is usually not directly detectable and understandable by a human to a form that is. Thus, an analytical instrument can be viewed as a communication device between the system under study and the scientist." This development can be regarded as a practical combination of the two divisions of analytical chemistry, classical or "wet" analysis and instrumental analysis.

REFERENCES

1. United States Environmental Protection Agency, Solid Waste and Emergency Response (OS-420), Field Measurements: Dependable Data When You Need It, EPA/530/UST-90–003, September, 1990.
2. Roberts, R.M., Khalaf, A.A., *Friedel-Crafts Alkylation Chemistry: A Century of Discovery*, M. Dekker, Inc., 1984.
3. Shriner, R.L., Fuson, R.C., Curtin, D.Y., Morrill, T.C., *The Systematic Identification of Organic Compounds*, John Wiley &Sons, New York, 1980.
4. Sweedler, J.V., Ratzlaff, K.L., Denton, M.B., *Charge Transfer Devices in Spectroscopy*, VCH Publishers, Inc., 1994.
5. Fox, M.A., Whitesell, J.K., *Organic Chemistry*, Jones and Bartlett Publishers, 1994.

EFFECTS OF PRESERVATION AGENTS ON DETERMINATIONS OF VOCS AND HAAS IN DRINKING WATER

E.E. Chang,[1] Ya-Wen Ko,[2] Pen-Chi Chiang,[2] and Chu-Fang Wang[3]

[1] Department of Analytical Chemistry
Taipei Medical College
Taipei, Taiwan, ROC
[2] Graduate Institute of Environmental Engineering
National Taiwan University
Taipei, Taiwan, ROC
[3] Graduate Institute of Nuclear Science
National Tsing-Hua University
Hsinchu, Taiwan, ROC

ABSTRACT

This research was performed to evaluate effects of inhibition/preservation agents on determinations of chlorinated volatile organic compounds (VOCs) and haloacetic acids (HAAs) in drinking water. Three typical preservation agents: ascorbic acid (25 mg/40 ml); $NaHSO_4$ (25 mg/40 ml); HCl (4 drops 1 N HCl/ 40 ml sample) were introduced. Among them, ascorbic acid was capable of recovering VOCs species within three days of preservation, and proved to be the best preservation agent for VOCs determined over the 60 day holding study. Another holding study was conducted to test the effects of three preservation techniques: Na_2SO_3 (100 mg/L), Na_2SO_3 (700 mg/L), and NH_4Cl (1.5 g/L) on determinations of HAAs in water samples. After 70 a day holding study, the stability of HAAs under various preservation techniques was in the sequence of: NH_4Cl 1.5 g/L> Na_2SO_3 700 mg/L> Na_2SO_3 100 mg/L.

1. INTRODUCTION

Since low concentrations of trace organics commonly occur in drinking water, it is necessary to use the recommended analytical methods followed by the QA/QC programs to obtain more precise and accurate data. However, prior to making any further analyses for the water samples, the sample storage which requires selecting the appropriate inhibition and preservation agents against the chemical interference caused by oxidation/reduc-

Chemistry for the Protection of the Environment 2, edited by Pawłowski et al.
Plenum Press, New York, 1996

tion and biochemical degradation has become one of the major operational procedures to fulfill with the QA/QC requirements.

Preservation of samples is highly dependent on the target compounds and the sample matrix.[1] Ongoing research indicates the following areas of concern: rapid biodegradation of aromatic compounds, even at low temperatures; dehydrohalogenation reactions such as conversion of pentachloroethane to tetrachloroethane; reaction of alkylbenzenes in chlorinated samples, even after acidification; and possible interactions among preservatives and reductants when dechlorination is used to prevent artifact formation, especially in samples potentially containing many target compounds.

According to suggestions of Standard Method,[1] ascorbic acid (25 mg/40 ml) appears to be a suitable preservation agent for volatile organic compounds (VOCs) that contain residual chlorine, but it needs to be demonstrated that this reductant is appropriate for the specific sample matrix. However, for samples collected soon after chlorination, quenching with reducing agent may not be sufficient to prevent formation of chlorinated VOCs completely because of hydrolysis of intermediates.

On the other hand, previous studies[2,3] have shown that the addition of ammonium chloride (NH_4Cl) does not dechlorinate a sample; rather, it converts the active free chlorine to chloramines. Krasner et al.[4] evaluated NH_4Cl as a preservative for haloacetic acids (HAAs) and found that it is acceptable for all HAAs; while both 100 mg/L and 700 mg/L of sodium sulfite (Na_2SO_3), as a dechlorinating agent, cannot preserve well MBAA (monobromoacetic acid) for the USA Weymouth Plant's chlorinated filter effluent.

Therefore, this research was performed to evaluate the effects of three different preservation agents on determination of chlorinated VOCs, including chloroform, 1,1-dichloroethene, 1,1,1-trichloroethane, carbon tetrachloride, trichloroethene and 1,2-dichloroethane, and HAAs, including MBAA (monobromoacetic acid), DCAA (dichloroacetic acid), TCAA (trichloroacetic acid) and DBAA (dibromoacetic acid), in Taipei Chang-Hsin Water Treatment Plant's chlorinated effluent.

2. MATERIAL & METHODS

2.1 VOCs Measurements

2.1.1 Sample Preparations. Tap water (with 0.4 mg/L residual chlorine) obtained from Taipei Chang-Hsin Water Treatment Plant and spiked with VOC stock standards in our laboratory was used as the principal test samples. Samples were collected in nominal 40-ml amber vials with Teflon-faced septum and screw caps. Samples were divided into three groups by introducing: ascorbic acid (25 mg/40 ml), HCl (4 drops 1 N HCl/40 ml), and $NaHSO_4$ (25 mg/40 ml), respectively. A total of 36 vials provided two replicates for analysis at the respective holding times (0, 3, 7, 14, 28 and 60 days). The sample vials were filled headspace-free and stored at 4°C.

2.1.2 Sample Analysis. Purge-and-trap gas chromatography with an electron capture detector (P & T/GC/ECD) technique was used for determinations of VOCs in water samples, based on U.S. EPA Method 500. At the beginning of each day, a reagent water blank was injected to confirm that there were no interferences. The reagent water blank for VOC analysis was prepared by boiling deionized water (Milli-Q SP) for 15 min. and then bubbling N_2 (99.99% purity) through the water for one hour, based on the Standard Method.[1] The detec-

Table 1. Detection limits and recoveries of VOCs and HAAs

Group	Species	Detection limits (μg/L)	Recoveries (%)
VOCs			
	1,1-dichloroethene	0.84	103.55
	chloroform	0.61	100.53
	1,1,1-trichloroethane	0.80	91.07
	carbon tetrachloride	0.11	105.65
	trichloroethene	0.15	92.13
	1,2-dichloroethane	0.97	90.08
HAAs			
	monobromoacetic acid	0.46	93.66
	dichloroacetic acid	0.83	112.05
	trichloroacetic acid	0.65	93.20
	dibromoacetic acids	0.95	88.27

tion limits and recoveries of VOCs and HAAs are summarized in Table 1. We also analyzed sample duplicates in order to meet the precision criterion, 15% of relative error percentage between duplicates.

2.2 HAAs Measurements

2.2.1 Sample Preparations. The principal test samples and the sample vials were designated as previous VOCs preparations, with the exception of the spiked HAAs stock standards. A total of 48 vials were divided into three groups: NH_4Cl (1.5 g/L sample), Na_2SO_3 (100 mg/L sample), and Na_2SO_3 (700 mg/L sample) and were held at 4°C. The duplicate samples from each group were analyzed on day 0, 6, 13, 18, 29, 43, 57 and 70.

2.2.2 Sample Analysis. HAAs were analyzed by an acidic, salted ether liquid-liquid extraction method, and they required esterification with diazomethane prior to GC/ECD analysis, based on Standard Method 6233B.[1] The internal standard calibration procedure is used to quantify each compound. Relative response is defined as the ratio of peak area for each compound to the internal standard peak area and used to construct the calibration curve, versus standard concentration. For a set of samples analyzed, the acceptable relative error percentages for internal standards and surrogate standards are 20% and 30%, respectively. The detection limits and recoveries of HAAs are summarized in Table 1.

3. RESULTS & DISCUSSIONS

3.1 Results of VOCs Holding Study

Figure 1 presents the analytical results for VOCs determinations of chloroform ($CHCl_3$), 1,1-dichloroethene, 1,1,1-trichloroethane, carbon tetrachloride (CCl_4), trichloroethene and 1,2-dichloroethane in the prepared samples. These water samples were preserved by hydrochloric acid (HCl, 4 drops 1 N HCl/40 ml sample), ascorbic acid (25 mg/40 ml) and sodium hydrogen sulfate ($NaHSO_4$, 25 mg/40 ml) for 60 days. The concentration levels of these VOCs were observed to be initially reduced then gradually increased during the first two weeks of preservation time. However, the concentrations of

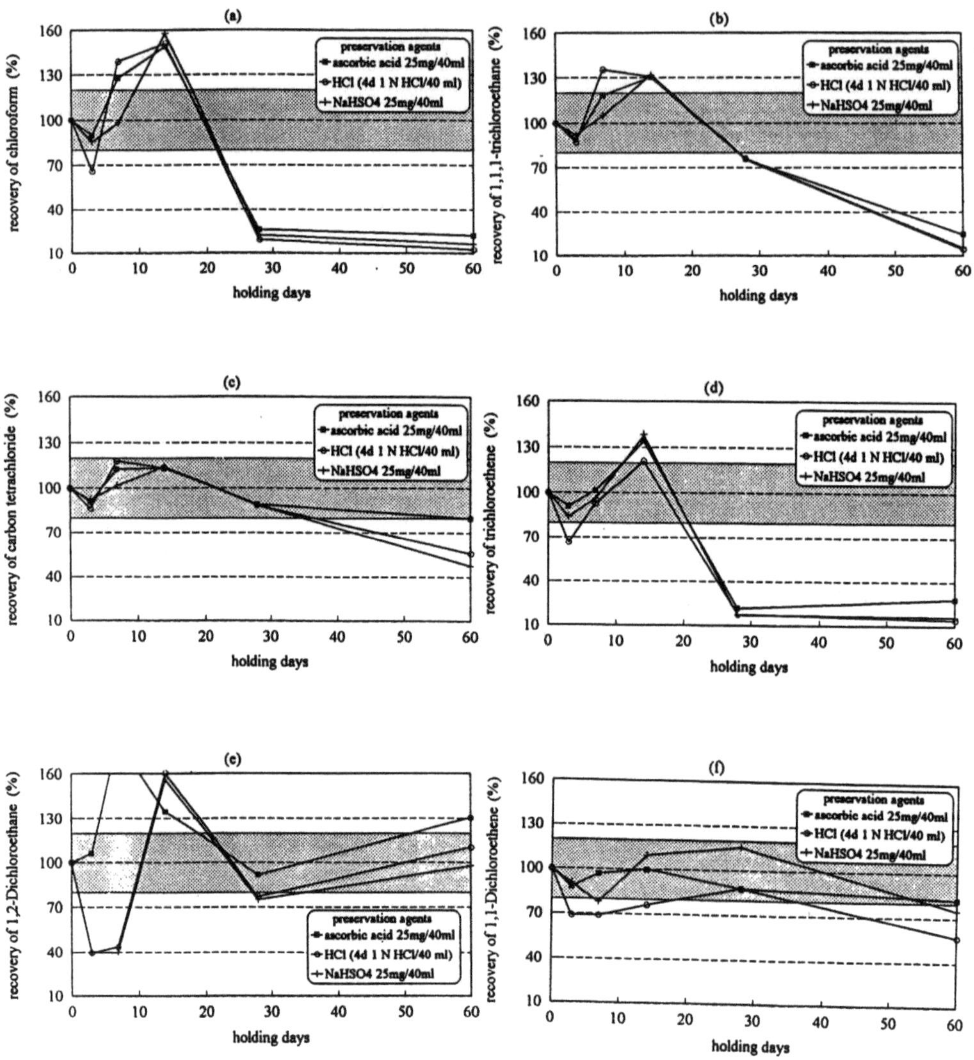

Figure 1. Analytical results of VOC holding study: (a) chloroform; (b) 1,1,1-tricholorethane; (c) carbon tetrachloride; (d) trichloroethene; (e) 1,2-dichloroethane; (f) 1,1-dichloroethene.

$CHCl_3$, 1,1,1-trichloroethane, carbon tetrachloride and trichloroethene exhibit a tendency to decrease after 2 weeks of preservation time.

To compare the performances of three inhibition/preservation agents for VOCs analyses, it is necessary to make further studies on the reaction kinetics of $CHCl_3$, 1,1,1-trichloroethane, CCl_4 and trichloroethene under two separate intervals, i.e., 0 to 14 days and 14 to 28 days. Table 2 summarizes the chemical reaction rate constants associated with the determined VOCs and the selected preservation agents during the 4 weeks of preservation time. In Table 2, it was clearly demonstrated that the measured concentrations can be expressed by the first-order reaction rate in which the "positive" and "negative" rate constants are found to occur during the first 2 weeks and 2 to 4 weeks of preservation time, respectively. Obviously, ascorbic acid has a slight advantage over HCl and $NaHSO_4$ in response to its relatively lower reaction rate constants shown in Table 2.

Table 2. The first-order reaction rate constants for VOC holding studies (units: day^{-1})

	Preservation agents					
	Ascorbic acid 25 mg/40 ml		HCl 4 d 1 N HCl/40 ml		$NaHSO_4$ 25 mg/40 ml	
Species	Day 0 to 14	Day 14 to 28	Day 0 to 14	Day 14 to 28	Day 0 to 14	Day 14 to 28
Chloroform	0.029	−0.124	0.029	−0.148	0.033	−0.139
1,1,1-Trichloroethane	0.019	−0.039	0.019	−0.039	0.020	−0.039
Carbon tetrachloride	0.009	−0.017	0.009	−0.018	0.010	−0.018
Trichloroethene	0.021	−0.130	0.014	−0.138	0.023	−0.150

This hypothesis is confirmed by Figure 1 and Figure 2. Figure 2 indicates that the ascorbic acid, in most of cases, possesses a reasonable recovery, i.e., 80 to 120%, which is better than HCl and $NaHSO_4$. The determinations of $CHCl_3$, 1,1,1-trichloroethane, carbon tetrachloride and trichloroethene shown in Figure 2 (a) to (d) exhibit very similar variation patterns regardless of the preservation agents introduced during the preservation period. The results of three reaction kinetic studies were clearly summarized in our previous discussion (Table 2). It was noted that the recoveries of 1,2-dichloroethane were extremely low and high, i.e., 40% for selecting HCl and $NaHSO_4$ agents and 180% for using ascorbic acid. In other words, the suggested three chemical agents in the literature are not suitable to determine the concentration of 1,2-dichloroethane once the preservation period is over 7 days. However, the recovery ratio of 1,1-dichloroethene can be held from 70 to 120% if the three selected preservation agents were introduced and preserved for 60 days.

Finally, a confirmation test was performed to assess the chemical interferences caused by the three preservation agents after 100 preservation days by examination of the gas chromatograms shown in Figure 3. In comparing the analytical results between the original sample (without preservation) and preserved samples, the preserved samples, obviously, had relatively higher unknown peaks and amount of noise than the original sample. It is noted that the sample preserved by ascorbic acid had fewer unknown peaks than the others.

Table 3 shows the performances of three preservation agents on VOC species during the 60 day holding study. Three ranking classes concerning the recovery of VOCs measurements were developed and shown in Table 3 to assess their respective performances associated with the selected preservation agents throughout this investigation. Shaded areas are given for the stable samples, which are defined as concentrations within 80–120% of the initial values. It was noted that samples "rehabilitated" after deviating from acceptable values were not included in shaded area. It was also concluded that ascorbic acid was capable of recovering the species in three days of preservation, and proved to be the most suitable one among three selected preservation agents, i.e., ascorbic acid (25 mg/40 ml); $NaHSO_4$ (25 mg/40 ml); HCl (4 drops 1 N HCl/40 ml sample), for the VOCs studied over a 60 day holding study.

3.2 Results of HAAs Holding Study

U.S. EPA suggests the use of 100 mg/L Na_2SO_3 for dechlorinating and preserving samples containing HAAs. However, the amount of Na_2SO_3 dosage is too low to prepare, especially for the 40 ml sample vial utilized in this investigation. As a result, we intended

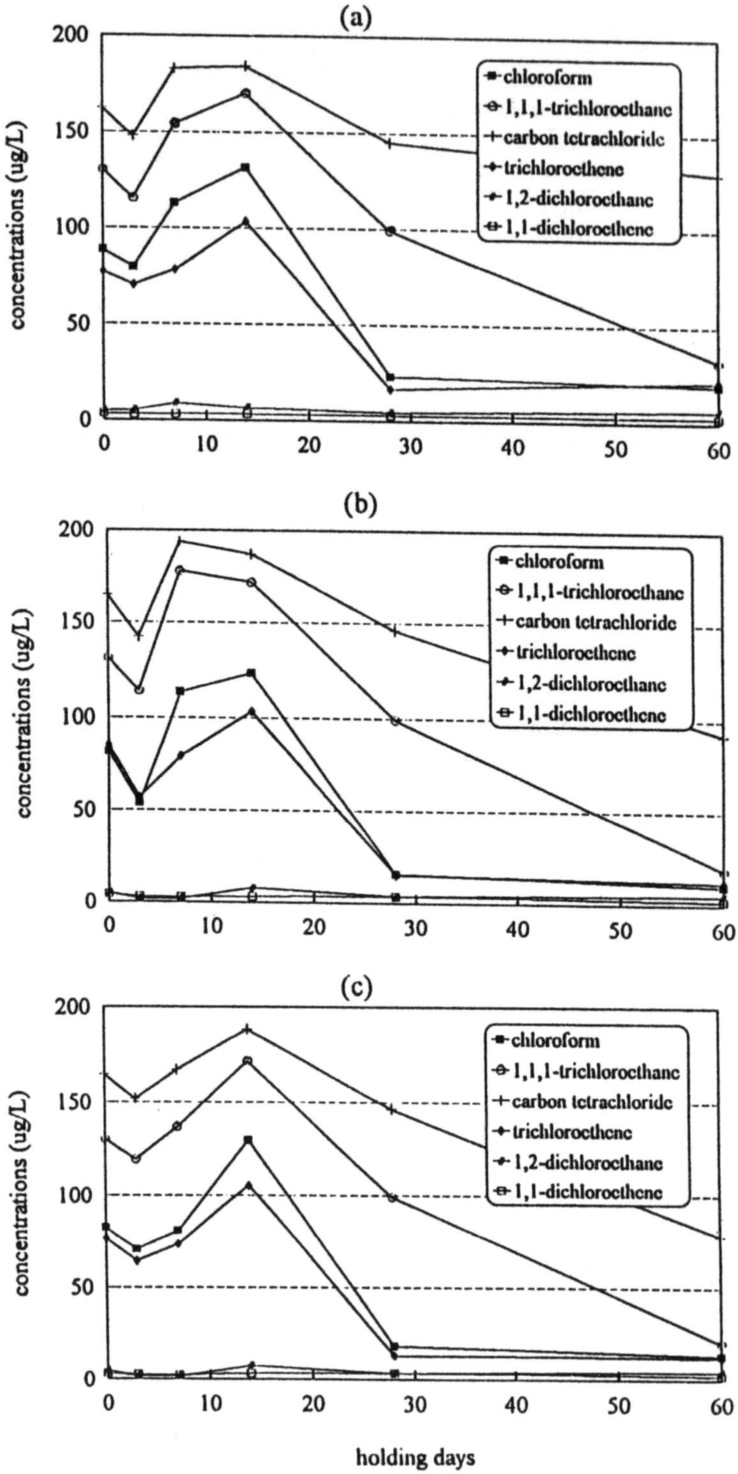

Figure 2. Analytical results of VOC holding study with preservation agents: (a) ascorbic acid (25 mg/40 ml sample); (b) HCl (4 drops 1 N HCl/40 ml sample); (c) $NaHSO_4$ (25 mg/40 ml sample).

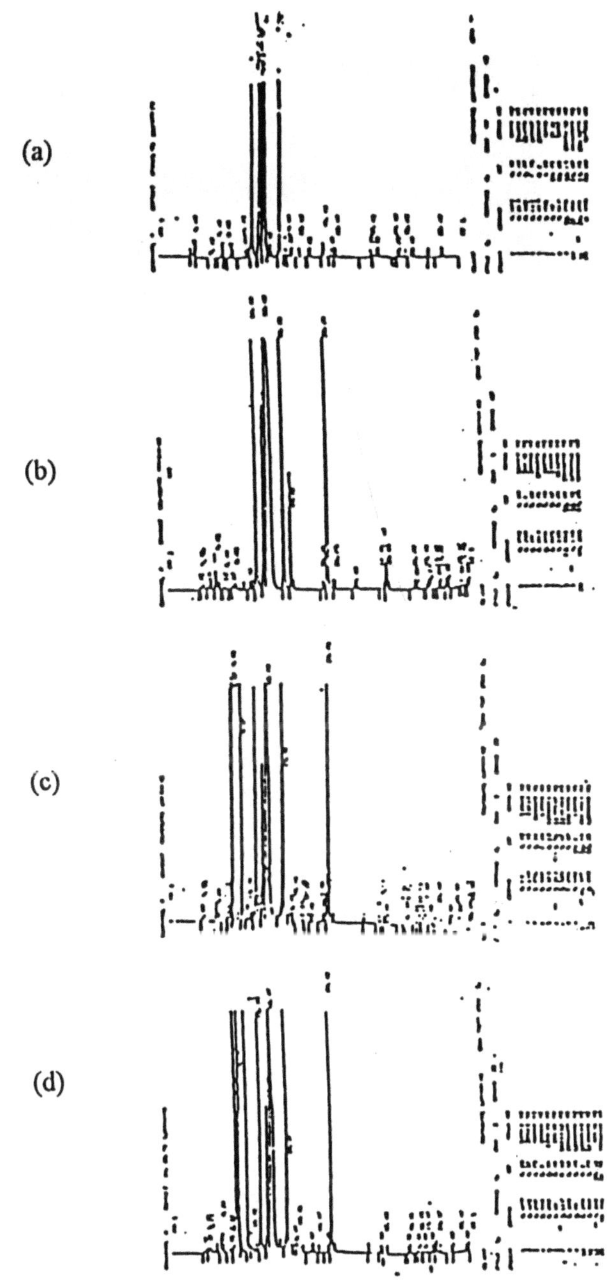

Figure 3. Gas chromatogram of VOC holding study after 100 preservation days: (a) original sample; (b) using ascorbic acid as a preservation agent; (c) using HCl as a preservation agent; (d) using $NaHSO_4$ as a preservation agent.

Table 3. Summary of six VOC Species with three preservation agents

	Preservation agents														
	Ascorbic acid 25 mg/40 ml holding days					HCl 4 d 1 N HCl/40 ml holding days					$NaHSO_4$ 25 mg/40 ml holding days				
Species	3	7	14	28	60	3	7	14	28	60	3	7	14	28	60
1,1-Dichloroethene						C	C	C	B	C		C	A	B	C
Chloroform		C	C	C	C	C	C	C	C	C			C	C	C
1,1,1-Trichloroethane			C	C	C		C	C	C	C			C	C	C
Carbon tetrachloride					C					C					C
Trichloroethene			C	C	C	C	A	C	C	C			C	C	C
1,2-Dichloroethane		C	C	A	C	C	C	C	C	B	C	C	C	C	A

A denotes sample concentrations ranges from 90 to 110 % of the initial values.
B denotes sample concentrations ranges from 80 to 120 % of the initial values.
C denotes sample concentrations out of the range 80–120% of the initial values.

to introduce a higher amount of Na_2SO_3 dosage, i.e., 700 mg/L, considered as an alternative for the HAAs holding study.

Figure 4(a) to (d) shows results of the holding studies using MBAA, DCAA, TCAA, and DBAA, respectively. In Figure 4(a), the measured MBAA concentrations can hold 80–120% of their initial concentration up to day 29 for NH_4Cl preservation samples. Similar recoveries (80–120%) can be made up to day 6 for the 100 mg/L of Na_2SO_3 preservation sample. After that, MBAA concentrations dropped gradually to the level of 55% and 20% at the end of holding studies (70 days) for the NH_4Cl and 100 mg/L Na_2SO_3 preservation samples, respectively. On the other hand, the use of 700 mg/L of Na_2SO_3 caused significant degradation of MBAA concentration after 6 days and continued to decay at the remaining holding study.

DCAA samples held their initial value within 80–120 percent until day 18 for all three preservation agents samples. As shown in Figure 4(b), the recovery levels of DCAA for those NH_4Cl preservation samples rose from day 6 to day 29, which is similar to the findings conducted by Krasner et al.[4] The variation pattern of DCAA with 700 mg/L of Na_2SO_3 is quite similar to that of DCAA with NH_4Cl. Furthermore, DCAA concentrations were stable (i.e., the variation of recovery level was within ± 20 percent of original concentration) over a 29-day holding time.

In Figure 4(c), the determination of TCAA in the presence of NH_4Cl remained within 20% of the time-zero concentration at the 70 holding day period; while the measured TCAA concentrations were almost acceptable if 700 mg/L of Na_2SO_3 preservation agent was added, except for the slight out-of-range points which in days 43 and 70. However, TCAA measurements for 100 mg/L of Na_2SO_3 preservation samples showed a loss of more than 20% after 43 days.

Figure 4(d) showed that DBAA was stable for the whole 70 day holding period when NH_4Cl was employed. In the course of a 70-day holding study, the Na_2SO_3-preserved (700 mg/L) samples held 80–120% of their initial concentration for at least 29 days, however, samples treated with 100 mg/L of Na_2SO_3 as a dechlorination and preservation agent were unstable after day 6.

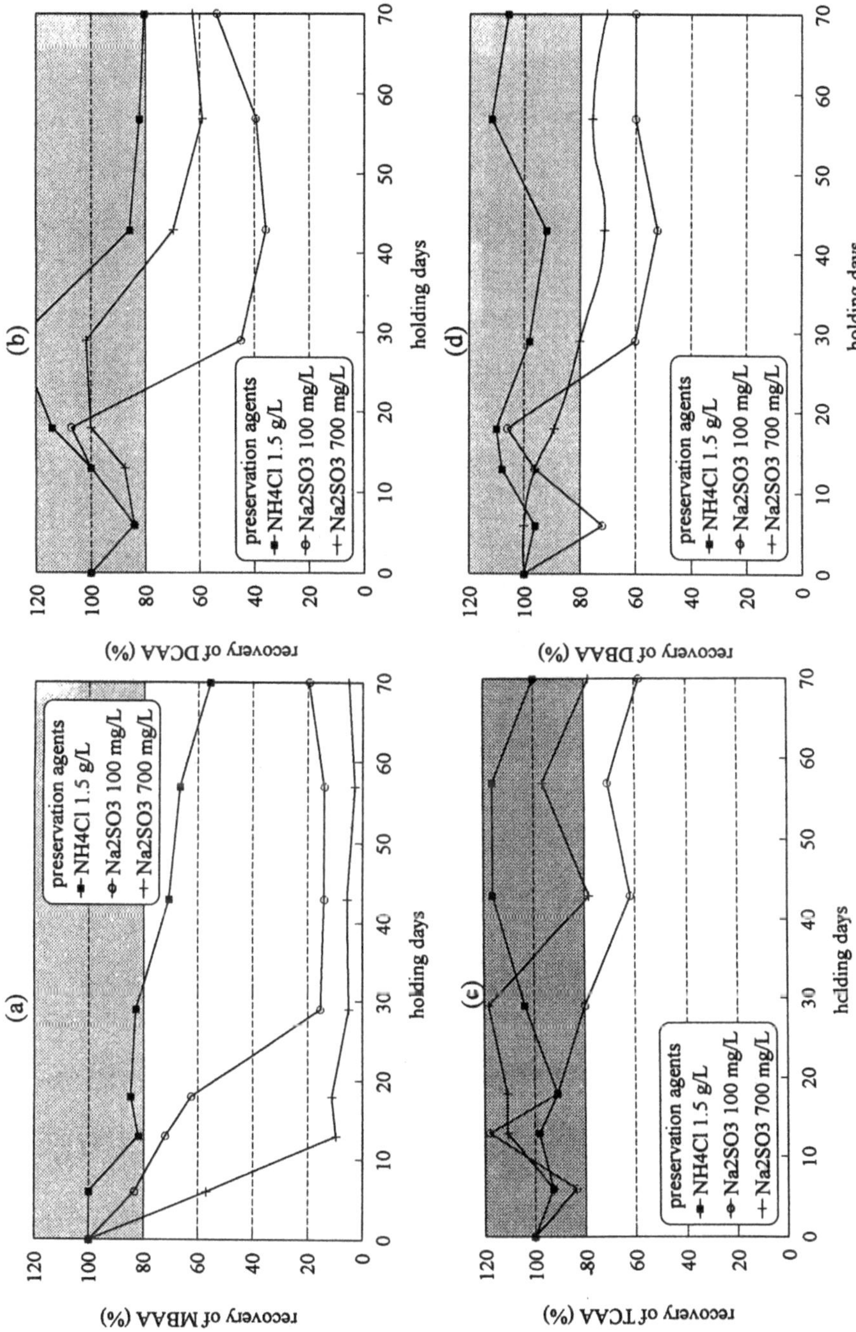

Figure 4. Analytical results of HAA holding study: (a) MBAA; (b) DCAA; (c) TCAA; (d) DBAA.

Table 4. Summary of four HAAs species with three preservation agents

	Preservation agents																				
	NH_4Cl 1.5 g/L holding days							Na_2SO_3 100 mg/L holding days							Na_2SO_3 700 mg/L holding days						
Species	6	13	18	29	43	57	70	6	13	18	29	43	57	70	6	13	18	29	43	57	70
Monobromoacetic acid					C	C	C		C	C	C	C	C	C	C	C	C	C	C	C	C
Dichloroacetic acid				C	B	B	B				C	C	C	C					C	C	C
Trichloroacetic acid												C	C	C					C	C	C
Dibromoacetic acids								C	A	A	C	C	C	C					C	C	C

A denotes sample concentrations ranges from 90 to 110% of the initial values.
B denotes sample concentrations ranges from 80 to 120% of the initial values.
C denotes sample concentrations out of the range 80–120% of the initial values.

It is possible to evaluate the performances of three preservation techniques for storage of HAAs through comparisons of relative error percentages of their concentration levels. From Figure 5, it was observed that relative error percentages of samples with 1.5 g/L of NH_4Cl as a preservation agent were generally less than those of the Na_2SO_3-preserved samples. On the other hand, Table 4 summarizes results of the HAA holding studies, which suggests that 1.5 g/L of NH_4Cl be capable of preserving four HAAs in 18 days of holding study and preserving TCAA and DBAA in whole holding period (70 days). As a result, it was concluded that the stability of HAAs under various preservation agents was in the sequence of: NH_4Cl 1.5 g/L> Na_2SO_3 700 mg/L> Na_2SO_3 100 mg/L after the 70 day holding study.

4. CONCLUSIONS

The analytical results for VOCs, including chloroform, 1,1-dichloroethylene, 1,1,1-trichloro-ethane, carbon tetrachloride, trichloroethylene, and 1,2-dichloroethane, showed that the decay of the concentrations of these species can be described by first-order decay reaction kinetics. In view of reaction kinetics, ascorbic acid has a slightl advantage over HCl and $NaHSO_4$ with respect to its relatively lower reaction rate constants shown in Table 2. From Table 3, it is concluded that among the three preservation methods studied in this research, ascorbic acid was capable of recovering these species within three days of preservation, and proved to be the most suitable one among the three preservation agents for the compounds studied over a 60 day holding study.

Another holding study was conducted to test the effectiveness of three preservation techniques on four HAAs, including MBAA, DCAA, TCAA and DBAA. Results shown in Table 4 suggest that 1.5 g/L of NH_4Cl was capable of preserving four HAAs within a 18 day holding study and preserving TCAA and DBAA within the entire holding period of 70 days.

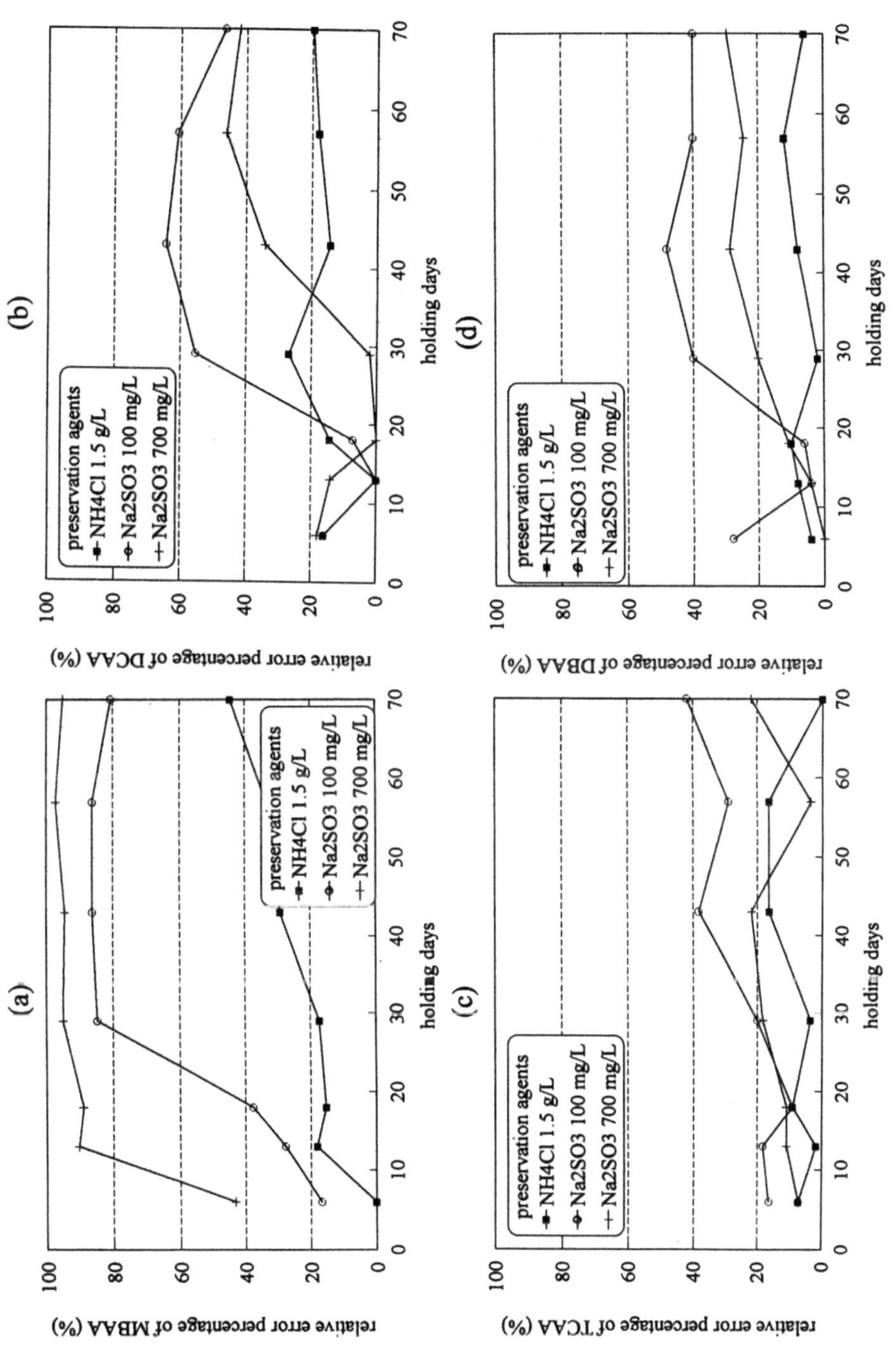

Figure 5. Results of relative error percentages of HAA holding study: (a) MBAA; (B) DCAA; (C) TCAA; (d) DBAA.

5. REFERENCES

1. *Standard Method for the Examination of Water and Wastewater* (1992), 18th ed., Prepared and published by APHA, AWWA, and WEF
2. Hodgeson, J., Cohen, A. and Collins, J., "Analytical Methods for Measuring Organic Chlorination By-Products", *Disinfection By-Products: Current Perspectives*, AWWA, 1989.
3. Koch, B., Crofts, E.W., Schimpff, W.K. and Davis, M.K., "Analysis of Halogenated Disinfection By-Products by Capillary Chromatography", *Disinfection By-Products: Current Perspectives*, AWWA, 1989.
4. Krasner, S., Sclimenti, M.J. and Hwang, C.J., "Experiences With Implementing A Laboratory Program to Sample and Analyze for Disinfection By-Products in a National Study", *Disinfection By-Products: Current Perspectives*, AWWA, 1989.

45

A COMPARISON OF SOIL- AND WATER CHEMISTRY IN A CATCHMENT IN CHINA WITH SITES IN POLAND AND NORWAY

Thorjørn Larssen, Rolf D. Vogt, and Hans M. Seip

Department of Chemistry
University of Oslo
P.O. Box 1033, Blindern, 0315 Oslo, Norway

ABSTRACT

The results from a small catchment outside Guiyang in southwestern China are compared to results from catchments in Poland and Norway. Deposition, soil water, stream water and soils have been sampled and analyzed in all these catchments, using similar techniques.

The sulfur deposition is high in the Chinese catchment, as in some of the Polish catchments. Base cations, especially calcium, is an important part of the cations in deposition in Guiyang; in most of the Polish catchments ammonium is important together with calcium. At the Norwegian sites ions originating from sea salts are very important.

The concentration of monomeric inorganic aluminum complexes (Ali) in soil water is high in the Guiyang catchment, as in the Polish catchments. However, the high calcium concentrations keep the Ali/(Ca^{2+}+Mg^{2+}) ratio relatively low in the Guiyang catchment; only in one plot the molar ratio is commonly observed above one, which often is considered as potentially harmful. Surface waters in two of the Polish sites (Ratanica and Brenna) and in the lower part of the Guiyang catchment have high sulfate and calcium concentrations and almost neutral pH. At the Czerniawka catchment in western Poland, and at the Norwegian sites, the pH in surface water is generally low, indicating lack of neutralization of the soil water before it enters the stream.

1. INTRODUCTION

Acid deposition is an existing or potential problem in many parts of the world (Rodhe, 1989). During the 1970s extensive research concerning ecosystem effects of acid deposition was carried out in Europe and in North America. Later, in the early eighties,

Chemistry for the Protection of the Environment 2, edited by Pawłowski et al.
Plenum Press, New York, 1996

Table 1. Characteristics of the different catchment involved in the comparison

Catchment	Climate	Altitude m.a.s.l.	Temp. °C (monthly avg.) Warmest	Temp. °C (monthly avg.) Coldest	Precip. mm	Major soil type	Texture class	Bedrock/parent material
Guiyang	Subtropical	1320–1400	24	5	1175	yellow, dystric Cambisol	Loams	Sandstone
Janow	Temperate	200	18.4	–3.8	550	Podzol	Sand	Sand
Ratanica	Temperate	270–427	17.7	0.4	850	Podzol/Luvisol	Sand	Sandstone, shales
Czerniawka	Temperate / sub-alpine	650–1050	14.6	–2.9	1200	Podzol	silty sand	Gneiss
Brenna	Temperate	640–730	14.2	–2.3	1300	dystric Cambisol/ Cambic podzol		Sandstone
Birkenes	Temperate	205–300	15.5	–3.7	1500	Podzol		Granite
Ingabekken	Temperate	280–370	13.7	–7.1	1300	Podzol		Granite
HUMEX	Temperate	150	13.5	–2.5	2560	Histosol/Podzol	Sand	Granite

References:

Guiyang: Larssen, 1994

Poland: Vogt et al., 1994a

Birkenes and Ingabekken: Mulder et al., 1992 (soils); Aune, B., 1993 (temperatures); Seip et al., 1990 (precipitation and altitude)

HUMEX: Vogt et al., 1992 (soils); Aune, B., 1993 (temperatures); SFT, 1993 (precipitation).

the fast industrialization led to an increased interest and worry about acid rain in other parts of the world, for instance in China. Norway was among the first countries in the world to start acidification research (Overrein et al., 1980). Parts of Poland has been among the worst affected areas, with extensive forest damage. China, with its fast industrial growth, experiences today large emissions of acidifying pollutants; effects have already been observed some places. In contrast to Europe, the emissions and hence the deposition of acidifying pollutants are increasing fast and what will happen to ecosystems in the near future is uncertain.

In this paper we compare some results from a catchment near Guiyang in China with different catchments in Norway and Poland.

2. THE CATCHMENTS

A summary of important characteristics of the different catchments is given in table 1. Locations of the different catchments are shown in Fig. 1.

2.1 China

The *Guiyang* catchment in Southwestern China (See Fig. 1) is located about 10 km. north-east of the city center of Guiyang, the capital of the Guizhou province (26°34'N, 106°38'E) (Seip et al., 1995; Larssen, 1994; Larssen et al., 1996). This 7 ha. catchment was in 1992 equipped with instruments for integrated monitoring of deposition, soils, soil water and surface water; probably as the first catchment in China. The catchment is lo-

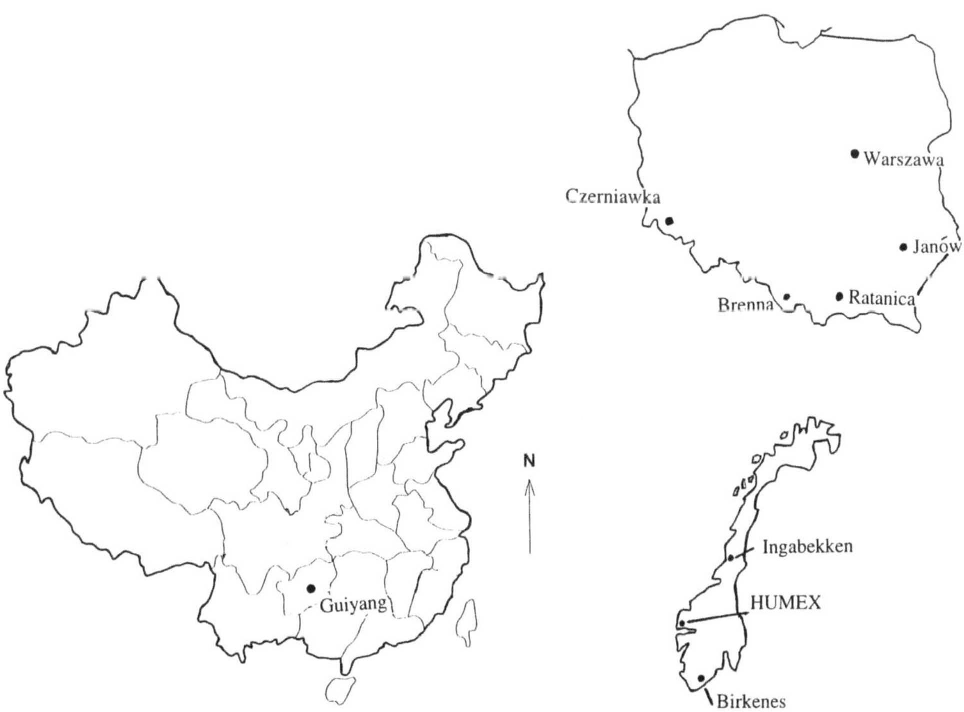

Figure 1. Maps showing location of the catchments discussed in this paper.

cated in the subtropical region at an altitude of 1320–1400 m. Monsoons result in high precipitation during summer and much less during winter although the humidity is often high most of the year. The soils are mainly dystric cambisols, the vegetation is Chinese fur (*Cunninghemia lancelota*), pine (*Pinus massoniana*) and wild rose bushes. No forest damage is apparent.

Precipitation has been sampled just outside the catchment. Throughfall has been sampled at 5 plots in the catchment, and soils and soil water at 7 plots. Two small first order streams join to form a second order stream in the lower part of the catchment, ending up in a dam at the lower end of the catchment. Surface water has been collected from the two first order streams and from the dam. For details about the catchment see Larssen (1994) and Larssen et al. (1996); about the soils see Liao et al. (1994) and Liao et al. (1996).

2.2 Poland

The four Polish catchments are presented in Vogt et al. (1994a). In all catchments samples of deposition, soil water, stream water and soils have been collected. The four catchments are relatively different with respect to deposition, and to some extent also with respect to the soils.

The *Janow* site is located about 70 km south of Lublin in southeastern Poland in the Lasy Janowskie forest (50°42'N, 22°24'E). The vegetation is mainly pine (*Pinus sylvestris*) with sparse blueberry (*Vaccinium myrtillus*) undergrowth. Lysimeters have been installed in two plots.

The *Ratanica* catchment is located about 40 km south of Kraków at an altitude ranging from 270 to 427 m.a.s.l. (49°51'N, 20°02'E). The climate is typically temperate with an annual precipitation of 850 mm. The vegetation is mainly beech (*Fagus sylvatica*) and pine (*Pinus sylvestris*) with blackberry (*Rubus hirtus*) and blueberry (*Vaccinium myrtillus*) ground vegetation. The soils are mainly luvisols and podzols developed in sand or sandy loams. Throughfall and soil water are sampled in three transects in the catchment at 29 plots, in addition to deposition and surface water.

The *Czerniawka* site is located in south-west Poland near the boarder of the Czech republic (50°48'N, 15°35'E). The annual amount of precipitation is about 1200 mm; about half falls as snow. The dominating tree species is Norway spruce (*Picea abies*), with a sparse ground vegetation of mainly grass and moss (*Polytrichum formosum*). Forest damage is serious at the higher altitudes of the catchment. Podzolic soils dominate in the catchment, with some leptosols at the high altitudes and fluvisols and histosols along the stream banks (Skotte, 1995). In addition to deposition and stream water, soil water has been sampled from 11 plots in the catchment.

The *Brenna* catchment is located about 70 km. south of Katowice, close to the Czech and Slovakian republics (49°40'N, 19°56'E). The dominating tree species is Norway Spruce (*Picea abies*) and the dominating soils are dystric cambisols and cambic podzols. Throughfall and soil water have been sampled at 21 plots along two transects, precipitation at two plots and surface water at the outlet in the lower part of the catchment.

2.3 Norway

The Birkenes catchment is located in southernmost Norway (58°15'N, 08°15'E) and has been intensively studied since 1972 (Overrein et al., 1980). This catchment is situated in an area where the population of trout (*Salmo trutta*) experienced a serious dieback since the 1950's. Norway spruce (*Picea abies*) and pine (*Pinus sylvestris*) are the dominating tree species and blueberry (*Vaccinium myrtillus*) the main undergrowth vegetation in the catchment. The major soil type is podzol over a bedrock of granite. Soil water has been sampled from 8 plots, the deposition and stream water have been monitored in detail.

The *Ingabekken* site is located in mid Norway (64°38'N, 12°18'E), and was selected in the latter part of the 1980's as a pristine catchment to study natural hydrogeochemical properties. The lower part of the catchment is mainly covered with pine forest, while the upper part has an alpine character with several small bogs and bare rock. The mineral soil is mainly shallow podzolic soils on a granite bedrock (Christophersen at al., 1990). The dominating middle part is covered by ombrogenic bogs.

The *HUMEX* catchment is located in the relatively pristine western part of Norway (61°26' N, 6°26'E). The two main vegetation species in this catchment are moss (*Sphangnum sp.*) and pine (*Pinus sylvestris*) (Røren, 1993). Podzols are the dominating mineral soil in the catchment, but a considerable part of the catchment is bogs (histosols) (Vogt et al., 1994b).

In this comparison study only the mineral soils in the catchments are discussed.

3. DEPOSITION

Volume-weighted average concentrations of the major ions in precipitation for the selected sites are shown in Fig. 2. The total concentration is highest in the Ratanica catchment, followed by Guiyang and the Polish catchments. In the Birkenes catchment the total ion concentration is similar to those in the Czerniawka and Brenna catchments. However, if we only look at the ions with possible anthropogenic origin, the concentrations in Birke-

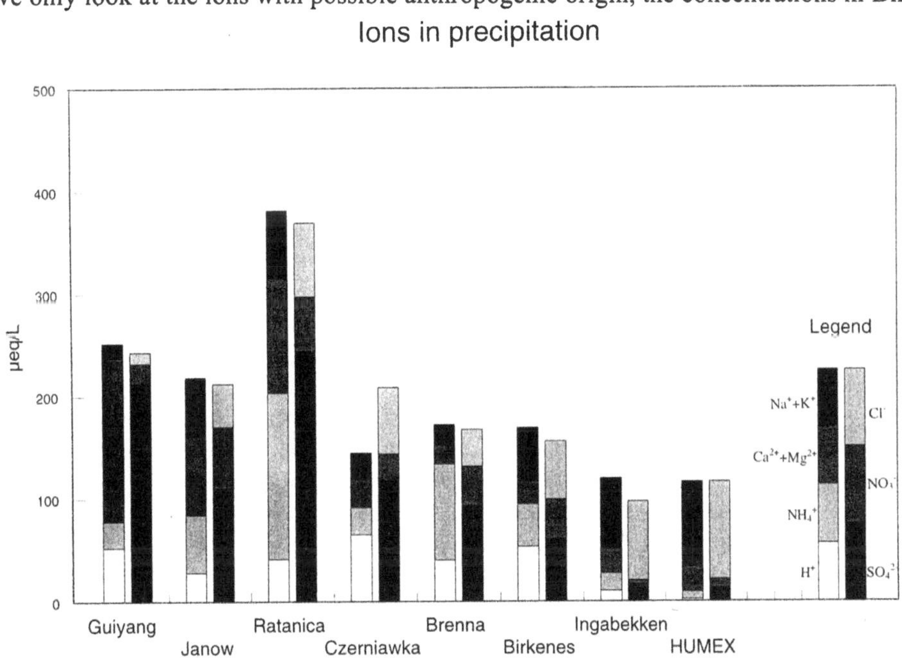

Figure 2. Volume weigthed annual precipitation chemistry in the catchments. Cations in left bar, anions in right bar. (References: Guiyang: Larssen, 1994; Poland: Vogt et al., 1994a; Birkenes and Ingabekken: Seip et al., 1990; HUMEX (data from Nausta meteorological station): SFT, 1993).

nes are markedly lower, while concentrations of seasalts are high due to the vicinity to the sea. The sea-salt dominance is even more pronounced at the two other Norwegian catchments, which are located near the ocean, and further away from any large anthropogenic emission sources. These two catchments (Ingabekken and HUMEX) are classified as relatively pristine sites. In Guiyang the concentration of seasalt in the precipitation is negligible because of the very long distance from the ocean. In the Polish catchments, especially Ratanica, the concentrations of chloride and sodium are surprisingly high in spite of the long distance from the sea (see: Farrell, 1995). Because of relatively dry conditions during parts of the year, soil dust may contribute considerably to the deposition in Guiyang and some of the Polish catchments.

The ion concentrations in throughfall samples are compared in Fig. 3. The concentration in throughfall samples is very sensitive to the vegetation type and density above the sampler, hence several samplers are used giving an average for the catchment. The differences between the sampling plots are large in Guiyang, where the median sulfate concentration in one plot was 1440 μeq/L and in an other 315 μeq/L (Larssen , 1994). The average of all the medians in Guiyang is 640 μeq/L. This large variation within the same catchment is explained by differences in vegetation density at the sampling sites combined with a high dry deposition.

The dry deposition decreases relatively fast when moving away from the emission sources. The Guiyang catchment is located close to large emission sources, hence the dry deposition is large. In the Norwegian catchments, located very far from the main emission sources, dry deposition is generally about 15% of the total deposition.

The sulfur deposition, as shown in table 2, is calculated from the SO_4^{2-} concentration in precipitation and the SO_2 concentration in air. The deposition velocity (V_d) used to calculate deposition from air SO_2 concentration varies with many factors, including vegetation type and density. We used V_d = 0.6 cm/s, as an estimate of annual average in southern Poland. In addition to the wet deposited SO_4^{2-} and the dry deposited SO_2, dry

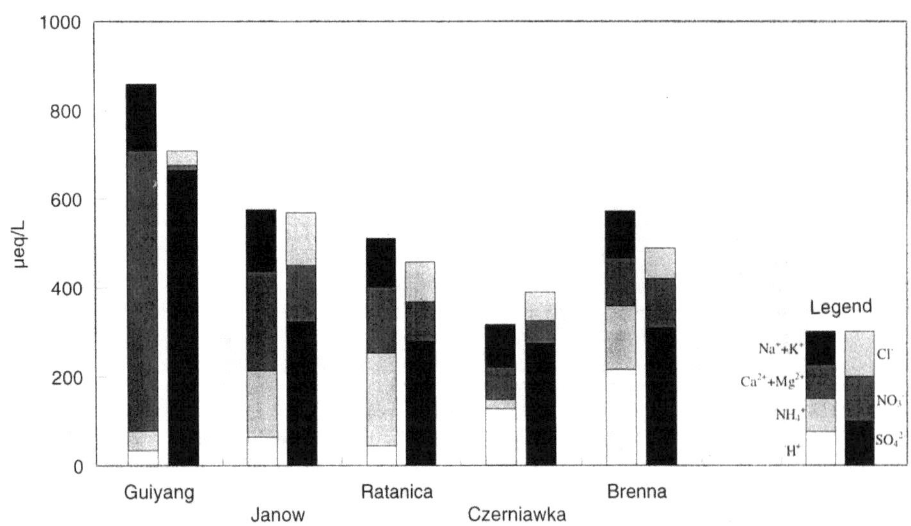

Figure 3. Median concentrations of ions in throughfall in the Chinese and Polish catchments. Cations in left bar, anions in right bar. (References: Guiyang: Larssen, 1994; Poland: Vogt et al., 1994a; Birkenes and Ingabekken: Seip et al., 1990; HUMEX (data from Nausta meteorological station): SFT, 1993).

Table 2. Total sulfur deposition for the different catchments

Site	Wet deposition $gSm^{-2}yr^{-1}$	Throughfall deposition $gSm^{-2}yr^{-1}$	SO_2 air conc. $\mu gS/m^3$	SO_2 deposition[a] $gSm^{-2}yr^{-1}$	Wet deposition + dry deposited SO_2 $gSm^{-2}yr^{-1}$	Reference
Guiyang	4.01	—	22	4.16	8.2	Larssen et al., 1996
Janow	0.97	2.71	5.5[b]	1.04	2.0[c]	Vogt et al., 1994c
Ratanica	3.33	2.45	10.5	1.99	5.3	Vogt et al., 1994b
Czerniawka	2.25	4.10	11.5	2.18	4.4	Vogt et al., 1994b
Brenna	1.94	4.44	13	2.46	4.4	Vogt et al., 1994b
Birkenes	1.47	—	—	0.18	1.64	SFT, 1992
Ingabekken	0.35	—	—	—	0.35	Seip et al., 1990
HUMEX	0.55	—	—	—	0.6	SFT, 1993

— denotes no information available.

[a]Assumed deposition velocity in Guiyang and the Polish sites is 0.6 cm/s (see: Iversen et al., 1991)

[b]At Jarzew EMEP station, NE of Lublin, data from Schaug et al., 1994.

[c]The total deposition is probably somewhat higher (as the figure for throughfall indicates) due to high deposition of sulfate in dust.

deposited SO_4^{2-} from aerosols may contribute significantly to the total sulfur deposition in certain dry areas. Based on the concentration in throughfall it is likely that dry deposited sulfate is of special importance at Janow, which also would be reasonable because of the small amount of precipitation.

The total deposition is highest in the Guiyang catchment, followed by Ratanica, Czerniawka and Brenna. The Janow catchment receives considerably less sulfur, but still higher than the Birkenes catchment. In the two pristine Norwegian catchments, Ingabekken and HUMEX, a considerable part of the total sulfur originates from sulfate in sea-spray.

4. SOILS

In Fig. 4, average mineral soil properties for the different catchments are shown. The upper, organic soil horizon in the Guiyang catchment has the lowest effective cation exchange capacity (CEC_E) compared to all the other catchments, while for the B horizon the Guiyang catchment has the highest. The CEC is generally strongly correlated to the amount of organic matter and a variable organic content is probably the main reason for the different CEC_E values in the OA horizons in these catchments. In the B horizons particle size is more important. The Norwegian catchments and Janow have more than 60% base cations and less than 20% aluminum on the ion exchange sites in the OA horizon, while at the Guiyang site aluminum and base cations account for 40% and 50%, respectively. In the B horizon the aluminum saturation (AlS) is between 80% and 90% in all the acidified catchments except Brenna, where it is 74%. The pristine Norwegian catchments have an average AlS of 75% and 65%.

5. SOIL WATER

Soil water concentrations for the different catchments are shown in Fig. 5. The concentrations presented are averages of the median values for the B-horizon from each plot.

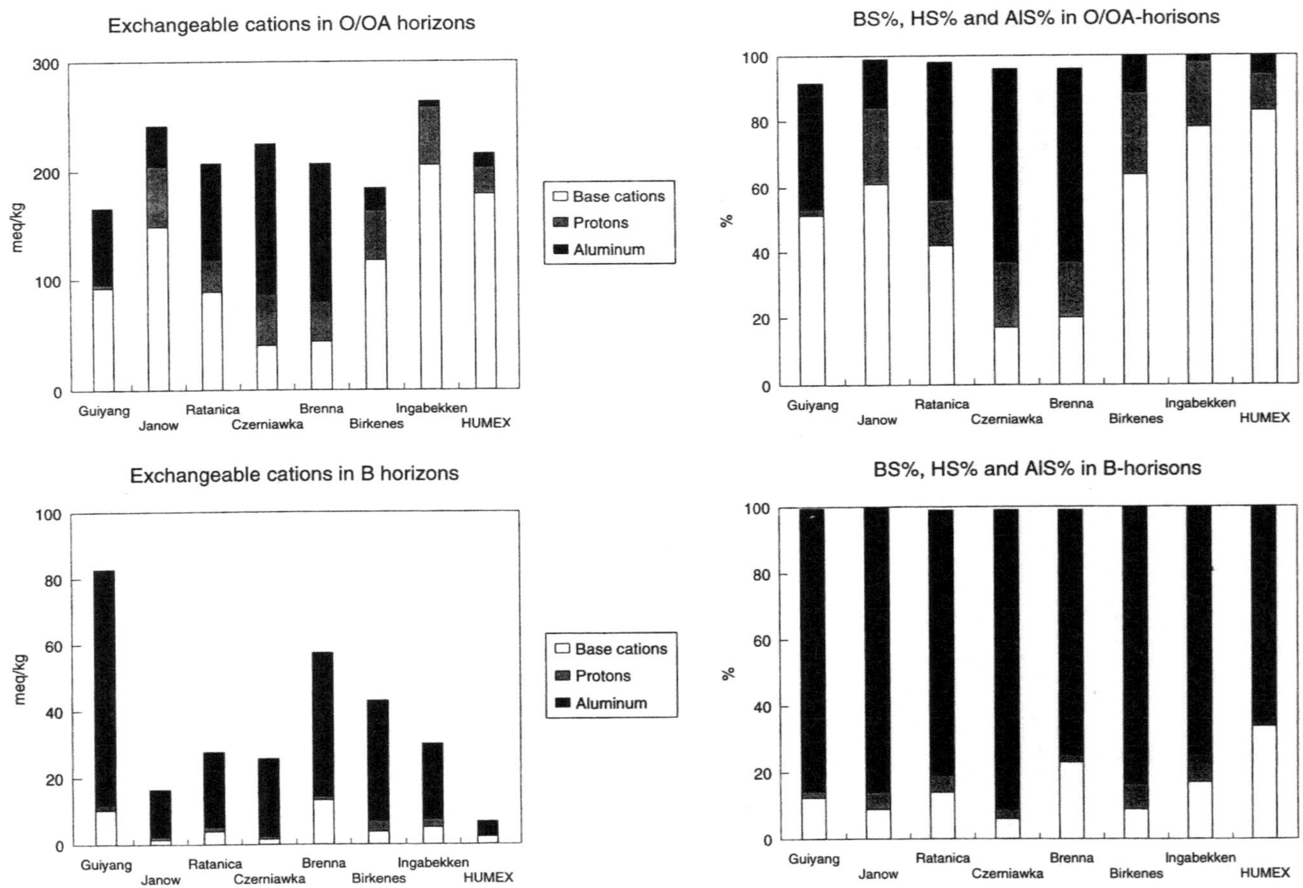

Figure 4. Cation exchange capacity (CEC_E), base saturation (BS%) and aluminum saturation (AlS%) for the catchments. (References: Guiyang: Larssen et al., 1996; Poland: Vogt et al., 1994a; Birkenes and Ingabekken: Mulder et al., 1992; HUMEX: Vogt et al., 1992.)

The large spatial and temporal variability in the different catchments is not shown in the figure, but should be kept in mind.

The total concentration of ions in soil water is highest in the Guiyang catchment, though hardly significantly higher than in Ratanica and Janow. The two other Polish catchments, Czerniawka and Brenna have on the average lower total ion concentrations in soil water, but still higher than the Norwegian catchments (Brenna is not shown in Fig. 5 because aluminum data is lacking).

Among the anions, sulfate is dominating in the Guiyang catchment and at the Polish sites. In the Norwegian catchments chloride of marine origin is the dominating anion, even though the sulfate concentration is considerable at Birkenes.

For the cations we see the same trend with seasalts in the Norwegian catchments; Na^+ is totally dominating in the "pristine" catchments, while in addition the aluminum concentration is considerable in the acidified Birkenes catchment. In all the Polish catchments and in Guiyang aluminum is very important. In addition calcium is high in some of them, giving rise to very different Ali/(Ca^{2+}+Mg^{2+}) ratios (see figure 6). Although the basis is weak, this ratio is commonly used in critical load studies as an indicator of possible harmful effects on vegetation when exceeding unity (Sverdrup and de Vries, 1994; Downing et al., 1993). Because of the high concentration of calcium in Guiyang, the ratio is considerably lower than in the Czerniawka catchment where serious forest decline is observed. This illustrates the importance of calcium at the Guiyang site. A considerable part of the calcium is supplied from the deposition. Therefore it is a cause of concern, in terms of acidification, what will happen if the emission of base cations is reduced much faster than the SO_2 emission, as is the case for present plans to limit air pollution in Guiyang (Xiong, 1995, pers. comm.).

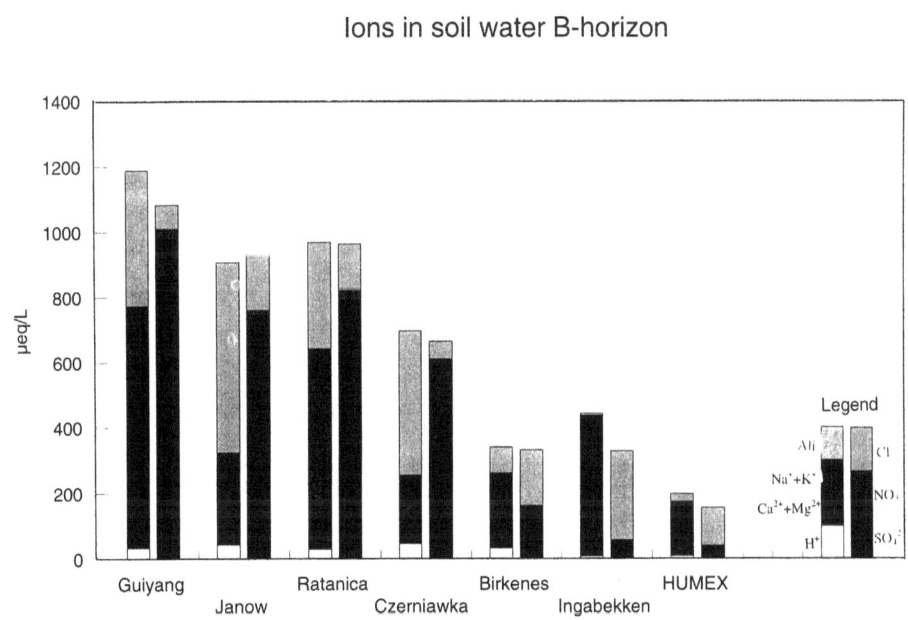

Figure 5. Concentration of ions in soil water collected from the B-horizons in the catchments. Cations in left bar, anions in right bar. (References: Guiyang: Larssen, 1994; Poland: Vogt et al., 1994a; Birkenes and Ingabekken: Mulder et al., 1995; HUMEX: Vogt et al., 1994b.)

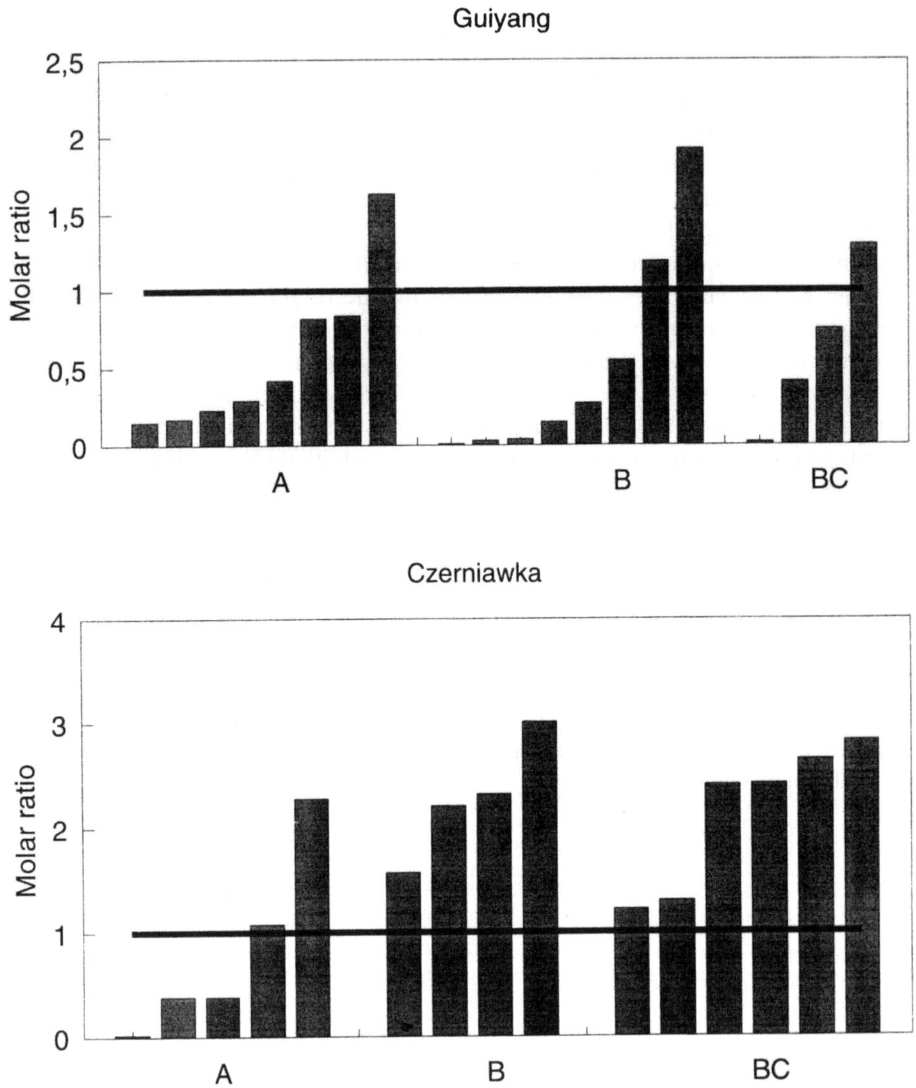

Figure 6. The molar ratio of Ali/(Ca^{2+} + Mg^{2+}) for soil water from the Guiyang and Czerniawka catchments. Each bar corresponds to median values for one lysimeter.

6. SURFACE WATER

The chemistry of the surface waters is very different in the different catchments (Fig. 7). In Ratanica and Brenna, and to some extent in the Guiyang catchment, neutralizing soil processes are taking place in deeper soil layers (and bedrock) giving rise to surface water with approximately neutral pH together with high sulfate and calcium concentrations. Hence it seems unlikely that surface water acidification will become a serious problem in these catchments. To what extent the results for the Guiyang catchment are representative for this part of China is uncertain. However the conclusions of Zhao and Seip (1991) and Xue and Schnoor (1994) indicate the same.

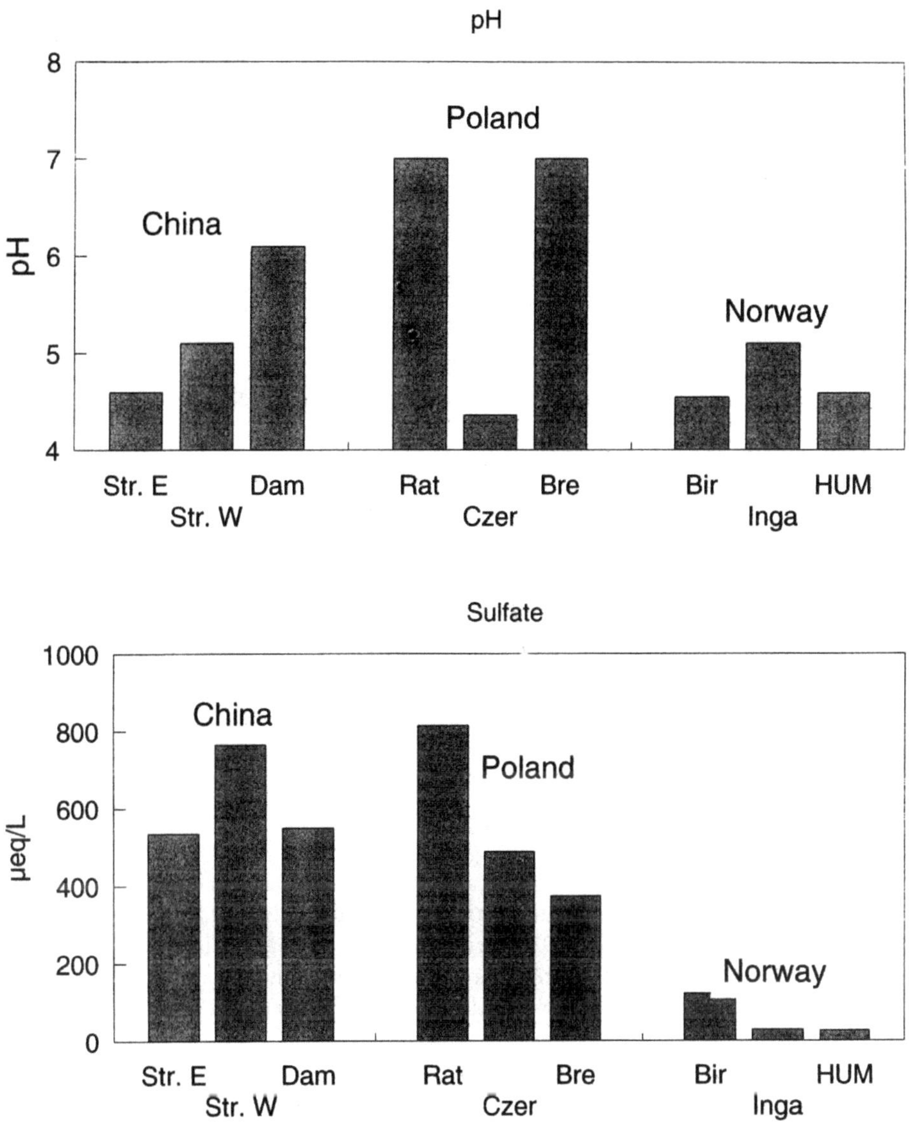

Figure 7. Median values of pH and SO_4^{2-} concentration in surface water. (References: Guiyang: Larssen et al., 1996; Poland: Vogt et al., 1994a; Birkenes and Ingabekken: Seip et al., 1990; HUMEX: Gjessing, 1994.)

In the Norwegian catchments and in Czerniawka the pH is low because the bedrock is of poorly weatherable gneiss or granite.

7. ALUMINUM CONTROL

The concentration of aluminum in soil- and surface water is of great importance when discussing effects on ecosystems by acidification. Hence it is of interest to find a model describing the aluminum chemistry in the catchment. A simple model involving a gibbsite ($Al(OH)_3$) mineral phase has been, and still is, widely used in acidification mod-

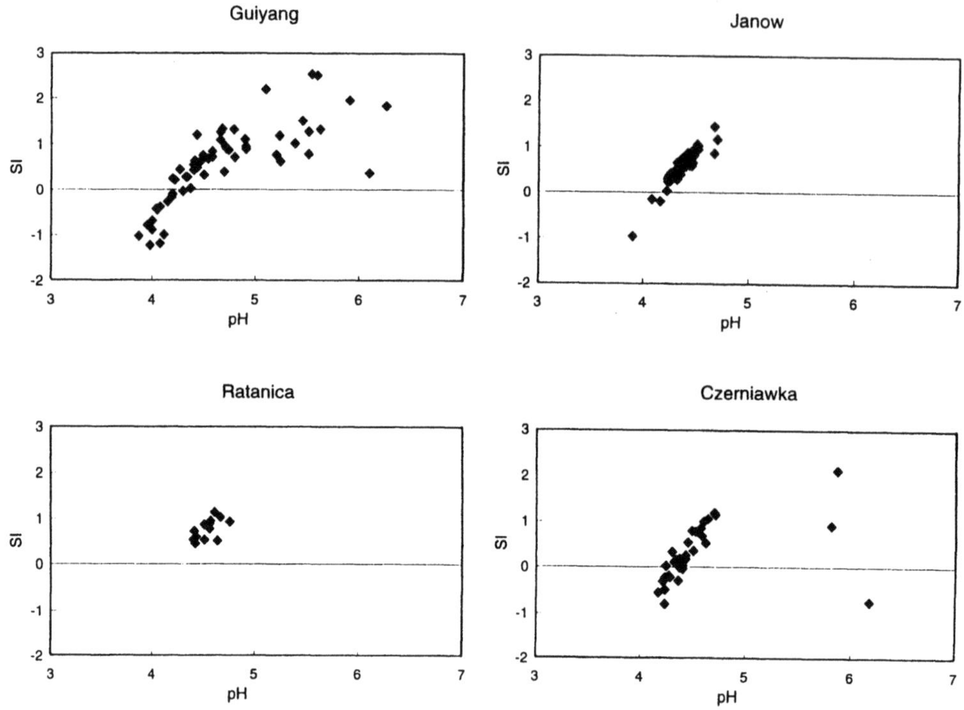

Figure 8. Saturation indices (SI = log Q/K_{sp}) for syntetic gibbsite for the Chinese and Polish catchments. (References: Guiyang: Larssen, 1994: Poland: Vogt et al., 1994a).

els. Several authors have pointed out that this model actually does not fit with the observed aluminum data (Neal et al., 1987; Seip et al., 1989; Vogt et al., 1994b). Here we illustrate that this is the case also for the Guiyang catchment and for the Polish catchments (Fig. 8). It is clear that better models should be found, further investigation is needed to develop such models.

8. CONCLUSIONS

Sulfur deposition in Guiyang is the highest among the selected catchments. In Guiyang the calcium deposition is high, while ammonium is more important in some of the Polish catchments.

Concentration of inorganic, monomeric aluminum in soil water is very high in Guiyang, as in the most acidified Polish catchments.

The ratio of $Ali/(Ca^{2+}+Mg^{2+})$ is lower in Guiyang than in Czerniawka, where serious forest decline has occurred. However, in one plot in Guiyang the ratio exceeds unity.

Soil acidification is likely, and may become a major problem in China.

Water acidification is not likely to become a serious problem in SW China, but more studies should be carried out.

Future trends will depend on emission of base cations as well as of sulfur- and nitrogen oxides.

REFERENCES

Aune, B., 1993. Air temperature normals, normal period 1961–1990. Report 02/93, The Norwegian Meteorol. Inst., Oslo, Norway.

Christophersen, N., Vogt, R.D., Neal, C., Anderson, H.A., Ferrier, R.C., Miller, J.D. and Seip, H.M., 1990. Controlling mechanisms for stream water chemistry at the Pristine Ingabekken site in mid-Norway. Water Resour. Res., 26:59–67.

Downing, R.J., Hettelingh, J-P., de Smet, P.A.M. (eds.), 1993. Calculation and mapping of critical loads in Europe: Status report 1993. Coordination Center for Effects, National Inst. of Publ. Healt and Environ. Protection (CCE, RIVM), Bilthoven, Netherlands.

Farrell, E.P., 1995. Atmospheric deposition in maritime environments. Water Air Soil Pollut. In press.

Gjessing, E.T., 1992. The HUMEX project: Experimental acidification of a catchment and its humic lake. Environ. Int., 18: 535–544.

Gjessing, E.T., 1994. HUMEX: Chemistry, hydrology and meteorology. Environ. Int., 20:267–276.

Iversen T., Halvorsen, N.E., Mylona, S., Sandnes, H., 1991. Calculated budgets for airborne acidifying components in Europe, 1985, 1987, 1988, 1989 and 1990. EMEP report1/91. The Norwegian Meteorological institute, Oslo.

Larssen, T., Xiong, J., Vogt, R.D., Seip, H.M., Liao, B. and Zhao, D., 1996. Studies of soils, soil water and stream water at a small catchment near Guiyang, China. In prep.

Larssen T., 1994. Acid deposition in southwestern China. Thesis, University of Oslo, Dept. of Chemistry, Norway.

Liao, B., Larssen, T., Seip, H.M. and Vogt, R.D., 1994. Anion adsorption and aluminum release from Chinese soils treated with different concentrations of $(NH_4)_2SO_4$ and NH_4NO_3. J. Ecol. Chem., 3:281–301.

Liao, B., Seip, H.M. and Larssen, T., 1996. Response of two Chinese forest soils to acidic inputs: Leaching experiment. In prep.

Mulder, J., Christophersen, N., Kopperud, K., Fjelldal, P.H., 1995. Water flow paths and the spatial distribution of soils as a key to understand differences in streamwater chemistry between the three catchments (Norway). Water Air Soil Pollut. 81:67–91.

Neal, C., Skeffington, R.A., Williams, R. and Roberts, D.J., 1987. Aluminium solubility controls in acid waters: the need for a reappraisal. Earth Planet. Sci. Lett., 86:105–112.

Overrein L.N., Seip, H.M. and Tollan, A., 1980. Acid precipitation - effects on forest and fish, final report SNSF project, 175 pp., Norw. Inst. for Water Res., Oslo.

Rodhe, H., 1989. Acidification in a global perspective. Ambio 18:155–160.

Røren, V., 1993. Vegetation map of the catchment area of Lake Skjervatjern. Newsletter 2/1993. HUMOR, HUMEX.

Schaug, J., Pedersen, U., Skjelmoen, J.E., Arnesen, K., Bartonova, A., 1994. EMEP Data Report 1992. Norwegian Institute of Air Research, Kjeller, Norway

Seip, H.M., Andersen, D.O., Christophersen, N., Sullivan, T.J. and Vogt, R.D., 1989. Variations in concentrations of aqueous aluminum and other chemical species during hydrological episodes at Birkenes, southernmost Norway. J. Hydrol., 108:387–405.

Seip, H. M., Blakar, A. B., Christophersen, N., Grip, H. and Vogt, R., 1990. Hydrochemical studies in Scandinavian catchments. In Mason, B.J. (ed.) The surface waters acidification programme, Cambridge University Press, pp. 19–29.

Seip, H.M., Zhao, D., Xiong, J., Zhao, D., Larssen, T., Liao, B. and Vogt, R.D., 1995. Acidic deposition and its effects in southwestern China. Water Air Soil Pollut. In press.

SFT, 1992. Overvåkning av langtransportert forurenset luft og nedbør. (Monitoring of long range transported air and precipitation.) Annual report 1991. Norwegian State Pollution Control Authority, Oslo. Report 507/92.

SFT, 1993. Overvåkning av langtransportert forurenset luft og nedbør. (Monitoring of long range transported air and precipitation.) Annual report 1992. Norwegian State Pollution Control Authority, Oslo. Report 533/93.

Skotte, G.C., 1995. Virkninger av sur nedbør på jord- og vannkjemi i tre polske nedbørfelter. Thesis, University of Oslo, Dept. of Chemistry, Norway. (in Norwegian, with english abstract)

Sverdrup, H. and de Vries, W., 1994. Calculating critical loads for acidity with the simple mass balance method. Water Air Soil pollut., 72:143–162.

Vogt, R.D., Seip, H.M. and Ranneklev, S.B., 1992. Soil and soil water studies at the HUMEX site. Environ. Int., 18: 555–564.

Vogt, R.D., Godzik, S., Kotowski, M., Niklinska, L., Pawlowski, L., Seip, H.M., Sienkiewicz, J., Skotte, G., Staszewski, T., Szarek, G., Tyszka, J. and Aagaard, P., 1994a. Soil soil water and stream water chemistry at some Polish sites with varying acid deposition. J. Ecol. Chem., 3:325–356.

Vogt, R.D., Ranneklev, S.B. and Mykkelbost, T.C., 1994b. The impact of acid treatment on soilwater chemistry at the HUMEX site. Environ. Int. 20:277–286.

Vogt, R.D., Seip, H.M., Pawlowski, L., Kotowski, M., Ødegård, S., Horváth, A. and Andersen S., 1994c. Potential acidification of soil and soil water: a monitoring study in the Janow Forest, southeastern Poland. Ecol. Engin., 3: 255–266.

Xiong, Jiling, 1995. Personal communication. Guizhou Inst. of Environ. Sci., 148 Xinhua rd., Guiyang, China.

Xue, H.B., and Schnoor, J.L., 1994. Acid deposition and lake chemistry in Southwest China. Water Air Soil Pollut., 75:61–78.

Zhao Dianwu and Seip, H.M., 1991. Assessing effects of acid deposition in southwestern China using the MAGIC model. Water Air Soil Pollut., 60:83–97.

THE TRACE METAL ELEMENTS IN WATER AND SEDIMENTS OF HORNÁD RIVER BASIN

N. Pliešovská,[1] K. Flórián,[1] and J. Seszták[2]

[1] Department of Chemistry, Faculty of Metallurgy
Technical University, Letná 9, SK-042 00 Košice, Slovakia
[2] Bodrog and Hornád River basin, state company
Ďumbierska 14, SK-041 59 Košice, Slovakia

1. INTRODUCTION

The natural concentration of metal cations in water varies between 10^{-3} and 10^{-8} $g.cm^{-3}$. This content of metal cations is required for biochemical processes to take place in organisms.

An increase of this content in water may become a problem from serious contamination. Historically most water quality investigations have attempted to assess trace elements in aquatic systems by analyzing water samples. However in most aquatic systems, the concentrations of trace elements in suspended sediment and the top few centimeters of bottom sediment are far greater than the concentration of trace elements in the water column. Bottom sediments can act as a reservoir for many trace elements.

An undisturbed sediment sink contains a historical record of chemical conditions. Under a changing environmental of physicochemical conditions (e.g. pH, Eh, dissolved oxygen, bacterial action), sediment-bound trace elements can dissolve into the water column, possibly enter the food chain and have significant environmental effects [1].

The area of the Hornád River basin in eastern Slovakia has been considered to be a heavily contaminated area mainly due to industrial activities and concentrated communal factors. The contaminants affected all the basic environmental media: air, water and soil and the related ecological sub-systems like stream sediments.

In the Hornád River basin the water reservoir at Ružín is characterized by a high rate of sedimentation. It is assumed that at present the Ružín reservoir contains approximately 4.5 million m^3 of sediments (8% of the volume of the reservoir[2]).

Research applying mainly spectrochemical methods for the investigation of metal elements in the environment of the Hornád River area began many years ago at the authors' laboratory. Spectrochemical methods[3,4,5,6,7] (D.C.ARC-OES) have been applied in observation of the surface water and sediments and these have been parallelly followed.

Both types of samples were treated as dry evaporates and 15 metallic cations in water were observed.

2. EXPERIMENTAL SECTION AND RESULTS

Simultaneous sampling of water and sediments were done at chosen sampling spots during long-lasting observations of the river Hornád.

Large amounts of data were obtained over a 4 years (1986–1990) period with monthly samples water and sediments analyzed by chemometric procedures.

The state standard classifies surface water according to its quality into five water classes. The surface waters classified in the first two classes can be used as sources of drinking water. The water quality is estimated on the basis of results of the analyses carried out during an uninterrupted period ranging from one to five years. Characteristic values are calculated from all obtained results, the minimum number of which should be 24. The characteristic value of the selected set is considered to be the value showing the water quality (in our case C_{90} concentration of heavy metal in $\mu g.dm^{-3}$ in water and $mg.kg^{-1}$ in sediment) with the a priori stated probability value not to be exceeded (P = 90%). This value is estimated according to the distribution line[8].

Sediments were evaluated by the so-called accumulations factors. This factor is calculated for first time as a ratio of concentration of the given element in "background" which was of following composition[8] to that estimated in the water sample C_{90}.

Cd	Hg	Co	Pb	Cu	Ni	Cr	Zn	Mn	
0.3	0.4	19	20	45	68	90	95	850	($mg.kg^{-1}$)

For the second time accumulation factor is calculated as a ratio $C_{90\ sediment} : C_{90\ H_2O}$ in the same sampling station. The results of this calculation are in Tables 2–4.

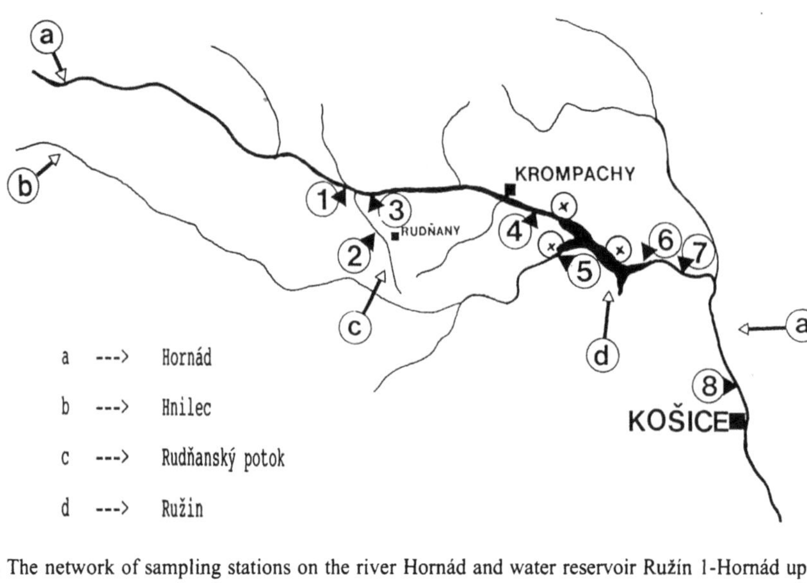

Figure 1. The network of sampling stations on the river Hornád and water reservoir Ružín 1-Hornád up Rudňany creek, 2-Rudňany creek-mouth, 3-Hornád-Rudňany creek, 4-Hornád below Krompachy, 5-Inlet of Hnilec to Ružín water reservoir, 6-The middle of the Ružín water reservoir, 7-Malá Lodina, 8-Ťahanovce.

Table 1. Classification of water and sediments

Element	Class (μg.dm^{-3}) I	II	III	IV	V	Sediments (mg.kg^{-1})
Hg	<0.1	<0.2	<0.5	<1.0	>1.0	1
Cd	<3	<5	<10	<20	>20	0.8
Pb	<10	<20	<50	<100	>100	65
As	<10	<20	<50	<100	>100	15
Cu	<20	<50	<100	<200	>200	50
Cr	<20	<100	<200	<500	>500	13
Co	<10	<20	<50	<100	>100	20
Ni	<20	<50	<100	<200	>200	20
Zn	<20	<50	<100	<500	>500	100
V	<10	<20	<50	<100	>100	
Ag		<10	<20	<50	>50	

The sediment deep samplings were carried out in the year 1993. The depth of the sediments in the reservoir was 130 cm. The sediments were sampled in three sampling localities:

1. the inlet of the Hornád river in to Ružín water reservoir,
2. the inlet of the Hnilec river in to Ružín water reservoir,
3. the middle of the Ružín water reservoir.

The results of the total concentration of mercury estimations as dependent on the depth of the sediment are given in graphs No. 2.

Mercury was found to be the heavy metal having the highest accumulation coefficient.

The estimated values confirm the assumption of mercury accumulation in the sediments and "memory" of the sediments. The highest mercury concentration at the river inlets is at depths of 120 to 150 cm, where the slowest hydraulic processes can be assumed and where sediments show not only natural mercury content in the various soils deposited in the river mouths but also a sharp increase in the metallic mercury production in the mining and dressing complex at Rudňany. In 1969, the mercury production was 14668 kg and in 1972, when there was the most intense production growth, it was 226561 kg.

Table 2. Accumulation factor for selected metal contents in location the middle of the Ružín water reservoir

Element	C_{90} (μg.dm^{-3})	C_{90} (mg.kg^{-1})	K_1	K_2
Co	0.18	13.8	$1.0 \cdot 10^5$	$0.77 \cdot 10^5$
Cr	12.4	215	$7.2 \cdot 10^3$	$1.7 \cdot 10^4$
Cu	27.7	683	$1.6 \cdot 10^3$	$2.5 \cdot 10^4$
Hg	17.4	8.2	$0.2 \cdot 10^2$	$0.5 \cdot 10^3$
Mn	361	1008	$2.3 \cdot 10^3$	$2.8 \cdot 10^3$
Ni	47	469	$1.4 \cdot 10^3$	$1.0 \cdot 10^4$
Pb	3	170	$6.7 \cdot 10^3$	$5.7 \cdot 10^4$
Zn	4.9	2700	$1.9 \cdot 10^4$	$5.5 \cdot 10^5$

Table 3. Accumulation factor for selected metal contents in location the inlet of the Hornád river in to Ružín water reservoir

Element	C_{90} (mg.dm^{-3})	C_{90} (mg.kg^{-1})	K_1	K_2
Co	—	28.4	—	—
Cr	10	138	$9.0 \cdot 10^3$	$1.4 \cdot 10^4$
Cu	300	1070	$1.5 \cdot 10^2$	$3.6 \cdot 10^3$
Hg	14	350	$0.2 \cdot 10^2$	$2.5 \cdot 10^4$
Mn	1060	1667	$0.8 \cdot 10^3$	$1.6 \cdot 10^3$
Ni	134	222	$0.5 \cdot 10^3$	$1.7 \cdot 10^3$
Pb	4	91	$5.0 \cdot 10^3$	$2.2 \cdot 10^4$
Zn	148	3794	$0.6 \cdot 10^3$	$2.6 \cdot 10^4$

As the depth of the sediment decreases, the concentration of the mercury decreases as well. This phenomena can be explained either by the mercury conversion in to methylmercury, dissolvable in water, or by the deposition of substances free of mercury.

The sediments of the Ružín reservoir analysis confirm the slight rise in the mercury concentration in the direction from the reservoir bottom to the sediment outer layers: The mercury content at depths of 120 cm and 90 cm is identity with the natural mercury content in this region (0.06 to 0.25 mg/kg).

The mercury content increases up to 5.5 mg/kg in the top layer of the sediments.

Due to the fact that the sediments of the Ružín reservoir adsorb surface heavy metals, the water quality in the reservoir water column is temporary improved.

3. DISCUSSION

The collected data enable evaluation of the water and sediment quality up to a certain degree.

As the present study deals with contamination by metallic ions it is not possible to evaluate absolutely the total contamination of this component of the environment. The major polluting metals in the Hornád river basin are copper and mercury. The levels of the above metals at the corresponding sampling locations are given in Fig. 3 and 4.

Table 4. Accumulation factors for selected metal contents in location the inlet of the Hnilec river in to Ružín water reservoir

Element	C_{90} (mg.dm^{-3})	C_{90} (mg.kg^{-1})	K_1	K_2
Co	9	217	$2.0 \cdot 10^3$	$2.4 \cdot 10^4$
Cr	18.5	118	$4.8 \cdot 10^3$	$1.8 \cdot 10^4$
Cu	147	3150	$3.0 \cdot 10^2$	$2.2 \cdot 10^4$
Hg	19	89	$0.2 \cdot 10^2$	$1.6 \cdot 10^4$
Mn	939	1884	$0.9 \cdot 10^3$	$2.0 \cdot 10^3$
Ni	12	80	$5.7 \cdot 10^3$	$6.9 \cdot 10^3$
Pb	3	77	$7.0 \cdot 10^3$	$2.5 \cdot 10^4$
Zn	103	5093	$0.9 \cdot 10^3$	$4.9 \cdot 10^4$

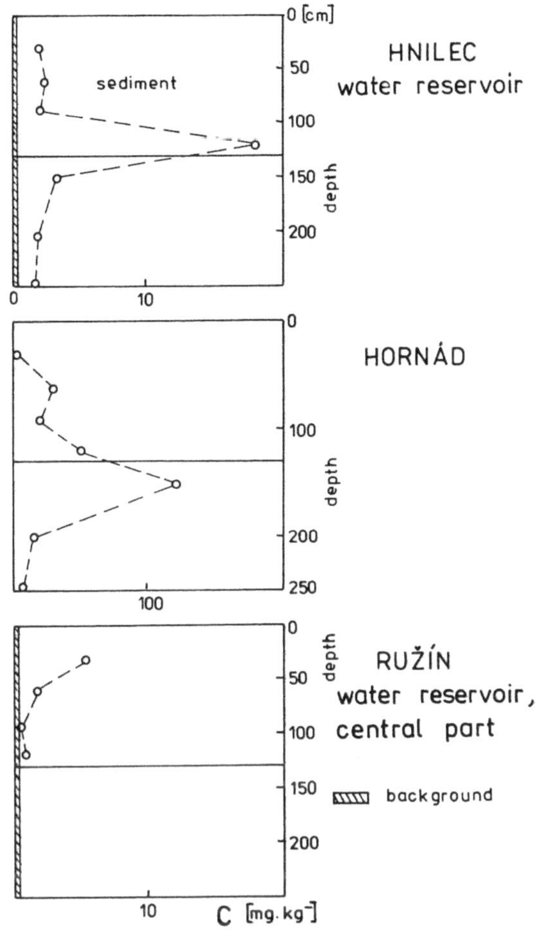

Figure 2. Mercury content alternation with depth.

In the case of copper (Fig. 3), it is evident that its concentration in water samples at three locations exceeds the permitted value for the surface water, i.e.100 $\mu g.dm^{-3}$. The excedence at location 4 is significant and is the result of the anthropogenic activity as this sampling spot is near the Krompachy Copper Smelter. The increased copper content at location 5 (discharge of the Hnilec river in to the Ružín dam) can be connected with the geological bottom quality.

According to the Hg content (Fig. 4), the water quality at all locations ranks it in the fifth class and means that water of the given basin cannot be used for long period as a source of drinking water.

The sediment quality study in the related basin has confirmed an assumed need to synoptically investigate the entire water system, i.e. to analyse water and sediment at the same sampling location at the same time.

Two methods for calculation of accumulation factors have showed that obtained results enable evaluation of the degree of metals accumulation in the sediments only relatively. We assume that the second method is more appropriate and while the first method exploits the geological bottom analysis for all sampling locations in the same manner this might not be properly suited for all sampling locations.

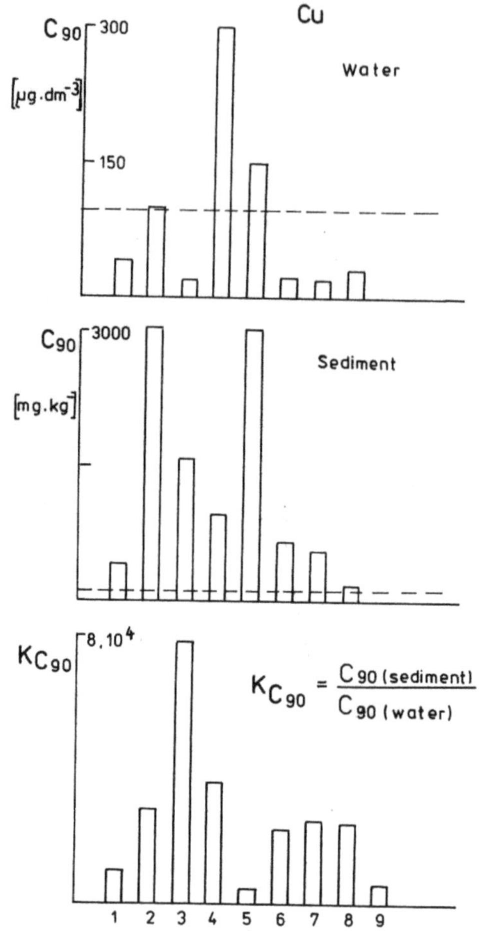

Figure 3. Results of analysis for Cu (1–8 the similar sampling station as Fig. 1).

The obtained accumulation coefficients confirm the manner of metals accumulation in the sediments and the analysis of the sediments sample from the deep sampling show that sediments "posses" memory and serve as a databank of the anthropogenie activity.

4. CONCLUSION

The long-term observations of 15 metallic elements in water and sediments have enabled the creation of a database to study the possibilities of risks related to the introduction of these heavy metals in the food chain, and to propose the possibilities for removal of the sediments from the water reservoir Ružín. The zone sampling of the sediments has proved their ability to retain "memory" and this fact could be utilized when creating a model background for acumulation factor calculations for the case of the individual heavy metals.

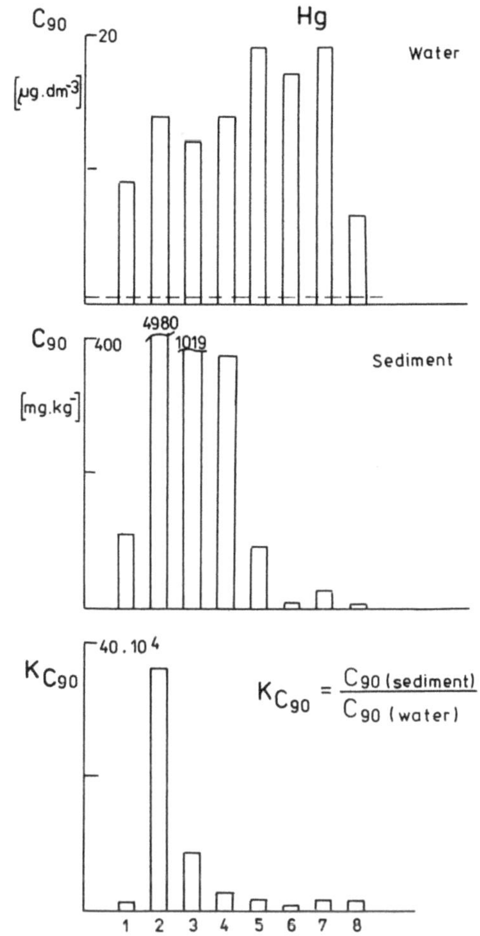

Figure 4. Results of analysis for Hg (1–8 the similar sampling place as Fig. 1).

REFERENCES

1. Horovith A.J.: *Sediment-Trace Element Chemistry*, Lewis Publishers Inc.19911 p.1
2. Stanèíková S.: *Proceedings of conference "Problems of complex efficient utilisation and protection of small rivers"*, Žilina (1985), pp.52–55
3. Heltai Gy., Flórián K., Pliešovská N.: *Ann. Ist. Sup. Sanita*, 1983, 19, pp.549–554
4. Flórián K., Pliešovská N.: *Acta Chim. Hung.*, 1980, 103, pp.301–322
5. Heltai Gy., Zimmer K., Flórián K.: *Acta Chim.Hung.*, 1981, 106, pp.239–255
6. Pliešovská N, Seszták J.: *Trans. TU Košice*, 1992, 2, pp. 79–84
7. Flórián K., Matherny M., Pliešovská N.: *Microchemical Journal 51*, 1995, pp.26–33,
8. ČSN No 757221, Praha, 1990

ABOUT THE EDITORS

LUCJAN PAWŁOWSKI

Lucjan Pawłowski was born in Poland in 1946. He studied chemistry at the Marie Curie–Skłodowska University and got his Ph.D. in 1976 and D.Sc. (habilitation) in 1980, both at the Technical University of Wrocław. In 1986 the President of Poland nominated L. Pawłowski to a full professor. He started research on application of ion exchange for water and wastewater treatment. As a result, he together with B. Bolto from CSIRO Australia, has published a book, "Wastewater Treatment by Ion Exchange," in which they summarized their own results and experience of the ion exchange area. In 1980 L. Pawłowski was elected President of the International Committee "Chemistry for Protection of the Environment." In 1980–1984 he was Chairman of the Environmental Chemistry Division of the Polish Chemical Society. In 1994 he was elected the Deputy President of the Polish Chemical Society and in the same year, the Deputy President of the Presidium Polish Academy of Science Committee "Men and Biosphere." He is member of editorial board of international journals: "Reactive Polymers," "The Science of the Total Environment," "Environment International," and "Journal of Ecological Chemistry." In 1990 he served as an adviser to President Lech Wałesa. In 1991 he was elected as the Deputy Rector of the Technical University of Lublin, a post he still holds. He has published 12 books, over 100 papers, and authored 51 patents.

WILLIAM J. LACY

Dr. William J. Lacy, a Chemical Engineer, is President of Lacy & Co., an international environmental consulting firm to industries and governments. He received his B.S. from the University of Connecticut, completed his Masters degree at New York University College of Engineering, and studied at the Oak Ridge Institute for Nuclear Studies, University of Michigan and Michigan State University. The School for Advanced Chemistry at Paul Sabatier University, France, awarded him the University Medal in 1983. He has an honorary Doctorate of Science in environmental engineering.

Dr. Lacy has authored 198 technical publications, including 14 textbooks. He has three patents and serves on the editorial advisory board of five technical journals. He has chaired or co-chaired 34 National International Conferences.

For his efforts, Dr. Lacy has received numerous awards and medals from Thailand, India, Egypt, Russia, Belgium, France, Poland, and Italy. Other honors include American Defense Preparedness Association award, Secretary of Defense Special Service Award, Presidential Recognition Award, EPA, Bronze Medal, and the U.S. Governmental Distinguished Service Medal.

CHRISTOPHER G. UCHRIN

Christopher G. Uchrin received a B.E. in Civil Engineering in 1972 and an M.E. in Environmental Engineering in 1974, both from Manhattan College, N.Y. He served as an Environmental Engineer for the United States Environmental Protection Agency from 1972 through 1977. He received a Ph.D. from the University of Michigan, Ann Arbor, in 1980. He was hired as an Assistant Professor by Rutgers University, New Brunswick, N.J., in 1980, was promoted to Associate Professor in 1986, and Full Professor in 1990. He served as the Director of the Graduate Program in Environmental Science from 1986 to 1991 and Department Chair from 1991 to 1994. He has been a licensed Professional Engineer in New York State since 1977. He is a member of Tau Beta Pi and Chi Epsilon honor societies. He is a member and has served on national committees for ASCE and ASTM.

MARZENNA R. DUDZIŃSKA

Marzenna R. Dudzińska received an M.Sc. in chemistry in 1983 from Marie Curie–Skłodowska University in Lublin, Poland. In 1983–1988 she was employed as research worker and teacher at the Physical Chemistry Department at M. Curie–Skłodowska University. In 1987 she worked in Institute of Spectroscopy and Applied Spectrochemistry in Dortmund, FRG, on a project combining chromatography and spectroscopy. In 1988 she was hired by Technical University of Lublin, Poland, and worked for the Water and Wastewater Treatment Department. She got a Fulbright Scholarship in 1989, and spent 1989/1990 at University of Houston, Texas, in the Civil and Environmental Enginering Department. She received a Ph.D. in environmental chemistry from Marie Curie–Skłodowska University in 1992. Now she works for the Environmental Protection Department at Technical University of Lublin. She was awarded by the Polish Chemical Society for her master thesis in 1983, by the Polish Ministry of Science and Education in 1987 and in 1992 and 1994 by the Rector of Technical University of Lublin. She is a member of the Polish Chemical Society and served as a President of the Student Section of the Society in 1984–1988.

AUTHOR INDEX

SUBJECT INDEX

MIX
Papier aus verantwortungsvollen Quellen
Paper from responsible sources
FSC® C105338

If you have any concerns about our products,
you can contact us on
ProductSafety@springernature.com

In case Publisher is established outside the EU,
the EU authorized representative is:
Springer Nature Customer Service Center GmbH
Europaplatz 3, 69115 Heidelberg, Germany

Printed by Libri Plureos GmbH
in Hamburg, Germany